SIMONE HÄRRI

Ecological Communities

To gain a more complete understanding of plant-based ecological community structure requires knowledge of the integration of direct and indirect effects in plant–herbivore systems. Trait modification of plants as a result of herbivory is very common and widespread in terrestrial plants, and this initiates indirect interactions between organisms that utilize the same host plant. This book argues that food webs by themselves are inadequate models for understanding ecological communities, because they ignore important indirect, nontrophic links. This subject is of great importance in understanding not only community organization but also in identifying the underlying mechanisms of maintenance of biodiversity in nature. This book will be an invaluable resource for researchers and graduate students interested in community and population ecology, evolutionary biology, biodiversity, botany, and entomology.

TAKAYUKI OHGUSHI is a Professor at the Center for Ecological Research at Kyoto University.

TIMOTHY P. CRAIG is an Associate Professor of Biology at the University of Minnesota in Duluth.

PETER W. PRICE is Regents' Professor Emeritus in the Department of Biological Sciences at Northern Arizona University.

Ecological Communities: Plant Mediation in Indirect Interaction Webs

Edited by

TAKAYUKI OHGUSHI
Kyoto University, Japan

TIMOTHY P. CRAIG
University of Minnesota Duluth, USA

PETER W. PRICE
Northern Arizona University, USA

CAMBRIDGE UNIVERSITY PRESS
Cambridge, New York, Melbourne, Madrid, Cape Town, Singapore, São Paulo

Cambridge University Press
The Edinburgh Building, Cambridge CB2 8RU, UK

Published in the United States of America by Cambridge University Press, New York

www.cambridge.org
Information on this title: www.cambridge.org/9780521850391

© Cambridge University Press 2007

This publication is in copyright. Subject to statutory exception
and to the provisions of relevant collective licensing agreements,
no reproduction of any part may take place without the
written permission of Cambridge University Press.

First published 2007

Printed in the United Kingdom at the University Press, Cambridge

A catalogue record for this publication is available from the British Library

ISBN-13 978-0-521-85039-1 hardback
ISBN-10 0-521-85039-8 hardback

Cambridge University Press has no responsibility for the persistence or accuracy
of URLs for external or third-party internet websites referred to in this publication,
and does not guarantee that any content on such websites is, or will remain,
accurate or appropriate.

Contents

List of contributors viii
Preface xiii

Part I   Introduction   1

1   Indirect interaction webs: an introduction   3
    TAKAYUKI OHGUSHI, TIMOTHY P. CRAIG, AND PETER W. PRICE

    Part II   Interaction linkages produced by plant-mediated indirect effects   17

2   Plant-mediated interactions in herbivorous insects: mechanisms, symmetry, and challenging the paradigms of competition past   19
    ROBERT F. DENNO AND IAN KAPLAN

3   Going with the flow: plant vascular systems mediate indirect interactions between plants, insect herbivores, and hemi-parasitic plants   51
    SUSAN E. HARTLEY, KATHY A. BASS, AND SCOTT N. JOHNSON

4   Plant-mediated effects linking herbivory and pollination   75
    JUDITH L. BRONSTEIN, TRAVIS E. HUXMAN, AND GOGGY DAVIDOWITZ

5   Trait-mediated indirect interactions, density-mediated indirect interactions, and direct interactions between mammalian and insect herbivores   104
    JOSÉ M. GÓMEZ AND ADELA GONZÁLEZ-MEGÍAS

6   Insect–mycorrhizal interactions: patterns, processes, and consequences   124
    ALAN C. GANGE

Part III Plant-mediated indirect effects in multitrophic systems 145

7 Plant-mediated interactions between below- and aboveground processes: decomposition, herbivory, parasitism, and pollination 147
Katja Poveda, Ingolf Steffan-Dewenter, Stefan Scheu, and Teja Tscharntke

8 Bottom-up cascades induced by fungal endophytes in multitrophic systems 164
Enrique J. Chaneton and Marina Omacini

9 Ecology meets plant physiology: herbivore-induced plant responses and their indirect effects on arthropod communities 188
Maurice W. Sabelis, Junji Takabayashi, Arne Janssen, Merijn R. Kant, Michiel van Wijk, Beata Sznajder, Nayanie S. Aratchige, Izabela Lesna, Belen Belliure, and Robert C. Schuurink

Part IV Plant-mediated indirect effects on communities and biodiversity 219

10 Nontrophic, indirect interaction webs of herbivorous insects 221
Takayuki Ohgushi

11 Effects of arthropods as physical ecosystem engineers on plant-based trophic interaction webs 246
Robert J. Marquis and John T. Lill

12 Host plants mediate aphid–ant mutualisms and their effects on community structure and diversity 275
Gina M. Wimp and Thomas G. Whitham

13 Biodiversity is related to indirect interactions among species of large effect 306
Joseph K. Bailey and Thomas G. Whitham

Part V Evolutionary consequences of plant-mediated indirect effects 329

14 Evolution of plant-mediated interactions among natural enemies 331
Timothy P. Craig

15 Linking ecological and evolutionary change in multitrophic interactions: assessing the evolutionary consequences of herbivore-induced changes in plant traits 354
David M. Althoff

Part VI Synthesis 377

16 **Indirect interaction webs propagated by herbivore-induced changes in plant traits** 379
    TAKAYUKI OHGUSHI, TIMOTHY P. CRAIG, AND PETER W. PRICE

Taxonomic index 411
Author index 421
Subject index 431

Contributors

David M. Althoff
Department of Biology, Syracuse University, Syracuse, NY 13244, USA

Nayanie S. Aratchige
Section Population Biology, Institute for Biodiversity and Ecosystem Dynamics, University of Amsterdam, 1098 SM Amsterdam, The Netherlands

Joseph K. Bailey
Department of Biological Sciences and the Merriam–Powell Center for Environmental Research, Northern Arizona University, Flagstaff, AZ 86011, USA

Kathy A. Bass
Department of Biology and Environmental Sciences, School of Life Sciences, University of Sussex, Falmer, Brighton BN1 9QG, UK

Belen Belliure
Section Population Biology, Institute for Biodiversity and Ecosystem Dynamics, University of Amsterdam, 1098 SM Amsterdam, The Netherlands

Judith L. Bronstein
Department of Ecology and Evolutionary Biology, University of Arizona, Tucson, AZ 85721, USA

Enrique J. Chaneton
IFEVA-Departamento de Recursos Naturales y Ambiente, Facultad de Agronomia, Universidad de Buenos Aires, 1417 Buenos Aires, Argentina

Timothy P. Craig
Department of Biology, University of Minnesota Duluth, Duluth,
MN 55812, USA

Goggy Davidowitz
Department of Ecology and Evolutionary Biology, University of Arizona,
Tucson, AZ 85721, USA

Robert F. Denno
Department of Entomology, University of Maryland, College Park,
MD 20742, USA

Alan C. Gange
School of Biological Science, Royal Holloway University of London,
Egham, TW20 0EX, UK

Jóse M. Gómez
Departamento de Ecología, Universidad de Granada, E-18071 Granada,
Spain

Adela González-Megías
Departmento de Biología Animal, Universidad de Granada, E-18071
Granada, Spain

Susan E. Hartley
Department of Biology and Environmental Sciences, School of Life
Sciences, University of Sussex, Falmer, Brighton BN1 9QG, UK

Travis E. Huxman
Department of Ecology and Evolutionary Biology, University of Arizona,
Tucson, AZ 85721, USA

Arne Janssen
Section Population Biology, Institute for Biodiversity and Ecosystem
Dynamics, University of Amsterdam, 1098 SM Amsterdam,
The Netherlands

Scott N. Johnson
School of Human and Environmental Sciences, Department of Soil
Science, University of Reading, Reading, RG6 6DW, UK

Merijn R. Kant
Section Population Biology, Institute for Biodiversity and Ecosystem
Dynamics, University of Amsterdam, 1098 SM Amsterdam,
The Netherlands

Ian Kaplan
Department of Entomology, University of Maryland, College Park, MD 20742, USA

Izabela Lesna
Section Population Biology, Institute for Biodiversity and Ecosystem Dynamics, University of Amsterdam, 1098 SM Amsterdam, The Netherlands

John T. Lill
Department of Biological Science, George Washington University, Washington, DC 20052, USA

Robert J. Marquis
Department of Biology, University of Missouri–St. Louis, St. Louis, MO 63121, USA

Takayuki Ohgushi
Center for Ecological Research, Kyoto University, Otsu, 520–2113, Japan

Marina Omacini
IFEVA-Departamento de Recursos Naturales y Ambiente, Facultad de Agronomia, Universidad de Buenos Aires, 1417 Buenos Aires, Argentina

Katja Poveda
Agroecology, University of Göttingen, D-37073 Göttingen, Germany

Peter W. Price
Department of Biological Sciences, Northern Arizona University, Flagstaff, AZ 86011, USA

Maurice W. Sabelis
Section Population Biology, Institute for Biodiversity and Ecosystem Dynamics, University of Amsterdam, 1098 SM Amsterdam, The Netherlands

Stefan Scheu
Technische Universitat Darmstadt, Fachbereich 10, Biologie, D-64287 Darmstadt, Germany

Robert C. Schuurink
Section Plant Physiology, Swammerdam Institute of Life Sciences, University of Amsterdam, 1098 SM Amsterdam, The Netherlands

Ingolf Steffan-Dewenter
Agroecology, University of Göttingen, D-37073 Göttingen, Germany

Beata Sznajder
Section Population Biology, Institute for Biodiversity and Ecosystem Dynamics, University of Amsterdam, 1098 SM Amsterdam, The Netherlands

Junji Takabayashi
Center for Ecological Research, Kyoto University, Otsu, 520–2113, Japan

Teja Tscharntke
Agroecology, University of Göttingen, D-37073 Göttingen, Germany

Michiel van Wijk
Section Population Biology, Institute for Biodiversity and Ecosystem Dynamics, University of Amsterdam, 1098 SM Amsterdam, The Netherlands

Gina M. Wimp
Department of Entomology, University of Maryland, College Park, MD 20742, USA

Thomas G. Whitham
Department of Biological Sciences, Northern Arizona University, Flagstaff, AZ 86011, USA

Preface

There is rapidly increasing appreciation of the community consequences of trait-mediated indirect effects in various ecosystems. Nevertheless, we know little about the underlying mechanisms. This book highlights the plant trait responses to herbivores as the mechanistic basis of plant-mediated indirect effects that are ubiquitous in terrestrial systems. Recent development of research on plant–herbivore interactions has revealed that herbivore-induced trait modification, such as defensive chemicals, nutritional status, and subsequent growth, is very common and widespread in terrestrial plants. Modification of plants in ways other than the simple removal of tissue can have complex impacts on not only other herbivores but also higher trophic levels through bottom–up cascades. Thus, the structure and biodiversity of plant-based terrestrial communities can be strongly influenced by the herbivore-initiated interaction linkages. This book represents the first major synthesis to integrate information on how trait-mediated indirect effects structure ecological communities through changes in plant traits in terrestrial systems.

This book is based on a symposium entitled "The impact of trait-mediated indirect effects of plants on the structure of herbivorous insect communities," which was held in Brisbane, Australia, at the International Congress of Entomology in August 2004. The symposium brought together a collection of contributors well qualified to evaluate and expand our understanding of how trait-mediated indirect effects of terrestrial plants structure ecological communities from the bottom up. We also invited authors who did not attend the conference to submit chapters in their fields of expertise in order to offer a broader view of the field. We have asked all contributors to review thoroughly the literature in their area, and to discuss their own research, with an emphasis on the novel insights gained through an understanding of plant-trait mediation of interactions as they influence ecological communities. This volume comprises an introduction and 15 chapters by international experts in this emerging field of plant-mediated indirect interactions. A majority of the chapters focus on plant–insect interactions because they are among the richest contributors to interaction biodiversity, and because the diverse array of herbivore feeding strategies induces many well-understood plant responses. It also includes work on ungulates, birds, and endophytic and mycorrhizal fungi. Thus, this book provides a comprehensive overview of the impact of herbivore-induced indirect effects on the ecology and evolution of plant–herbivore interactions, community organization, and biodiversity. There are contributions from Argentina, Britain, Germany, Japan, Spain, the Netherlands, and the United States. All of the contributors have achieved excellent work in this field, and have established outstanding reputations in the academic community. Hence, this provides a truly international view of ecological interactions and community ecology that increases the exposure of new concepts on community ecology and biodiversity. This book provides current syntheses of this emerging field and a guide for future research for professional biologists, graduate students, and advanced undergraduates in the various fields of ecology, evolution, botany, entomology, mycology, mammalogy, biochemistry, biological control, conservation biology, and biodiversity. It is also relevant to the scientists interested in plant–herbivore interactions in the basic and applied aspects of agriculture, forestry, and horticulture.

Finally, we thank the authors for their hard work and keeping up with the exacting schedule for preparation and submission of their chapters. In particular, we greatly appreciated their dealing with the numerous bothersome requests from the editors for clarification of their written texts. We would like to thank the many ecologists who helped improve the quality by critically reviewing the chapters. It has been a pleasure to work with Cambridge University Press, and we particularly thank Dominic Lewis, Ward Cooper, Wendy Phillips, and Emma Pearce for their assistance with the production of this volume. It has been extremely exciting for us to be involved in production of a book in such an emerging, important research area.

Takayuki Ohgushi
Timothy P. Craig
Peter W. Price
July 2006

Part I   INTRODUCTION

1

Indirect interaction webs: an introduction

TAKAYUKI OHGUSHI, TIMOTHY P. CRAIG, AND PETER W. PRICE

The modification of plant traits by species utilizing a plant can be of crucial importance in determining the ecological and evolutionary interactions among community members centered on the plant. The authors in this volume review studies that integrate direct and indirect effects of plant-mediated traits on the ecology and evolution of plant–herbivore and other plant-mediated interactions. They synthesize these data to provide a more complete understanding of plant-based ecological community structure, which will provide a new perspective on the organization of these communities. Also, they suggest directions for future study. The richness of plant-trait-mediated effects results in the production of far more linkages among species than food web interactions alone. This richness also strengthens the argument that insect species on plants constitute real communities rather than noninteracting assemblages of individual species. Herbivore modification of plant traits is very common and widespread in terrestrial plants, and this initiates indirect interactions between species utilizing the same host plant. These interactions can be mediated by herbivore-induced changes in plant defensive chemicals, nutritional status, and the subsequent growth of plants attacked. Modification of plants in ways other than the simple removal of tissue can have complex impacts on other herbivores, and produce effects that cascade upward to higher trophic levels including predators and parasitoids. Thus, the structure and biodiversity of plant-based terrestrial communities can be strongly influenced by interaction linkages initiated by herbivory.

Indirect effects occur when the impacts of one species on another are influenced by one or more intermediate species. Indirect effects are important

Ecological Communities: Plant Mediation in Indirect Interaction Webs, ed. Takayuki Ohgushi, Timothy P. Craig, and Peter W. Price. Published by Cambridge University Press. © Cambridge University Press 2007.

because they are common in biological systems, and may change outcomes of interactions among species. Indirect effects are diverse, but they can be classified as either (1) density-mediated or (2) trait-mediated (Abrams *et al.* 1996). Density-mediated indirect effects result in a numerical response of the induced species, while trait-mediated indirect effects influence traits such as behavior, morphology, and life histories of the induced species. Density-mediated indirect effects, such as apparent competition and top–down trophic cascades, have been studied much more thoroughly than trait-mediated indirect effects (Holt and Lawton 1994, Pace *et al.* 1999). Recently there has been an increased appreciation of the importance of trait-mediated indirect effects on community organization (Werner and Peacor 2003). For example, a special feature section on "Linking individual-scale trait plasticity to community dynamics" in the journal *Ecology* (volume 84:1081–1157 in 2003) focused on this issue. However, the authors focused on trait-mediated indirect effects in prey–predator interactions, such as changes in prey behavior to avoid predation risk, and they paid little attention to the plant-mediated indirect effects on community composition and structure. We argue that density-mediated indirect effects predominate in herbivore–predator interactions, while trait-mediated indirect effects are of primary importance in plant–herbivore interactions. This is because predators usually kill individuals in lower trophic levels while herbivores predominately alter traits of plants attacked, without killing them. Our emphasis is on the importance of herbivore modification of terrestrial plant traits in producing indirect effects that link multiple interactions and shape community structure.

"Indirect interaction webs" (Ohgushi 2005) is a new concept used in this book to evaluate the functioning of nontrophic, indirect links in ecological communities. Food webs that measure only direct and trophic interactions are the traditional tool used to evaluate the forces structuring ecological communities. We argue that food webs by themselves are inadequate models for understanding ecological communities because they ignore important indirect, nontrophic links. On the other hand, indirect interaction webs contrast them from traditional food webs that fail to capture the frequency and importance of trait-mediated effects. This novel approach of incorporating indirect, nontrophic links into the web of biotic interactions offers a new perspective on how species interactions evolve in communities, and how biodiversity originates, and is maintained, in terrestrial systems. Plants, as primary producers, are at the base of interaction webs as well as food webs so it is logical to develop our understanding of these systems from the "bottom up." The importance of bottom–up forces on community structure has recently gained extensive empirical and theoretical support that challenges the traditional view that plant-centered communities are structured from the top down by the influence of natural enemies (Hunter and Price 1992, Whitham *et al.* 2003). The inclusion of non-trophic and indirect interactions provides a new perspective on plant–herbivore interactions that is not found in earlier reviews on this subject. The book, therefore, provides a comprehensive overview of the impact of herbivore-induced indirect effects on the ecology and evolution of plant–herbivore interactions, community organization, and biodiversity. This volume is the first to integrate information on how trait-mediated indirect effects influence the structure of plant-based ecological communities.

This book comprises an introduction and 15 chapters by international experts in the rapidly developing field of indirect effects produced by plant-mediated interactions. The authors focus on four questions of fundamental importance in research on the indirect effects of plant-mediated interactions: (1) What novel interaction linkages are produced by plant-mediated indirect effects? (2) What complex interactions are generated by plant-mediated indirect effects in multi-trophic systems? (3) What are the effects of plant-mediated indirect effects on community structure and biodiversity? and (4) What are the evolutionary consequences of plant-mediated indirect effects?

What novel interaction linkages are produced by plant-mediated indirect effects?

Interaction linkages among organisms produced by plant-mediated effects have important, but poorly understood, roles in structuring ecological communities. The effect of one organism can be propagated through a community to other organisms as the result of changes in plants that we term plant-mediated indirect effects. Ecological research has long focused on studies of how single, direct interactions shape ecological communities. Recent research, however, indicates that complex interactions among multiple species are often more important than interactions among pairs of species (Tscharntke and Hawkins 2002, Stanton 2003). Indirect effects frequently structure these multispecies interactions, but only recently have studies been conducted that can evaluate these effects in terrestrial communities. Density-mediated indirect effects, such as trophic cascades and apparent competition, are one kind of indirect interaction that has been recognized to be much more common than previously thought (Holt and Lawton 1994, Pace *et al.* 1999).

Trait-mediated indirect effects are another, potentially very important, kind of indirect interaction that has been studied less frequently than density-mediated indirect effects. Research on plant–herbivore systems has provided increasing evidence of the importance of these plant-trait-mediated interactions. Substantial, indirect interactions caused by herbivore-induced changes in terrestrial plants occur frequently among temporally separated, spatially separated, and taxonomically distinct species (Ohgushi 2005). The importance of such interactions has not been previously recognized, because of the traditional view that interactions usually occur only among closely related species or among species that simultaneously use the same plant resources. Herbivory on terrestrial plants is usually nonlethal, and it can induce a wide variety of changes in plant traits that can influence interactions among very different species. Measuring these changes provides a mechanistic understanding of how plants mediate indirect interactions among herbivores, and other members of an ecological community. An understanding of plant-mediated indirect interactions among herbivores will allow us to establish linkages between previously isolated areas of inquiry such as: above- and belowground ecology, herbivory and pollination ecology, and insect and microbial ecology. This will result in a much more complete picture of how multiple species are connected in ecological communities through plant-mediated indirect effects. We believe that the indirect interactions are ubiquitous in plant–herbivore interactions in terrestrial systems and that inadequate knowledge of such interactions has impeded progress in understanding how ecological communities function.

There is increasing evidence of plant-mediated indirect interactions among insect herbivores. In particular, feeding-induced plant resistance is a well-documented phenomenon for both sap-feeding and leaf-chewing insects. Denno and Kaplan (Chapter 2) provide a thorough overview on plant-mediated indirect interactions between herbivorous insects, and they identify possible underlying mechanisms responsible for them. Plant-mediated indirect interactions are highly asymmetric, and they occur among temporally and spatially separated insects, and among phylogenetically distinct herbivores. They also explore the life-history traits of herbivorous insects that contribute to competitive superiority in plant-mediated interactions. They demonstrate that studying plant-mediated indirect interactions among insect herbivores provides novel conclusions that challenge the traditional paradigms of community organization that have been based on competition theory.

Hartley *et al.* (Chapter 3) focus on the asymmetric interactions between aphids and other organisms that exploit the plant vascular system. First, they examine the indirect interactions between phloem-feeding aphids and leaf miners that physically disrupt the flow of nutrients within the vascular system. The leaf miners significantly decreased aphid survival, but the aphids did not affect the leaf miners. Second, they address how plants mediate interactions between taxonomically separated organisms: phloem-feeding aphids and hemi-parasites that obtain nutrients from the xylem of the host plant roots. The aphids improved performance of the parasite because the aphids increased root biomass, and provided more opportunities for the parasite to connect to its host. In contrast, the parasite had a negative effect on the aphid performance. Hartley *et al.* stress that the indirect interactions can result in the physical alteration of the structures within plants, and that such interactions have been poorly studied.

Studies of plant–herbivore and plant–pollinator interactions have long developed in isolation. Bronstein *et al.* (Chapter 4) explore how herbivory indirectly affects pollination through changes in reproductive traits of shared plants. Herbivore damage can alter flower appearance, reduce flower numbers and size, affect both nectar quantity and chemistry, alter flowering phenologies, and restrict potential resource allocation to seeds. Each of these phenomena, in turn, can deter pollinators, and/or alter the efficacy of individual pollinator visits. To develop a predictive framework, they propose a new physiologically based graphical model designed to explore how plant resources are apportioned to pollinator attraction/reward and to herbivore deterrence, and how allocation schedules are altered by both activities. Bronstein *et al.* apply this model to a pollination system in which a major pollinator (a hawkmoth, *Manduca sexta*) is also a damaging herbivore of the same plant (*Datura wrightii*) while in the larval stage.

Plant-mediated indirect effects can produce interactions between phylogenetically distant herbivores. Mammalian herbivores are ubiquitous and important plant consumers in most terrestrial communities, and they can affect plants by modifying their life history, overall architecture, leaf morphology, chemistry, and spatial distribution. Gómez and González-Megías (Chapter 5) provide a variety of examples of how mammalian herbivores influence directly and indirectly performance or population size of herbivorous insects. Specifically, they assess the importance of the trait-mediated indirect effects that influence the interaction between these two kinds of herbivores. They stress that plant-trait-mediated indirect effects are crucial for insects since the main effect of ungulates is not a change in plant density but a modification of the chemical, physiological, morphological, and architectural traits of plants. Thus, future research needs to consider jointly the effects of both herbivorous vertebrates and invertebrates to draw a more accurate picture of the dynamics of terrestrial communities.

Plant-mediated indirect interactions extend to microorganisms. The use of a common host plant indirectly links insects and mycorrhizal fungi, because both require carbon from the host. Gange (Chapter 6) reviews examples of how indirect interactions between insects and mycorrhizal fungi link below- and aboveground processes. In general, insect herbivory reduces mycorrhizal colonization of plants, through competition for carbon. The most important effect of herbivory is on the mycorrhizal species diversity. Insect host plant choice is unaffected by the mycorrhizal status of a plant, but mycorrhizae can dramatically increase or decrease herbivorous insect growth. Ectomycorrhizal fungi seem to have fewer effects than do arbuscular mycorrhizae. Positive effects of mycorrhizae on insects are caused by alterations in plant morphology or nutritional quality, while negative effects are the result of the modification of plant defenses. Gange also explores the potential role of these interactions in structuring the insect and mycorrhizal communities.

What complex interactions are generated by plant-mediated indirect effects in multitrophic systems?

Plant-mediated indirect effects can affect not only herbivores, but also their natural enemies as the result of bottom–up cascades through trophic levels (Price et al. 1980, Hunter et al. 1992). Indirect effects can influence the relative importance of top–down effects (control exerted by predators on lower trophic levels) and bottom–up effects (control exerted by resources available to each trophic level). In the past top–down effects have often been considered to be more important than bottom–up effects in multitrophic interactions (cf. Berryman 2002). We believe this is primarily due to the use of food webs to analyze community structure. Food web analysis focuses on the top–down effects of predators, and thus does not reveal effects exerted by lower trophic levels on higher trophic levels. The importance of indirect effects in structuring complex systems has been indicated by recent work on interactions that include more than three trophic levels (Gange and Brown 1997, Tscharntke and Hawkins 2002). Trophic cascades from higher to lower trophic levels are well-known indirect effects in aquatic multitrophic systems (Carpenter et al. 1985, Carpenter and Kitchell 1988), in which predators decrease herbivore abundance, and the decreased herbivory in turn increases primary production. Thus, trophic cascades initiated by predators have chiefly density-mediated indirect effects (but see Schmitz et al. 2004). In contrast, herbivore-induced changes in plant traits cascade upwards from plants to predators via herbivores. These are indirect effects because plants do not respond numerically, but instead change traits in response to herbivory. Interaction linkages of herbivores through plant-mediated indirect effects have the potential to influence the third trophic level, which enhances the bottom–up cascades in multitrophic systems. Thus, interactions between trophic levels can be bidirectional, with each trophic level exerting influences on the other. Bidirectionality creates feedback loops from terrestrial plants that receive herbivory to higher trophic levels. We argue that plant-mediated indirect effects have great impacts not only on herbivores, but also on their natural enemies. These bottom–up effects from plants may alter the patterns and intensity of top–down effects on lower trophic levels. Thus, to understand the relative importance of bottom–up and top–down forces in plant-based multitrophic systems, we need to know much more about how feedbacks induced by changes in plant quality and architecture in response to herbivory influence higher trophic levels.

Recent studies of multitrophic interactions including plants, herbivores, and their natural enemies have highlighted the relative importance of top–down and bottom–up forces and that of direct and indirect effects in forming ecological communities. Thus, the multitrophic level approach addresses the complexity of ecological communities much more realistically than we have recognized from the study of pair-wise interactions alone. Traditionally, above- and belowground communities have been investigated separately even though both systems are highly interdependent. Poveda et al. (Chapter 7) review the evidence of interactions between above- and belowground components. They argue that the relative importances, and the combined effects, of belowground and aboveground herbivores on plant performance and on plant–herbivore, herbivore–parasitoid, and plant–pollinator interactions are almost unknown. They describe a system consisting of multitrophic interactions with organisms including root herbivores and soil decomposers underground, and aphids and their parasitoids and flower visitors aboveground. There are two positive plant-mediated indirect interactions between below- and aboveground processes: the positive effect of root herbivores on flower visitors, and the positive impact of root herbivores and decomposers on aphid densities that indirectly affect the number of parasitized aphids.

Endophytic fungi cause morphological, physiological, and chemical changes in their host plants, including the synthesis of alkaloids that may confer protection against herbivores and pathogens. Thus, they may indirectly affect insect herbivores and natural enemies through changes in these plant traits. Chaneton and Omacini (Chapter 8) provide an overview of indirect effects of fungal endophytes on herbivorous insects and their natural enemies to illustrate the role of mutualistic fungal endophytes as elicitors of bottom–up indirect effects transmitted to higher trophic levels. They describe direct and indirect effects involved in a system consisting of Italian ryegrass, *Neotyphodium* endophyte, aphids, and their primary and secondary parasitoids. They illustrate how the symbiotic interactions result in the propagation of bottom–up cascades to higher trophic levels in multitrophic interaction webs.

Plant secondary chemical compounds, those that apparently lack primary roles in basic functions such as growth and reproduction, are thought to play a role in direct defense against herbivores. These may be produced constitutively or induced by wounding or herbivory. These secondary compounds may also have indirect effects at the second or higher trophic levels. Sabelis et al. (Chapter 9) review how secondary plant compounds from plants under herbivore attack indirectly affect other herbivores, predators, hyperpredators, and omnivores. Such indirect effects are of fundamental importance in understanding the structure of arthropod communities on plants.

What are the effects of plant-mediated indirect effects on community structure and biodiversity?

The role of herbivore-induced changes in plant traits in shaping arthropod communities and in maintaining biodiversity on terrestrial plants has only recently been appreciated (Waltz and Whitham 1997, Martinsen et al. 2000). Adaptations and counter-adaptations in plant–herbivore interactions produce temporal and spatial resource heterogeneity that increases species richness and interaction diversity. Most of the earth's biodiversity is in its interaction diversity: the tremendous variety of ways in which species are linked together into constantly interacting networks (Thompson 1996). Plant-mediated interaction linkages strongly influence interaction diversity and shape the network structure of interacting species in ecological communities. Knowledge of both the trophic and nontrophic interaction linkages produced by herbivore-induced effects is critical if we are to understand how the network of species interactions alters biodiversity components in arthropod communities. Herbivory changes plant quality and architecture, and it increases the food and habitat heterogeneity provided by terrestrial plants. Herbivore-induced changes can generate bottom–up cascading effects that influence entire arthropod communities, thereby altering both species richness and abundance. Moreover, positive indirect interactions often increase arthropod biodiversity. For example, one herbivore can facilitate the success of other herbivores by inducing compensatory regrowth that increases the nutritional status and/or plant biomass. This may, in turn, result in an increase of species and interaction diversity. Ecosystem engineers create physical structures and they initiate a web of nontrophic interactions (Lill and Marquis 2003). They can also increase species and interaction diversity by creating habitats for other arthropods.

Herbivorous insects rarely kill terrestrial plants, but they frequently induce changes in a range of plant traits. Therefore herbivore-induced indirect interactions are common and widespread in plant–insect interactions. Ohgushi (Chapter 10) illustrates how herbivore-induced indirect interactions create multiple plant–insect linkages mediated by plant- and ant-mediated effects in willow and goldenrod. He proposes the use of indirect interaction webs, which include nontrophic and indirect links, to depict interaction linkages. He demonstrates that these indirect interaction webs revealed more than three times the number of direct and indirect interactions in these systems than would have been detected using traditional plant-based food webs. Recognizing the importance of the indirect interaction linkages produced by herbivore-induced changes in plant traits will alter our understanding of the organization of arthropod community structure on terrestrial plants.

Numerous insect herbivore species act as physical ecosystem engineers by modifying plant architecture to create shelters. Plant manipulation by these engineers through leaf-shelter building, galling, or stem-boring, provides habitat for other species, and has a positive effect on organism abundance and species diversity in arthropod communities. Marquis and Lill (Chapter 11) provide a number of examples of how insect ecosystem engineers are widespread on terrestrial plants, and identify several pathways through which ecosystem engineers interact with other herbivorous and predatory arthropods, thereby altering trophic structure of the associated arthropod fauna of the host plant. One example of ecosystem engineering is insects that build leaf shelters. They not only create habitats for other organisms but they can also modify leaf quality by reducing the production of secondary compounds and lignin that are positively correlated with light availability. Thus the leaf-shelter builder has an indirect positive effect on habitat quality for herbivores inhabiting the shelter. To fully understand the impact of insect ecosystem engineers on the entire arthropod community, future research needs to focus on both the interactions within the shelters and on the impact of the engineer outside of the shelter.

The role of mutualism in broader levels of community organization is poorly understood because much of the study of mutualisms has focused on just a few species. Wimp and Whitham (Chapter 12) explore how plant traits determine the distribution and intensity of ant–aphid mutualisms, which in turn affect arthropod community structure. Ant–aphid mutualisms negatively impact other herbivores, generalist predators, and sub-dominant species of tending ants, but they attract a specialist community of parasites and predators that have unique adaptations for evading ant defenses. In addition to their effects on species richness and the relative abundance of community members, ant–aphid mutualisms can also enhance community stability in terms of species turnover rate and spatial heterogeneity of species diversity. Wimp and Whitham propose that future studies should: (1) consider the important role of bottom–up effects of host plants in determining the distribution of the mutualists, (2) examine interactions between the mutualism and the surrounding community on a variety of different scales, (3) encompass multiple trophic levels, and (4) explore the effects of mutualisms on species composition as well as diversity.

Bailey and Whitham (Chapter 13) examine how herbivores can indirectly impact biodiversity through plant-mediated impacts on dependent community members. In an aspen and cottonwood system, arthropod species diversity is driven by the synergistic impact of plant-mediated indirect interactions among common herbivores, including a gall-forming sawfly, elk, and beavers. Specifically, the individual and combined effects of common, dominant, and keystone herbivores create a gradient of habitats or conditions that increase the probability of unique indirect interactions, increase the heterogeneity of potential niches for other species, and increase the possibility of synergistic outcomes, all of which can positively affect biodiversity.

What are the evolutionary consequences of plant-mediated indirect effects?

It is important to understand the evolutionary as well as the ecological consequences of plant-mediated indirect effects in a community. For example, the evolution of a plant–insect–natural-enemy interaction is modified by the presence of other species, and the evolutionary changes in one interaction can indirectly affect the evolution of other interactions within a community. Evolutionary changes in insect–plant interactions can influence, and be influenced by, complex direct and indirect interactions with other members of ecological communities (Thompson 1994, Stanton 2003). An examination of indirect plant-mediated effects may lead to a better understanding of coevolutionary processes in plant–herbivore–natural-enemy interactions. Coevolution, the reciprocal evolutionary change in interacting species (Thompson 1994), has been proposed as an important process in generating and maintaining biodiversity (Ehrlich and Raven 1964). However, documenting the extent of coevolution in plant–herbivore interactions has a long and controversial history (Jermy 1993, Thompson 1994, Bernays 1998, Zangerl and Berenbaum 2003). One difficulty in assessing these coevolutionary forces has been the past focus on studying putatively coevolving species pairs in isolation from the rest of the community. We know little about how interactions with other community members influences the evolution of interacting species pairs. Indirect effects may substantially influence evolutionary interactions between species. Recent work has suggested that indirect effects should be particularly influential where organisms are faced with multiple selective agents in nature (Agrawal and Van Zandt 2003). Changes in the abundance of the third species may alter either the patterns of selection imposed by a focal pair of species on each other or the response of those species to that selection. For example, the patterns of natural selection on plant resistance to one herbivore depend on the presence or absence of other herbivores in the community (Pilson 1992). The future of community ecology will demand much more knowledge about how communities are shaped by constantly evolving networks of species interactions, and how these networks of different kinds of interactions, in turn, shape community dynamics and ongoing natural selection on these interactions.

The indirect effects that increase the ecological complexity of multitrophic communities may also be important in directing the evolution of members of these communities. Craig (Chapter 14) proposes that to understand community structure we need to study evolutionary interaction webs as well as ecological interaction webs. Ecological interaction webs involve alterations in mediating species that influence other species in the current generation, while in evolutionary interaction webs selection to alter the mediating species affects other species in subsequent generations. He provides an example of an evolutionary interaction web by showing that selection by one species can alter plant characteristics that indirectly influence the evolution of characters of a third species. A study of geographic variation in the goldenrod–galling-fly–natural-enemy community shows that indirect interactions between natural enemies mediated by the host plant and fly can influence the evolution of ovipositor length of a parasitoid. Craig also reviews interactions in other gall-centered insect communities to address the importance of evolutionary indirect effects on several gall characteristics. He also points out some problems with detecting and evaluating indirect effects, and suggests multivariate statistical techniques, such as multiple regression and path analysis, as a powerful tool for analyzing indirect selection.

The evolutionary consequences of herbivore-induced plant volatiles used by parasitoids to locate their hosts in multitrophic systems have been largely unstudied. Althoff (Chapter 15) outlines three approaches to investigating these evolutionary changes in multitrophic interactions: (1) selection experiments, (2) comparisons among closely related species, and (3) phylogenetic surveys across taxa in which herbivore-induced changes in plants might be important evolutionarily. He argues that, at the microevolutionary scale, demonstrating fitness differences for individuals that vary in their response to herbivore-induced changes is required. At the macroevolutionary scale the incorporation of phylogenetics into a comparative approach provides a powerful tool to identify the origins of traits that have evolved in response to herbivore-induced changes.

This volume therefore provides an overview and current perspectives which integrate direct and indirect effects to gain a more complete understanding of plant-centered ecological community structure. In particular, it brings together several different areas of study in plant–herbivore interactions, involving multitrophic interactions, diffuse coevolution, top–down vs. bottom–up effects, induced plant responses following herbivory, ecosystem engineering, and chemically mediated interactions. This book will also offer an interface in linking different disciplines in ecology, which have been conducted largely in isolation. These include aboveground and belowground ecology, herbivory and pollination ecology, mammal, insect, and microbial ecology, and chemical and community ecology. Finally, we hope that this book stimulates an interesting, new way of looking at direct and indirect interactions in ecological communities.

References

Abrams, P. A., B. A. Menge, G. G. Mittelbach, D. A. Spiller, and P. Yodzis. 1996. The role of indirect effects in food webs, pp. 371–395 in G. A. Polis and K. O. Winemiller (eds.) *Food Webs: Integration of Patterns and Dynamics*. New York: Chapman and Hall.

Agrawal, A. A., and P. A. Van Zandt. 2003. Ecological play in the coevolutionary theatre: genetic and environmental determinants of attack by a specialist weevil on milkweed. *Journal of Ecology* **91**:1049–1059.

Bernays, E. 1998. Evolution of feeding behavior in insect herbivores. *BioScience* **35**:35–43.

Berryman, A. (ed.) 2002. *Population Cycles: The Case for Trophic Interactions*. Oxford, UK: Oxford University Press.

Carpenter, S. R., and J. F. Kitchell. 1988. Consumer control in lake productivity. *BioScience* **38**:764–769.

Carpenter, S. R., J. F. Kitchell, and J. R. Hodgson. 1985. Cascading trophic interactions and lake productivity. *BioScience* **35**:634–639.

Ehrlich, P. R., and P. H. Raven. 1964. Butterflies and plants: a study in coevolution. *Evolution* **18**:586–608.

Gange, A. C., and V. K. Brown (eds.) 1997. *Multitrophic Interactions in Terrestrial Systems*. Oxford, UK: Blackwell Science.

Holt, R. D., and J. H. Lawton. 1994. The ecological consequences of shared natural enemies. *Annual Review of Ecology and Systematics* **25**:495–520.

Hunter, M. D., and P. W. Price. 1992. Playing chutes and ladders: heterogeneity and the relative roles of bottom-up and top-down forces in natural communities. *Ecology* **73**:724–732.

Hunter, M. D., T. Ohgushi, and P. W. Price (eds.) 1992. *Effects of Resource Distribution on Animal–Plant Interactions*. San Diego, CA: Academic Press.

Jermy, T. 1993. Evolution of insect–plant relationships: a devil's advocate approach. *Entomologia Experimentalis et Applicata* **66**:3–12.

Lill, J. T., and R. J. Marquis. 2003. Ecosystem engineering by caterpillars increases insect herbivore diversity on white oak. *Ecology* **84**:682–690.

Martinsen, G. D., K. D. Floate, A. M. Waltz, G. M. Wimp, and T. G. Whitham. 2000. Positive interactions between leafrollers and other arthropods enhance biodiversity on hybrid cottonwoods. *Oecologia* **123**:82–89.

Ohgushi, T. 2005. Indirect interaction webs: herbivore-induced effects through trait change in plants. *Annual Review of Ecology, Evolution, and Systematics* **36**:81–105.

Pace, M. L., J. J. Cole, S. R. Carpenter, and J. F. Kitchell. 1999. Trophic cascades revealed in diverse ecosystems. *Trends in Ecology and Evolution* **14**:483–488.

Pilson, D. 1992. Aphid distribution and the evolution of goldenrod resistance. *Evolution* **46**:1358–1372.

Price, P. W., C. E. Bouton, P. Gross, *et al.* 1980. Interactions among three trophic levels: influence of plants on interactions between insect herbivores and natural enemies. *Annual Review of Ecology and Systematics* **11**:41–65.

Schmitz, O. J., V. Krivan, and O. Ovadia. 2004. Trophic cascades: the primacy of trait-mediated indirect interactions. *Ecology Letters* **7**:153–163.

Stanton, M. L. 2003. Interacting guilds: moving beyond the pairwise perspective on mutualisms. *American Naturalist* **162**:S10–S23.

Thompson, J. N. 1994. *The Coevolutionary Process*. Chicago, IL: University of Chicago Press.

Thompson, J. N. 1996. Evolutionary ecology and the conservation of biodiversity. *Trends in Ecology and Evolution* **11**:300–303.

Tscharntke, T., and B. A. Hawkins (eds.) 2002. *Multitrophic Level Interactions*. Cambridge, UK: Cambridge University Press.

Waltz, A. M., and T. G. Whitham. 1997. Plant development affects arthropod communities: opposing impacts of species removal. *Ecology* **78**:2133–2144.

Werner, E. E., and S. D. Peacor. 2003. A review of trait-mediated indirect interactions in ecological communities. *Ecology* **84**:1083–1100.

Whitham, T. G., W. P. Young, G. D. Martinsen, *et al.* 2003. Community and ecosystem genetics: a consequence of the extended phenotype. *Ecology* **84**:559–573.

Zangerl, A. R., and M. R. Berenbaum. 2003. Phenotype matching in wild parsnip and parsnip webworms: causes and consequences. *Evolution* **57**:806–815.

Part II INTERACTION LINKAGES PRODUCED BY PLANT-MEDIATED INDIRECT EFFECTS

2

Plant-mediated interactions in herbivorous insects: mechanisms, symmetry, and challenging the paradigms of competition past

ROBERT F. DENNO AND IAN KAPLAN

Introduction

Interspecific interactions between insect herbivores can be either negative (competitive) or positive (facilitative) (Damman 1993, Denno *et al.* 1995). In the context of traditional community ecology, however, negative interactions have received the most attention (e.g., Lawton and Strong 1981, Schoener 1982, Strong *et al.* 1984, Denno *et al.* 1995) until quite recently (e.g., Lill and Marquis 2003, Nakamura *et al.* 2003). Nonetheless, the importance of interspecific competition as a factor structuring communities of insect herbivores has experienced a controversial history to say the least (Strong *et al.* 1984, Damman 1993, Denno *et al.* 1995). During the 1960s and 1970s, competition was revered as a central organizing force structuring communities of phytophagous insects (Denno *et al.* 1995). During these decades, field investigations into interspecific competition were heavily dominated by observational studies of resource partitioning as evidence for reduced competition and thus coexistence (e.g., McClure and Price 1976, Rathcke 1976, Waloff 1979). Notably, experimental field studies documenting the occurrence of interspecific competition between insect herbivores were scarce (but see McClure and Price 1975).

In the 1980s, the role of competition in structuring phytophagous insect communities was challenged severely, and within a few years it fell from a position of prominence to the status of a weak and infrequent process (Lawton

Ecological Communities: Plant Mediation in Indirect Interaction Webs, ed. Takayuki Ohgushi, Timothy P. Craig, and Peter W. Price. Published by Cambridge University Press. © Cambridge University Press 2007.

and Strong 1981, Lawton 1982, Lawton and Hassell 1984, Strong et al. 1984). Two lines of criticism led to its downfall. The first had its roots in a theoretical paper by Hairston et al. (1960) who argued that because defoliation was infrequent, food must be rarely limiting for herbivores, and that natural enemies were largely responsible for maintaining herbivore densities below competitive levels. A second avenue of criticism that led to the demise of competition stemmed from the analysis of phytophagous insect distributions and co-occurrences. Cases of positive interspecific association and the presence of vacant niches led many ecologists to question the importance of competition (Strong 1981, Lawton 1982). During these times, it is noteworthy that the studies responsible for both the veneration and demise of interspecific competition were overwhelmingly observational, and nonexperimental (reviewed in Denno et al. 1995). By the mid 1980s the scientific community had responded to a plea for a more experimental approach, and many more manipulative investigations of competition between insect herbivores began to appear in the literature (e.g., McClure 1980, 1989, 1990, Kareiva 1982, Stiling and Strong 1984, Karban 1986, 1989, Crawley and Pattrasudhi 1988, Denno and Roderick 1992). Along with such studies came the rejuvenated perception that herbivorous insects frequently compete and subsequent reviews reflect this reversal in thinking (Faeth 1987, Damman 1993).

In the mid 1990s, Denno et al. (1995) published a review based on experimental studies of interspecific competitive interactions between insect herbivores. This review examined 193 pair-wise interactions, and found strong evidence for competitive effects on performance and fitness. There was clear evidence for competitive effects in 93% of the studies involving sap-feeding herbivores (e.g., aphids, scale insects, planthoppers, and leafhoppers) and in 78% of the studies involving chewing herbivores (e.g., grasshoppers, caterpillars, beetles, and sawflies). Most surprising was the finding that over half of the cases of interspecific competition between mandibulate herbivores (52%) were examples of plant-mediated competition in which previous feeding by one species induced either nutritional or allelochemical changes in the plant that adversely affected the performance of another species feeding later in the season. Thus, one of the major outcomes of this review (Denno et al. 1995) and other contemporaneous assessments (Damman 1993) was that competitive effects were often plant-mediated via induced resistance and that failure to investigate such a mechanism may have vastly underestimated the possible importance of competition as a structuring force in communities of insect herbivores.

In the 1990s and thereafter, many more studies began to appear documenting that plant-mediated interactions were widespread and that insect herbivores frequently compete via induced resistance (e.g., Moran and Whitham 1990, Masters and Brown 1992, Rausher et al. 1993, Inbar et al. 1995, Salt et al. 1996,

Wold and Marquis 1997, Agrawal 1998, Heard and Buchanan 1998, Denno *et al.* 2000, Redman and Scriber 2000, Nykänen and Koricheva 2004, Van Zandt and Agrawal 2004b). Moreover, books on induced resistance also appeared during this time further highlighting that plants mediate interactions among insect herbivores (Tallamy and Raupp 1991, Karban and Baldwin 1997, Agrawal *et al.* 1999c). Thus, competition has re-entered the mix as a potentially important force influencing the structure of insect communities on plants.

Not only did reviews reveal that plants often mediate negative interactions between insect herbivores (competition via induced resistance or altered risk of natural enemy attack), but they also documented that plants can promote positive interactions (induced susceptibility) via altered plant nutrition (feeding facilitation), allelochemistry (leaf volatiles), architecture (leaf flush), or protective housing (leaf shelters) (Damman 1993, Denno *et al.* 1995, Nykänen and Koricheva 2004). Thus, in the context of a rapidly growing literature that documents the prevalence of plant-mediated interactions between insect herbivores, our objectives for this chapter are to: (1) identify the mechanisms underlying plant-mediated interactions between insect herbivores including induced resistance, induced susceptibility, and induced leaf volatiles and extrafloral nectar that alter risk of enemy attack, (2) outline herbivore traits (e.g., early colonization, tolerance of induced resistance, aggregation, dispersal, and feeding guild) that affect the symmetry of plant-mediated interactions between insect herbivores (e.g., competitive superiority), (3) explore the constraints that superiority in plant-mediated interactions (e.g., competitive dominance) may place on other life-history traits such as dispersal that ultimately affect landscape-scale population dynamics, and (4) discuss how plant-mediated interactions among insect herbivores increase the chances for interspecific interaction, yet challenge the paradigms of traditional competition theory. Last, we argue for a more all-encompassing approach to future studies that extends beyond a focus on pair-wise interactions and examines plant-mediated effects on multiple species at multiple spatial and temporal scales. Overall, we suggest that by considering plant-mediated interactions among insect herbivores, forces such as competition and facilitation will play an increasingly important role in structuring communities of phytophagous insects, albeit in nontraditional ways.

Mechanisms underlying plant-mediated interactions between insect herbivores

The causal mechanisms of plant-mediated interaction between insect herbivores are diverse and include induced plant allelochemistry, nutrition, morphology, and altered natural-enemy attack. In this section we first consider mechanisms underlying negative interspecific interactions (competition via induced resistance) and then we address those mechanisms promoting positive interactions (facilitation via induced susceptibility). In several cases of interspecific interaction, the precise mechanism of plant mediation is indeed known. However, much of the mechanistic work on induced resistance, especially earlier investigations, focused on intraspecific effects whereby previous feeding by one species had later consequences for conspecifics. On occasion, we also call on such intraspecific studies to provide a mechanistic understanding of induced effects, but emphasize that such mechanisms likely underlie interspecific effects as well.

Induced resistance and plant-mediated competition between insect herbivores
Induced allelochemistry

When herbivorous insects feed on plants, numerous allelochemicals can be induced with putative adverse effects on the performance or fitness of other herbivores that feed either contemporaneously or at a later time (Karban and Baldwin 1997, Constabel 1999). Allelochemicals known to be induced by herbivore feeding or mechanical wounding include defensive proteins (e.g., proteinase inhibitors, oxidative enzymes, cell-wall proteins, phenylpropanoid enzymes, lectins, and carbohydrate-binding enzymes), phenolics (phenolic acids, phenolic glycosides, furanocoumarins, coumarins, condensed and hydrolyzable tannins, lignin, and total phenolics), terpenoids (monoterpenes, diterpene aldehydes, phytoecdysteroids, and cucurbitacins), alkaloids (nicotine, quinolizidine, tropane, and hydroxamic acids such as DIMBOA), and indole glucosinolates (Karban and Baldwin 1997, Constabel 1999).

Induction of phytochemicals by herbivory can take place in two ways, either by the release of "preformed chemicals" or by "activated synthesis," both of which can have adverse consequences for other herbivores (Karban and Baldwin 1997). Unlike most activated phytochemicals, the induction of "preformed defenses" does not result exclusively from changes in synthesis or degradation (Karban and Baldwin 1997). Rather, the release of preformed chemicals results from the disruption of tissues where such chemicals are compartmentalized or from the mixing of locally separated substrates and enzymes during tissue damage. Examples of preformed/compartmentalized allelochemicals include the monoterpenes stored in the resin ducts of conifers (Raffa 1991), the cardenolide-containing latex borne in the canal systems of a diversity of plant taxa (Dussourd and Denno 1991), and the furanocoumarins housed in oil tubes of wild parsnip plants (Berenbaum and Zangerl 1999, Zangerl 1999). When herbivores puncture such ducts, canals, or oil tubes during feeding, the compartmentalized allelochemicals are released, often with adverse effects. Examples of preformed allelochemicals that are induced by the mixing of separated substrates when plant tissue is damaged include the production of hydrogen cyanide in cyanogenic plants (Conn 1979), the hydrolysis of glucosinolates to form thiocyanates in crucifers (Chew 1988), and the conversion of phenolic glycosides to more active feeding deterrents such as salicin in poplars (Clausen et al. 1989).

In the "activated" class of allelochemicals are proteinase inhibitors and many alkaloids that are synthesized following herbivory and result in an actual increase in the allelochemical pool (Karban and Baldwin 1997, Constabel 1999). The induced increase in allelochemistry can occur over a range of time intervals from a few hours or days (rapid induced response) to the next season (delayed induced response) (Karban and Baldwin 1997). Regardless, the distinction between "preformed" and "activated" induced responses is best viewed as a continuum (Karban and Baldwin 1997), because some allelochemicals have elements of both. For example, while furanocoumarins, monoterpenes, and glucosinolates are expressed constitutively (i.e., are preformed), they can be also induced several-fold by herbivory, suggesting induced synthesis (Lewinsohn et al. 1991, Agrawal et al. 1999b, Zangerl 1999). Rapid induced allelochemicals and the quick release of preformed allelochemicals are more likely to affect herbivores that feed contemporaneously on plants, whereas delayed plant responses offer the possibility of mediating interactions between herbivores that are temporally displaced.

Although there are numerous studies documenting the induction of specific allelochemicals by insect herbivores (reviewed in Karban and Baldwin 1997, Constabel 1999) and many others demonstrating plant-mediated competition between different herbivore species (Denno et al. 1995, I. Kaplan and R.F. Denno unpublished data), there are few reports linking the induction of a specific allelochemical by one species with a measured negative consequence for another herbivore species. Thus, the allelochemical mechanism underlying plant-mediated interspecific competition remains largely unknown for most interactions. Moreover, even when there is an association between an induced chemical and altered herbivore performance, the causal link is often ambiguous because only selected allelochemicals were measured (see Constabel 1999). Nonetheless, there are a few cases identifying the probable allelochemical involved in induced resistance.

Several examples involve the induction of defensive proteins. For instance, previous feeding by the fruitworm *Helicoverpa zea* induces increased levels of proteinase inhibitors and polyphenol oxidases in tomato *Lycopersicon esculentum*, which have adverse effects on the growth of a leaf-chewing lepidopteran and the density of aphids and a dipterous leaf miner (Stout and Duffey 1996, Stout et al. 1998). Similarly, feeding by the whitefly *Bemisia argentifolii* induces increases in defensive proteins in cabbage *Brassica oleracea* with negative consequences for the growth and survival of the cabbage looper *Trichoplusia ni*, and in *L. esculentum* with negative effects on the oviposition, density, and survival of the leaf miner *Liriomyza trifolii* (Inbar et al. 1999a, 1999b, Mayer et al. 2002). Also, early season feeding by lepidopterans on oak induces increases in protein-binding capacity with negative density effects on a guild of lepidopteran species that feed in mid- to-late season (Wold and Marquis 1997).

There are other cases of induced resistance that implicate terpenoids, glucosinolates, and latex. For instance, root-feeding by the wireworm *Agriotes lineatus* induces increases in aboveground terpenoids in cotton *Gossypium herbaceum* which adversely affect the growth of the leaf-chewing lepidopteran *Spodoptera exigua* (Bezemer et al. 2003). Also, glucosinolates are induced by larvae of the imported cabbage worm *Pieris rapae* feeding on wild radish *Raphanus raphanistrum* that negatively affect the weight, colonization, and density of lepidopterans, aphids, and a dipterous leaf miner (Agrawal 1998, 1999, 2000). Likewise, feeding by monarch butterfly larvae *Danaus plexippus* induces latex production in milkweed *Asclepias syriaca* with adverse effects on the larval mass of the chrysomelid beetle *Labidomera clivicollis* and colonization by the aphid *Aphis nerii* (Van Zandt and Agrawal 2004b). Regardless of the specific mechanism, however, an important generality that emerges from this small collection of studies is that induced resistance conferred by allelochemistry can result in substantial competitive effects between insect herbivores that are taxonomically discrepant, temporally displaced, and spatially separated on the same plant.

Induced morphology

Early colonizing herbivores can induce changes in gross plant morphology (e.g., leaf flush, bud burst, branching architecture, floral traits, and leaf shelters) as well as small-scale surface structure (e.g., trichomes) that can have significant effects on the performance, survival, and density of other herbivores that feed later in the colonization sequence (Karban and Baldwin 1997). The induced morphological change can have either negative or positive effects (see below) on subsequent feeding herbivores. Two induced morphological changes that are known to adversely affect herbivores are changes in leaf trichome density (Agrawal 1998, 1999, Traw and Dawson 2002) and altered flower size and number (Strauss 1997). Because induced changes in plant morphology are expressed well after the initial bout of herbivory, their effects are likely to be felt by other herbivores that are temporally displaced from the inducing species, and significantly so in some cases. For example, when attacked by the chrysomelid beetle *Agelastica alni*, it is the reflushed leaves of grey alder *Alnus incana* that have a higher density of nonglandular trichomes (Baur et al. 1991). Also, the resprouted leaves of stinging nettle *Urtica dioica* carried a higher density of trichomes following damage than did noninduced leaves (Pullin and Gilbert

1989). Similarly, induced changes in floral and seed morphology often occur well after the initial bout of herbivory (Hendrix 1988, Strauss et al. 1996, Thalmann et al. 2003).

Notably, feeding by larvae of *Pieris rapae* induced increased trichome density in black mustard *Brassica nigra*, which had very adverse effects on the growth of conspecifics that fed several weeks later (Traw and Dawson 2002). Evidence from a similar study system links feeding-induced increases in leaf trichome density with the reduced performance of an assemblage of other herbivore species. Specifically, larvae of *P. rapae* induced trichomes in *R. raphanistrum* that had negative consequences for the growth and density of leaf-chewing lepidopterans, aphids, and dipterous leaf miners (Agrawal 1998, 1999, 2000).

Folivory-induced changes in floral traits are also likely to mediate interspecific interactions between insect herbivores with negative consequences for later-feeding species. For example, foliar herbivory can result in reduced flower size and number (Hendrix and Trapp 1989, Strauss et al. 1996, Lehtilä and Strauss 1997), and in pollen and nectar production (Strauss et al. 1996, Krupnick et al. 1999). Such induced changes have been viewed in the context of herbivore–pollinator interactions (Strauss 1997, Ohgushi 2005). However, one could argue that many pollinators (e.g., syrphid flies and bees) are indeed herbivores. The traditional focus has been on how folivory-induced changes in floral traits subsequently affect pollinator services to the plant (Strauss et al. 1996, Lehtilä and Strauss 1997, Krupnick et al. 1999). From the pollinator's perspective, it is less clear how herbivory-induced changes in floral morphology feed back to alter foraging strategies and ultimately pollinator fitness. Thus, the extent to which such interactions might represent plant-mediated competition remains plausible but open.

Most assuredly though, foliar herbivory can affect other herbivores that require a particular floral morphology for concealment and/or development. For instance, foliar herbivory can drastically reduce seed size (Hendrix 1988, Thalmann et al. 2003), and reduced seed size can have dire consequences for the growth and survival of seed-infesting herbivores (Szentesi and Jermy 1995, Moegenburg 1996, Willott et al. 2000). However, only one study involving weevils on musk thistle *Carduus nutans* provides a mechanistic link between folivore-induced changes in floral or seed morphology and their effects on an inflorescence/seed-feeder (Milbrath and Nechols 2004). Early-season feeding by the folivorous weevil *Trichosirocalus horridus* induced shorter flower stems, significantly fewer flower heads, and delayed flowering by one week, all of which had adverse effects (colonization, oviposition, and survival) on the seed-head-feeding weevil *Rhinocyllus conicus*.

Altered nutrition and source–sink dynamics

When phloem-tapping herbivores such as aphids feed, they alter the source–sink dynamics of phloem transport (Way and Cammal 1970, Larson and Whitham 1991). Using $^{14}CO_2$-labeling experiments, it has been shown that aphids act as nutrient sinks by diverting assimilates from neighboring leaves and drawing them toward feeding sites where aphid performance is dramatically enhanced (Way and Cammal 1970, Larson and Whitham 1991, Inbar et al. 1995). The strength of the sink, and thus aphid performance, is influenced by the position of the feeding site such that aphids occupying the basal portion of the leaf (on the midrib) divert more assimilates than more distally located aphids (Larson and Whitham 1991, Inbar et al. 1995).

The manipulation of the phloem transport system provides the opportunity for plant-mediated exploitative competition between two phloem feeders because of where they feed on the plant with respect to one another. For example, the phloem-feeding aphid *Geoica* sp. forms galls at the base of wild pistachio *Pistacia palaestina* leaves where it diverts assimilates from the distal margins of leaves where another gall-forming aphid *Forda formicaria* develops (Inbar et al. 1995). The diversion of assimilates and the creation of basal nutrient sinks benefits *Geoica* but results in 84% mortality in distal *Forda* galls (Inbar et al. 1995).

Other phloem-feeding herbivores also compete via induced changes in plant nutrition, but the mechanism is quite different. For instance, the planthoppers *Prokelisia dolus* and *P. marginata* co-occur on intertidal marshes along the Atlantic coast of North America where they feed exclusively on the cordgrass *Spartina alterniflora* (Denno and Roderick 1992, Denno et al. 2000). Densities can be high and they often share the same plant where they feed together on the upper surfaces of the leaf blades (Denno et al. 2000). Previous feeding by *P. dolus* reduces the concentration of essential amino acids in cordgrass (Olmstead et al. 1997), which has very adverse effects on the performance and survival of *P. marginata* (Denno et al. 2000). Reciprocal effects are far less severe due to the ability of *P. dolus* to better tolerate low plant nitrogen via feeding compensation (Denno et al. 2000).

Induced reductions in plant nitrogen are also implicated in the interspecific triggering of emigrants (macropterous forms or alates) in both planthoppers (Denno and Roderick 1992, Matsumura and Suzuki 2003) and aphids (Itô 1960, Lamb and MacKay 1987). As populations grow, both crowding and induced changes in plant nutrition combine to stimulate nymphs to develop into dispersing morphs (Denno and Roderick 1992). Such interactions are competitive because one species stimulates the production of migrants in another, which promotes emigration from a shared plant resource (Denno and Roderick 1992).

Leaf-chewing herbivores also compete via induced changes in plant nutrition. For example, early-season feeding by lepidopterans induces reductions in foliar nitrogen that negatively affect the survival and performance of the late-season leaf miners and other free-living leaf-chewers on *Quercus robur* (West 1985, Hunter 1992). Likewise, previous feeding by the gypsy moth *Lymantria dispar* decreases plant nitrogen that has negative consequences for the growth and survival of the tiger swallowtail *Papilio canadensis* on quaking aspen *Populus tremuloides* (Redman and Scriber 2000). Like induced changes in plant allelochemistry and morphology, induced nutrition provides the opportunity for spatially and temporally displaced herbivores to compete.

Although feeding-induced changes in plant nutrition and source–sink dynamics can promote interactions between herbivore species, they should be viewed as "induced responses" (*sensu* Karban and Baldwin 1997) because the nutrient sinks created by aphids in fact compete with plants for assimilates by diverting nutrients away from developing fruits (Larson and Whitham 1991). Thus, from the plant's perspective, whereas some induced allelochemicals and trichomes may benefit the plant and act as "induced defenses" (Agrawal 1998, 1999), known cases of induced nutrition do not clearly benefit the plant (Weis *et al.* 1988, Larson and Whitham 1991).

Altered risk of enemy attack

Feeding-induced changes in plant volatiles, extrafloral nectar, and morphology can also affect herbivore–herbivore interactions via altered risk of attack from natural enemies (Paré *et al.* 1999, Thaler 1999, Ness 2003). When herbivores feed on plants, volatiles such as terpenoids, green-leaf lipoxygenase products, and indole are either released from storage structures or are synthesized (Paré *et al.* 1999). Induced volatiles are then used by predators and parasitoids as cues to locate their habitats and hosts (Vinson 1998, Paré *et al.* 1999, Thaler 2002a). As a consequence, parasitism rates of herbivores on induced plants can as much as double (Thaler 1999). In some cases of induction, plants produce a herbivore-specific blend of volatiles that allow specialist parasitoids to distinguish between their host and close non-host relatives (De Moraes *et al.* 1998). For such cases involving specialists, parasitoids attack only the inducing herbivore and interspecific effects on other herbivore species are precluded. However, in other instances, damage by one herbivore species attracts generalist parasitoids (Thaler 2002a) or predators (Turlings *et al.* 1991, Paré *et al.* 1999) that have the potential to attack other co-occurring herbivore species. For example, damage of aspen leaves by gypsy moth larvae attracts generalist parasitoids that then parasitize larvae of swallowtail larvae as well as other caterpillar species (Redman and Scriber 2000). Apparently, volatiles emitted by wounded trees attract a wide range of parasitoid species with broad host preferences (Schultz 1999, Redman and Scriber 2000). In another example, prior feeding by the mirid bug *Tupiocoris notatus* causes native tobacco plants to release volatile organic compounds that attract the generalist predator *Geocoris pallens*, which then selectively attacks caterpillars of the less mobile and co-occurring herbivore *Manduca quinquemaculata* (Kessler and Baldwin 2004).

Herbivory can also induce the flow of extrafloral nectar that subsequently attracts generalist natural enemies with negative effects on herbivores. For example, feeding by catalpa sphinx moth larvae *Ceratomia catalpae* induces two-fold increases in the volume of extrafloral nectar, which then results in a ten-fold increase in the density of attending ants *Forelius pruinosus* with subsequent reductions in the density of sphinx moth larvae (Ness 2003). Because such ants are generalist predators, the possibility remains that the density of other herbivore species was similarly reduced, but this interaction was not examined.

By altering plant morphology via building leaf shelters, herbivores can also enhance the density of natural enemies, an effect that cascades to adversely affect other herbivore species. For instance, several shelter-building caterpillars on willow *Salix miyabeana* promote the occurrence of specialist aphids *Chaitophorus saliniger* that feed preferentially within leaf shelters (induced susceptibility); the aphids in turn attract predaceous ants that then reduce the survival of the co-occurring leaf beetle larvae *Plagiodera versicolora* by 60% (indirect induced resistance) (Nakamura and Ohgushi 2003, Ohgushi 2005).

One of the frequent outcomes of induced resistance for insect herbivores is delayed development (Inbar *et al.* 1999b, Denno *et al.* 2000, Wise and Weinberg 2002), a result that may prolong exposure to natural enemies with increased risk of predation (Benrey and Denno 1997). Although the slow growth–high mortality hypothesis has been explored in the context of single predator–prey interactions (see review by Williams 1999), it remains largely untested in an interspecific framework whereby one herbivore induces a plant change that results in delayed growth and increased risk of attack in another herbivore species. Two studies, however, provide support (Kessler and Baldwin 2004, I. Kaplan *et al.* unpublished data). First, previous feeding by the mirid bug *Tupiocoris notatus* induces the accumulation of proteinase inhibitors in wild tobacco that are correlated with developmental delays in larvae of *Manduca quinquemaculata* making them more susceptible to attack by the generalist predator *Geocoris pallens* (Kessler and Baldwin 2004). In this system, the risk of predation in *Manduca* is further exacerbated because mirids also induce the production of volatile organic compounds that attract *Geocoris* to plants also occupied by *Manduca*. In a second study, previous feeding by potato leafhoppers *Empoasca fabae* induces chemical and morphological changes in potato plants that result in a significant developmental delay for later-feeding larvae of Colorado potato beetles *Leptinotarsa decemlineata* (I. Kaplan *et al.* unpublished data). Moreover, slow-developing larvae on induced plants incur much higher mortality from the predaceous stinkbug *Podisus maculiventris* than larvae feeding on undamaged plants. Thus, developmental delays imposed on one herbivore by another via induced resistance can increase significantly risk of predation.

To assess the overall interspecific effect of induced resistance on a herbivore, one must know both the magnitude and direction of the direct (allelochemistry, nutrition, and morphology) and indirect effects (natural enemies). For example, in the potato system, feeding by leafhoppers has a negative effect on the growth and survival of potato beetles, an effect that is further exacerbated in the presence of *Podisus* predators. By contrast, if induced responses in plants affect herbivores adversely but also reduce the quality of herbivores for natural enemies (e.g., result in small host size), then the direct negative effect may be offset in part by a relaxed indirect effect from higher trophic levels (Thaler 2002b). Thus, to understand fully the consequences of interspecific herbivore interactions mediated via induced resistance, interactions among all the major players in the system must be taken into account. Nonetheless, there are clear cases where induced responses in plants enhance enemy effects on other herbivores in the system resulting in instances of diffuse competition (Thaler 1999, Redman and Scriber 2000, Kessler and Baldwin 2004, I. Kaplan *et al.* unpublished data).

Induced susceptibility and plant-mediated facilitation between insect herbivores
Induced allelochemistry

When previous feeding by one herbivore induces allelochemicals that are used by another to locate their host plants, induced susceptibility can occur. Induced susceptibility is evident when larvae of *Pieris rapae* feed on wild radish and induce glucosinolates and perhaps other volatile compounds that are used by the specialist flea beetle *Phyllotreta* sp. to locate its host, a response that results in increased beetle colonization, oviposition, and plant damage (Agrawal and Sherriffs 2001).

Induced susceptibility can also arise when one herbivore deactivates the pre-formed defenses of a plant thus providing another herbivore the opportunity to feed. For instance, feeding by jack pine budworms *Choristoneura pinus* severs resin canals thereby reducing resin flow and the concentration of monoterpenes (Wallin and Raffa 2001). Deactivation of the resin defense system by budworms promotes colonization by bark beetles *Ips grandicollis* and pine sawyers *Monochamus carolinensis* resulting in positive density associations among the three herbivores. A similar situation occurs on plants bearing latex canals. The vein-cutting and trenching behaviors exhibited by a wide variety of insect herbivores depressurizes the latex-bearing canal system (Dussourd and Eisner 1987, Dussourd and Denno 1994). By feeding distal to the cuts, not only does the inducing herbivore benefit, but so can other species that feed on the undefended leaf tissue (Dussourd and Denno 1991, 1994). For example, armyworm larvae *Spodoptera ornithogalli* grow much more rapidly on the leaves of the latex bearing plant *Lactuca* following trenching by looper larvae *Trichoplusia ni* (Dussourd and Denno 1994). Given the prevalence of herbivores with behaviors that deactivate canal-borne defenses, and the abundance of plants with preformed defenses (Dussourd and Denno 1991, 1994), future research will undoubtedly discover more species of herbivores that take advantage of deactivated defenses and feed in association with trenching and vein-cutting species.

Induced morphology

Herbivores can also induce changes in plant morphology, such as increased branching and the re-flush of leaves, or alter architecture by building leaf shelters and leaf rolls that favor other herbivore species. Notably, induced changes in plant morphology usually occur well after the initial bout of herbivory or shelter construction providing the opportunity for temporally displaced herbivores to interact (Ohgushi 2005).

Regarding induced changes in plant morphology, damage to the apical meristem of goldenrod *Solidago altissima* by a guild of early-season gall-formers causes more branching that subsequently promotes increased colonization and densities of several aphids and spittlebugs (Pilson 1992). Similarly, there are studies showing that early-season herbivory by caterpillars on oaks and the gall-makers on willow stimulates a secondary leaf flush on which aphids, leaf beetles, and leaf rollers can flourish (Hunter 1992, Nakamura *et al.* 2003). In these examples, the mechanisms underlying facilitation extend beyond simply altered plant architecture. For instance, the regrowth leaves on which sap-feeders thrive are higher in nitrogen (Nakamura *et al.* 2003). Moreover, re-flushed leaves are tender and are easier for chewing herbivores to consume or construct leaf shelters with (Hunter 1992, Nakamura *et al.* 2003).

By constructing leaf shelters (rolls, folds, and ties) or forming leaf galls, herbivores create new habitats for other herbivores that often generate positive associations (Cappuccino 1993, Damman 1993, Martinsen *et al.* 2000, Fukui 2001, Lill and Marquis 2003, Ohgushi 2005). Positive interactions result because the shelters built by these so-called "ecosystem engineers" provide other colonizing herbivores with enemy-free space (Cappuccino 1993, Damman 1993), improved microclimate (Larsson *et al.* 1997), or improved food quality (Fukui 2001). Importantly, primary shelter-makers spend significant time and energy constructing their leaf structures, expenditures that later shelter-users do not incur (Fukui 2001). The experimental addition of artificial leaf shelters designed to mimic the structures of early-season leaf-tying and leaf-rolling lepidopterans results in population increases in other shelter-building caterpillars (Lill and Marquis 2003) and non-shelter-building herbivores such as aphids, lepidopterans, sawflies, and beetles (Cappuccino 1993, Nakamura and Ohgushi 2003). In one instance, recruitment of previously occupied artificial ties by secondary leaf tyers is double that compared to never-occupied ties implicating feeding-related volatiles (Lill and Marquis 2003). Conversely, the removal of natural leaf shelters results in remarkable reductions in the densities of other insect herbivores (Cappuccino and Martin 1994, Martinsen et al. 2000, Lill and Marquis 2003). It is important to realize that the initiating shelter-builder can induce positive interactions with other shelter colonists (aphids), but negative associations with folivores (leaf beetles) that are adversely affected by ants that tend the aphids (Nakamura and Ohgushi 2003).

Altered nutrition and source–sink dynamics

When herbivores feed, particularly phloem feeders, they can induce nutrient sinks by diverting nutrients from surrounding leaves to their feeding site (Larson and Whitham 1991, Inbar et al. 1995). If other herbivores colonize and exploit surrounding sites, they are deprived of nutrients and their performance is often adversely affected (Inbar et al. 1995). On the other hand, if other herbivores co-occur locally along with the inducers at the site of the nutrient sink, they can benefit tremendously from the enhanced levels of assimilates. For instance, numerous aphid species either selectively colonize the specific feeding sites occupied by other aphid species or shift their feeding site to co-occur with another aphid species (Shearer 1976, Salyk and Sullivan 1982). The result is a pattern of positive interspecific association either within a plant or across plants in the field (Kidd et al. 1985, Formusoh et al. 1992, Montandon et al. 1993, Waltz and Whitham 1997). Similarly, leafhoppers *Idiocerus* sp. occur far more abundantly on poplar branches with *Pemphigus* gall aphids compared to branches where gall aphids are excluded, and leafhoppers are frequently observed feeding directly on galls (Waltz and Whitham 1997, G. Wimp personal communication). By occurring in close proximity to an induced nutrient sink, sap-feeders experience enhanced growth, body size, and survival (Forrest 1971, Shearer 1976, Kidd et al. 1985, Formusoh et al. 1992, Montandon et al. 1993).

Feeding-induced changes in source–sink dynamics can also benefit spatially separated herbivores on the same plant. Root-feeding coleopterans, for example, induce increases in aboveground leaf nitrogen that promotes increased growth, fecundity, and density of foliar-feeding aphids (Gange and Brown 1989, Masters and Brown 1992) and inflorescence-feeding tephritid flies (Masters *et al.* 2001). Notably, most cases of plant-mediated facilitation via altered nutritional dynamics involve phloem-feeders, either as the inducing species or even more frequently as the beneficiary (Denno *et al.* 1995). The sensitivity of phloem-feeders to changes in amino nitrogen (Cook and Denno 1994) undoubtedly contributes to their overrepresentation in cases of plant-mediated facilitation involving altered source–sink dynamics.

Altered risk of natural-enemy attack

It is conceivable that one herbivore might induce allelochemicals that are then sequestered by another species for use in defense against natural enemies. This would represent a case of plant-mediated facilitation involving induced defenses and reduced enemy attack. Although no such cases are explicitly known, the scenario could result between leaf beetle species (*Galerucella lineola* and *Phratora vitellinae*) that co-occur on phenolic-glycoside-containing willows. Leaf beetles induce salicylates (Ruuhola 2001) and salicylates are the precursors of sequestered salicyl aldehyde that is excreted by *Phratora* as an effective defense against some invertebrate predators (Denno *et al.* 1990). Despite the possibility, however, pieces of this hypothetical interspecific scenario have not been linked.

Life-history traits promoting competitive superiority in plant-mediated interactions

Plant-mediated interactions between insect herbivores, both negative and positive, are overwhelmingly asymmetric (Denno *et al.* 1995). Competitive interactions, the ones we emphasize here, are particularly one-sided, with one species being clearly the superior competitor. Admittedly, however, many experimental designs examining plant-mediated effects between two herbivore species lack reciprocity (e.g., only examine the effects of spring feeders on late-season herbivores) and it is therefore not possible to separate competition from amensalism. Nonetheless, where designs have tested for reciprocal effects, these have been found even though interactions are strongly asymmetric (Tomlin and Sears 1992, Agrawal 2000). In this section we explore various herbivore traits that contribute to competitive superiority. It goes without saying that the various traits conferring competitive superiority may not be mutually exclusive and in fact may act in concert in some cases. In a forthcoming section, however, we provide evidence that trade-offs exist between certain traits (e.g., feeding compensation and dispersal), and that such trade-offs may constrain competitive superiority.

Early-season exploitation of plant resources (breaking diapause and colonization)

Early-season arrival/exploitation often provides a distinct advantage in plant-mediated competitive interactions (Denno et al. 1995). This advantage frequently involves the early breaking of diapause coupled with the rapid colonization and monopolization of optimal feeding sites (McClure 1980, Hunter 1992). The early exploitation of plant resources by one herbivore provides the opportunity for induced resistance with imposed effects on other herbivores that feed later (Denno et al. 1995, 2000). For instance, early colonization and feeding by nymphs of the scale *Fiorinia externa* significantly reduces the nitrogen available in young hemlock *Tsuga canadensis* foliage, which in turn dramatically decreases the survival of the co-occurring scale *Nuculaspis tsugae* (McClure 1980). Notably, when the advantage of early arrival is experimentally removed, competitive superiority is lost for *F. externa* (McClure 1980). The asymmetry of interspecific interactions between gall aphids (*Geoica* and *Forda*) results because competitive domination in *Geoica* is associated with the spring occupation of optimal leaf sites (leaflet base) where phloem transport is easily manipulated, nutrients are monopolized via nutrient sinks, and the distal galls of *Forda* suffer dramatically (Inbar et al. 1995). Although early-season feeding is clearly associated with competitive superiority in many plant-mediated interactions (Denno et al. 1995), there are admittedly no studies that have examined possible carry-over effects of late-season inducers on spring feeders. However, feeding by the spring defoliator *Epirrita autumnata* induces higher concentrations of total phenolic compounds in mountain birch (*Betula pubescens*) that carry over to the next season with adverse intraspecific effects on fecundity (Kaitaniemi et al. 1998, 1999). Thus, the possibility exists for induced interannual carry-over effects on other herbivore species.

Tolerance of allelochemicals

It goes without saying that insect herbivores induce resistance characteristics in plants that feed back to have adverse growth and fitness consequences for the inducing herbivore (Tallamy and Raupp 1991, Karban and Baldwin 1997, Agrawal et al. 1999c). In the context of interspecific interactions and competitive superiority, however, the question becomes one of differential tolerance. Specifically, competitive superiority can result because colonizing herbivores not only experience or create optimal feeding sites, but they are also relatively more tolerant of feeding-induced changes in plant physiology compared to later-feeding or co-occurring species. For instance, early-season feeding by the leaf beetle *Phratora vitellinae* induces increases in phenolic glycosides that it tolerates, but which adversely affect generalist herbivores such as *Operophtera brumata* (Ruuhola 2001). Similarly, the asymmetry of plant-mediated competition between *Prokelisia* planthoppers is due to the ability of *P. dolus* to better tolerate feeding-depleted levels of plant nitrogen compared to its co-occurring congener *P. marginata* (Denno et al. 2000).

The mechanisms underlying tolerance, however, are diverse. Tolerance of induced allelochemistry involves detoxification, excretion, sequestration, and/or behavioral avoidance (Karban and Agrawal 2002). For example, furanocoumarins are expressed both constitutively but can be also induced several-fold by herbivory (Berenbaum and Zangerl 1999, Zangerl 1999). Adapted swallowtail larvae *Papilio polyxenes* that encounter furanocoumarins detoxify these compounds using cytochrome P450 monooxygenases (Cohen et al. 1992). Prior ingestion of furanocoumarins induces cytochrome P450 activity which then allows detoxification and leaf consumption (Cohen et al. 1989). Likewise, larvae of the specialist swallowtail *Battus philenor* induce high levels of aristolochic acid without physiological cost or increased mortality, but the mechanism underlying tolerance is unknown (Fordyce 2001).

Differences in tolerance to feeding-induced reductions in amino nitrogen between *Prokelisia* species result from dissimilarities in cibarial muscle mass (musculature associated with ingestion) that influence feeding compensation (Denno et al. 2000). The competitive superiority of *P. dolus* over *P. marginata* results from its greater commitment to cibarial muscle mass and increased ability to compensate for declining plant nitrogen via increased throughput of cell sap. Thus, in mixed-species crowds, not only does *P. dolus* contribute more to declines in plant nitrogen but it also tolerates such declines better and experiences fewer performance and fitness costs (Olmstead et al. 1997, Denno et al. 2000).

Also, feeding by the whitefly *Bemisia argentifolii* induces defensive proteins (e.g., chinases, peroxidases, lysozymes) that have little effect on the whitefly but adversely affect the development and survival of the leaf-mining dipteran *Liriomyza trifolii* and the foliar-feeding lepidopteran *Trichoplusia ni* (Inbar et al. 1999a, 1999b, Mayer et al. 2002). Competitive superiority in this case results because sap-feeders, by virtue of their feeding style, are less exposed to defensive proteins (Raven 1983).

Regardless of the mechanism conferring tolerance, however, competitive superiority in several cases results from the selective ability of the inducing herbivore to tolerate or avoid induced plant responses better than other co-occurring species. Given that both generalists (*Bemisia*) and specialists (*Phratora* and *Prokelisia*) can dominate in plant-mediated competitive interactions, it seems unlikely that diet breadth will explain much of the variation underlying competitive superiority. On the other hand, it makes intuitive sense that generalists might be more likely to suffer from induced plant responses than adapted specialists and thus be relegated to the role of inferior competitor, but this hypothesis remains untested (see Karban and Baldwin 1997).

Aggregation

Herbivores that aggregate are overrepresented in cases of demonstrated interspecific competition between insect herbivores (Damman 1993, Denno *et al.* 1995). Moreover, aggregation appears to confer competitive dominance in plant-mediated interactions, especially in small sap-feeding herbivores such as aphids and psyllids that compete via the formation of nutrient sinks that divert nutrients from other "competitors" (e.g., Shearer 1976, Salyk and Sullivan 1982, Inbar *et al.* 1995, Heard and Buchanan 1998). The ability of aphids to form nutrient sinks and benefit from enhanced nutrition is positively density dependent up to a certain aggregation size beyond which resource depletion occurs (Way and Cammell 1970). Optimal group size in some aphids, and therefore maximal nutrient manipulation, is maintained to some extent by a balance between on-site reproduction and emigration (Way and Cammell 1970).

Although aggregation appears to confer competitive superiority in sap-feeders that form nutrient sinks, it is unclear whether aggregation in chewing herbivores provides the same advantage, especially if allelochemicals are concerned. For instance, many plants become resistant after only small amounts of leaf tissue are removed (Karban and Baldwin 1997). In other cases, resistance and damage are positively related (Karban and Baldwin 1997). In these latter cases, aggregation may enhance resistance and thus confer competitive superiority if the inducing species is relatively more tolerant of the induced allelochemical. Because several of the chewing herbivores involved in cases of plant-mediated competition are both competitively superior and feed in groups (Denno *et al.* 1995), the possibility remains that aggregation imparts a competitive edge.

Dispersal ability

Although early colonization of plant resources confers competitive superiority in numerous interactions between insect herbivores, dispersal ability per se does not appear to do so (Denno *et al.* 1995). In fact, in a survey of interspecific competitive interactions at large among insect herbivores, the loser was a better disperser than the superior competitor in 10 of 13 interactions (Denno *et al.* 1995). Many of these interactions involved pairs of sap-feeders (aphids and planthoppers) in which competition was likely mediated via induced resistance (Itô 1960, Tamaki and Allen 1969, Addicott 1978, McClure 1980, 1990, Edson 1985, Denno and Roderick 1992). It has been suggested elsewhere that immobile insects owe their success to their ability to manipulate plant physiology, particularly the mechanisms by which plants establish source–sink relationships (Karban and Baldwin 1997). Existing data for wing-dimorphic sap-feeders (aphids and planthoppers) suggest a phenotypic trade-off between competitive ability and dispersal capability, an issue that we develop in a forthcoming section.

Feeding guild

There is reason to expect that feeding guild (e.g., sap-feeder, free-living chewer, leaf miner, or root-feeder) could promote competitive superiority in plant-mediated interactions. This expectation arises because allelochemicals are much less concentrated in vascular tissue compared to leaf tissue (Raven 1983, Cook and Denno 1994). Thus, feeding and probing by sap-feeders may induce allelochemicals in tissues that they themselves do not experience or are exposed to in lower concentrations (Inbar *et al.* 1999a, 2003, de Ilarduya *et al.* 2003). Accordingly, sap-feeders may be advantaged over chewing herbivores that consume whole-leaf tissues. However, data at hand do not support this expectation. Indeed, there are three reports showing that sap-feeders (aphids and whiteflies) do impose adverse effects on the survival and performance of chewing herbivores (caterpillars and leaf-mining flies) (Mattson *et al.* 1989, Inbar *et al.* 1999a, 1999b, Mayer *et al.* 2002). There are conflicting data, namely five case studies, demonstrating that chewing herbivores induce plant responses that negatively affect sap-feeders (Ajlan and Potter 1992, Tomlin and Sears 1992, Stout and Duffey 1996, Agrawal 1998, Stout *et al.* 1998, Van Zandt and Agrawal 2004b). Moreover, the mechanisms underlying induced resistance in these studies are diverse, not guild specific, and include trichomes, defensive proteins, proteinase inhibitors, and glucosinolates.

Examples of competitive superiority in interactions between other feeding guilds are scant but no clear pattern prevails. For example, the belowground root-feeding coleopteran *Agriotes lineatus* imposed adverse effects on the aboveground leaf-chewer *Spodoptera exigua* in one case (Bezemer *et al.* 2003), but the reverse occurred in another instance where aboveground feeding by the leaf miner *Chromatomyia syngensiae* negatively affected the root-feeding scarab beetle *Phyllopertha horticola* (Masters and Brown 1992). Although the stem-boring weevil *Rhyssomatus lineaticollis* imposes adverse fitness effects on a variety of other feeding guilds (leaf-chewers, leaf miners, and sap-feeders) (Van Zandt and Agrawal 2004b), reciprocal effects have not been examined. Admittedly there are too few studies to draw a strong conclusion, but the data at hand do not suggest that feeding guild per se strongly confers competitive superiority in plant-mediated interactions involving insect herbivores.

Competitive superiority, trade-offs, and constraints imposed on other life-history traits

Competitive superiority may constrain other life-history traits such as dispersal due to underlying phenotypic trade-offs, and thus have widespread consequences for population dynamics (Denno et al. 1995). Physiological trade-offs among life-history traits such as fecundity, age to first reproduction, egg size, and dispersal are common in herbivorous insects (Zera and Denno 1997). Here we argue that competitive superiority, especially for sap-feeders that interact via induced changes in plant nutrition, may limit dispersal. Our suspicion stems from the observation that competitive dominance and dispersal ability are inversely related in several species of sap-feeders (Denno et al. 1995).

For planthoppers, competitive superiority arises because of their ability to compensate for feeding-induced reductions in amino nitrogen by increasing their intake of phloem sap (Denno et al. 2000). This is made possible by a large commitment to the musculature associated with ingestion. Planthoppers and leafhoppers feed by inserting their stylets into phloem and xylem tissues respectively (Backus 1985, Cook and Denno 1994). Then using a cibarial pump (modified esophagus), they ingest cell sap. The cibarial pump is driven by a series of dilator muscles that insert on the interior of the face, and face size is positively related to the cross-sectional mass and thus the power of the cibarial muscles to ingest cell sap (Backus 1985). For *Prokelisia* planthoppers, *P. dolus* has a much broader face and commitment to subtending cibarial musculature than *P. marginata* (Denno et al. 2000). Thus, *P. dolus* is more capable of increasing food uptake in response to any reductions in plant nitrogen, either natural or those induced by previous feeding, than *P. marginata*. Furthermore, feeding by *P. dolus* may deplete plant nitrogen more than feeding by *P. marginata*, further enhancing the consequences for *P. marginata*. This supposition may explain why previous feeding by *P. dolus* has much more dire consequences on the fitness of *P. marginata* than the reverse situation. Moreover, the broad face and associated large commitment to cibarial muscles in *Delphacodes penedetecta* (R.F. Denno unpublished data) likely explains why this planthopper does not suffer in plant-mediated interactions with *P. dolus* (Ferrenberg and Denno 2003).

Data for *Prokelisia* planthoppers suggest a trade-off between cibarial musculature and flight musculature and thus between competitive ability and dispersal. *Prokelisia dolus* is a very sedentary species with a high ratio of cibarial to flight muscle mass, which contrasts with the low ratio characteristic of the very dispersive *P. marginata* (Denno et al. 2000, A.F. Huberty and R.F. Denno unpublished data). Thus, competitive superiority is associated with feeding compensation and tolerance of low plant nitrogen, whereas poor competitive ability is linked with intolerance of depleted nitrogen and high mobility (Denno *et al.* 2000). The two planthoppers cope with their nitrogen requirements in two very different ways, *P. dolus* by feeding compensation and *P. marginata* by dispersal to nitrogen-rich habitats elsewhere (Denno *et al.* 1996). We hypothesize that such trade-offs explain the inverse relationship between competitive ability and dispersal in sap-feeders. For sap-feeders, dispersal is also negatively correlated with other life-history traits such as fecundity and age to first reproduction (Denno *et al.* 1989, 1995). Thus, habitat factors (nonpersistence) that select for a high incidence of dispersal in populations should adversely affect both reproduction and competitive ability (Denno *et al.* 1996, 2000).

One can ask how widespread trade-offs between competitive ability and dispersal might be. The trade-off we report here has not been explored in species other than *Prokelisia*. However, if competitive ability is linked to feeding musculature, then there is every reason to expect such a trade-off in not only sap-feeders but in mandibulate herbivores as well.

Plant-mediated interactions challenge the past paradigms of competition theory

Traditional competition theory posits that: (1) populations of herbivores frequently reach densities at which interspecific competition occurs, (2) species interact symmetrically by engaging in reciprocal struggles either via exploitation or interference over shared resources, (3) niche divergence (resource partitioning) diminishes competition, and (4) closely related species are more likely to compete because they share the same niche (reviewed in Denno *et al.* 1995). We suggest that when plants mediate "competitive interactions" between insect herbivores via induced resistance these traditional paradigms of community organization are often violated.

Herbivore densities and competition

Although plants generally become more resistant as herbivore damage increases, even very low amounts of damage can induce substantial levels of resistance in plants (reviewed in Karban and Baldwin 1997). Consequently, herbivores may interact with dire consequences even if their densities are not particularly high and apparent levels of herbivory are low. It is noteworthy that one of the major observations used by Hairston *et al.* (1960) to argue against the importance of competition between insect herbivores was that "defoliation" was rare, that food was not often limiting, and that natural enemies maintain herbivores below competitive levels. That insect herbivores can induce resistance, and therefore compete at low to moderate densities, challenges the basis for this argument.

Symmetry of interaction

Plant-mediation pre-adapts competitive interactions for asymmetric outcomes. In fact, 84% of interactions between insect herbivores, many of which are plant-mediated, are asymmetric (Denno et al. 1995). Interference competition among insect herbivores, whereby individuals fight or vie for access to resources, tends to encourage more symmetric interactions (Denno et al. 1995). Plant-mediated interactions are fundamentally more exploitative in nature such that one herbivore influences the quality and thus quantity of available resources for another. Moreover, herbivores that compete via induced resistance are often spatially or temporally isolated and may never meet to vie for resources. Thus, early-season exploitation coupled with induced resistance and a diversity of tolerance mechanisms in the inducing herbivore often leads to competitive superiority.

Niche divergence and resource partitioning

Induced responses can be rapid (within a single herbivore generation) or quite delayed (across multiple generations) (Karban and Baldwin 1997), providing the opportunity not only for contemporaneous herbivores to interact but also those that occur in different seasons or perhaps even different years (Denno et al. 1995, Ohgushi 2005). Also, induced responses can be local, occurring in the damaged tissue, or they can be systemic where the response often carries beyond the damaged tissue to other nondamaged tissue, and in several cases even to different nondamaged tissue types (Inbar et al. 1995, Karban and Baldwin 1997, Karban and Kuć 1999). Thus, herbivores that co-occur locally as well as those that are spatially or temporally separated on the same plant can interact via feeding induced responses (Damman 1993, Denno et al. 1995, Ohgushi 2005). Needless to say, the conventional paradigm that "resource partitioning" minimizes "competitive interactions" is in serious jeopardy in the context of induced responses.

Phylogenetic relatedness and competition

Recent evidence from plant-mediated systems suggests that phylogenetically discrepant taxa of herbivores often compete intensely on the same plant (Tomlin and Sears 1992, Stout and Duffey 1996, Agrawal 1998, Stout et al. 1998, Inbar et al. 1999a, 1999b, Van Zandt and Agrawal 2004a, 2004b). Several characteristics of induced plant responses undoubtedly contribute to this pattern. First, many different cues can elicit an induced response in the same plant including feeding from a diversity of herbivore taxa (Karban and Kuć 1999, Stout and Bostock 1999). Second, although some induced responses are quite specific, many are active against a wide variety of organisms including insect herbivores in multiple feeding guilds and orders (Karban and Baldwin 1997, Agrawal et al. 1999a, 2000, Karban and Kuć 1999, Stout and Bostock 1999). Although jasmonic acid is the primary pathway by which induced resistance mediates interactions between insect herbivores, and likewise the salicylic pathway for plant pathogens (Agrawal et al. 1999c, Karban and Kuć 1999), there are numerous cases whereby herbivores stimulate the salicylic pathway and pathogens the jasmonic (reviewed in Karban and Kuć 1999). Thus, a possible mechanistic explanation exists not only for interactions between diverse taxa of insect herbivores, but also for interphyletic interactions between arthropods and plant pathogens (Karban and Baldwin 1997, Stout and Bostock 1999, Ohgushi 2005).

All considered, plant mediation via induced resistance enhances significantly the probability for interspecific interaction between insect herbivores, a view corroborated by recent reviews (Denno et al. 1995, Ohgushi 2005). However, plant-mediated competition also defies the traditional views of competition theory in that interactions between insect herbivores occur even at low densities, are frequently asymmetric, are not often diminished by niche divergence on the same plant, and frequently occur between unrelated taxa.

Looking ahead to more holistic approaches of community dynamics

Almost all studies of interactions between insect herbivores, plant-mediated or otherwise, involve pair-wise assessments (reviewed in Damman 1993, Denno et al. 1995, Nykänen and Koricheva 2004). There are only a handful of studies that have considered multiple species interactions among the major co-occurring herbivores in any one system (Hunter 1992, Agrawal 2000, Van Zandt and Agrawal 2004b, Ohgushi 2005) and most of these (Ohgushi 2005 excepted) are very focused on plant-mediated lateral effects without much regard for higher trophic levels. Moreover, most studies assessing the relative strength of top–down and bottom–up forces on herbivores have largely excluded lateral effects (e.g., Stiling and Rossi 1997, Forkner and Hunter 2000, Denno et al. 2002). Despite the daunting experimental challenge, we must move beyond simple pair-wise assessments to a broader approach that includes not only interactions with multiple herbivores but also incorporates important top–down and bottom–up variables. Taking a keystone species perspective may simplify matters to some extent and make any experimental design more tractable (e.g., Hunter 1992, González-Megías and Gómez 2003). Moreover, because environmental factors (e.g., soil nutrient availability and stress) influence plant quality, a plant's ability to induce defenses against herbivores, and thus the relative strength of bottom–up controls (Karban and Baldwin 1997, Denno et al. 2002), it will be critical to examine interaction strengths (both direct and indirect) and how environmental factors alter them. Only such a holistic approach will provide insights into the factors organizing community structure. Any predictive model of community structure will likely find plant-mediated responses at its core because plants mediate interactions not only among insect herbivores as shown here but also between herbivores and their natural enemies (Hunter and Price 1992, Denno *et al.* 2002, Ohgushi 2005).

Unraveling the food web ramifications of plant-mediated responses is further complicated by issues of temporal and spatial scaling. For instance, although induced resistance has been implicated in the cyclic fluctuations of forest defoliators and other insect herbivores, there is very little empirical evidence rigorously isolating the long-term carry-over effect of induced resistance as it might influence interspecific interactions and thus population dynamics across years (Karban and Baldwin 1997, Kaitaniemi *et al.* 1998, 1999, Underwood 1999). Similarly, spatial variation in plant responses to herbivory and its extended consequences is poorly understood in any context, let alone in a community-wide one. Future research on the spatial ecology of plant-mediated interactions would benefit from the broad experimental approaches used to explore spatially explicit variation in the strength of top–down and bottom–up controls and food web dynamics at large (Hacker and Bertness 1995, Uriarte and Schmitz 1998, Fagan and Bishop 2000, Denno *et al.* 2002).

Acknowledgments

Deborah Finke, Rick Karban, Peter Price, Gina Wimp, and an anonymous reviewer provided comments on an earlier draft of this article, and we hope to have incorporated their many insightful suggestions. This research was supported by National Science Foundation Grants (DEB-9527846 and DEB-9903601) and a USDA Grant (00-35302-9334) to RFD.

References

Addicott, J. F. 1978. Niche relationships among species of aphids feeding on fireweed. *Canadian Journal of Zoology* **57**:558–569.

Agrawal, A. A. 1998. Induced responses to herbivory and increased plant performance. *Science* **279**:1201–1202.

Agrawal, A. A. 1999. Induced responses to herbivory in wild radish: effects on several herbivores and plant fitness. *Ecology* **80**:1713–1723.

Agrawal, A. A. 2000. Specificity of induced resistance in wild radish: causes and consequences for two specialist and two generalist caterpillars. *Oikos* **89**:493–500.

Agrawal, A. A., and M. F. Sherriffs. 2001. Induced plant resistance and susceptibility to late-season herbivores of wild radish. *Annals of the Entomological Society of America* **94**:71–75.

Agrawal, A. A., C. Kobayashi, and J. S. Thaler. 1999a. Influence of prey availability and induced host plant resistance on omnivory by western flower thrips. *Ecology* **80**: 518–523.

Agrawal, A. A., S. Y. Strauss, and M. J. Stout. 1999b. Costs of induced responses and tolerance to herbivory in male and female fitness components of wild radish. *Evolution* **53**:1093–1104.

Agrawal, A. A., S. Tuzun, and E. Bent (eds.) 1999c. *Induced Plant Defenses against Pathogens and Herbivores*. St. Paul, MN: American Phytopathological Society Press.

Agrawal, A. A., R. Karban, and R. Colfer. 2000. How leaf domatia and induced plant resistance affect herbivores, natural enemies and plant performance. *Oikos* **89**:70–80.

Ajlan, A. M., and D. A. Potter. 1992. Lack of effect of tobacco mosaic virus-induced systemic acquired-resistance on arthropod herbivores in tobacco. *Phytopathology* **82**:647–651.

Backus, E. A. 1985. Anatomical and sensory mechanisms of leafhopper and planthopper feeding behavior, pp. 163–194 in L. R. Nault and J. G. Rodriguez (eds.) *The Leafhoppers and Planthoppers*. New York: John Wiley.

Baur, R., S. Binder, and G. Benz. 1991. Nonglandular leaf trichomes as short-term inducible defense of the grey alder, *Alnus incana* (L.), against the chrysomelid beetle *Agelastica alni* L. *Oecologia* **87**:219–226.

Benrey, B., and R. F. Denno. 1997. The slow growth–high mortality hypothesis: a test using the cabbage butterfly. *Ecology* **78**:987–999.

Berenbaum, M. R., and A. R. Zangerl. 1999. Coping with life as a menu option: inducible defenses of wild parsnip, pp. 10–32 in R. Tolrian and C. D. Harvell (eds.) *The Ecology and Evolution of Inducible Defenses*. Princeton, NJ: Princeton University Press.

Bezemer, T. M., R. Wagenaar, N. M. van Dam, and F. L. Wäckers. 2003. Interactions between above- and belowground insect herbivores as mediated by the plant defense system. *Oikos* **101**:555–562.

Cappuccino, N. 1993. Mutual use of leaf-shelters by lepidopteran larvae on paper birch. *Ecological Entomology* **8**:287–292.

Cappuccino, N., and M. A. Martin. 1994. Eliminating early-season leaf-tiers of paper birch reduces abundance of midsummer species. *Ecological Entomology* **19**:399–401.

Chew, F. S. 1988. Biological effects of glucosinolates, pp. 155–181 in H. G. Cutler (ed.) *Biologically Active Natural Products: Potential Use in Agriculture*. Washington, DC: American Chemical Society.

Clausen, T. P., P. B. Reichardt, J. P. Bryant, *et al.* 1989. Chemical model for short-term induction in quaking aspen (*Populus tremuloides*) foliage against herbivores. *Journal of Chemical Ecology* **15**:2335–2346.

Cohen, M. B., M. R. Berenbaum, and M. A. Schuler. 1989. Induction of cytochrome P450-mediated detoxification of xanthotoxin in the black swallowtail. *Journal of Chemical Ecology* **15**:2347–2355.

Cohen, M. B., M. A. Schuler, and M. R. Berenbaum. 1992. A host-inducible cytochrome P450 from a host-specific caterpillar: molecular cloning and evolution. *Proceedings of the National Academy of Sciences of the USA* **89**:10920–10924.

Conn, E. E. 1979. Cyanide and cyanogenic glycosides, pp. 387–412 in G. A. Rosenthal and D. H. Janzen (eds.) *Herbivores: Their Interaction with Secondary Plant Metabolites*. New York: Academic Press.

Constabel, C. P. 1999. A survey of herbivore-inducible defensive proteins and phytochemicals, pp. 137–166 in A. A. Agrawal, S. Tuzan, and E. Bent (eds.) *Induced Plant Defenses against Pathogens and Herbivores*. St. Paul, MN: American Phytopathological Society Press.

Cook, A., and R. F. Denno. 1994. Planthopper/plant interactions: feeding behavior, plant nutrition, plant defense and host plant specialization, pp. 114–139 in R. F. Denno and T. J. Perfect (eds.) *Planthoppers: Their Ecology and Management*. New York: Chapman and Hall.

Crawley, M. J., and P. Pattrasudhi. 1988. Interspecific competition between insect herbivores: asymmetric competition between cinnabar moth and the ragwort seed-head fly. *Ecological Entomology* **13**:243–249.

Damman, H. 1993. Patterns of herbivore interaction among herbivore species, pp. 132–169 in N. E. Stamp and T. M. Casey (eds.) *Caterpillars: Ecological and Evolutionary Constraints on Foraging*. New York: Chapman and Hall.

de Ilarduya, O. M., Q. G. Xie, and I. Kaloshian. 2003. Aphid-induced defense responses in Mi-1-mediated compatible and incompatible tomato interactions. *Molecular Plant–Microbe Interactions* **16**:699–708.

De Moraes, C. M., J. Lewis, P. W. Paré, H. T. Alborn, and J. H. Tumlinson. 1998. Herbivore-infested plants selectively attract parasitoids. *Nature* **393**:570–573.

Denno, R. F., and G. K. Roderick. 1992. Density-related dispersal in planthoppers: effects of interspecific crowding. *Ecology* **73**:1323–1334.

Denno, R. F., K. L. Olmstead, and E. S. McCloud. 1989. Reproductive cost of flight capability: a comparison of life history traits in wing dimorphic planthoppers. *Ecological Entomology* **14**:31–44.

Denno, R. F., S. Larsson, and K. L. Olmstead. 1990. Host plant selection in willow-feeding leaf beetles (Coleoptera: Chrysomelidae): role of enemy-free space and plant quality. *Ecology* **71**:124–137.

Denno, R. F., M. S. McClure, and J. R. Ott. 1995. Interspecific interactions in phytophagous insects: competition revisited and resurrected. *Annual Review of Entomology* **40**:297–331.

Denno, R. F., G. K. Roderick, M. A. Peterson, et al. 1996. Habitat persistence underlies the intraspecific dispersal strategies of planthoppers. *Ecological Monographs* **66**:389–408.

Denno, R. F., M. A. Peterson, C. Gratton, et al. 2000. Feeding-induced changes in plant quality mediate interspecific competition between sap-feeding herbivores. *Ecology* **81**:1814–1827.

Denno, R. F., C. Gratton, M. A. Peterson, et al. 2002. Bottom–up forces mediate natural-enemy impact in a phytophagous insect community. *Ecology* **83**:1443–1458.

Dussourd, D. E., and R. F. Denno. 1991. Deactivation of plant defense: correspondence between insect behavior and secretory canal architecture. *Ecology* **72**:1383–1396.

Dussourd, D. E., and R. F. Denno. 1994. Host range of generalist Lepidoptera: larval trenching permits feeding on plants with secretory canals. *Ecology* **75**:69–78.

Dussourd, D. E., and T. Eisner. 1987. Vein-cutting behavior insect counterploy to the latex defense of plants. *Science* **237**:898–901.

Edson, J. L. 1985. The influences of predation and resource subdivision on the coexistence of goldenrod aphids. *Ecology* **66**:1736–1743.

Faeth, S. 1987. Community structure and folivorous insect outbreaks: the role of vertical and horizontal interactions, pp. 135–171 in P. Barbosa and J. C. Schultz (eds.) *Insect Outbreaks*. New York: Academic Press.

Fagan, W. F., and J. G. Bishop. 2000. Trophic interactions during primary succession: herbivores slow a plant reinvasion at Mount St. Helens. *American Naturalist* **155**:238–251.

Ferrenberg, S. M., and R. F. Denno. 2003. Competition as a factor underlying the abundance of an uncommon phytophagous insect, the salt-marsh planthopper *Delphacodes penedetecta*. *Ecological Entomology* **28**:58–66.

Fordyce, J. A. 2001. The lethal plant defense paradox remains: inducible host plant aristolochic acids and the growth and defense of the pipevine swallowtail. *Entomologia Experimentalis et Applicata* **100**:339–346.

Forkner, R. E., and M. D. Hunter. 2000. What goes up must come down? Nutrient addition and predation pressure on oak herbivores. *Ecology* **81**:1588–1600.

Formusoh, E. S., G. E. Wilde, and J. C. Reese. 1992. Reproduction and feeding-behavior of greenbug biotype-E (Homoptera: Aphididae) on wheat previously fed upon by aphids. *Journal of Economic Entomology* **85**:789–793.

Forrest, J. M. S. 1971. The growth of *Aphis fabae* as an indicator of the nutritional advantage of galling to the apple aphid *Dysaphis devecta*. *Entomologia Experimentalis et Applicata* **14**:447–483.

Fukui, A. 2001. Indirect interactions mediated by leaf shelters in animal–plant communities. *Population Ecology* **43**:31–40.

Gange, A. C., and V. K. Brown. 1989. Effects of root herbivory by an insect on a foliar-feeding species, mediated through changes in the host plant. *Oecologia* **81**:38–42.

González-Megías, A., and J. M. Gómez. 2003. Consequences of removing a keystone herbivore for the abundance and diversity of arthropods associated with a cruciferous shrub. *Ecological Entomology* **28**:299–308.

Hacker, S. D., and M. D. Bertness. 1995. A herbivore paradox: why salt marsh aphids live on poor-quality plants. *American Naturalist* **145**:192–210.

Hairston, N. G., F. E. Smith, and L. B. Slobodkin. 1960. Community structure, population control, and competition. *American Naturalist* **44**:421–425.

Heard, S. B., and C. K. Buchanan. 1998. Larval performance and association within and between two species of hackberry nipple gall insects, *Pachypsylla* spp. (Homoptera: Psyllidae). *American Midland Naturalist* **140**:351–357.

Hendrix, S. D. 1988. Herbivory and its impact on plant reproduction, pp. 246–266 in J. Lovett-Doust and L. Lovett-Doust (eds.) *Plant Reproductive Ecology: Patterns and Strategies*. Oxford, UK: Oxford University Press.

Hendrix, S. D., and E. J. Trapp. 1989. Floral herbivory in *Pastinaca sativa*: do compensatory responses offset reductions in fitness? *Evolution* **43**:891–895.

Hunter, M. D. 1992. Interactions within herbivore communities mediated by the host plant: the keystone herbivore concept, pp. 287–325 in M. D. Hunter, T. Ohgushi, and P. W. Price (eds.) *Effects of Resource Distribution on Animal–Plant Interactions*. San Diego, CA: Academic Press.

Hunter, M. D., and P. W. Price. 1992. Playing chutes and ladders: heterogeneity and the relative roles of bottom-up and top-down forces in natural communities. *Ecology* **73**:724–732.

Inbar, M., A. Eshel, and D. Wool. 1995. Interspecific competition among phloem-feeding insects mediated by induced host-plant sinks. *Ecology* **76**:1506–1515.

Inbar, M., H. Doostdar, G. L. Leibee, and R. T. Mayer. 1999a. The role of plant rapidly induced responses in asymmetric interspecific interactions among insect herbivores. *Journal of Chemical Ecology* **25**:1961–1979.

Inbar, M., H. Doostdar, and R. T. Mayer. 1999b. Effects of sessile whitefly nymphs (Homoptera: Aleyrodidae) on leaf-chewing larvae (Lepidoptera: Noctuidae). *Environmental Entomology* **28**:353–357.

Inbar, M., R. T. Mayer, and H. Doostdar. 2003. Induced activity of pathogenesis related (PR) proteins in aphid galls. *Symbiosis* **34**:293–300.

Itô, Y. 1960. Ecological studies on population increase and habitat segregation among barley aphids. *Bulletin of the National Institute of Agricultural Science, Series C* **11**:45–130.

Kaitaniemi, P., K. Ruohomaki, V. Ossipov, E. Haukioja, and K. Pihlaja. 1998. Delayed induced changes in the biochemical composition of host plant leaves during an insect outbreak. *Oecologia* **116**:182–190.

Kaitaniemi, P., K. Ruohomaki, T. Tammaru, and E. Haukioja. 1999. Induced resistance of host tree foliage during and after a natural insect outbreak. *Journal of Animal Ecology* **68**:382–389.

Karban, R. 1986. Interspecific competition between folivorous insects on *Erigeron glaucus*. *Ecology* **67**:1063–1072.

Karban, R. 1989. Community organization of *Erigeron glaucus* folivores: effects of competition, predation, and host plant. *Ecology* **70**:1028–1039.

Karban, R., and A. A. Agrawal. 2002. Herbivore offense. *Annual Review of Ecology and Systematics* **33**:641–664.

Karban, R., and I. T. Baldwin. 1997. *Induced Responses to Herbivory*. Chicago, IL: University of Chicago Press.

Karban, R., and J. Kuć. 1999. Induced resistance against pathogens and herbivores: an overview, pp. 1–19 in A. A. Agrawal, S. Tuzun, and E. Bent (eds.) *Induced Plant Defenses against Pathogens and Herbivores*. St. Paul, MN: American Phytopathological Society Press.

Kareiva, P. 1982. Exclusion experiments and the competitive release of insects feeding on collards. *Ecology* **63**:696–704.

Kessler, A., and I. T. Baldwin. 2004. Herbivore-induced plant vaccination. I. The orchestration of plant defenses in nature and their fitness consequences in the wild tobacco *Nicotiana attenuata*. *Plant Journal* **38**:639–649.

Kidd, N. A. C., G. B. Lewis, and C. A. Howell. 1985. An association between two species of pine aphid, *Schizolachnus pineti* and *Eulachnus agilis*. *Ecological Entomology* **10**:427–432.

Krupnick, G. A., A. E. Weis, and D. R. Campbell. 1999. The consequences of floral herbivory for pollinator service to *Isomeris arborea*. *Ecology* **80**:125–134.

Lamb, R. J., and P. A. MacKay. 1987. *Acyrthosiphon kondoi* influences alata production by the pea aphid, *A. pisum*. *Entomologia Experimentalis et Applicata* **45**:195–198.

Larson, K. C., and T. G. Whitham. 1991. Manipulation of food resources by a gall-inducing aphid: the physiology of sink–source interactions. *Oecologia* **88**:15–21.

Larsson, S., H. E. Häggström, and R. F. Denno. 1997. Preference for protected feeding sites by larvae of the willow-feeding leaf beetle *Galerucella lineola*. *Ecological Entomology* **22**:445–452.

Lawton, J. H. 1982. Vacant niches and unsaturated communities: a comparison of bracken herbivores at sites on two continents. *Journal of Animal Ecology* **51**:573–595.

Lawton, J. H., and M. P. Hassell. 1984. Interspecific competition in insects, pp. 451–495 in C. B. Huffaker and R. L. Rabb (eds.) *Ecological Entomology*. New York: John Wiley.

Lawton, J. H., and D. R. Strong. 1981. Community patterns and competition in folivorous insects. *American Naturalist* **118**:317–338.

Lehtilä, K., and S. Y. Strauss. 1997. Leaf damage by herbivores affects attractiveness to pollinators in wild radish, *Raphanus raphanistrum*. *Oecologia* **111**:396–403.

Lewinsohn, E., M. Gijzen, and R. Croteau. 1991. Defense mechanisms of conifers: differences in constitutive and wound-induced monoterpene biosynthesis among species. *Plant Physiology* **96**:44–49.

Lill, J. T., and R. J. Marquis. 2003. Ecosystem engineering by caterpillars increases insect herbivore diversity on white oak. *Ecology* **84**:682–690.

Martinsen, G. D., K. D. Floate, A. M. Waltz, G. M. Wimp, and T. G. Whitham. 2000. Positive interactions between leafrollers and other arthropods enhance biodiversity on hybrid cottonwoods. *Oecologia* **123**:82–89.

Masters, G. J., and V. K. Brown. 1992. Plant-mediated interactions between two spatially separated insects. *Functional Ecology* **6**:175–179.

Masters, G. J., T. H. Jones, and M. Rogers. 2001. Host-plant mediated effects of root herbivory on insect seed predators and their parasitoids. *Oecologia* **127**:246–250.

Matsumura, M., and Y. Suzuki. 2003. Direct and feeding-induced interactions between two rice planthoppers, *Sogatella furcifera* and *Nilaparvata lugens*: effects on dispersal capability and performance. *Ecological Entomology* **28**:174–182.

Mattson, W. J., R. A. Haack, R. K. Lawrence, and D. A. Herms. 1989. Do balsam aphids (Homoptera: Aphididae) lower tree susceptibility to spruce budworm? *Canadian Entomologist* **121**:93–103.

Mayer, R. T., M. Inbar, C. L. McKenzie, *et al*. 2002. Multitrophic interactions of the silverleaf whitefly, host plants, competing herbivores, and phytopathogens. *Archives of Insect Biochemistry and Physiology* **51**:151–169.

McClure, M. S. 1980. Competition between exotic species: scale insects on hemlock. *Ecology* **61**:1391–1401.

McClure, M. S. 1989. Biology, population trends, and damage of *Pineus boerneri* and *P. coloradensis* (Homoptera: Adelgidae) on red pine. *Environmental Entomology* **18**:1066–1073.

McClure, M. S. 1990. Cohabitation and host species effects on the population growth of *Matsucoccus resinosae* (Homoptera: Margarodidae) and *Pineus boerneri* (Homoptera: Adelgidae) on red pine. *Environmental Entomology* **19**:672–676.

McClure, M. S., and P. W. Price. 1975. Competition among sympatric *Erythroneura* leafhoppers (Homoptera: Cicadellidae) on American sycamore. *Ecology* **56**:1388–1397.

McClure, M. S., and P. W. Price. 1976. Ecotope characteristics of coexisting *Erythroneura* leafhoppers (Homoptera: Cicadellidae) on sycamore. *Ecology* **57**:928–940.

Milbrath, L. R., and J. R. Nechols. 2004. Indirect effect of early-season infestations of *Trichosirocalus horridus* on *Rhinocyllus conicus* (Coleoptera: Curculionidae). *Biological Control* **30**:95–109.

Moegenburg, S. M. 1996. Sabal palmetto seed size: causes of variation, choice of predators, and consequences for seedlings. *Oecologia* **106**:539–543.

Montandon, R., J. E. Slosser, and W. A. Frank. 1993. Factors reducing the pest status of the Russian wheat aphid (Homoptera: Aphididae) on wheat in the rolling plains of Texas. *Journal of Economic Entomology* **86**:899–905.

Moran, N. A., and T. G. Whitham. 1990. Interspecific competition between root-feeding and leaf-galling aphids mediated by host-plant resistance. *Ecology* **71**:1050–1058.

Nakamura, M., and T. Ohgushi. 2003. Positive and negative effects of leaf shelters on herbivorous insects: linking multiple herbivore species on a willow. *Oecologia* **136**:445–449.

Nakamura, M., Y. Miyamoto, and T. Ohgushi. 2003. Gall initiation enhances the availability of food resources for herbivorous insects. *Functional Ecology* **17**:851–857.

Ness, J. H. 2003. *Catalpa bignonioides* alters extrafloral nectar production after herbivory and attracts many bodyguards. *Oecologia* **134**:210–218.

Nykänen, H., and J. Koricheva. 2004. Damage-induced changes in woody plants and their effects on insect herbivore performance: a meta-analysis. *Oikos* **104**:247–268.

Ohgushi, T. 2005. Indirect interaction webs: herbivore-induced effects through trait change in plants. *Annual Review of Ecology, Evolution, and Systematics* **36**:81–105.

Olmstead, K. L., R. F. Denno, T. C. Morton, and J. T. Romeo. 1997. Influence of *Prokelisia* planthoppers on the amino acid composition and growth of *Spartina alterniflora*. *Journal of Chemical Ecology* **23**:303–321.

Paré, P. W., W. J. Lewis, and J. H. Tumlinson. 1999. Induced plant volatiles: biochemistry and effects on parasitoids, pp. 167–180 in A. A. Agrawal, S. Tuzun, and E. Bent (eds.) *Induced Defenses against Pathogens and Herbivores*. St. Paul, MN: American Phytopathological Society Press.

Pilson, D. 1992. Aphid distribution and the evolution of goldenrod resistance. *Evolution* **46**:1358–1372.

Pullin, A. S., and J. E. Gilbert. 1989. The stinging nettle, *Urtica dioica*, increases trichome density after herbivore and mechanical damage. *Oikos* **54**:275–280.

Raffa, K. F. 1991. Induced defensive reactions in conifer–bark beetle systems, pp. 245–276 in D. W. Tallamy and M. J. Raupp (eds.) *Phytochemical Induction by Herbivores*. New York: Academic Press.

Rathcke, B. J. 1976. Competition and coexistence within a guild of herbivorous insects. *Ecology* **57**:76–87.

Rausher, M. D., K. Iwao, E. L. Simms, N. Ohsaki, and D. Hall. 1993. Induced resistance in *Ipomoea purpurea*. *Ecology* **74**:20–29.

Raven, J. A. 1983. Phytophages of xylem and phloem: a comparison of animal and plant sapfeeders. *Advances in Ecological Research* **13**:135–234.

Redman, A. M., and J. M. Scriber. 2000. Competition between the gypsy moth, *Lymantria dispar*, and the northern tiger swallowtail, *Papilio canadensis*: interactions mediated by host plant chemistry, pathogens, and parasitoids. *Oecologia* **125**:218–228.

Ruuhola, T. 2001. *Dynamics of salicylates in willows and its relation to herbivory*. Ph.D. dissertation, University of Joensuu, Finland.

Salt, D. T., P. Fenwick, and J. B. Whittaker. 1996. Interspecific herbivore interactions in a high CO_2 environment: root and shoot aphids feeding on *Cardamine*. *Oikos* **77**:326–330.

Salyk, R. P., and D. J. Sullivan. 1982. Comparative feeding behavior of two aphid species: bean aphid (*Aphis fabae* Scopoli) and pea aphid (*Acyrthosiphon pisum* Harris) (Homoptera: Aphididae). *Journal of the New York Entomological Society* **90**:87–93.

Schoener, T. W. 1982. The controversy over interspecific competition. *American Scientist* **70**:586–595.

Schultz, J. C. 1999. Discussion, p. 19 in D. J. Chadwick and J. A. Goode (eds.) *Insect–Plant Interactions and Induced Plant Defense*. New York: John Wiley.

Shearer, J. W. 1976. Effect of aggregations of aphids (*Periphyllus* spp.) on their size. *Entomologia Experimentalis et Applicata* **20**:179–182.

Stiling, P., and A. M. Rossi. 1997. Experimental manipulations of top-down and bottom-up factors in a tri-trophic system. *Ecology* **78**:1602–1606.

Stiling, P. D., and D. R. Strong. 1984. Experimental density manipulation of stem-boring insects: some evidence for interspecific competition. *Ecology* **65**:1683–1685.

Stout, M. J., and R. M. Bostock. 1999. Specificity of induced responses to arthropods and pathogens, pp. 183–209 in A. A. Agrawal, S. Tuzun, and E. Bent (eds.) *Induced Defenses against Pathogens and Herbivores*. St. Paul, MN: American Phytopathological Society Press.

Stout, M. J., and S. S. Duffey. 1996. Characterization of induced resistance in tomato plants. *Entomologia Experimentalis et Applicata* **79**:273–283.

Stout, M. J., K. V. Workman, R. M. Bostock, and S. S. Duffey. 1998. Specificity of induced resistance in the tomato, *Lycopersicon esculentum*. *Oecologia* **113**:74–81.

Strauss, S. Y. 1997. Floral characteristics link herbivores, pollinators, and plant fitness. *Ecology* **78**:1640–1645.

Strauss, S. Y., J. K. Conner, and S. L. Rush. 1996. Foliar herbivory affects floral characteristics and plant attractiveness to pollinators: implications for male and female plant fitness. *American Naturalist* **147**:1098–1107.

Strong, D. R. 1981. The possibility of insect communities without competition: hispine beetles on *Heliconia*, pp. 183–194 in R. F. Denno and H. Dingle (eds.) *Insect Life History Patterns: Habitat and Geographic Variation*. New York: Springer-Verlag.

Strong, D. R., J. H. Lawton, and T. R. E. Southwood. 1984. *Insects on Plants*. Cambridge, MA: Harvard University Press.

Szentesi, A., and T. Jermy. 1995. Predispersal seed predation in leguminous species: seed morphology and bruchid distribution. *Oikos* **73**:23–32.

Tallamy, D. W., and M. J. Raupp. 1991. *Phytochemical Induction by Herbivores*. New York: John Wiley.

Tamaki, G., and W. W. Allen. 1969. Competition and other factors influencing the population dynamics of *Aphis gossypii* and *Macrosiphoniella sanborni* on greenhouse chrysanthemums. *Hilgardia* **39**:447–505.

Thaler, J. S. 1999. Jasmonate-inducible plant defenses cause increased parasitism of herbivores. *Nature* **399**:686–688.

Thaler, J. S. 2002a. Jasmonate-deficient plants have reduced direct and indirect defenses against herbivores. *Ecology Letters* **5**:764–774.

Thaler, J. S. 2002b. Effect of jasmonate-induced plant responses on the natural enemies of herbivores. *Journal of Animal Ecology* **71**:141–150.

Thalmann, C., J. Freise, W. Heitland, and S. Bacher. 2003. Effects of defoliation by horse chestnut leafminer (*Cameraria ohridella*) on reproduction in *Aesculus hippocastanum*. *Trees* **17**:383–388.

Tomlin, E. S., and M. K. Sears. 1992. Indirect competition between the Colorado potato beetle (Coleoptera: Chrysomelidae) and the potato leafhopper (Homoptera: Cicadellidae) on potato: laboratory study. *Environmental Entomology* **21**:787–792.

Traw, M. B., and T. E. Dawson. 2002. Reduced performance of two specialist herbivores (Lepidoptera: Pieridae, Coleoptera: Chrysomelidae) on new leaves of damaged black mustard plants. *Environmental Entomology* **31**:714–722.

Turlings, T. C. J., J. H. Tumlinson, R. R. Heath, A. T. Proveaux, and R. E. Doolittle. 1991. Isolation and identification of allelochemicals that attract the larva parasitoid, *Cotesia marginiventris* (Gesson), to the microhabitat of one of its hosts. *Journal of Chemical Ecology* **17**:2235–2251.

Underwood, N. 1999. The influence of induced plant resistance on herbivore population dynamics, pp. 211–229 in A. A. Agrawal, S. Tuzan, and E. Bent (eds.) *Induced Plant Defenses against Pathogens and Herbivores*. St. Paul, MN: American Phytopathological Society Press.

Uriarte, M., and O. J. Schmitz. 1998. Trophic control across a natural productivity gradient with sap-feeding herbivores. *Oikos* **82**:552–560.

Van Zandt, P. A., and A. A. Agrawal. 2004a. Specificity of induced plant responses to specialist herbivores of the common milkweed *Asclepias syriaca*. *Oikos* **104**:401–409.

Van Zandt, P. A., and A. A. Agrawal. 2004b. Community-wide impacts of herbivore-induced plant responses in milkweed (*Asclepias syriaca*). *Ecology* **85**:2616–2629.

Vinson, B. 1998. The general host selection behavior of parasitoid hymenoptera and a comparison of initial strategies utilized by larvaphagous and oophagous species. *Biological Control* **11**:79–96.

Wallin, K. F., and K. F. Raffa. 2001. Effects of folivory on subcortical plant defenses: can defense theories predict interguild processes? *Ecology* **82**:1387–1400.

Waloff, N. 1979. Partitioning of resources by grassland leafhoppers (Auchenorrhyncha, Homoptera). *Ecological Entomology* **4**:379–385.

Waltz, A. M., and T. G. Whitham. 1997. Plant development affects arthropod communities: opposing impacts of species removal. *Ecology* **78**:2133–2144.

Way, M. J., and M. Cammell. 1970. Aggregation behavior in relation to food utilization by aphids, pp. 229–247 in A. Watson (ed.) *Animal Populations in Relation to their Food Resources*. Oxford, UK: Blackwell Scientific Publications.

Weis, A. E., R. Walton, and C. L. Crego. 1988. Reactive plant tissue sites and the population biology of gall makers. *Annual Review of Entomology* **33**:467–486.

West, C. 1985. Factors underlying the late seasonal appearance of the lepidopterous leaf-mining guild on oak. *Ecological Entomology* **10**:111–120.

Williams, I. S. 1999. Slow-growth, high-mortality: a general hypothesis, or is it? *Ecological Entomology* **24**:490–495.

Willott, S. J., S. G. Compton, and L. D. Incoll. 2000. Foraging, food selection and worker size in the seed harvesting ant *Messor bouvieri*. *Oecologia* **125**:35–44.

Wise, M. J., and A. M. Weinberg. 2002. Prior flea beetle herbivory affects oviposition preference and larval performance of a potato beetle on their shared host plant. *Ecological Entomology* **27**:115–122.

Wold, E. N., and R. J. Marquis. 1997. Induced defense in white oak: effects on herbivores and consequences for the plant. *Ecology* **78**:1356–1369.

Zangerl, A. R. 1999. Locally-induced responses in plants: the ecology and evolution of restrained defense, pp. 231–249 in A. A. Agrawal, S. Tuzun, and E. Bent (eds.) *Induced Plant Defenses against Pathogens and Herbivores*. St. Paul, MN: American Phytopathological Society Press.

Zera, A. J., and R. F. Denno. 1997. Physiology and ecology of dispersal polymorphisms in insects. *Annual Review of Entomology* **42**:207–231.

3

Going with the flow: plant vascular systems mediate indirect interactions between plants, insect herbivores, and hemi-parasitic plants

SUSAN E. HARTLEY, KATHY A. BASS, AND SCOTT N. JOHNSON

Introduction

Plant-mediated indirect interactions between phytophagous insects

There is increasing interest in the consequences of indirect interactions for community structure and function (Wootton 1994). Herbivory by one phytophagous species has the potential to affect other herbivores exploiting the same plant, hence plants are able to mediate indirect interactions between organisms that exploit them, even if these organisms are spatially or temporally separated (Masters and Brown 1997). For example, root-feeding herbivores may impact on the performance of foliar feeding insects (Gange and Brown 1989, Masters and Brown 1992), while herbivores feeding early in the season affect the growth and development of those feeding later (West 1985, Harrison and Karban 1986). Many such interactions are mediated by damage-induced changes in the chemical composition of the shared host plant (Hartley and Jones 1997, Karban and Baldwin 1997), particularly increases in secondary compounds (Hartley and Lawton 1987, Haukioja et al. 1990), but there are also cases where alterations in the nutrient levels within the host explain the impact of one insect herbivore on another (McClure 1980, Denno et al. 2000). Thus both changes in nutrient and in secondary compounds have been associated with detrimental effects on other phytophagous insects and may underpin competitive indirect interactions between herbivores (Denno et al. 1995).

Ecological Communities: Plant Mediation in Indirect Interaction Webs, ed. Takayuki Ohgushi, Timothy P. Craig, and Peter W. Price. Published by Cambridge University Press. © Cambridge University Press 2007.

The importance of competitive interactions between phytophagous insects has been re-evaluated in recent years. Despite examples of induced responses to herbivores by plants, the prevailing view over a decade ago was that plant resources were rarely limiting for insect herbivores, hence competitive interactions also were thought to be rare (Strong et al. 1984). A turning point was 1995, when a review by Denno et al. (1995) found evidence of interspecific competition in 76% of interactions studied and that these occurred across all feeding guilds of phytophagous insects. Denno et al. (1995) concluded that host plants mediate competitive interactions between phytophagous insects more frequently than natural enemies or physical factors (e.g., temperature, rainfall). This shift in the importance of plant-based factors on insect population studies has resulted in much new research focusing on indirect interactions between phytophagous insects, including several reviews (Masters and Brown 1997, Blossey and Hunt-Joshi 2003) and many field-based examples (Denno et al. 2000, Fisher et al. 2000, Petersen and Sandstrom 2001).

Consistent features emerge from these studies of plant-mediated indirect interactions between insect herbivores. First, temporal separation of the herbivores means many of these interactions are asymmetric. Thus early feeding herbivores have an impact on late-feeding ones but not vice versa. For example, feeding by a cecidomyid on the buds of red pine (*Pinus resinosa*) resulted in modified stamens, preventing a weevil from feeding later in the season (Mattson 1986). The review by Denno et al. (1995) found that, of 193 pair-wise interactions, 84% were asymmetric interactions (e.g., McClure 1980, Karban 1986, Denno and Roderick 1992). Second, many of the interactions are based on the manipulation of plant nutrients, particularly the alteration of source–sink relationships within the plant (Larson and Whitham 1991). Gall-formers provide several examples of this phenomenon because they can create very strong sinks within susceptible host plants (Larson and Whitham 1997). Inbar et al. (1995) found an exploitative interspecific interaction between two gall-forming aphids mediated by host-plant sinks. Both species formed galls on the host leaf, but *Geoica* sp. formed a gall on the leaflet midrib and diverted the normal phloem transport away from *Forda formicaria*, which made galls on the leaf margin. *Geoica* sp. initiated early leaf senescence, which increased *F. formicaria* mortality to 84%.

Another type of plant modification by insect herbivores, which is overlooked in comparison to the manipulation of secondary compounds and nutrients, is the structural modification of plant tissues in a way that affects other phytophagous insects. Until recently this idea has generally been disregarded (Masters and Brown 1997) or reported as idiosyncratic phenomena (e.g., Mattson 1986), but it is now realized that insect herbivores can create shelters (e.g., Fukui 2001), change the morphology of plants, or even, if they attack the apical meristem for example, alter plant architecture, all of which can alter resource availability for other herbivores. Gall-forming herbivores often have profound impacts on the morphology and architecture of their host plants (e.g., Price and Louw 1996, Price et al. 1997). Interactions mediated by morphological changes in a shared host plant often occur between herbivores that are temporally separated, and they are almost always asymmetric. For example, Nakamura and Ohgushi (2003) demonstrated how morphological changes produced by shelter-making lepidopteran larvae in the host *Salix miyabeana* increased the abundance of both aphids and ants and decreased the survival rate of leaf beetle larvae. Hence physical modifications of the host plant by one organism can have wide-ranging impacts on other organisms in the community.

Indirect interactions based on the plant vascular system

Phytophagous insects exploit and compete for host resources in a variety of ways, but the focus of this chapter is competition between organisms that utilize the host's vascular system. In some cases this is mediated by changes in the nutritional quality of the host plant. For example, Denno et al. (2000) described an asymmetric interaction between two species of phloem-feeding planthoppers on cordgrass in a salt marsh. Prior feeding by *Prokelisia dolus* significantly reduced the concentration of essential amino acids within the host, so had a negative impact on development time, body size and survival of *Prokelisia marginata*. However, some insect herbivores physically modify plant tissues in a way that disrupts the vasculature (Price and Louw 1996). Many different types of phytophagous insects do this because several benefits derive from modifying vasculature, including the avoidance of plant defense compounds (Carroll and Hoffman 1980, Dussourd and Eisner 1987, Dussourd and Denno 1994) and nutrient accumulation in tissues no longer connected to the vascular system (Forcella 1982, White 1984). Again gall-forming insects are particularly good examples of herbivores that improve plant nutritional quality by physically modifying vasculature (Hartley 1998, Wool et al. 1999). Gall-formers cause short-term structural modification of plant vessels, as well as longer-term changes in plant growth and architecture that could potentially affect other insect species feeding on shared leaves.

However, insect herbivores are not the only type of organism that exploit the plant vascular system. Parasitic angiosperms remove resources from the vascular system of host plants by means of haustorial connections. Hemi-parasitic plants contain chlorophyll and are capable of photosynthesis, removing only xylem resources from the host (Press et al. 1998). In some ecosystems, such as the subarctic, parasitic plants can make up to 25% of the plant community. In such environments they are a clearly defined functional group that has the potential to influence the structure and function of the communities in which they occur

(Pennings and Callaway 2002), because they can alter the rates of community processes such as decomposition (Quested *et al.* 2002). In addition, parasitic angiosperms can be important agricultural weeds and the most significant are in the genus *Striga* (Press *et al.* 1998). For example, *Striga hermonthica* is a major biotic factor limiting grain production of millet, sorghum, and maize in the semi-arid African tropics (Graves *et al.* 1990, Cechin and Press 1993).

Removal of host resources by parasitic angiosperms can alter the rate of host metabolic functions such as photosynthesis, respiration, and uptake of water and solutes (Graves 1995). Hemi-parasites usually have a negative impact on photosynthesis, and this in turn reduces the biomass of the host (Press and Seel 1996, Davies and Graves 2000) and may also potentially impact on the chemical composition of the phloem (Jiang *et al.* 2004). The impacts of hemi-parasitic plants on water and solute uptake mean that they alter the composition of the xylem sap in their hosts by decreasing amino acid concentration and increasing the concentration of inorganic ions (Seel and Jeschke 1999).

Despite the impacts of parasitic angiosperms on their hosts, very little information is available on the occurrence of indirect interactions with other organisms mediated through these parasite-induced changes in host quality. One of the few experiments to assess such interactions was undertaken by Puustinen and Mutikainen (2001), who investigated a multispecies interaction between the hemi-parasite *Rhinanthus serotinus*, the host *Trifolium repens*, and a generalist snail herbivore. Snails were fed with leaves from host plants that had grown with or without parasitism from *R. serotinus*. Snail growth rate was reduced when eating leaves from plants that were infected with the hemi-parasite. However, in this experiment no attempt was made to determine effects of snail herbivory on hemi-parasite growth or the reasons for reduced growth of the snail. This type of study suggests that interactions between parasitic angiosperms and insect herbivores are possible, but it also illustrates the need for more research in this field, both to identify such interactions, and determine the mechanisms behind them.

Aims of this chapter

In this chapter we consider interactions between organisms that exploit the plant vascular system, focusing on two separate experimental systems. First, we examine the indirect interactions between phloem-feeding aphids and another guild of insects, leaf miners, which physically disrupt the flow of nutrients within the vascular system. Second, we consider the plant-mediated interactions between two completely different organisms that exploit the vascular system, namely aphids and hemi-parasitic plants, one of which removes resources from the phloem and one from the xylem. This is a relatively new area in which to provide a synthesis, particularly for interactions between hemi-parasitic plants and insect herbivores of which there are very few examples in the literature. However, for the two examples we consider here we show how these interactions are asymmetric and reflect a variety of plant-based mechanisms including changes in allocation patterns, physical changes in the roots and vascular structure, and changes in the amount and quality of plant resources.

Interactions based on the physical modification of the vascular system

Interactions between aphids and leaf miners

The negative effect of one insect herbivore on the survivorship of another of a different feeding guild is usually considered in terms of nutritional or allelochemical changes induced by one species that subsequently affect the other (Hartley and Lawton 1987, Masters and Brown 1997). However, different feeding guilds might also interact via physical changes to their shared host plant, particularly when one guild has a feeding mechanism which causes a structural alteration to the leaves of the host.

We examined how a free-living aphid, *Euceraphis betulae* Koch (Homoptera: Drepanosiphinae), is affected by co-occurring leaf-mining moths of the genus *Eriocrania* spp. Zeller (Lepidoptera: Eriocraniidae). Both these species feed on leaves of the deciduous tree *Betula pendula* Roth (Betulaceae) (silver birch). The two species are spatially separated and do not compete directly for the same resource; *Eriocrania* feeds internally within leaves between the upper and lower epidermis, whereas *E. betulae* feeds on phloem sap from the basal leaf midrib and petiole (Hajek and Dahlsten 1986). *Eriocrania* spp. have the potential to indirectly affect *E. betulae*, however, if the feeding of their larvae disrupts the vascular system. While most attention to date has focused on the impacts of gall-forming insects on leaf vasculature, leaf-mining insects may also impact on the plant vascular system (Connor and Taverner 1997) though in a rather different way. Gall insects induce the development of new vasculature from meristematic tissues which feeds in to the outer gall tissue (Shorthouse and Rohfritsch 1992, Wool et al. 1999); hence they alter source–sink relationships but they feed on the lining of the gall cavity so do not sever the new specially formed vasculature. Leaf miners feed by chewing on the unmodified inner mesophyll of the leaf, including the vasculature, which they cut through. Phloem-feeding insects such as aphids might be particularly susceptible to vasculature manipulation by other insects, as they are reliant on the high flux of phloem sap (Raven 1983, Hartley and Jones 1997). In contrast, disruption of the vasculature may have beneficial effects for the leaf-mining moth by improving the nutritional quality of leaves (Johnson et al. 2002).

The impact of leaf miners on aphid performance

In 1999, we conducted field surveys of 300 young B. pendula trees in north east Scotland and found that E. betulae were significantly less abundant on mined leaves with midrib damage than on mined leaves with just lamina damage, or mine-free leaves (ANOVA: $F_{2,1427} = 22.00$; $p < 0.0001$) (see Johnson et al. 2002 for further details). We also found that when E. betulae was forced to feed on mined leaves, by caging aphids on leaves with Eriocrania leaf miners, their mortality was significantly increased (GLM: $F_{1,70} = 14.76$; $p < 0.001$). However, it was the actual region of the leaf mined, rather than leaf miner presence per se, that was strongly correlated with E. betulae survivorship. Specifically, when Eriocrania mines impinged on the leaf midrib, aphid survivorship was significantly lower than when leaf mining was restricted to the lamina (GLM: $F_{1,32} = 12.97$; $p = 0.001$) (Fig. 3.1a) (see Johnson et al. 2002).

That E. betulae were equally abundant in the field on leaves with damaged and undamaged lamina, versus on leaves with damaged leaf vasculature, reinforces the suggestion that it is a particular consequence of leaf mining, namely damage to the primary vasculature, that makes a leaf unsuitable as a resource to the aphid, rather than leaf-miner presence as such. This was supported by the results of experiments simulating leaf-miner damage on B. pendula leaves. When E. betulae were reared on leaves with simulated leaf-miner damage (by cauterizing the abaxial lamina with a soldering iron), aphid performance was significantly lower on leaves with midrib damage compared to either lamina damaged or undamaged leaves (GLM: $F_{2,116} = 65.04$; $p < 0.001$) (Fig. 3.1b). Thus midrib damage was pinpointed as being the sole factor associated with higher E. betulae mortality.

A second part of the same study (see Johnson et al. 2002) investigated whether midrib damage had beneficial effects on leaf-miner performance. Larval mass on emergence was recorded from mines with just lamina damage, i.e., midrib intact, from mines where the midrib was damaged by the miner ("natural" damage) and from mines where the midrib was manually severed (using a mounted dissection pin). Eriocrania mass was significantly greater when the midrib was damaged compared to when it was intact (ANOVA: $F_{2,40} = 4.22$; $p = 0.022$). Post hoc analysis indicated that the only significant difference was between leaf miners on leaves with manually severed midribs and leaf miners that only damaged the lamina ($t = -2.83$; $p = 0.0193$). A possible mechanism for the enhanced larval mass of Eriocrania feeding on leaves with manually severed midribs compared with those with intact midribs may be an increase in leaf nutritional value. In addition to restricting water supply to leaves by damaging the xylem, nutrient export from leaves may be curbed by disrupting phloem

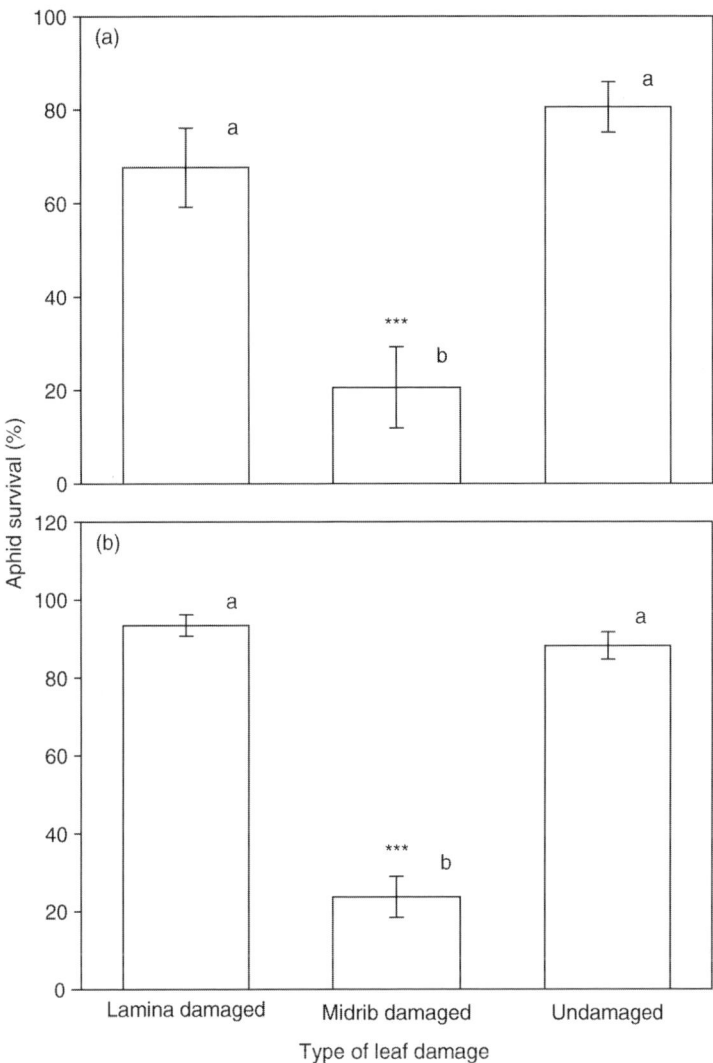

Figure 3.1. *Euceraphis betulae* survivorship (% of total number of aphids in experiment which survived under each treatment) when caged (a) with *Eriocrania* leaf miners that had damaged either the lamina or the midrib, and mine-free leaves, and (b) on leaves that had simulated (cauterized tissue) leaf-miner damage representing lamina and midrib damage, compared with undamaged leaves. Mean values +/− standard errors are shown. Lowercase superscripts indicate significant differences between leaf damage types. Survivorship was significantly lower on leaves with the midrib damaged in both cases. *** indicates $p < 0.001$ (GLM; see text for details).

vessels. Accelerated protein degradation due to water stress may therefore result in the accumulation of nitrogenous compounds in leaves with midrib damage. However, it remains unclear why this mechanism does not produce significant improvement in larval performance when the midrib is naturally damaged, particularly since leaves with both naturally and manually damaged midribs have a higher nitrogen content than mine-free leaves (Johnson et al. 2002).

The impact of leaf miners on the vascular system

The competitive interaction between *Eriocrania* and *E. betulae*, like many direct (Lawton and Hassell 1981) and indirect (Bonsall and Hassell 1997) interactions between insects, is asymmetric. All *Eriocrania* larvae were alive at the end of the caging experiments, suggesting that *E. betulae* has little discernible impact on *Eriocrania* and, in a previous study, we found no impact of aphids on miner performance, despite the presence of aphids reducing the resources available for the miners by 33% (Fisher et al. 2000).

Disruption of phloem hydraulics is proposed as the mechanism underpinning the negative impacts of the miners on the aphid. Such disruption could involve a drop in phloem pressure, or reduced total flow into the leaf. It may also encompass change in the nitrogen content of the sap. However, it seems unlikely that *Eriocrania* caused changes in the nutritional chemistry of phloem sap only in leaves with midrib damage and not in lamina-damaged leaves. Although not an ideal indication of phloem sap quality, we found higher nitrogen concentrations in mined leaves with midrib damage than mine-free leaves (Johnson et al. 2002). A more probable explanation for the negative effects on the aphid is that damage to the midrib disrupts the rate or amount of phloem flow on which phloem-feeders are dependent (Dixon 1998). Similar effects have been noted, particularly in systems involving gall-formers (Hartley 1998). For example, when an aphid gall (*Geoica* sp.) was situated on the midrib of a pistachio leaf it diverted nutrients away from a second aphid gall (*Forda formicaria*), adversely affecting its performance (Inbar et al. 1995). A previous study (Hartley 1998) clearly demonstrated that the relative position of galls on the vascular system is critical in determining the impact they have on one another. This intraspecific interaction was also asymmetric: the removal of *Cynips divisa* galls attached to veins near the midrib had a more beneficial effect on performance of the remaining galls than removing galls that were drawing their resources from the vascular system at the edge of the leaf.

Euceraphis betulae might be particularly susceptible to interference with phloem hydraulics because of its specific feeding sites on the larger primary veins, namely the basal midrib and petiole. Although the flow of phloem sap is greater in the larger veins, the concentration and quality tends to be lower than in neighboring smaller vessels (Fellows and Geiger 1974), suggesting that

E. betulae is adapted to feeding on lower-quality sap but at higher flux rates. Any interference with the phloem hydraulics might then be especially detrimental to *E. betulae*, compared to an aphid that is adapted to feed on vessels with a lower flow of phloem sap. Indeed, Prestidge and McNeil (1982) have suggested that the availability of phloem sap has selected for two discrepant life-history styles in phloem-feeding insects. Those phloem-feeders that have highly specific demands for phloem sap (e.g., high flux) are highly mobile so as to meet those demands, whereas phloem-feeders that are more tolerant of fluctuating phloem sap availability tend to be less mobile. Under such circumstances, vasculature damage by leaf-mining insects might disproportionately affect large and mobile aphids such as *E. betulae*.

Interactions based on the resources in the vascular system

Interactions between aphids and hemi-parasitic plants

We investigated the indirect interactions between two organisms, a hemi-parasitic angiosperm *Rhinanthus minor* L. (Scrophulariaceae) and a phloem-feeding insect *Sitobion avenae* F. (Aphididae) on a shared host, *Poa annua* L. (Poaceae). These two organisms are spatially separated when feeding, removing resources from different parts of the host, but they both attack the plant vascular system. Through haustorial connections, *R. minor* penetrates host roots below ground to remove xylem contents, while *S. avenae* uses a rostrum to penetrate the phloem vessels in leaves above ground. Both *R. minor* and *S. avenae* may modify plant quality by creating sinks and diverting the movement of nutrients within the plant (Awmack and Leather 2002). When sharing the same host, these spatially separated organisms may compete indirectly for host resources; the interaction is mediated by modifications in nutrient allocation, biomass, or allometry within the host plant.

Studies that have investigated the effects of the removal of resources by *R. minor* on the host plant all reported a reduction in host biomass (Seel and Press 1996, Puustinen and Salonen 1999, Jiang *et al.* 2004). For example, Davies and Graves (2000) found that *R. minor* parasitism significantly suppressed host biomass by between 32% and 44%. The impacts of *R. minor* parasitism are largest on younger hosts and persist until the following year (Seel and Press 1996). Indeed, *R. minor* can have such a large impact on its graminaceous hosts that it can decrease the productivity and change the species composition of grassland communities (Pywell *et al.* 2004).

Aphids feed on phloem sap and hence act as sinks, removing photosynthetic products and manipulating host physiology and morphology (Crawley 1983,

Louda *et al.* 1990, Hartley and Jones 2004). Aphids also can influence host chemical composition by injecting saliva, hastening senescence, modifying the balance of the carbohydrate sink strengths, and transmitting viruses (Kennedy 1951). The grain aphid *S. avenae* has been shown to significantly reduce *P. annua* above- and belowground biomass by 21% and 32% respectively (Fraser and Grime 1999).

The impact on the shared host by each of these organisms will be important in determining indirect interactions between them. Pennings and Callaway (2002) suggested that, in comparison to herbivores, parasitic plants might have greater effects on their hosts in proportion to the mass of the consumer and the amount of resources removed. They argued that the disproportionate impact of parasitic plants on their hosts is mostly caused by parasite-induced changes in host allocation and physiology. These large impacts, including a reduction of grassland productivity (Pywell *et al.* 2004), lead us to suggest that *R. minor* parasitism may have a larger impact on the host biomass than *S. avenae*, and hence may be the stronger competitor for host resources, at least in natural systems. In crop systems, high densities of *S. avenae* can impact on host biomass and reduce yields, although effects are most severe at early crop stages (Voss *et al.* 1997, Wangai *et al.* 2000). If *R. minor* and *S. avenae* do have different abilities to acquire host resources, it will result in an asymmetric interaction between them on a shared host.

Changes in the host as a result of resource removal by *R. minor* and *S. avenae* will mediate the indirect interaction between *R. minor* and *S. avenae*. Specifically, changes in host root:shoot ratio and biomass will alter the availability of resources for each organism. *Rhinanthus minor* may reduce host photosynthesis by removing essential material from the host xylem vessels. That could reduce phloem quality for *S. avenae*. In addition, to compensate for the removal of xylem contents by root hemi-parasites, host root:shoot ratio has been observed to increase in many infected host plants (e.g., Graves *et al.* 1990, 1992). For example, Ramlan and Graves (1996) found *R. minor* parasitism increased *Lolium perenne* root biomass by 50%. The reduction in aboveground biomass compared to root biomass will reduce resource availability for *S. avenae* and is likely to have a negative impact on its development time and fecundity. At the same time, *S. avenae* removes and excretes host photosynthetic products taken from the phloem. This could reduce total host biomass and result in the host allocating proportionally less resources to belowground biomass. Aphids have been shown in a range of studies to reduce root biomass of their hosts (Choudhury 1984, Fraser and Grime 1999, Hartley and Jones 2004). The proportional reduction in root biomass may reduce the haustorial connectivity of *R. minor* with the host. It could also affect xylem sap quality for *R. minor* by reducing the energy supply for nutrient uptake by host roots (Bowling and Dunlop 1978). However, when both organisms are sharing the same host, both xylem and phloem resources are being removed from the host vascular system simultaneously. The induced host response is less easy to predict and will be determined by which nutrients are the most limiting for the host and which organism is the strongest sink for host nutrients.

The impact of hemi-parasites on aphid performance

We investigated the plant-mediated interaction between R. minor and S. avenae in a greenhouse experiment. Poa annua seedlings were grown in pots containing washed potting sand and Osmocote fertilizer at 1/16 of the manufacturer's recommended amount, equivalent to 94 mg N of fertilizer per liter of sand. Three R. minor seedlings were transferred to each pot and 7 weeks later 20 S. avenae were added to half the replicates and all pots were caged. The experiment was harvested 21 days later, 10 weeks into the experiment. The numbers of R. minor flowers were counted and aphids were collected, counted and weighed. The heights of both plant species were measured before being separated into above- and belowground biomass, dried and weighed. Haustorial attachments were clearly visible and counted in the rehydrated roots.

Biomass of R. minor was significantly increased by the presence of S. avenae on a shared host ($F_{1,64} = 7.82$; $p = 0.007$) (Fig. 3.2). The same relationship was observed with the number of R. minor flowers ($F_{1,64} = 12.36$; $p = 0.001$) (Fig. 3.2),

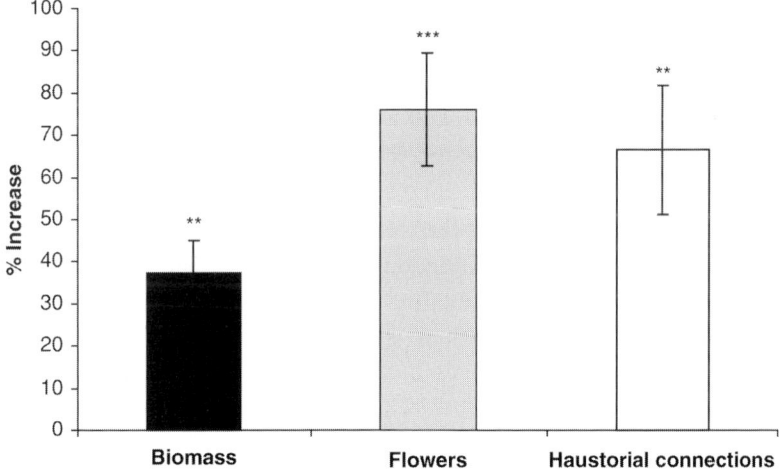

Figure 3.2. The increase in biomass, flowers (% dry weight), and haustorial connections (number) of Rhinanthus minor when sharing a host with the phloem feeding aphid Sitobion avenae, compared to R. minor growing alone with the host (Poa annua). Mean values ± standard errors are shown. ** indicates $p < 0.01$; *** indicates $p < 0.001$ (ANOVA; see text for details).

which was positively correlated with R. minor biomass ($r^2 = 0.49$; $p < 0.001$). Although the number of R. minor flowers at the time of harvesting was not a direct indicator of the hemi-parasite's fecundity, it does suggest an increase in potential seed production in R. minor when growing on a host experiencing aphid herbivory. The improved performance of the hemi-parasite when aphids were present may result from the fact that aphid herbivory significantly increased the number of haustorial connections between R. minor and its host ($F_{1,41} = 8.53$; $p = 0.006$) (Fig. 3.2). Further analysis revealed that there was a strong positive correlation between R. minor biomass and the number of haustorial connections within the host root system ($r^2 = 0.33$; $p < 0.001$). However, the increase in haustorial connections in R. minor attacking hosts experiencing aphid herbivory was not simply related to changes in root biomass: no correlation between the number of connections and host root biomass was found ($r^2 = 0.02$; $p = 0.4$).

There was a positive correlation between P. annua aboveground biomass and the number of S. avenae ($r^2 = 0.27$; $p = 0.001$). There was a marginal reduction in aphid numbers ($F_{1,32} = 3.97$; $p = 0.05$) and total biomass ($F_{1,32} = 3.68$; $p = 0.059$) on a shared host when R. minor was present, but the weight of individual aphids was unaffected by the presence of R. minor ($F_{1,32} = 0.12$; $p = 0.73$) (Fig. 3.3). The marginal effects observed may reflect the large variability in the data. In addition, although R. minor parasitism reduced aboveground biomass of the host, the number of aphids per unit area of leaf was significantly increased on a shared host ($F_{1,32} = 4.70$; $p = 0.033$). Three weeks of aphid herbivory on an unparasitized host did not alter host height, biomass, or allometry, but R. minor parasitism did significantly reduce host biomass over this time period ($F_{1,152} = 42.23$; $p < 0.001$), though root:shoot ratio was unaffected ($F_{1,152} = 1.60$; $p = 0.209$). Parasitism by R. minor alone reduced host biomass by around 37% on average, i.e., almost to the same degree as when both aphid and parasite were present on the same host (Fig. 3.3). Although the combined effect of both herbivory and parasitism resulted in a marked reduction in host height and aboveground biomass (Fig. 3.3), there was high variability in the belowground response, with 85% of plants showing no change in root:shoot ratio.

The impacts of hemi-parasites and aphids on the vascular system

Contrary to our original expectation that hemi-parasites and aphids might compete indirectly for the vascular resources of the host, the presence of aphids on the same host greatly increased R. minor biomass and fecundity. This increase in biomass was facilitated by an increase in haustorial connections between R. minor and the host roots. This resulted in a growth in parasite sink strength by increasing the surface area across which xylem contents were

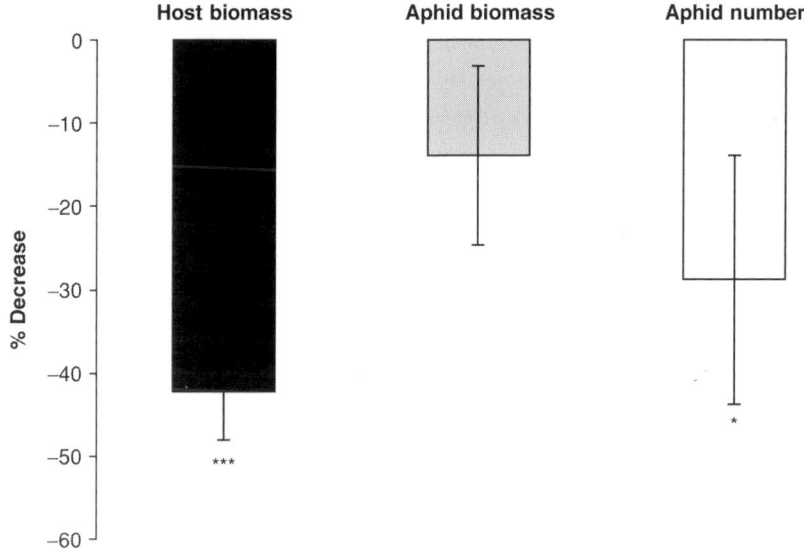

Figure 3.3. The impact of R. minor parasitism on the host plant and on S. avenae. The figure shows the decrease in host (P. annua) aboveground biomass (% dry weight) when experiencing both herbivory by S. avenae and parasitism by R. minor; and the decrease in the total biomass and size of populations of S. avenae on a host experiencing parasitism compared to populations grown on the host alone. Mean values ± standard errors are shown. * indicates $p < 0.05$; *** indicates $p < 0.001$ (ANOVA; see text for details).

transferred (Seel et al. 1992). This increase in connectivity was not linked with an associated increase in host root biomass and may be related to a change in the structure rather than biomass of the host roots. For example, phloem-feeding by the aphid may have reduced the ability of the host to defend against penetration from parasite roots, perhaps mediated by some sort of change in root structure. Also, once attached to the host, the parasite and host compete for resources (Graves 1995) and the effect of aphid herbivory on the host may have altered the host's sink strength. In contrast, our expectation that R. minor parasitism on a shared host would have a negative impact on aphid populations was supported. Aphid herbivory alone did not have an effect on host biomass, which is contrary to many earlier studies, such as that of Fraser and Grime (1999) who showed that herbivory by S. avenae on the host P. annua reduced the host's root:shoot ratio as well as reducing overall biomass. This may have been related to the very low nutrient levels used in this experiment, which resulted in small aphid populations. At a higher nutrient level or over a longer timescale, aphid populations may have become large enough to impact on host biomass and allocation patterns.

These results were considered in relation to the suggestion by Pennings and Callaway (2002) that in comparison to herbivores, parasitic plants might have greater effects on their hosts in proportion to the mass of the consumer and the amount of resources removed. The results superficially suggest that *R. minor* parasitism did have a much larger effect on the host biomass compared to *S. avenae*. However, we found a strong correlation between the reduction in aboveground host biomass and the biomass of *R. minor* and *S. avenae*, suggesting that impact on the host was related to biomass of the consumer rather than the type of consumer. This means that the effects of the insects on the host may have been less significant because insect biomass was much smaller than the biomass of the parasitic plant, rather than supporting Pennings and Callaway's theory. Clearly further examples are required to establish any generality in this pattern.

The outcome of the indirect interaction between *R. minor* and *S. avenae* on a shared host was asymmetric. The interaction was positive for *R. minor* and negative for *S. avenae*, an example of contramensalism. The positive effect on *R. minor* biomass in the presence of aphid herbivory on the shared host was mediated by an increase in haustorial connections with the host roots. The reduction in the size of the aphid population was correlated with the reduction in the host aboveground biomass. This experiment is the first to demonstrate an asymmetric interaction between a parasitic plant and a phloem-feeding insect herbivore. We suggest that the reduction in host aboveground biomass and, potentially, the modification of host phloem resources by parasitic plants, could be more widespread than previously reported. With approximately 400 species of parasitic plants distributed in habitats all over the world from the Arctic to Australia, this type of interaction could be an important community process. It is suggested that *R. minor* plays a central role in determining the structure and dynamics of grassland communities because of its impact on grass productivity and species richness (Pywell *et al.* 2004) and, since plant diversity is an important driver of insect diversity in many ecosystems (Hartley *et al.* 2003), there could be significant knock-on effects of *R. minor* presence on insect community structure. It is certain that the potential indirect interactions between parasitic plants and insect herbivores have been under-reported thus far and that such interactions could potentially be of significance in natural communities. Other indirect interactions between very different types of organisms have been shown to have important and wide-ranging effects, particularly so in the case of plant-mediated interactions between insects and fungi, whether pathogenic (Hatcher 1995, Johnson *et al.* 2003), mycorrhizal (Gange and Bower 1997), or endophytic (Clay 1988).

Discussion

Herbivores, parasites, and the plant vascular system

Interactions between diverse organisms based on the plant vascular system have been little studied but are likely to be of much wider significance than currently believed. They involve widespread and abundant insect herbivores and hemi-parasites, which are a significant component of many ecosystems. Aphids rely on the vascular system, specifically the phloem, for their resources. Hemi-parasites also acquire much of their nutrients from the plant vascular system, this time from the xylem. *Eriocrania* leaf miners are the only organisms featured in the interactions presented here that do not directly exploit the resources in the plant vascular system. However, their feeding action does impact on the phloem in such a way that the performance of organisms that do require vascular resources is adversely affected. The impacts of these organisms on each other are not trivial. Leaf miners that cut the midrib reduce the survivorship of aphids feeding on these leaves by an average of 60–70% and in natural communities aphids avoid these leaves. There are around six times more aphids per leaf on unmined leaves than on those with a damaged midrib (Johnson *et al.* 2002). Similarly, the presence of the hemi-parasite *R. minor* reduced aphid population size by 36% on average, as well as slightly reducing their biomass, although the data are variable. The presence of aphids improves the performance of the hemi-parasite to such an extent there is a potential increase in fecundity: the number of haustorial connections, the biomass, and the amount of flowering were all increased.

Organisms that exploit the vascular system will alter source–sink relationships within the host plant and this underpins some of the interactions described in this chapter. For example, parasitic plants reduce host biomass and may increase root:shoot ratio, and in a previous study (Seel and Press 1996), the impacts of *R. minor* on the host were related to biomass partitioning. In this study, we found that root:shoot ratio was not significantly affected by *R. minor* parasitism overall, but the response was variable and some individual plants did respond to infection by *R. minor* with an increase in root:shoot ratio. Seel and Press (1996) also found that *R. minor* parasitism did not have a large impact on root:shoot ratio of the host. The impact of *R. minor* on the root biomass of the host seems to depend on nutrient levels. Davies and Graves (2000) found that parasitism caused a decrease in host root:shoot ratio when phosphorus levels were high but not when they were low, possibly because when nutrient availability is low, the host is unable to respond to xylem resource removal by increasing root biomass.

Some insect herbivores, particularly gall-formers, also are able to manipulate source–sink relationships within the host plant (e.g., Hartley 1998). The creation of additional sinks by insect herbivores can both increase (Abrahamson and McCrea 1986) or decrease (Hartley and Lawton 1992, Masters and Brown 1992) the nitrogen levels in nearby foliage. Impacts of aboveground insect herbivores on root : shoot ratios have also been demonstrated (e.g., Fraser and Grime 1999), but the impacts on this parameter seem surprisingly variable (Hartley and Jones 2004), even for herbivores that create strong sinks aboveground (Abrahamson and McCrea 1986).

Comparisons and contrasts

Aphids vs. other organisms

It is very clear that, as with many indirect interactions, the effects are asymmetric. Denno *et al.* (1995) noted that interspecific competition was recorded in 76% of interactions between insect herbivores and 84% of these were asymmetric. The aphids are the "victims" in both interactions highlighted in this chapter in that they are negatively affected by both leaf miners and hemi-parasitic plants, but they have either no or positive effects on their competitors (Fig. 3.4). Previous studies of the competitive interactions between haustellate and mandibular herbivores found that haustellate herbivores were superior competitors about 50% of the time (Denno *et al.* 1995). A study of the herbivores on goldenrod *Solidago altissima* found that xylem-feeding spittlebugs had a far greater impact on the host plant, and hence on potential competitors, than either leaf-feeding beetles or aphids (Meyer 1993). Although the effects of competitors on aphids reported here have been negative, other outcomes have been observed in interactions between aphids and other types of root-feeding organisms. For instance, when aboveground aphids are indirectly affected by root-feeding insects the effects on aphid performance and abundance tend to be positive (Masters and Brown 1992; but see Blossey and Hunt-Joshi 2003).

One possible explanation for the direction of the asymmetric interactions we describe is that the biomass of the organism concerned is more significant than the type of organism in terms of determining its impact on the shared host plant, and hence on the other organisms. Aphids are small so need high abundance to have a significant impact on the host plant in a way that produces an adverse impact on their competitors. Although such impacts have been found in some other studies (e.g., Fraser and Grime 1999), in our studies we did not find that aphids had a significant adverse effect on either miners (Fisher *et al.* 2000), hemi-parasites, or the host plant in the case of the grasses. Parasitism by *R. minor* had the greater impact on the host than the insect herbivores. Three weeks of *S. avenae* herbivory had no effect on the host plant, while *R. minor* parasitism

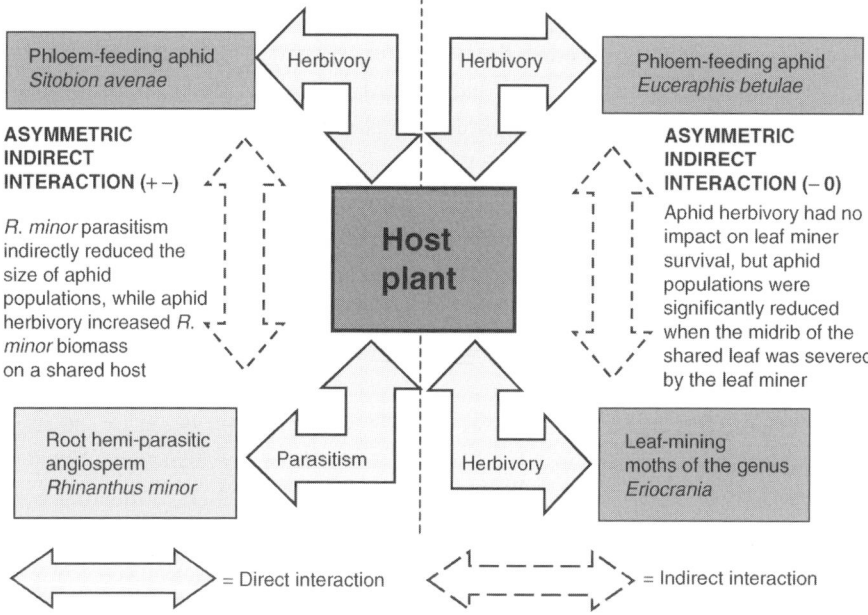

Figure 3.4. A schematic representation of two different indirect interactions between organisms, mediated by a shared host, illustrating the results obtained in this chapter. The outcomes of these interactions are asymmetric and can either be positive (+), negative (−), or nonsignificant (0). The directions of the outcomes (+, −, 0) of the interactions shown may vary according to the relative densities of aphids, miners, and parasitic plants.

significantly reduced both above- and belowground biomass. In our experiments, the relative reduction in host biomass compared with control plants was correlated with the biomass of R. minor and S. avenae which were removing resources from the host ($r^2 = 0.18$; $p < 0.001$). Consequently, the lack of significant reduction in host biomass by aphid herbivory was related to the low biomass of S. avenae, rather than simply to their lack of impact.

Insect herbivores vs. hemi-parasites

Despite the fact that they are from different kingdoms, many parallels can be drawn between parasitic plants and insect herbivores (Pennings and Callaway 2002). Both types of consumer have the potential to reduce host biomass (Seel and Press 1996, Fraser and Grime 1999), alter host allometry (Crawley 1983, Davies and Graves 2000), and modify community interactions (Brown and Gange 1992, Bigger and Marvier 1998, Pywell et al. 2004). The relationship between insect herbivores and their host plants is considered to be closer to parasite–host than predator–prey relationships (Strauss and Zangerl

2002). However, there are differences in mobility, in the level of impact and interaction with the host, and in the effect the organisms have on the hormonal and chemical composition of the host plant (Pennings and Callaway 2002). Pennings and Callaway (2002) suggest that in comparison to herbivores, parasitic plants might have greater effects on their hosts in proportion to the mass of the consumer and the amount of resources removed. Hence in our studies, R. minor had a bigger impact on the host because it was removing host resources for longer and R. minor biomass was greater than the biomass of S. avenae.

Denno et al.'s (1995) review of indirect interactions between phytophagous insects concluded that insect herbivores were most likely to compete under specific conditions, i.e., if they are closely related, sessile, aggregative, feed on discrete resources, and feed on grasses or forbs. *Sitobion avenae* is a grass specialist and grasses are the main host for R. minor. Organisms are also more likely to compete if they are aggregated, and a field survey (Bass 2004) found that both R. minor and grass-feeding Hemiptera have aggregated distributions in natural communities. Discrete resources, due to their limited nature, are likely to result in competition. However, in contrast to the general conclusion that closely related organisms are more likely to interact (Denno et al. 1995), the results presented here, and those from studies of plant-mediated interactions involving insects and fungi (e.g., Johnson et al. 2003), indicate that strong interactions between organisms from different kingdoms are possible.

Mechanical vs. chemical effects

Some of the interactions we describe here are based predominantly on mechanical disruption of the vascular system, but other cases may involve both mechanical and chemical effects. Indeed in the study reported here, leaves that were mined by *Eriocrania* also had elevated levels of phenolic compounds (see Johnson et al. 2002), a major group of defensive compounds in birch, but this seems less likely as the causal mechanism for the negative impacts on aphids. As phloem-feeders, many aphids circumvent the effects of phenolics because they usually occur in low concentrations in the phloem (Hartley and Jones 1997). Several aphid species (Martin et al. 1994), including *E. betulae* (Johnson et al. 2002), do not appear to be adversely affected by phenolic compounds, but insects of different feeding guilds may be negatively affected by this change in leaf chemistry. The disruption in phloem hydraulics in the system we studied is a more feasible explanation, especially given the greater dependence of aphids on the flux of phloem sap rather than its nutritional quality per se (Dixon 1998). Whether mechanical or chemical effects predominate in a particular interaction might therefore be determined by the feeding guild of the insect being affected by the change in host plant anatomy and/or chemistry. No studies, to our knowledge, have directly measured the changes in phloem hydraulics that may result from leaf-miner damage, so this remains an issue that needs to be addressed.

Parasitic plants may alter both the quantity and quality of the phloem and xylem sap in the host vascular system. The reduction in host biomass produced by *R. minor* will have reduced the number of phloem vessels available to *S. avenae*. We did not measure any changes in phloem quality in this experiment, but many hemi-parasites have a negative impact on host photosynthesis (Graves *et al.* 1989), so it is not inconceivable that there were major changes in the quality and quantity of phloem resources available for the aphid on parasitized hosts. Some previous work supports this possibility. Jiang *et al.* (2004) compared the model of net flow of nutrients around the host *Hordeum* when infected and uninfected by *R. minor*. They found that total nitrogen uptake was lower in infected plants and 18% of root xylem flow was redirected towards the parasite. As a consequence, 66% less nitrogen was available for xylem flow to the shoot, but despite this, phloem retranslocation of nitrogen was relatively increased in response to parasitism. Nitrogen is the limiting nutrient for phloem-feeding aphids so changes in nitrogen content of the phloem sap would alter the quality of the resource for the aphids (Raven 1983). This mechanism is consistent with our results: *R. minor* parasitism did reduce host biomass which resulted in a reduction in aphid numbers, but we also observed that the presence of *R. minor* increased the number of aphids per unit of leaf area, which may reflect enhanced phloem quality.

Conclusion

The results presented in this chapter clearly demonstrate that plants have the potential to mediate interactions between taxonomically separated organisms: plant–insect interactions have consequences for hemi-parasitic plants and vice versa. Furthermore, we found that these interactions may be based on physical alteration of the structures within plants (the midrib, the roots), rather than solely the changes in the quality and quantity of resources that we usually associate with indirect interactions. All the interactions were asymmetric but could be positive as well as negative, even when organisms seem to be exploiting similar resources (Fig. 3.4). We may need to consider a broader range of both organisms and mechanisms than we have in the past when examining the impacts of plant–insect interactions on the wider ecological community.

Acknowledgments

We are grateful to Angela Douglas for help with the work on *Eriocrania* leaf miners, particularly the chemical analysis of mined leaves. We thank Libby John and Malcolm Press for their advice on the experiments with parasitic plants and aphids. We are grateful to the Natural Environment Research Council, the University of Sussex, and the Royal Society for financial support.

References

Abrahamson, W. G., and K. D. McCrea. 1986. Nutrient and biomass allocation in *Solidago altissima*: effects of two stem gall-makers, fertilisation and ramet isolation. *Oecologia* **68**:174–180.

Awmack, C. S., and S. R. Leather. 2002. Host plant quality and fecundity in herbivorous insects. *Annual Review of Entomology* **47**:817–844.

Bass, K. A. 2004. Resource partitioning in the multi-species interaction between a host plant, a parasitic angiosperm and an insect herbivore. Ph.D. dissertation, University of Sussex, Brighton, UK.

Bigger, D. S., and M. A. Marvier. 1998. How different would a world without herbivory be? A search for generality in ecology. *Integrative Biology* **1**:60–67.

Blossey, B., and T. R. Hunt-Joshi. 2003. Belowground herbivory by insects: influence on plants and aboveground herbivores. *Annual Review of Entomology* **48**:521–547.

Bonsall, M. B., and M. P. Hassell. 1997. Apparent competition structures ecological assemblages. *Nature* **388**:371–373.

Bowling, D. J. F., and J. Dunlop. 1978. Uptake of phosphate by white clover. I. Evidence for an electrogenic phosphate pump. *Journal of Experimental Botany* **29**:1139–1146.

Brown, V. K., and A. C. Gange. 1992. Secondary plant succession: how is it modified by insect herbivory? *Vegetatio* **101**:3–13.

Carroll, C. R., and C. A. Hoffman. 1980. Chemical feeding deterrent mobilized in response to insect herbivory and counteradaptation by *Epilachna tredecimnotata*. *Science* **209**:414–416.

Cechin, I., and M. C. Press. 1993. Nitrogen relations of the sorghum–*Striga hermonthica* host–parasite association: growth and photosynthesis. *Plant, Cell and Environment* **16**:237–247.

Choudhury, D. 1984. Aphids and plant fitness: a test of Owen and Weigert's hypothesis. *Oikos* **43**:401–402.

Clay, K. 1988. Fungal endophytes of grasses: a defensive mutualism between plants and fungi. *Ecology* **69**:10–16.

Connor, E. F., and M. P. Taverner. 1997. The evolution and adaptive significance of the leaf-mining habit. *Oikos* **79**:6–25.

Crawley, M. J. 1983. *Herbivory: The Dynamics of Animal–Plant Interactions*. Oxford, UK: Blackwell Science.

Davies, D. M., and J. D. Graves. 2000. The impact of phosphorus on interactions of the hemi-parasitic angiosperm *Rhinanthus minor* and its host *Lolium perenne*. *Oecologia* **124**:100–106.

Denno, R. F., and G. K. Roderick. 1992. Density-related dispersal in plant hoppers: effects of interspecific crowding. *Ecology* **73**:1323–1334.

Denno, R. F., M. S. McClure, and J. R. Ott. 1995. Interspecific interactions in phytophagous insects: competition re-examined and resurrected. *Annual Review of Entomology* **40**:297–331.

Denno, R. F., M. A. Peterson, C. Gratton, J. Cheng, G. A. Langellotto, A. F. Huberty, and D. L. Finke. 2000. Feeding-induced changes in plant quality mediate interspecific competition between sap-feeding herbivores. *Ecology* **81**:1814–1827.

Dixon, A. F. G. 1998. *Aphid Ecology: An Optimization Approach*, 2nd edn. London: Chapman and Hall.

Dussourd, D. E., and R. F. Denno. 1994. Host range of generalist caterpillars: trenching permits feeding on plants with secretory canals. *Ecology* **75**:69–78.

Dussourd, D. E., and T. Eisner. 1987. Vein cutting behavior: insect counterploy to the latex defense of plants. *Science* **237**:898–901.

Fellows, R. J., and D. R. Geiger. 1974. Structural and physiological changes in sugar beet leaves during sink to source conversion. *Plant Physiology* **54**:877–885.

Fisher, A. E. I., S. E. Hartley, and M. Young. 2000. Direct and indirect competitive effects of foliage feeding guilds on the performance of the birch leaf-miner *Eriocrania*. *Journal of Animal Ecology* **69**:165–176.

Forcella, F. 1982. Why twig-girdling beetles girdle twigs. *Naturwissenschaften* **69**:398–400.

Fraser, L. H., and J. P. Grime. 1999. Aphid fitness on 13 grass species: a test of plant defence theory. *Canadian Journal of Botany* **77**:1783–1789.

Fukui, A. 2001. Indirect interactions mediated by leaf shelters in animal–plant communities. *Population Ecology* **43**:31–40.

Gange, A. C., and E. Bower. 1997. Interactions between insects and mycorrhizal fungi, pp. 115–132 in A. C. Gange and V. K. Brown (eds.) *Multitrophic Interactions in Terrestrial Systems*. Oxford, UK: Blackwell Science.

Gange, A. C., and V. K. Brown. 1989. Effects of root herbivory by an insect on a foliar-feeding species, mediated through changes in the host plant. *Oecologia* **81**:38–42.

Graves, J. D. 1995. Host-plant responses to parasitism, pp. 207–225 in M. C. Press and J. D. Graves (eds.) *Parasitic Plants*. London: Chapman and Hall.

Graves, J. D., M. C. Press, and G. R. Stewart. 1989. A carbon balance model of the sorghum–*Striga hermonthica* host–parasite association. *Plant, Cell and Environment* **12**:101–107.

Graves, J. D., A. Wylde, M. C. Press, and G. R. Stewart. 1990. Growth and carbon allocation in *Pennisetum typhoides* infected with the parasitic angiosperm *Striga hermonthica*. Plant, Cell and Environment **13**:367–373.

Graves, J. D., M. C. Press, S. Smith, and G. R. Stewart. 1992. The carbon economy of the association between cowpea and the parasitic angiosperm *Striga gesnerioides*. Plant, Cell and Environment **13**:319–328.

Hajek, A. E., and D. L. Dahlsten. 1986. Discriminating patterns of variation in aphid (Homoptera, Drepanosiphidae) distribution on *Betula pendula*. *Environmental Entomology* **15**:1145–1148.

Harrison, S., and R. Karban. 1986. Effects of an early season folivorous moth on the success of a later season species, mediated by a change in quality of the shared host, *Lupinus arboreus* Sims. *Oecologia* **69**:354–359.

Hartley, S. E. 1998. The chemical composition of plant galls: are levels of nutrients and secondary compounds controlled by the gall-former? *Oecologia* **113**:492–501.

Hartley, S. E., and C. G. Jones. 1997. Plant chemistry and herbivory, or why the world is green, pp. 284–324 in M. J. Crawley (ed.) *Plant Ecology*. Oxford, UK: Blackwell Science.

Hartley, S. E., and T. H. Jones. 2004. Insect herbivores, nutrient cycling and plant productivity: a review, pp. 27–52 in W. W. Weisser and E. Siemann (eds.) *Insects and Ecosystem Function*. Berlin, Germany: Springer-Verlag.

Hartley, S. E., and J. H. Lawton. 1987. Effects of different types of damage on the chemistry of birch foliage, and the responses of birch feeding insects. *Oecologia* **74**:432–437.

Hartley, S. E., and J. H. Lawton. 1992. Host-plant manipulation by gall insects: a test of the nutrition hypothesis. *Journal of Animal Ecology* **61**:113–119.

Hartley, S. E., S. M. Gardner, and R. J. Mitchell. 2003. Indirect effects of grazing and nutrient addition on the hemipteran community of heather moorlands. *Journal of Applied Ecology* **40**:793–803.

Hatcher, P. E. 1995. Three-way interactions between plant pathogenic fungi, herbivorous insects and their host plants. *Biological Reviews* **70**:639–694

Haukioja, E., K. Ruohomaki, J. Senn, J. Suomela, and M. Walls. 1990. Consequences of herbivory in the mountain birch (*Betula pubescens* ssp. *tortuosa*): importance of the functional organization of the tree. *Oecologia* **82**:238–247.

Inbar, M., A. Eshel, and D. Wool. 1995. Interspecific competition among phloem-feeding insects mediated by induced host-plant sinks. *Ecology* **76**:1506–1515.

Jiang, F., W. D. Jeschke, and W. Hartung. 2004. Solute flows from *Hordeum vulgare* to the hemiparasite *Rhinanthus minor* and the influence of infection on host and parasite nutrient relations. *Functional Plant Biology* **31**:633–643.

Johnson, S. N., P. J. Mayhew, A. E. Douglas, and S. E. Hartley. 2002. Insects as leaf engineers: can leaf-miners alter leaf structure for birch aphids? *Functional Ecology* **16**:575–584.

Johnson, S. N., A. E. Douglas, S. Woodward, and S. E. Hartley. 2003. Microbial impacts on plant–herbivore interactions: the indirect effects of a birch pathogen on a birch aphid. *Oecologia* **134**:388–396.

Karban, R. 1986. Interspecific competition between folivorous insects on *Erigeron glaucus*. *Ecology* **67**:1063–1072.

Karban, R., and I. T. Baldwin. 1997. *Induced Responses to Herbivory*. Chicago, IL: University of Chicago Press.

Kennedy, J. S. 1951. Benefits to aphids from feeding on galled and virus-infected leaves. *Nature* **168**:825–826.

Larson, K. C., and T. G. Whitham. 1991. The manipulation of food resources by a gall-forming aphid: the physiology of sink–source interactions. *Oecologia* **88**:15–21.

Larson, K. C., and T. G. Whitham. 1997. Competition between gall aphids and natural plant sinks: plant architecture affects resistance to galling. *Oecologia* **109**:575–582.

Lawton, J. H., and M. P. Hassell. 1981. Asymmetrical competition in insects. *Nature* **289**:793–795.

Louda, S. M., K. H. Keeler, and R. D. Holt. 1990. Herbivore influences on plant performance and competitive interactions, pp. 413–444 in J. B. Grace and D. Tilman (eds.) *Perspectives in Plant Competition*. London: Academic Press.

Martin, M. A., N. Cappuccino, and D. Ducharme. 1994. Performance of *Symydobius americanus* (Homoptera, Aphididae) on paper birch grazed by caterpillars. *Ecological Entomology* **19**:6–10.

Masters, G. J., and V. K. Brown. 1992. Host plant mediated interactions between two spatially separated insects. *Functional Ecology* **6**:175–179.

Masters, G. J., and V. K. Brown. 1997. Host-plant mediated interactions between spatially separated herbivores: effects on community structure, pp. 217–237 in A. C. Gange and V. K. Brown (eds.) *Multitrophic Interactions in Terrestrial Systems*. Oxford, UK: Blackwell Science.

Mattson, W. J. 1986. Competition for food between two principal cone insects of red pine, *Pinus resinosa*. *Environmental Entomology* **15**:88–92.

McClure, M. S. 1980. Competition between exotic species: scale insects on hemlock. *Ecology* **61**:1391–1401.

Meyer, G. A. 1993. A comparison of leaf and sap feeding insects on photosynthetic rates of goldenrod. *Oecologia* **92**:480–489.

Nakamura, M., and T. Ohgushi. 2003. Positive and negative effects of leaf shelters on herbivorous insects: linking multiple herbivore species on willow. *Oecologia* **136**:445–449.

Pennings, S. C., and R. M. Callaway. 2002. Parasitic plants: parallels and contrasts with herbivores. *Oecologia* **131**:479–489.

Petersen, M. K., and J. P. Sandstrom. 2001. Outcome of indirect competition between two aphid species mediated by responses in their common host plant. *Functional Ecology* **15**:525–534.

Press, M. C., and W. E. Seel. 1996. Interactions between hemiparasitic angiosperms and their hosts in the subarctic. *Ecological Bulletins* **45**:151–158.

Press, M. C., J. D. Scholes, and J. R. Watling. 1998. Parasitic plants: physiological and ecological interactions with their hosts, pp. 174–197 in M. C. Press, J. D. Scholes, and M. G. Barker (eds.) *Physiological Plant Ecology*. Oxford, UK: Blackwell Science.

Prestidge, R. A., and S. McNeil. 1982. The role of nitrogen in the ecology of grassland Auchenorrhyncha (Homoptera). *Symposium of the British Ecological Society* **22**:257–281.

Price, P. W., and S. Louw. 1996. Resource manipulation through architectural modification of the host plant by a gall-forming weevil *Urodontus scholtzi* Louw (Coleoptera: Anthribidae). *African Entomology* **4**:103–110.

Price, P. W., G. W. Fernandes, and R. DeClerck-Floate. 1997. Gall-inducing insect herbivores in multitrophic systems, pp. 239–255 in A. C. Gange and V. K. Brown (eds.) *Multitrophic Interactions in Terrestrial Systems*. Oxford, UK: Blackwell Science.

Puustinen, S., and P. Mutikainen. 2001. Host–parasite–herbivore interactions: implications of host cyanogenesis. *Ecology* **82**:2059–2071.

Puustinen, S., and V. Salonen. 1999. The effect of host defoliation on hemiparasitic-host interactions between *Rhinanthus serotinus* and two *Poa* species. *Canadian Journal of Botany* **77**:523–530.

Pywell, R. F., J. M. Bullock, K. J. Walker, S. J. Coulson, S. J. Gregory, and M. J. Stevenson. 2004. Facilitating grassland diversification using the hemiparasitic plant *Rhinanthus minor*. *Journal of Applied Ecology* **41**:880–887.

Quested, H. M., M. C. Press, T. V. Callaghan, and J. H. C. Cornelissen. 2002. The hemiparasitic angiosperm *Bartsia alpina* has the potential to accelerate decomposition in sub-arctic communities. *Oecologia* **130**:88–95.

Ramlan, M. F., and J. D. Graves. 1996. Estimation of the sensitivity to photoinhibition in *Striga hermonthica*-infected sorghum. *Journal of Experimental Botany* **47**:71–78.

Raven, J. A. 1983. Phytophages of xylem and phloem: a comparison of animal and plant sap-feeders. *Advances in Ecological Research* **13**:135–234.

Seel, W. E., and W. D. Jeschke. 1999. Simultaneous collection of xylem sap from *Rhinanthus minor* and the hosts *Hordeum* and *Trifolium*: hydraulic properties, xylem sap composition and effects of attachment. *New Phytologist* **143**:281–298.

Seel, W. E., and M. C. Press. 1996. Effects of repeated parasitism by *Rhinanthus minor* on the growth and photosynthesis of a perennial grass, *Poa alpina*. *New Phytologist* **134**:495–502.

Seel, W. E., I. Cechin, C. A. Vincent, and M. C. Press. 1992. Carbon partitioning and transport in parasitic angiosperms and their hosts, pp. 199–223 in C. J. Pollock, J. F. Farrar, and A. J. Gordon (eds.) *Carbon Partitioning within and between Organisms*. Oxford, UK: Bios.

Shorthouse, J. D., and O. Rohfritsch (eds.) 1992. *Biology of Insect-Induced Galls*. New York: Oxford University Press.

Strauss, S. Y., and A. R. Zangerl. 2002. Plant–insect interactions in terrestrial ecosystems, pp. 77–106 in C. M. Herrera and O. Pellmyr (eds.) *Plant–Animal Interactions: An Evolutionary Approach*. Oxford, UK: Blackwell Science.

Strong, D. R., J. H. Lawton, and T. R. E. Southwood. 1984. *Insects on Plants: Community Patterns and Mechanisms*. Oxford, UK: Blackwell Science.

Voss, T. C., R. W. Kieckefer, B. W. Fuller, M. J. Mcleod, and D. A. Beck. 1997. Yield losses in maturing spring wheat caused by cereal aphids (Homoptera: Aphididae) under laboratory conditions. *Journal of Economic Entomology* **90**:1346–1350.

Wangai, A. W., R. T. Plumb, and H. F. van Emden. 2000. Effects of sowing date and insecticides on cereal aphid populations and barley yellow dwarf virus on barley in Kenya. *Journal of Phytopathology* **148**:33–37.

West, C. 1985. Factors underlying the late seasonal appearance of the lepidopterous leaf-mining guild on oak. *Ecological Entomology* **10**:111–120.

White, T. C. R. 1984. The abundance of invertebrate herbivores in relation to the availability of nitrogen in stressed food plants. *Oecologia* **63**:90–105.

Wool, D., R. Aloni, O. Ben-Zvi, and M. Wollberg. 1999. A galling aphid furnishes its home with a built-in pipeline to the host food supply. *Entomologia Experimentalis et Applicata* **91**:183–186.

Wootton, J. T. 1994. The nature and consequences of indirect effects in ecological communities. *Annual Review of Ecology and Systematics* **25**:443–466.

4

Plant-mediated effects linking herbivory and pollination

JUDITH L. BRONSTEIN, TRAVIS E. HUXMAN, AND GOGGY DAVIDOWITZ

Introduction

Over its lifetime, an individual plant must cope with a variety of challenges posed by other species. In fact, it may be interacting simultaneously with competitors, consumers, and a variety of mutualists, including pollinators, seed dispersers, and root symbionts. How the plant will fare in the presence of both beneficial and antagonistic species is at least in part a function of the resources it is able to devote to attracting and deterring them. Since resource availability is not unlimited, this sets up a situation in which allocation in the context of one set of interactions (such as attracting and rewarding mutualists) may force a trade-off with allocation in the context of another (such as defending against antagonists). While these trade-offs are well known, the vast majority of studies of interspecific interactions nevertheless focus on a single kind of interaction at a time. Hence, we know remarkably little about the nature and consequences of interactions *among interactions* experienced by a single organism.

One notable exception involves the topic of this chapter, interactions between herbivory and pollination. The literature on the ways in which herbivory alters pollination rates and hence reproductive success has grown rapidly in recent years. The large communities of researchers who focus on herbivory and on pollination now clearly recognize that they are not independent (Strauss 1997, Strauss and Armbruster 1997, Adler and Bronstein 2004, Irwin et al. 2004; see also Chapter 7, this volume). In light of the number of excellent case studies that

Ecological Communities: Plant Mediation in Indirect Interaction Webs, ed. Takayuki Ohgushi, Timothy P. Craig, and Peter W. Price. Published by Cambridge University Press. © Cambridge University Press 2007.

have accumulated over the past quarter-century, the question is no longer whether herbivore damage can reduce pollination success; the answer to that question clearly is yes. Rather, the focal questions have been these. By what routes does this reduction take place; does reduced pollination success necessarily lead to reduced fitness, or can plants compensate to some extent for their losses? Are defenses to herbivory costly in terms of pollination? Are herbivores ever *beneficial* to pollination? Finally, how does pollination alter herbivory (as opposed to the other way around)? We review the current state of knowledge in relation to each of these four issues.

While a great deal is already known about interactions between pollination and herbivory in a wide range of systems, we are at a much earlier stage in developing a mechanistic understanding of how and why those interactions occur. For this reason, few predictions are yet available as to when herbivory will in fact affect reproductive success via pollination, and when it will not. We present below a mechanistically based framework designed to elucidate physiological processes underlying links between pollination and herbivory. This approach has the added benefit of taking into consideration not only the plant's perspective on plant–animal interactions, but the animal's as well. We apply this approach to one interaction that we have recently begun to study, involving a plant and the insect that both pollinates it and feeds upon it. We will argue that this framework has value beyond the single system that we have been investigating, and has the potential to make predictions about interactions in a range of resource environments.

For the purposes of this chapter, we have chosen to include florivory, or consumption of flowers, in our definition of herbivory. Consumption of both foliage and flowers are well documented to have an impact on pollination and plant reproductive success, although the mechanisms underlying these effects may vary depending on the location of the damage. Furthermore, many generalist grazers feed relatively indiscriminately on flowers and foliage of a given plant (e.g., Bertness *et al.* 1987, Matter *et al.* 1999, Mothershead and Marquis 2000, Sharaf and Price 2004), making it difficult and perhaps somewhat artificial to separate the two phenomena.

Review: interactions between herbivory and pollination

Mechanisms by which herbivory can reduce reproductive success

Herbivory can reduce plant reproductive success through two general mechanisms. First, pollinators may avoid damaged plants or the plants on which herbivores reside. The negative effect can occur via a reduction in pollen receipt and/or removal that in turn reduces seed production and/or seed paternity.

Second, the effect of herbivory may occur via alterations in the plant's investment in reproduction. More precisely, herbivory may reduce plants' stored resources, leading through various routes to reduced production of flowers, nectar, and/or pollen, and ultimately of offspring. Below, we explore these mechanisms in greater depth. These mechanisms are not as distinct as they may at first appear. For example, herbivore-damaged plants may have fewer resources to invest in flower production (our second mechanism), leading pollinators to avoid these plants in preference to ones presenting more flowers and thus more nectar (our first mechanism). We consider below how one might quantify the cumulative effects on plants of both independent and interacting mechanisms. We also note that other authors have divided up the range of herbivory–pollination interactions somewhat differently than we do here. In particular, some define all negative effects of herbivory mediated by pollinator visitation as "indirect" effects of herbivory on pollination; negative effects mediated by investment in and allocation to reproduction independent of pollinator actions, as well as direct flower consumption, are then considered "direct" effects of herbivory (e.g., Mothershead and Marquis 2000).

Pollinator avoidance mechanisms
Pollinator avoidance of herbivore damage

A variety of pollinators avoid visiting plants whose flowers have been damaged by consumers. The most thoroughly documented case involves the andromonoecious shrub *Isomeris arborea*, whose flowers are attacked by the nitidulid beetle *Meligethes rufimanus* (Krupnick and Weis 1999, Krupnick *et al.* 1999). Beetle-attacked flowers have wilted or tattered petals and anthers, reduced nectar volumes, and corollas that fail to open completely. This damage results in reduced visitation by bee pollinators at the scale of the damaged flower, plant, and patch, leading to a 70% decrease in pollen deposition and a 65% decrease in pollen reaching other flowers. Other cases in which pollinators reject damaged flowers include butterfly pollinators and tephritid fly larval damage (Murawski 1987), bat pollinators and katydid damage (Cunningham 1995), and bee pollinators and thrips damage (Karban and Strauss 1993). Rejection may be stimulated by a reduction in flower numbers, not only by damage per se (Cunningham 1995, Adler *et al.* 2001; see p. 78).

Pollinator avoidance of herbivores

It seems likely that pollinators sometimes avoid visiting flowers on plants because they observe the presence there of florivore or herbivore individuals or aggregations. These antagonists may directly interfere with pollinators' access to flowers. Alternatively, their presence may signal either reduced floral rewards or an elevated risk of predation (if, for example, predators recruit to herbivore aggregations). Evidence that pollinators avoid herbivores/florivores is difficult to come by (but see Lohman et al. (1996) for a case in which pollinators cannot gain access to flowers mired in parsnip webworm damage). The paucity of evidence may be due to simple lack of study, or to the difficulty of distinguishing *why* pollinators avoid the flowers that they do (that is, due to herbivore damage or to herbivores per se). Certain pollinators avoid plants occupied by their own predators (Dukas and Morse 2003, Muñoz and Arroyo 2004), implying that they at least have the cognitive abilities to avoid herbivores should there be a reason to do so.

Plant-mediated effects of herbivory on pollination

Reduced flower numbers

Probably the most common effect that herbivory has been reported to have on pollination occurs via a reduction in flower numbers (e.g., Karban and Strauss 1993, Lehtilä and Strauss 1997, Krupnick et al. 1999, Mothershead and Marquis 2000, Adler et al. 2001, Rathcke 2001). This reduction comes about directly when herbivores fully consume flowers (e.g., Bertness et al. 1987). Alternatively, it can result from depletion of resources available for reproduction, or from reallocation of resources to other functions (e.g., regrowth). A reduction in flower numbers can result in lowered fruit and seed production in one of two ways: either through a direct reduction in the maximum numbers of fruit a plant can produce, or through pollinator preference for plants bearing larger numbers of flowers. For example, Adler et al. (2001) showed that in the hemi-parasitic plant *Castilleja indivisa*, bud herbivory reduced the number of open flowers; this in turn led to a reduction in pollinator visits and hence lowered seed set. Reduced flower numbers can have similar negative effects on the male component of reproductive function: the maximum amount of pollen produced is lower on plants with fewer flowers, and/or pollen export may be reduced due to a paucity of floral visitors (e.g., Quesada et al. 1995). In some cases, plants continue to produce fewer flowers one year or more after experimental herbivory treatments have been discontinued (Karban and Strauss 1993, Primack et al. 1994), suggesting long-term effects on the resource base available for reproduction.

Reduced flower size

In general, flower size is expected to show orders of magnitude less variability within species than do flower numbers (Harper 1977). However, in some cases, herbivore-attacked plants do produce smaller as well as fewer flowers. In one particularly detailed study, Mothershead and Marquis (2000) examined the effects of herbivory on *Oenothera macrocarpa*. When subjected to artificial herbivore attack, *O. macrocarpa* produces not only fewer flowers, but flowers with significantly smaller corolla diameters and shorter floral tube lengths. Flowers with smaller corolla diameters have reduced probabilities of setting fruit; furthermore, seed set and floral tube length decrease together. Mothershead and Marquis demonstrated that reproductive failure was not due to herbivory depleting resources that *O. macrocarpa* might have otherwise invested into fruit and seed maturation. Rather, the effects were indirect: leaf and floral bud damage led to reductions in floral size, which in turn led to reductions in hawkmoth pollinator visitation and reduced pollen delivery to individual flowers.

Causal relationships between herbivory, flower size, and pollinator visitation rates have as yet received minimal study. Different pollinator taxa are known to respond rather differently to flower size variation induced by herbivory. For example, in wild radish, *Raphanus raphanistrum*, a self-incompatible annual pollinated by a diversity of small insects, herbivore damage early in the growing season results in production of flowers of reduced petal length and width. Syrphid flies preferentially visit larger flowers, whereas visitation by solitary bees appears to be unrelated to flower size (Strauss *et al.* 1996, Conner and Rush 1996, Lehtilä and Strauss 1997).

Flower size variation may also be differentially detectable across plant species. Wolfe and Krstolic (1999) have shown that bilaterally symmetrical flowers are less variable in size than are radially symmetrical ones; they suggest that this pattern may be due to pollinators' relative ease of recognizing and avoiding deviations from bilateral as opposed to radial symmetry. Furthermore, pollinator avoidance of relatively small flowers within a population would seem more likely in species showing a high correlation between flower size and floral rewards (e.g., see Cresswell and Galen 1991). These observations suggest that predictions may be possible regarding which plant species are under stronger and weaker selection to hold flower size constant in the face of reduced resource pools.

Alteration of flowering phenology

Herbivory and florivory can cause a shift in an individual's flowering period. Flower consumption may take place early (e.g., Pilson 2000) or late (Louda and Potvin 1995) in the phenological period, leading to a delayed onset or early end to flowering. Additionally, early-season herbivory may deplete resources typically allocated to reproduction, and/or plants may simply not have time to replace damaged flowers before a climate-enforced end of the reproductive season. Either phenomenon would truncate the flowering period (Freeman *et al.* 2003). Phenological shifts have the potential to result in reproductive failure. This may occur due to a paucity of pollinators, or to a shift in pollinator identity and hence pollination quality. These effects can be relatively complex.

For example, Sharaf and Price (2004) studied ungulate browsing of flowering stalks in the monocarpic herb *Ipomopsis aggregata*. Clipping delayed flowering, leading plants to miss the peak of hummingbird pollinator activity and to have lower per-flower visitation rates than unclipped controls in one year, but to have greater overlap with hummingbirds and higher visitation rates in the subsequent year. Some researchers have argued that selection should shift flowering periods to times that do not coincide with peak floral and foliar consumption (e.g., Juenger and Bergelson 1998, Pilson 2000). However, the temporal variability evident in studies like that of Sharaf and Price (2004) suggests that at least in some species, selection may be rather inconsistent in direction and magnitude.

Effects on nectar quantity and quality

In a few cases, herbivore attack has been shown to reduce either the quantity or quality of nectar on the attacked plant. Krupnick *et al.* (1999) found that undamaged *Isomeris arborea* flowers produced over three times the amount of nectar as ones damaged by a flower-feeding beetle, *Meligethes rufimana*. Aizen and Raffaele (1996) found a tendency for reduced nectar sugar production on defoliated *Alstroemeria alata* and Strauss *et al.* (1996) for reduced nectar production on damaged *Raphanus raphanistrum*. In both cases, however, this effect was highly variable within the plant population.

There is abundant evidence that floral visitors are sensitive to even small variations in nectar volumes, and either shorten or eliminate visits to plants that produce less (Zimmerman 1983, Cresswell and Galen 1991, Mitchell and Waser 1992, Ackerman *et al.* 1994, Hodges 1995). Reduced pollinator visitation can lead to reduced pollen donation, fruit, and seed set. It can be difficult, however, to isolate the effects of nectar reduction, since many other floral traits can be altered simultaneously in attacked plants. For example, in *Isomeris arborea*, not only nectar volume but also flower numbers and anthers per flower are reduced by florivore attack (Krupnick *et al.* 1999). Furthermore, the number of pollinator visits may not be predictive of reproductive success. Increased visitation may result in the deposition of more self pollen (Hodges 1995), or plant reproductive success may be limited primarily by the availability of resources other than pollen (McDade and Davidar 1984).

Effects on pollen quantity and quality

Herbivore attack can also have a negative affect on pollen traits, presumably a consequence of reduced resource availability in attacked plants. In one particularly well-documented study, artificially defoliated branches of *Cucurbita texana* produced significantly fewer staminate flowers and fewer pollen grains per flower than undamaged branches on the same plant (Quesada *et al.*

1995). Pollen from staminate flowers on partially defoliated plants was also less likely to sire seeds than pollen from undamaged plants, implying that leaf damage not only decreased pollen production but pollen performance as well. Pollen grain size and post-pollination pollen performance is similarly reduced in herbivore-attacked *Alstroemeria alata*, although pollen quantity is unaffected (Aizen and Raffaele 1996, 1998). In *Lobelia siphilitica*, removal of 50% of the leaves during floral development causes a significant reduction in pollen-tube growth rates (Mutikainen and Delph 1996).

Alteration of sexual expression

Herbivory can have differential effects on female and male reproductive function (Mutikainen and Delph 1996, Strauss 1997, Strauss *et al.* 2001). In plant species with individuals that reproduce via both functions, herbivory can result in a shift in sexual expression: that is, they are relatively more male or more female than they would have been if not attacked. Such shifts in sexual expression have been demonstrated several times in andromonoecious plant species (in which individuals normally bear both male and hermaphroditic flowers). Herbivory early in the season leads to increased investment in the production of hermaphroditic flowers in *Depressaria pastinacella* (Hendrix 1984), and *Isomeris arborea* (Krupnick and Weis 1998). In contrast, Lowenberg (1997) documented that neither floral herbivory nor lack of early pollination had a significant effect on the ratio of the two floral morphs in the andromonoecious *Sanicula arctopodes*. Some flower-feeding herbivores preferentially feed on female or male tissue, which also can alter sexual expression at the level of either individuals or populations (Lowenberg 1997, Leege and Wolfe 2002).

Damage does not always lead to fitness effects

In the previous sections we have reviewed evidence that plant-mediated effects of herbivory can lead to reductions in reproductive success, either via forced reductions in reproductive investment or via pollinator rejection or avoidance of damaged plants. However, herbivore impacts on reproductive structures do not always lead to fitness reductions. Two very different phenomena may buffer plants against the effects of herbivores. First, as we have pointed out, visitation rates of some pollinators are rather unresponsive to traits that may be altered by herbivory, including flower size, flower numbers, and nectar volumes (e.g., Frazee and Marquis 1994, Conner and Rush 1996, Strauss *et al.* 1996). Second, shifts in resource allocation may lead to full or partial compensation for the damage (Lowenberg 1994, English-Loeb *et al.* 1999, Krupnick *et al.* 1999, Stowe *et al.* 2000). For example, English-Loeb *et al.* (1999) studied the effects of the strawberry bud weevil, *Anthonomus signatus*, on cultivated strawberries. Bud removal was found to have no effect on fruit set; plants compensated for bud loss by increasing the weight of fruits produced by the remaining buds, and by reducing bud abortion rates. There was no comparable compensation, however, when flowers rather than buds were removed. In two contrasting environments, the herb *Raphanus raphanistrum* had equivalent reproductive success whether herbivore-damaged or undamaged (Strauss *et al.* 2001). In the poorer-quality environment, however, damaged plants experienced relatively greater success as male parents, apparently due to changes in allocation to flowers versus seeds after damage.

Consequences of anti-herbivore traits

There is some evidence that pollination is affected not only by herbivory per se, but also by some of the traits that have apparently evolved to deter herbivores. Strauss *et al.* (1999) review evidence that plant secondary compounds can deter pollinators. This can occur directly by their presence in floral tissue or nectar; alternatively, the production of these herbivore deterrents may divert resources that might otherwise have been allocated towards pollinator attractants and rewards. On the other hand, investment into defenses can pay off if the consequent reductions of herbivory serve to make plants more attractive to pollinators (Adler *et al.* 2001, Irwin *et al.* 2004).

Can herbivory ever increase pollination success?

Until now we have considered the effect of herbivory on pollination to range from negative to neutral. However, there are two situations in which herbivory may act to *increase* pollination success.

Overcompensation

In a study that ignited extensive controversy, Paige and Whitham (1987) argued that herbivory to *Ipomopsis aggregata* plants led to increased reproductive success: browsing led to the production of multiple stalks and thus more inflorescences, ultimately leading to a 2.4-fold increase in fruit and seed production. Thus, *I. aggregata* not only compensated for damage, but overcompensated for it, leading to the suggestion that herbivores can actually be mutualists of the plants they consume. Overcompensation has subsequently been shown to occur in *I. aggregata* under a very narrow range of biotic and abiotic conditions, specifically, when browsing is very heavy and occurs early in the growing season. Under other conditions, consumed plants do worse or else no better than unconsumed ones (Bergelson and Crawley 1992, Paige 1994, Bergelson *et al.* 1996, Juenger and Bergelson 1997, 1998, Sharaf and Price 2004). Overcompensation to herbivory has been documented in a handful of other plant species as well (Lennartson *et al.* 1998, Juenger *et al.* 2000, Paige *et al.* 2001). This phenomenon has now been explored extensively in ecological, physiological, and evolutionary models

(Vail 1992, de Mazancourt et al. 1998, 2001, Jaremo et al. 1999, Agrawal 2000, de Mazancourt and Loreau 2000, Yamauchi and Yamamura 2004). While there are sets of conditions under which benefits of herbivory are likely to emerge, most researchers now consider it unlikely that plants are under selection to attract herbivores to obtain these benefits.

Herbivorous pollinators

Interactions between herbivory and pollination are particularly intriguing when pollinator species are themselves consumers of the same plant species. In these cases, plants experience an obvious conflict: should they attract more visitors, or fewer?

There are three distinct groups of plant-consuming pollinators. First, individuals of some animal species pollinate plants while they feed upon them. Notable among these are species that visit flowers in order to consume pollen (or to collect pollen to feed their offspring), incidentally transferring pollen among flowers while they do so. These include many of the best-studied pollinators, such as bees and beetles. Remarkably little attention has been paid to their simultaneous roles as antagonistic consumers, however (but see Westercamp 1996).

Second, certain highly specialized insects lay their eggs in or near the flowers they pollinate; their developing offspring then consume a subset of the seeds initiated by their mothers' actions. These systems include figs and fig wasps (Cook and Rasplus 2003), yuccas and yucca moths (Pellmyr 2003), and a growing list of other, less well-studied interactions involving moths, beetles, and flies (Holland and Fleming 1999, Jaeger et al. 2000, Dufaÿ and Anstett 2003, Kato et al. 2003, Meekijjaroenroj and Anstett 2003). These interactions manifest an array of specializations, including extreme specificity, floral traits that exclude all other visitors, and active pollination by adult females; they have extensive histories of coevolution and even cospeciation (e.g., Pellmyr 1997, Jousselin et al. 2003, Cook et al. 2004). There is an extensive literature on how conflicts between pollination and consumption within these interactions play out ecologically and evolutionarily (e.g., Herre et al. 1999, Ferdy et al. 2002, Holland and DeAngelis 2002, Yu et al. 2004).

Finally, some insects oviposit upon but also take nectar at and incidentally pollinate the same plant individuals (Adler and Bronstein 2004). The pollinators' offspring feed upon vegetative tissue. These interactions are considerably less specialized than the highly coevolved mutualisms mentioned above, but they have the potential to be no less damaging to the plant. Adler and Bronstein (2004) have reviewed the taxonomic distribution of these poorly known pollinating folivores. The best-documented examples are all hawkmoths (Lepidoptera: Sphingidae), although only one species, *Manduca sexta*, has as yet been studied in depth. This group of antagonistic pollinators is worth a closer look, since they offer particular advantages for studying the ecology and physiology of interactions between pollination and herbivory.

Pollination can affect herbivory too

In the preceding sections, we have focused on how herbivory can affect pollination rates. However, herbivory can also be a response to pollination rates. For example, in the perennial herb *Paeonia broteroi*, plants exposed to more pollinators suffer a disproportionate effect of herbivory, because herbivores preferentially graze large fruits (Herrera 2000). Floral traits that have apparently evolved to attract pollinators (visual and olfactory cues, as well as food rewards) often attract antagonistic species at the same time (Adler and Bronstein 2004). Examples of such plant antagonists include nectar robbers and thieves (Galen and Cuba 2001), florivores (Zangerl and Rutledge 1996), pre-dispersal seed predators (Brody 1997), and even fungal pathogens (Shykoff and Bucheli 1995). In particular, plants with large flowering displays can be particularly attractive not only to pollinators but to herbivores as well, notably florivores and pre-dispersal seed predators (Cunningham 1995, Brody and Mitchell 1997, Matter *et al.* 1999, Ehrlén 2002, Gómez 2003). For example, high-flowering individuals of the perennial herb *Lathyrus vernus* experience increased fruit set, but also an elevated risk of grazing damage (Ehrlén 2002). In these cases, rates of herbivory and pollination may increase together not because they are causally linked, but because each increases with plant size and flower numbers. Yet, the ecological effect is the same: certain plants simultaneously attract higher numbers of both pollinators and herbivores. Thus, the opportunity for, and direction of, selection on floral traits may vary with the herbivore environment (Gómez 1993, Herrera 2000, Adler *et al.* 2001, Ehrlén 2002, Herrera *et al.* 2002, Cariveau *et al.* 2004, Vázquez and Simberloff 2004).

Towards a predictive framework

It should be clear from the foregoing that extensive knowledge is now available about a myriad of ways in which herbivory and pollination interact, both via plant traits and via pollinator behavior. The picture is still incomplete, however. We believe that our understanding of the relationship between herbivory and pollination has been constrained by two characteristics of research to date.

First, the research approach has been heavily phytocentric. That is, the ecological and evolutionary consequences of herbivory/pollination interactions have been studied largely from the perspective of the shared plant. This is a critical perspective (and, of course, the one central to the focus of this volume). However, to understand these interactions, particularly in an evolutionary context, it seems equally important to consider the perspective of the pollinators and herbivores as well. As one extreme example, consider obligate mutualisms between plants and insects (such as fig wasps or yucca moths) that simultaneously pollinate and lay eggs within inflorescences. The dynamics of these interactions are clearly driven by an interplay between the evolutionary interests of plants (i.e., how to maximize pollination while minimizing seed loss to the pollinators' offspring) and pollinators (how to maximize offspring production). In fact, the cost to the plant of losing offspring translates more or less directly into the benefit to the insect of gaining offspring (Bronstein 2001). Clearly, in this case, studying the interaction from only the plant perspective would miss the larger picture. Coevolutionary interactions and conflicts of interest are likely to occur in many of the other interactions we have discussed in which plant antagonists also pollinate, including interactions considerably less specialized than the ones just discussed (e.g., Westercamp 1996, Adler and Bronstein 2004).

Second, we still lack a general predictive framework for when to expect relatively stronger and weaker interactions between pollinators and herbivores. It is clear that herbivory sometimes has a large impact on pollination success and subsequent reproduction (e.g., Krupnick and Weis 1999, Mothershead and Marquis 2000), and sometimes does not (Strauss *et al.* 2001). However, it is still an open question how we might have predicted those differences. Furthermore, similarity of certain study systems, as well as a reliance on a few tractable model systems, risk giving the illusion that more generalizations exist than really do. As one example, there is extensive evidence that floral herbivory can lead to shifts in plant sex expression, but to our knowledge, all studies of this phenomenon to date have involved andromonoecious plants. Andromonoecy is found in only about 2% of angiosperms (Richards 1997). As another example, much of what is known empirically about context-dependent outcomes of herbivory (from antagonistic to overcompensatory, i.e., mutualistic) comes from studies of a single plant species, *Ipomopsis aggregata*.

Knowledge of herbivory impacts on pollination derives in part from the use of multiple regression or path-analytic models to examine phenomena within particular networks of interactions. Path analysis is a useful tool for illustrating causal relationships thought to be important in driving some behavior and for ranking their strength (Sokal and Rohlf 1981). This technique is especially useful when the potential exists for nonindependence among variables (Li 1981). For example, Adler *et al.* (2001) used path analysis in order to determine the direct and indirect effects of herbivory and pollination on lifetime seed set in the hemi-parasitic plant *Castilleja indivisa*, and the direct and indirect effects of host-acquired alkaloids on seed set via herbivory and pollination. The authors point to the value of path analysis as a tool for measuring the significance of multiple interactions simultaneously when attempting to understand mechanisms underlying correlations between plant traits and fitness (see also Mothershead and Marquis 2000, Hamback 2001, Herrera et al. 2002, Cariveau et al. 2004).

The value of such structural models is predicated on a priori knowledge of the physical structure amongst model variables (Schemske and Horvitz 1988, Mitchell 1992). Often, these models are constructed by allowing interactions among all variables in an analysis where the structure of the model can alter usefulness (Smith et al. 1997). In the context of species interactions, these models often operate outside of known allometrically or physiologically linked plant trait relationships. The end result is knowledge based on correlative information from individual study systems, a valuable approach but one likely to limit a model's predictive value beyond the system for which it was developed.

The need to develop a predictive framework for describing behavior in many settings is an emerging theme in ecology. Plant physiological ecologists have made substantial progress in developing an understanding of how plants and ecosystems may respond to future global change by using a "resource-based" analysis of biological response (Strain and Bazzaz 1983, Bloom et al. 1985). This framework utilizes well-described physiological processes and the understanding of how variation is driven by resource availability (environmental variation), coupled with an understanding of how allocation of resources links small-scale physiology and whole-organism responses (Wolfe et al. 1998). Such an approach can be extended to provide a means of evaluating the relationship between herbivory and pollination, as mediated through the ecophysiology of plants. This approach can also be extended to incorporate the animal perspective that we have argued for above, because the currencies involved have important fitness consequences to these partners. Below, we provide a detailed example of how this method might be used to study the interaction between one plant and an insect that is both its major pollinator and major herbivore. We emphasize a framework that explicitly considers the costs and benefits of different patterns of herbivory to both the plant and insect, with the assumption that pollination behavior derives as a simple function of changes in plant size.

We stress that we are certainly not the first to advocate a physiologically based, whole-plant perspective on the relationship between pollination and herbivory (see, e.g., Frazee and Marquis 1994, Sharaf and Price 2004, Yamauchi and Yamamura 2004). We also note that we are at early stages of this study. At this point, we know considerably less about this interaction than we do about many that we have highlighted above.

Linked pollination and herbivory in the Manduca/Datura *interaction*

Manduca sexta is a large, widespread North and Central American hawk-moth. Despite its broad geographic range, its host range is exceptionally narrow. It feeds almost exclusively on plants in the family Solanaceae; in any one region, it has only a handful of native host plants (although it also commonly feeds upon solanaceous crop plants, giving rise to its common name, the tobacco hornworm). In the southwestern USA, where its agricultural hosts are not grown commercially, *M. sexta* lays its eggs only on the herbs *Datura wrightii* (Solanaceae) and (more rarely) *Proboscidea parviflora* (Martyniaceae) (Mechaber and Hildebrand 2000, Mira and Bernays 2002).

Nectaring and oviposition are tightly linked behaviorally in *M. sexta*. Before an oviposition event on *D. wrightii*, a female will typically make a nectaring visit to a flower on the same plant. *Manduca sexta* transfers copious amounts of pollen during these nectaring visits, and resulting fruit and seed set are quite high (J.L. Bronstein, T.E. Huxman, and G. Davidowitz unpublished data). *Datura wrightii* in fact shows a suite of classic traits suggesting reliance on pollination by hawkmoths. Flowers are white, fragrant, exceptionally large (up to 25 cm in length), and open for a single night; they secrete copious (65 ± 10 ml), concentrated (25% sucrose equivalents) nectar (Elle and Hare 2002, Raguso *et al.* 2003). This nectar appears to be a critical food resource for adult *M. sexta*. Raguso *et al.* (2003) have calculated that nectar within a single *D. wrightii* flower provides a 1.2 g *M. sexta* with 10–15 min of hovering capability; a short bout of five to ten flower visits provides a moth with enough energy to fly for an hour or more in search of a mate or oviposition site. As a metabolic reward in a pollination mutualism, this is almost unprecedented.

The benefit of *M. sexta* pollination, however, appears to come at a considerable cost to *Datura*. Female *M. sexta* can lay 300–400 eggs in their lifetime (Madden and Chamberlin 1945) and up to 100 each night (G. Davidowitz unpublished data). Provided that there is enough foliage, a larva will feed exclusively upon its natal plant (primarily on leaves, but also on flowers and fruits). In contrast to its alternate host *P. parviflora*, *D. wrightii* provides a high-quality food source for *M. sexta* (Mira and Bernays 2002). The few larvae that escape predation and parasitism grow to be so large that a single individual can defoliate its host by its last instar (McFadden 1968). An individual *M. sexta* larva can process 1400–1900 cm^2 of leaves (Casey 1976, Heinrich 1971), which is greater than the size of many *D. wrightii* in natural settings (T. Huxman unpublished data).

In this interaction, then, it seems likely that plant traits that favor increased pollination success will simultaneously lead to a higher probability of defoliation by the pollinators' offspring. Adler and Bronstein (2004) have in fact shown that in another *Datura* species, experimental augmentation of nectar volumes leads to both increased pollination (resulting in higher fruit set) and increased egg deposition by *M. sexta*. The physiological link between resource costs of herbivory and subsequent resource investment into reproduction remains unknown, however. Without this knowledge, observed relationships between pollination and herbivory cannot confidently be extrapolated across environments differing in resource availability. Below, we describe a mechanistically based conceptual model designed to highlight the physiological processes that underlie the links between herbivory and pollination, and between plants and insects, based on our research on this interaction in southern Arizona.

Resource allocation model of Datura/Manduca *interactions*

We have been exploring relationships between variation in herbivore behavior and pollination success, based in part on the morphological and physiological determinants of organismal size in *D. wrightii* and *M. sexta*. Figure 4.1 illustrates a set of hypothetical linkages between reproductive success of the two species as a function of differential herbivory on leaves of different sizes within a *D. wrightii* canopy. We stress that we are at an early stage in disentangling the phenomena represented in this figure. Here, we will focus on predictors of the costs of herbivory to *D. wrightii*, and how they translate into the benefits of consumption from the perspective of the herbivore.

The seasonal photosynthetic activity of *D. wrightii* is the foundation of energy resources available for use in this interaction. That is, *M. sexta* success depends heavily on the quality and quantity of the resource pool represented by *D. wrightii* leaves. At the same time, *D. wrightii* has a finite pool of photosynthates available to allocate to growth and reproduction. The removal of leaf material by herbivory is a loss of potential carbon gain by the plant; in fact, given the size, abundance, and voraciousness of *Manduca* larvae, there is the potential for carbon gain to disappear completely at some times of year. However, the actual cost of herbivory depends upon both plant and insect characteristics. Can differential larval herbivory across the range of sizes and ages of leaves of *D. wrightii* shift the cost and benefit relationships of the *Datura/Manduca* interaction through time, by affecting either the plant's or the insect's ability to produce offspring?

First, we will consider the plant perspective. *Datura wrightii* leaves differ in nutrient content, most likely as a result of the typical trend for differential allocation of resources to photosynthesis with leaf age (Mooney 1972, Chabot and Hicks 1982). Newly produced, small leaves have relatively high photosynthetic rates (Fig. 4.2), and are thus likely to have high nitrogen and phosphorus contents relative to older, larger leaves. In contrast, large *D. wrightii* leaves have

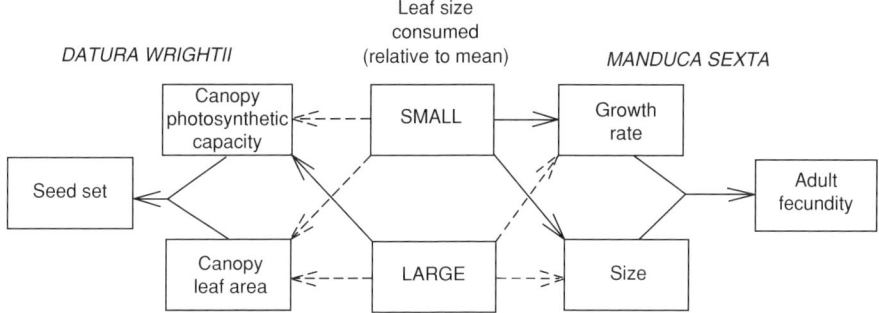

Figure 4.1. Conceptual linkages between *Manduca sexta* and *Datura wrightii* fecundity as a result of differential herbivory by leaf size. Solid lines represent positive effects between an increase in one character and the linked character; dashed lines are negative effects. For example, an increase in consumption of relatively small leaves may result in increased *Manduca* growth rate, but a relatively greater decrease in canopy photosynthetic capacity and seed set as compared to consumption of large leaves.

lower photosynthetic rates (Fig. 4.2) and thus relatively lower nutrient contents, as a result of retranslocation of resources to newly produced leaves and a dilution effect (higher carbon concentrations relative to nitrogen and phosphorus) associated with carbohydrate production in older leaves (Field 1983). Generally speaking, large and small leaves contribute differentially to total plant carbon gain as a function of these characteristic nutrient content/photosynthetic relationships and the frequency distribution of leaf size on a plant (Mooney 1972). Combined, these physiological and morphological characteristics can be used to understand potential whole-plant carbon gain. What emerges is that canopy (i.e., whole-plant) photosynthetic capacity and canopy leaf area display will have quasi-independent effects on plant size (e.g., Comstock and Ehleringer 1986, Hamerlynck et al. 2002). Because fecundity in this (like many) plant species is size dependent (Fig. 4.3), these parameters likely have strong control over *D. wrightii* reproductive success in a given season. These relationships imply that factors affecting whole-plant photosynthetic capacity and leaf area display can combine to produce different effects on plant fecundity.

In the absence of herbivory and under greenhouse conditions, *D. wrightii* produces leaves at the rate of 0.8 ± 0.2 leaves per day. Fifth instar (i.e., nearly full-grown) *M. sexta* larvae can consume considerably more than this during the same interval (see above). However, not all leaves are equivalent to feeding *Manduca*. The rate of consumption increases as leaf size increases (Fig. 4.4). Since larger leaves are of lower quality (Fig. 4.2), this suggests that larvae are eating a greater amount in order to obtain similar nutrient intake values.

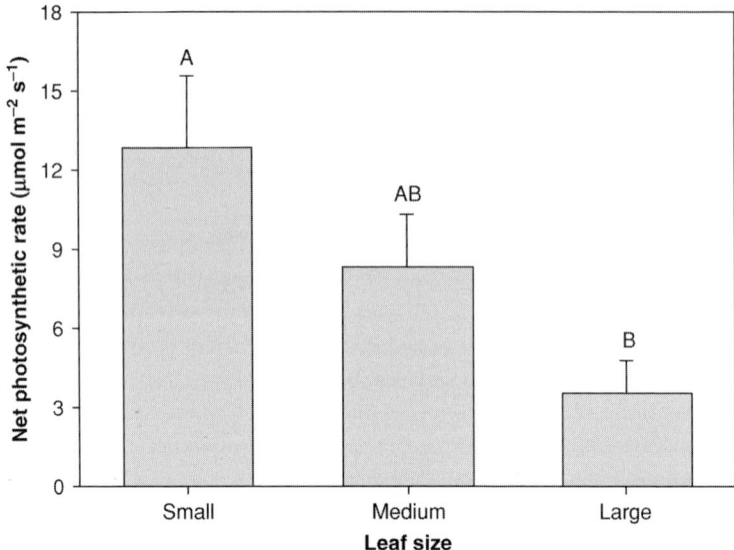

Figure 4.2. Net photosynthetic rate (CO_2 exchange) versus leaf size for *Datura wrightii* grown in a greenhouse. Small leaves are less than 6 cm in length, while medium leaves are between 7 and 13 cm, and large leaves are greater than 14 cm in length. $F = 5.1$; d.f. $= 2.6$; $p < 0.05$. Different letters represent groups that are significantly different (pair-wise comparison with $p < 0.05$).

We can predict that consumption of relatively small or relatively large leaves will have differing effects on *Datura* fecundity, and that these effects will operate through different mechanisms (Fig. 4.1). Consumption of relatively small leaves should have a large impact on canopy photosynthetic capacity, since small leaves have higher proportional photosynthetic rates (Fig. 4.2). However, this will have little effect on canopy leaf area (since small leaves contribute little to this measure). Conversely, consumption of larger leaves will have a large impact on canopy leaf area, but little impact on canopy photosynthetic capacity. Both kinds of consumption, then, will have negative effects on plant carbon gain. However, depending upon environmental conditions, these two patterns can compound to relatively larger or smaller costs to the plant. For example, as water stress develops, the loss of relatively small leaves will result not only in significant reductions in photosynthetic capacity, but in a shift in whole-plant water use, such that more water is used per unit carbon gain. This occurs as a result of the amount of low-photosynthetic-rate leaf area that remains on the plant. The plant will consequently have lower water-use efficiency at a whole-plant scale, adding to the relatively high cost of losing small leaves to herbivores. Thus, the spatial

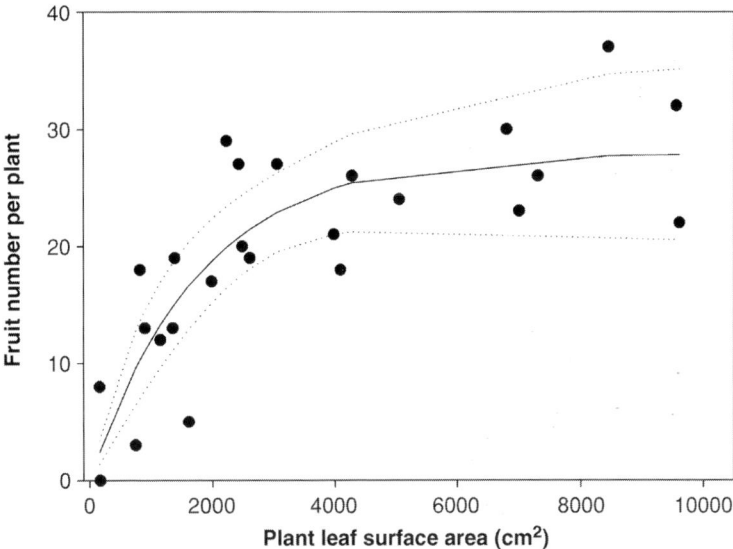

Figure 4.3. Fruit production as a function of leaf surface area in *Datura wrightii*. The nonlinear curve was generated from 25 plants collected at the Santa Rita Experimental Range, 45 km south of Tucson, Arizona, USA. The dotted lines represent 95% confidence intervals.

distribution of herbivory by *Manduca* larvae within the canopy of an individual plant should be an important predictor of *Datura* fitness. Further, the timing of herbivory (i.e., during rapid plant growth early in the season, following a defoliation event, or at any other time when most leaves in a canopy are small) will also determine its fitness impact, due in part to fewer resources being available for reproduction later in the exceptionally long (7-month) flowering period.

The interaction between herbivory and loss of photosynthetic capacity differs not only among leaves of a single plant, but among plants in contrasting resource environments. In particular, the loss of photosynthetic capacity due to herbivory may be greater on water- and nutrient-stressed plants. Individuals grown under stress show a reduction in photosynthetic rates of remaining leaf tissue during defoliation as compared to those grown in relatively resource-rich conditions (Fig. 4.5). This suggests that in addition to the reduction in photosynthetic capacity at the whole-plant scale occurring through herbivory, whole-plant photosynthetic performance is further reduced through adjustments in the photosynthetic characteristics of remaining leaves following a defoliation event.

We now turn to the herbivore's perspective on the same set of phenomena. *Manduca* larvae consuming small leaves have higher growth rates than do larvae feeding on large leaves (Fig. 4.6). As a result, the ratio of consumption rate to

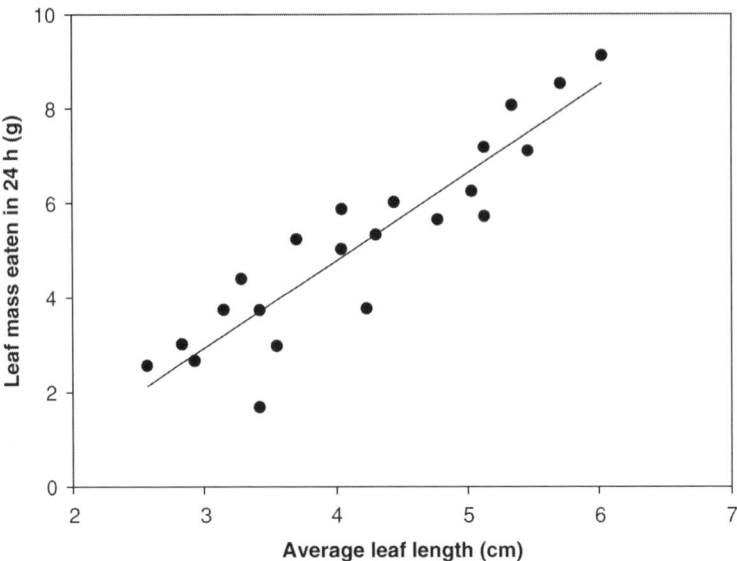

Figure 4.4. Consumption rate of *Datura wrightii* leaf material by last-instar *Manduca sexta* larvae reared from field-collected eggs, as a function of leaf length ($Y = 1.85x - 2.61$; $r^2 = 0.83$; $p < 0.05$).

growth rate is small, since the nutrient content of these small leaves is relatively high. The opposite is true for larvae that feed upon larger leaves: they consume greater amounts of leaf area, but have lower growth rates (Figs. 4.4 and 4.6).

Growth rate is a critical fitness variable for *M. sexta*. It interacts with the physiological mechanism that controls the duration of the growth period, and, as a result, influences both development time and body size (Davidowitz and Nijhout 2004). Changes in these life-history traits have implications for both larval (Mira and Bernays 2002) and adult (Davidowitz and Nijhout 2004, Davidowitz et al. 2004) components of fitness. Variation in the nutrient content of the larva's diet is also important in determining size at maturity (Davidowitz et al. 2003, 2004) which, in turn, can explain 67% of the variation in egg output in females (G. Davidowitz unpublished data). In summary, then, the fitness achieved by any given larva feeding within a *Datura* canopy can vary as a function of the size distribution of leaves consumed.

This analysis provides a valuable example of how a physiologically based approach to the determinants of fitness can provide insight into potential variation in the behavior of partners under different conditions. Admittedly, there are many gaps in our knowledge at this point. Our research approach does not yet provide details into organismally based homeostatic adjustments that can influence pollination. Furthermore, it currently relies on the assumption

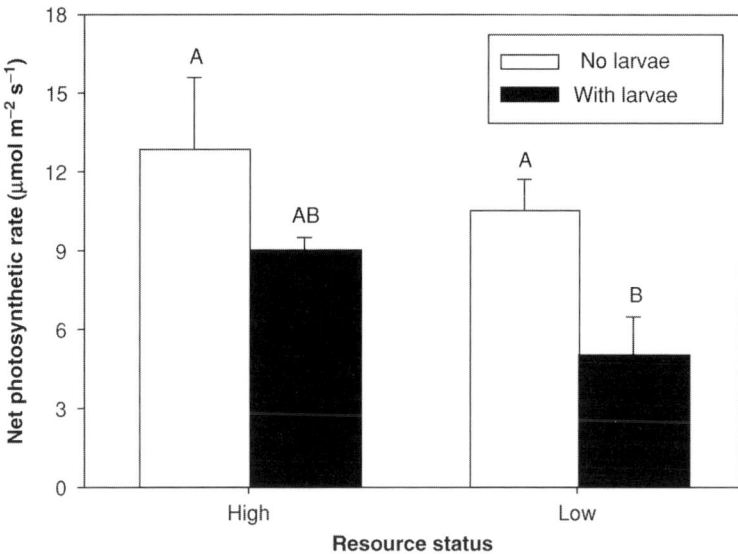

Figure 4.5. Net photosynthetic rates (CO_2 exchange) as a function of water and nutrient status with and without herbivory by *Manduca sexta* larvae. High-resource-status plants were grown in high water and high nutrient conditions. Low-resource-status plants were given 25% of the water and nutrients as plants in the high-resource treatment. Nutrient × herbivory interaction: $F = 3.42$; d.f. $= 1,3$; $p = 0.10$. Different letters represent groups that are significantly different by pair-wise comparison with $p < 0.05$.

that shifts in reproductive output are driven largely by changes in organism size. Larger *D. wrightii* show higher reproductive output (Fig. 4.3), but we do not yet know if the fraction of viable seeds produced by a plant, or ratios of fruit to mature seed, are constant functions of plant size or vary due to whole-plant resource constraints (e.g., Huxman and Loik 1997). If we do use this assumption, however, then several predictions can be made with regard to interactions between these plants and their pollinating herbivores. For example, exclusive consumption of small leaves by *Manduca* larvae should result in relatively faster-growing larvae, but relatively greater seasonal reductions in plant size. Since flower number is a function of plant size, such a strategy by larvae should result in reduced local rewards available for adult moths throughout the season and a reduction in seed set, due to reduced flowering later in the year, as well as, possibly, the departure of pollinators to richer patches.

Datura wrightii mediates an interaction between herbivory and pollination by the same organism. That is, insects feeding on *D. wrightii* are the same as those that pollinate flowers later in the flowering season or the following year – the

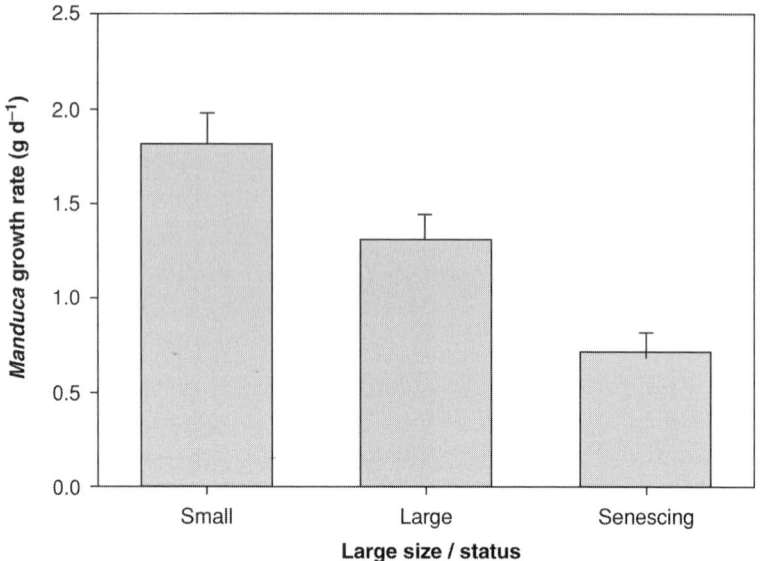

Figure 4.6. Growth rate (g d^{-1}) of last-instar *Manduca sexta* larvae as a function of *Datura wrightii* leaf size. Leaves were collected from plants in the field and fed to larvae hatched from field-collected eggs and reared in the laboratory.

positive and negative effects simply occur at two different points in their life cycle. This certainly simplifies our ability to identify the critical ecophysiological variables underlying links between pollination and herbivory. In particular, it allows us to explicitly consider the role of environment and resources on all components of the interaction, as the climatic features driving herbivore and pollinator behavior are the same (*M. sexta* initiates its seasonal activity in association with the southwestern monsoonal precipitation occurring in the months of July to September). However, our approach could be extended fairly straightforwardly to the more usual case in which pollinators and herbivores are different species.

Summary and conclusions

Herbivory and pollination offer a particularly compelling case in which plant-mediated phenomena can link interspecific interactions that are often treated as independent. A growing literature has provided a wealth of data related to the following four questions.

1. In what ways can herbivory reduce pollination success? Pollinators may avoid either herbivore damage or herbivores per se. Alternatively, the

effects may be plant-mediated: plants attacked by herbivores may produce fewer or smaller flowers, or nectar or pollen of reduced quantity or quality, or they may experience a shift in flowering time. Any of these phenomena, alone or (more commonly) in combination, can result in reduced pollination success. However, fitness effects are not inevitable. Plants exhibit a range of compensatory mechanisms that may ameliorate or even eliminate fitness reductions associated with herbivory. Furthermore, some pollinators are apparently insensitive to plant-mediated herbivory effects.

2. Are defenses against herbivory costly in terms of pollination? In some cases, investment into defenses against herbivory reduces resources that might otherwise be devoted to attracting or rewarding pollinators. On the other hand, these investments may still be beneficial, if reduced herbivore damage itself increases plant attractiveness to pollinators.

3. Is herbivory ever beneficial to pollination? There are two situations in which herbivores may augment pollination rates: in the rare case when plants are able to overcompensate for herbivory, i.e., to produce more reproductive units in response to being eaten; and in cases where herbivores themselves are effective pollinators of the plants they consume.

4. When can herbivory be a response to pollination rate (as opposed to the other way around)? Herbivores may be attracted by the presence of nectar, large flower displays, or large fruit displays, such that heavily pollinated plants are those most heavily attacked by consumers.

We have argued that despite a growing body of knowledge demonstrating how herbivory and pollination are intertwined in a wide range of interactions, a framework able to generate predictions about when and how these diverse effects will occur has been more elusive. As one attempt to fill this gap, we have posed a fifth question.

5. How is the relationship between herbivory and pollination mediated through plant and animal ecophysiology? We demonstrate the potential value of this approach using data from our studies of the herb *Datura wrightii*, which is pollinated by a hawkmoth (*Manduca sexta*) which lays eggs when it visits the flowers. In this case, pollination comes at a direct cost of herbivory by the pollinators' offspring. Our data suggest that the costs and the benefits of this interaction to both partners are functions of organism size. We have documented how feeding choices made by moth larvae translate physiologically into larval growth rate and ultimately moth fitness, and simultaneously into loss of

photosynthetic capacity (Fig. 4.5), reductions in plant size and, by extrapolation, plant fitness.

The general approach that we have taken to studying linkages between pollination and herbivory has the potential to shed light on interactions among other kinds of interactions taking place on a shared partner. As one example, it is well documented that outcomes of some protection mutualisms between ants and reward-producing herbivores are mediated by the resource status of the plants on which the herbivores feed (see Chapter 12, this volume). Our approach could be extended straightforwardly to studying this and other multitrophic interactions.

More generally, our focus on how the costs and benefits to each partner are a function of organism size allows us to integrate core areas of organismal biology with emerging themes from macroecology. Our understanding of the determinants of plant growth is rapidly reaching an exceptional level (Bazzaz and Grace 1997). The factors that control photosynthetic rate, as well as how that rate varies as a function of a changing environment, have been effectively formulated into models that can be translated across species (Chen and Reynolds 1997). Our knowledge of the role of growth rate in determining patterns of resource allocation across plant taxa, plant size, and environmental conditions is also growing (Grace 1997, Enquist et al. 2002, Poorter and Navas 2002). Parallel knowledge for animals, including herbivorous insects, lags far behind, but has begun to accumulate (e.g., Davidowitz and Nijhout 2004, Davidowitz et al. 2004). Since we are beginning to understand how organismal size distributions, density, and metabolic features behave in natural systems (e.g., Enquist et al. 2003, Brown et al. 2004), the ecophysiological approach to linked species interactions has the potential to produce knowledge that can be translated across both scales and systems.

Acknowledgments

We thank the editors for their invitation to contribute to this volume, and Rob Raguso for discussion and access to unpublished data, A. Levine and G. Barron-Gafford for assistance in the field and laboratory, and the reviewers for helpful comments on the manuscript. The research reported here was supported by the National Science Foundation DEB-0316205 to J. L. Bronstein, T. E. Huxman, and G. Davidowitz.

References

Ackerman, J. D., J. A. Rodriguez-Robles, and E. J. Meléndez. 1994. A meager nectar offering by an epiphytic orchid is better than nothing. *Biotropica* **26**:44–49.

Adler, L. S., and J. L. Bronstein. 2004. Attracting antagonists: does floral nectar increase leaf herbivory? *Ecology* **85**:1519–1526.

Adler, L. S., R. Karban, and S. Y. Strauss. 2001. Direct and indirect effects of alkaloids on plant fitness via herbivory and pollination. *Ecology* **82**:2032–2044.

Agrawal, A. A. 2000. Overcompensation of plants in response to herbivory and the by-product benefits of mutualism. *Trends in Plant Science* **5**:309–313.

Aizen, M. A., and E. Raffaele. 1996. Nectar production and pollination in *Alstroemeria aurea*: responses to level and pattern of flowering shoot defoliation. *Oikos* **76**:312–322.

Aizen, M. A., and E. Raffaele. 1998. Flowering-shoot defoliation affects pollen grain size and postpollination pollen performance in *Alstroemeria aurea*. *Ecology* **79**:2133–2142.

Bazzaz, F. A. and J. Grace (eds.) 1997. *Plant Resource Allocation.* San Diego, CA: Academic Press.

Bergelson, J., and M. J. Crawley. 1992. Herbivory and *Ipomopsis aggregata*: the disadvantages of being eaten. *American Naturalist* **139**:870–882.

Bergelson, J., T. Juenger, and M. J. Crawley. 1996. Regrowth following herbivory in *Ipomopsis aggregata*: compensation but not overcompensation. *American Naturalist* **148**:744–755.

Bertness, M. D., C. Wise, and A. M. Ellison. 1987. Consumer pressure and seed set in a salt marsh perennial plant community. *Oecologia* **71**:190–200.

Bloom, A. J., F. S. Chapin, and H. A. Mooney. 1985. Resource limitation in plants: an economic analogy. *Annual Review of Ecology and Systematics* **16**:363–392.

Brody, A. K. 1997. Effects of pollinators, herbivores, and seed predators on flowering phenology. *Ecology* **78**:1624–1631.

Brody, A. K., and R. J. Mitchell. 1997. Effects of experimental manipulation of inflorescence size on pollination and pre-dispersal seed predation in the hummingbird-pollinated plant *Ipomopsis aggregata*. *Oecologia* **110**:86–93.

Bronstein, J. L. 2001. The costs of mutualism. *American Zoologist* **41**:127–141.

Brown, J. H., J. F. Gillooly, A. P. Allen, V. M. Savage, and G. B. West. 2004. Towards a metabolic theory of ecology. *Ecology* **58**:1771–1789.

Cariveau, D., R. E. Irwin, A. K. Brody, L. S. Garcia-Mayeya, and A. von der Ohe. 2004. Direct and indirect effects of pollinators and seed predators to selection on plant and floral traits. *Oikos* **104**:15–26.

Casey, T. M. 1976. Activity patterns, body temperature and thermal ecology in two desert caterpillars (Lepidoptera: Sphingidae). *Ecology* **57**:485–497.

Chabot, B. F., and D. J. Hicks. 1982. The ecology of leaf life spans. *Annual Review of Ecology and Systematics* **13**:229–259.

Chen, J. L., and J. F. Reynolds. 1997. GePSI: a generic plant simulator based on object-oriented principles. *Ecological Modeling* **94**:53–66.

Comstock, J. P., and J. R. Ehleringer, Jr. 1986. Canopy dynamics and carbon gain in response to soil moisture availability in *Encelia frutescens* Gray, a drought deciduous shrub. *Oecologia* **68**:271–278.

Conner, J. K., and S. Rush. 1996. Effects of flower size and number on pollinator visitation to wild radish, *Raphanus raphanistrum*. *Oecologia* **105**:509–516.

Cook, J. M., and J.-Y. Rasplus. 2003. Mutualists with attitude: coevolving fig wasps and figs. *Trends in Ecology and Evolution* **18**:241–248.

Cook, J. M., D. Bean, S. A. Power, and D. J. Dixon. 2004. Evolution of a complex coevolved trait: active pollination in a genus of fig wasps. *Journal of Evolutionary Biology* **17**:238–246.

Cresswell, J. E., and C. Galen. 1991. Frequency-dependent selection and adaptive surfaces for floral character combinations: the pollination of *Polemonium viscosum*. *American Naturalist* **138**:1342–1353.

Cunningham, S. A. 1995. Ecological constraints on fruit initiation by *Calyptrogyne ghiesbreghtiana* (Arecaceae): floral herbivory, pollen availability, and visitation by pollinating bats. *American Journal of Botany* **82**:1527–1536.

Davidowitz, G., and H. F. Nijhout. 2004. The physiological basis of reaction norms: the interaction between growth rate, the duration of growth and body size. *Integrative and Comparative Biology* **44**:443–449.

Davidowitz, G., L. J. D'Amico, and H. F. Nijhout. 2003. Critical weight in the development of insect body size. *Evolution and Development* **5**:188–197.

Davidowitz, G., L. J. D'Amico, and H. F. Nijhout. 2004. The effects of environmental variation on a mechanism that controls insect body size. *Evolutionary Ecology Research* **6**:49–62.

de Mazancourt, C., and M. Loreau. 2000. Grazing optimization, nutrient cycling, and spatial heterogeneity of plant–herbivore interactions: should a palatable plant evolve? *Evolution* **54**:81–92.

de Mazancourt, C., M. Loreau, and L. Abbadie. 1998. Grazing optimization and nutrient cycling: when do herbivores enhance plant production? *Ecology* **79**:2242–2252.

de Mazancourt, C., M. Loreau, and U. Dieckmann. 2001. Can the evolution of plant defense lead to plant–herbivore mutualism? *American Naturalist* **158**:109–123.

Dufaÿ, M., and M.-C. Anstett. 2003. Conflicts between plants and pollinators that reproduce within inflorescences: evolutionary variations on a theme. *Oikos* **100**:3–14.

Dukas, R., and D. Morse. 2003. Crab spiders affect flower visitation by bees. *Oikos* **101**:157–163.

Ehrlén, J. 2002. Assessing the lifetime consequences of plant–animal interactions for the perennial herb *Lathyrus vernus* (Fabaceae). *Perspectives in Plant Ecology, Evolution and Systematics* **5**:145–163.

Elle, E., and J. D. Hare. 2002. Environmentally induced variation in floral traits affects the mating system in *Datura wrightii*. *Functional Ecology* **16**:79–88.

English-Loeb, G., M. Pritts, J. Kovach, R. Rieckenberg, and M. J. Kelly. 1999. Compensatory ability of strawberries to bud and flower removal: implications for managing the strawberry bud weevil (Coleoptera: Curculionidae). *Journal of Economic Entomology* **92**:915–921.

Enquist, B. J., J. P. Haskell, and B. H. Tiffney. 2002. General patterns of taxonomic and biomass partitioning in extant and fossil plant communities. *Nature* **419**:610–613.

Enquist, B. J., E. P. Economo, T. E. Huxman, et al. 2003. Scaling metabolism from organisms to ecosystems. *Nature* **423**:639–642.

Ferdy, J.-B., L. Després, and B. Godelle. 2002. Evolution of mutualism between globeflowers and their pollinating flies. *Journal of Theoretical Biology* **217**:219–234.

Field, C. 1983. Allocating leaf nitrogen for the maximization of carbon gain: leaf age as a control on the allocation program. *Oecologia* **56**:341–347.

Frazee, J. R., and R. J. Marquis. 1994. Environmental contribution to floral trait variation in *Chamaecrista fasciculata* (Fabaceae: Caesalpinoideae). *American Journal of Botany* **81**:206–215.

Freeman, R. S., A. K. Brody, and C. D. Neefus. 2003. Flowering phenology and compensation for herbivory in *Ipomopsis aggregata*. *Oecologia* **136**:394–401.

Galen, C., and J. Cuba. 2001. Down the tube: pollinators, predators, and the evolution of flower shape in the alpine skypilot, *Polemonium viscosum*. *Evolution* **55**:1963–1971.

Gómez, J. M. 1993. Phenotypic selection of flowering synchrony in a high mountain plant, *Hormathophylla spinosa* (Cruciferae). *Journal of Ecology* **81**:605–613.

Gómez, J. M. 2003. Herbivory reduces the strength of pollinator-mediated selection in the Mediterranean shrub *Erysimum mediohispanicum*: consequences for plant specialization. *American Naturalist* **162**:242–256.

Grace, J. 1997. Towards models of resource allocation in plants, pp. 279–281 in F. A. Bazzaz and J. Grace (eds.) *Plant Resource Allocation*. San Diego, CA: Academic Press.

Hamback, P. A. 2001. Direct and indirect effects of herbivory: feeding by spittlebugs affects pollinator visitation rates and seedset of *Rudbeckia hirta*. *Ecoscience* **8**:45–50.

Hamerlynck, E. P., T. E. Huxman, T. N. Charlet, and S. D. Smith. 2002. Effects of elevated CO_2 (FACE) on the functional ecology of the drought-deciduous Mojave Desert shrub, *Lycium andersonii*. *Environmental and Experimental Botany* **43**:93–106.

Harper, J. L. 1977. *Population Biology of Plants*. London: Academic Press.

Heinrich, B. 1971. The effect of leaf geometry on the feeding behavior of the caterpillar of *Manduca sexta* (Sphingidae). *Animal Behavior* **19**:119–124.

Hendrix, S. D. 1984. Reactions of *Heracleum lanatum* to floral herbivory by *Depressaria pastinacella*. *Ecology* **65**:191–197.

Herre, E. A., N. Knowlton, U. G. Mueller, and S. A. Rehner. 1999. The evolution of mutualisms: exploring the paths between conflict and cooperation. *Trends in Ecology and Evolution* **14**:49–53.

Herrera, C. M. 2000. Measuring the effects of pollinators and herbivores: evidence for non-additivity in a perennial herb. *Ecology* **81**:2170–2176.

Herrera, C. M., M. Medrano, P. J. Rey, *et al.* 2002. Interaction of pollinators and herbivores on plant fitness suggests a pathway for correlated evolution of mutualism- and antagonism-related traits. *Proceedings of the National Academy of Sciences of the USA* **99**:16823–16828.

Hodges, S. A. 1995. The influence of nectar production on hawkmoth behavior, self pollination, and seed production in *Mirabilis multiflora* (Nyctaginaceae). *American Journal of Botany* **82**:197–204.

Holland, J. N., and T. H. Fleming. 1999. Mutualistic interactions between *Upiga virescens* (Pyralidae), a pollinating seed-consumer, and *Lophocereus schottii* (Cactaceae). *Ecology* **80**:2074–2084.

Holland, J. N., and D. L. DeAngelis. 2002. Ecological and evolutionary conditions for fruit abortion to regulate pollinating seed-eaters and increase plant reproduction. *Theoretical Population Biology* **61**:251–263.

Huxman, T. E., and M. E. Loik. 1997. Reproductive patterns of two varieties of *Yucca whipplei* (Liliaceae) with different life histories. *International Journal of Plant Sciences* **158**:778–784.

Irwin, R. E., L. S. Adler, and A. K. Brody. 2004. The dual role of floral traits: pollinator attraction and defense. *Ecology* **85**:1503–1511.

Jaeger, N., I. Till-Bottraud, and L. Després. 2000. Evolutionary conflict between *Trollius europaeus* and its seed-parasite pollinators *Chiastocheta* flies. *Evolutionary Ecology Research* **2**:885–896.

Jaremo, J., J. Tuomi, P. Nilsson, and T. Lennartsson. 1999. Plant adaptations to herbivory: mutualistic versus antagonistic coevolution. *Oikos* **84**:313–320.

Jousselin, E., J. Y. Rasplus, and F. Kjellberg. 2003. Convergence and coevolution in a mutualism: evidence from a molecular phylogeny of *Ficus*. *Evolution* **57**:1255–1269.

Juenger, T., and J. Bergelson. 1997. Pollen and resource limitation of compensation to herbivory in scarlet gilia, *Ipomopsis aggregata*. *Ecology* **78**:1684–1695.

Juenger, T., and J. Bergelson. 1998. Pairwise versus diffuse natural selection and the multiple herbivores of scarlet gilia, *Ipomopsis aggregata*. *Evolution* **52**:1583–1592.

Juenger, T., T. Lennartsson, and J. Tuomi. 2000. The evolution of tolerance to damage in *Gentianella campestris*: natural selection and the quantitative genetics of tolerance. *Evolutionary Ecology* **14**:393–419.

Karban, R., and S. Y. Strauss. 1993. Effects of herbivores on growth and reproduction of their perennial host, *Erigeron glaucus*. *Ecology* **74**:39–46.

Kato, M., A. Takimura, and A. Kawakita. 2003. An obligate pollination mutualism and reciprocal diversification in the tree genus *Glochidion* (Euphorbiaceae). *Proceedings of the National Academy of Sciences of the USA* **100**:5264–5267.

Krupnick, G. A., and A. E. Weis. 1998. Floral herbivore effect on the sex expression of an andromonoecious plant, *Isomeris arborea* (Capparaceae). *Plant Ecology* **134**:151–162.

Krupnick, G. A., and A. E. Weis. 1999. The effect of floral herbivory on male and female reproductive success in *Isomeris arborea*. *Ecology* **80**:135–149.

Krupnick, G. A., A. E. Weis, and D. R. Campbell. 1999. The consequences of floral herbivory for pollinator service to *Isomeris arborea*. *Ecology* **80**:125–134.

Leege, L. M., and L. M. Wolfe. 2002. Do floral herbivores respond to variation in flower characteristics in *Gelsemium sempervirens* (Loganiaceae), a distylous vine? *American Journal of Botany* **89**:1270–1274.

Lehtilä, K., and S. Y. Strauss. 1997. Leaf damage by herbivores affects attractiveness to pollinators in wild radish, *Raphanus raphanistrum*. *Oecologia* **111**:396–406.

Lennartson, T., P. Nilsson, and J. Tuomi. 1998. Induction of overcompensation in the field gentian, *Gentianella campestris*. *Ecology* **79**:1061–1072.

Li, C. C. 1981. *Path Analysis: A Primer*, 3rd edn. Pacific Grove, CA: Boxwood.

Lohman, D. J., A. R. Zangerl, and M. R. Berenbaum. 1996. Impact of floral herbivory by parsnip webworm (Oecophoridae: *Depressaria pastinacella* Duponchel) on pollination and fitness of wild parsnip (Apiaceae: *Pastinaca sativa* L.). *American Midland Naturalist* **136**:407–412.

Louda, S.M., and M.A. Potvin. 1995. Effect of inflorescence-feeding insects on the demography and lifetime fitness of a native plant. *Ecology* **76**:229–245.

Lowenberg, G.J. 1994. Effects of floral herbivory on maternal reproduction in *Sanicula arctopoides* (Apiaceae). *Ecology* **75**:359–369.

Lowenberg, G.J. 1997. Effects of floral herbivory, limited pollination, and intrinsic plant characteristics on phenotypic gender in *Sanicula arctopoides*. *Oecologia* **109**:279–285.

Madden, A.H., and F.S. Chamberlin. 1945. Biology of the tobacco hornworm in the southern cigar tobacco district. *US Department of Agriculture Technical Bulletin* **896**.

Matter, S.F., J.B. Landry, A.M. Greco, and C.D. Lacourse. 1999. Importance of floral phenology and florivory for *Tetraopes tetraophthalmus* (Coleoptera: Cerambycidae): tests at the population and individual level. *Environmental Entomology* **28**:1044–1051.

McDade, L.A., and P. Davidar. 1984. Determinants of fruit and seed set in *Pavonia dasypetala* (Malvaceae). *Oecologia* **64**:61–67.

McFadden, M.W. 1968. Observations on feeding and movement of tobacco hornworm larvae. *Journal of Economic Entomology* **61**:352–356.

Mechaber, W.L., and J.G. Hildebrand. 2000. Novel, non-solanaceous hostplant record for *Manduca sexta* (Lepidoptera: Sphingidae) in the southwestern United States. *Annals of the Entomological Society of America* **93**:447–451.

Meekijjaroenroj, A., and M.-C. Anstett. 2003. A weevil pollinating the Canary Islands date palm: between parasitism and mutualism. *Naturwissenschaften* **90**:452–455.

Mira, A., and E.A. Bernays. 2002. Trade-offs in host use by *Manduca sexta*: plant characters vs. natural enemies. *Oikos* **97**:387–397.

Mitchell, R.J. 1992. Testing evolutionary and ecological hypotheses using path analysis and structural equation modelling. *Functional Ecology* **6**:123–129.

Mitchell, R.J., and N.M. Waser. 1992. Adaptive significance of *Ipomopsis aggregata* nectar production: pollination success of single flowers. *Ecology* **73**:633–638.

Mooney, H.A. 1972. The carbon balance of plants. *Annual Review of Ecology and Systematics* **3**:315–346.

Mothershead, K., and R.J. Marquis. 2000. Fitness impacts of herbivory through indirect effects on plant–pollinator interactions in *Oenothera macrocarpa*. *Ecology* **81**:30–40.

Muñoz, A.A., and M.T.K. Arroyo. 2004. Negative impacts of a vertebrate predator on insect pollinator visitation and seed output in *Chuquiraga oppositifolia*, a high Andean shrub. *Oecologia* **138**:68–73.

Murawski, D.A. 1987. Floral resource variation, pollinator response, and potential pollen flow in *Psiguria warscewiczii*. *Ecology* **65**:1273–1282.

Mutikainen, P., and L.F. Delph. 1996. Effects of herbivory on male reproductive success in plants. *Oikos* **75**:353–358.

Paige, K.N. 1994. Herbivory and *Ipomopsis aggregata*: differences in response, differences in experimental protocol – a reply to Bergelson and Crawley. *American Naturalist* **143**:739–749.

Paige, K.N., and T.G. Whitham. 1987. Overcompensation in response to mammalian herbivory: the advantage of being eaten. *American Naturalist* **129**:407–416.

Paige, K.N., B. Williams, and T. Hickox. 2001. Overcompensation through the paternal component of fitness in *Ipomopsis arizonica*. *Oecologia* **128**:72–76.

Pellmyr, O. 1997. Pollinating seed-eaters: why is active pollination so rare? *Ecology* **78**:1655–1660.

Pellmyr, O. 2003. Yuccas, yucca moths, and coevolution: a review. *Annals of the Missouri Botanical Garden* **90**:35–55.

Pilson, D. 2000. Herbivory and natural selection on flowering phenology in wild sunflower, *Helianthus annuus*. *Oecologia* **122**:72–82.

Poorter, H., and M.L. Navas. 2002. Plant growth and competition at elevated CO_2: on winners, losers and functional groups. *New Phytologist* **157**:175–198.

Primack, R.B., S.L. Miao, and K.R. Becker. 1994. Costs of reproduction in the pink lady's slipper orchid (*Cypripedium acaule*): defoliation, increased fruit production, and fire. *American Journal of Botany* **81**:1083–1090.

Quesada, M., K. Bollman, and A.G. Stephenson. 1995. Leaf damage decreases pollen production and hinders pollen performance in *Cucurbita texana*. *Ecology* **76**:437–443.

Raguso, R.A., C. Henzel, S.L. Buchmann, and G.P. Nabhan. 2003. Trumpet flowers of the Sonoran Desert: floral biology of *Peniocereusi* cacti and sacred *Datura*. *International Journal of Plant Science* **164**:877–892.

Rathcke, B.J. 2001. Pollination and predation limit fruit set in a shrub, *Bourreria succulenta* (Boraginaceae), after hurricanes on San Salvador Island, Bahamas. *Biotropica* **33**:330–338.

Richards, A.J. 1997. *Plant Breeding Systems*. New York: Chapman and Hall.

Schemske, D.W., and C.C. Horvitz. 1988. Plant-animal interactions and fruit production in a neotropical herb: a path analysis. *Ecology* **69**:1128–1137.

Sharaf, K.E., and M.V. Price. 2004. Does pollination limit tolerance to browsing in *Ipomopsis aggregata*? *Oecologia* **138**:396–404.

Shykoff, J.A., and E. Bucheli. 1995. Pollinator visitation patterns, floral rewards and the probability of transmission of *Microbotryum violaceum*, a venereal disease of plants. *Journal of Ecology* **83**:189–198.

Smith, F.A., J.H. Brown, and T.J. Valone. 1997. Path analysis: a critical evaluation using long-term experimental data. *American Naturalist* **149**:29–42.

Sokal, R.R., and F.J. Rohlf. 1981. *Biometry*, 2nd edn. San Fransisco, CA: W.H. Freeman.

Stowe, K.A., R.J. Marquis, C.G. Hochwender, and E.L. Simms. 2000. The evolutionary ecology of tolerance to consumer damage. *Annual Review of Ecology and Systematics* **31**:565–595.

Strain, B.R., and F.A. Bazzaz. 1983. Terrestrial plant communities, pp. 177–222 in E.R. Lemon (ed.) CO_2 *and Plants: The Response of Plants to Rising Levels of Atmospheric Carbon Dioxide*. Boulder, CO: Westview Press.

Strauss, S.Y. 1997. Floral characters link herbivores, pollinators, and plant fitness. *Ecology* **78**:1640–1655.

Strauss, S.Y., and W.S. Armbruster. 1997. Linking herbivory and pollination: new perspectives on plant and animal ecology and evolution. *Ecology* **78**:1617–1618.

Strauss, S.Y., J.K. Conner, and S.L. Rush. 1996. Foliar herbivory affects floral characters and plant attractiveness to pollinators: implications for male and female plant fitness. *American Naturalist* **147**:1098–1107.

Strauss, S.Y., D.H. Siemens, M.B. Decher, and T. Mitchell-Olds. 1999. Ecological costs of plant resistance to herbivores in the currency of pollination. *Evolution* **53**:1105–1113.

Strauss, S.Y., J.K. Connor, and K.P. Lehtilä. 2001. Effects of foliar herbivory by insects on the fitness of *Raphanus raphanistrum*: damage can increase male fitness. *American Naturalist* **158**:496–504.

Vail, S.G. 1992. Selection for overcompensatory plant responses to herbivory: a mechanism for the evolution of plant–herbivore mutualism. *American Naturalist* **139**:1–8.

Vázquez, D.P., and D. Simberloff. 2004. Indirect effects of an introduced ungulate on pollination and plant reproduction. *Ecological Monographs* **74**:281–308.

Westercamp, C. 1996. Pollen in bee-flower relations: some considerations on melittophily. *Botanica Acta* **109**:325–332.

Wolfe, D.W., R.M. Gifford, D. Hilbert, and Y. Luo 1998. Integration of photosynthetic acclimation to CO_2 at the whole-plant level. *Global Change Biology* **4**:879–893.

Wolfe, L.M., and J.L. Krstolic. 1999. Floral symmetry and its influence on variance in flower size. *American Naturalist* **154**:484–488.

Yamauchi, A., and N. Yamamura. 2004. Herbivory promotes plant production and reproduction in nutrient-poor conditions: effects of plant adaptive phenology. *American Naturalist* **163**:138–153.

Yu, D.W., J. Ridley, E. Jousselin, *et al.* 2004. Oviposition strategies, host coercion and the stable exploitation of figs by wasps. *Proceedings of the Royal Society of London Series B* **271**:1185–1195.

Zangerl, A.R., and C.E. Rutledge. 1996. The probability of attack and patterns of constitutive and induced defense: a test of optimal defense theory. *American Naturalist* **147**:599–608.

Zimmerman, M. 1983. Plant reproduction and optimal foraging: experimental nectar manipulations in *Delphinium nelsonii*. *Oikos* **41**:57–63.

5

Trait-mediated indirect interactions, density-mediated indirect interactions, and direct interactions between mammalian and insect herbivores

JOSÉ M. GÓMEZ AND ADELA GONZÁLEZ-MEGÍAS

Introduction

Ecologists have traditionally recognized the consequences that direct interactions between species have on the functioning of ecological communities and on the flow of energy through food webs (Pimm 2002). However, ecological communities are among the most complex natural systems, and thus the interactions between species are far from simple. In this respect, the importance of indirect effects (those effects transmitted from one species to another through one or more intermediate species) as a determinant of the structure and dynamics of ecological communities has been clearly acknowledged only in recent years (Wootton 1994, Abrams 1995, Abrams et al. 1996). Indirect effects are those transmitted from one species to another through one or more intermediate species. These indirect effects can be propagated through food webs as a consequence of changes in the density of the intervening species, a mechanism known as density-mediated indirect interaction (DMII) (Abrams 1995, Werner and Peacor 2003). Nevertheless, indirect effects can also occur through changes in the phenotypes of the interacting organisms, a mechanism known as trait-mediated indirect interaction (TMII) (Werner and Peacor 2003, Schmitz et al. 2004). Although DMIIs have been traditionally considered to be the main source of any variation in ecological communities, ecologists are progressively more conscious of the

Ecological Communities: Plant Mediation in Indirect Interaction Webs, ed. Takayuki Ohgushi, Timothy P. Craig, and Peter W. Price. Published by Cambridge University Press. © Cambridge University Press 2007.

essential role played by TMIIs (Bolker *et al.* 2003, Dill *et al.* 2003, Luttbeg *et al.* 2003, Trussell *et al.* 2003, Werner and Peacor 2003; and chapters throughout this volume).

Our knowledge of the importance of TMIIs is surely biased by the systems that have been traditionally studied. First, most experimental studies have been done on tritrophic systems, with predator presence affecting prey behavior and, therefore resource ingestion (Luttbeg *et al.* 2003). Second, most studies have been done with systems involving only animals (Raimondi *et al.* 2000, Bolker *et al.* 2003, Dill *et al.* 2003, Luttbeg *et al.* 2003, Trussell *et al.* 2003). By contrast, very few studies have been done in systems involving plants. This is quite paradoxical since interactions between plants and herbivores have two features that make them very suitable for the occurrence of TMIIs: nonlethality and phenotypic plasticity of one of the intervening species, the plant.

As a consequence of their strong effect on plants, it has been traditionally assumed that herbivores affect each other by decreasing the availability of shared resources, that is, through DMIIs (Schoonhoven *et al.* 1998, Speight *et al.* 1999). However, during the last two decades empirical evidence has accumulated showing the importance of competition among herbivores even in systems where resources are not limiting (Denno *et al.* 1995, Stewart 1996) or where competitors are temporally and spatially separated (Faeth 1992, Faeth and Wilson 1997, Master and Brown 1997, Denno *et al.* 2000, Fisher *et al.* 2000, Ferrenberg and Denno 2003). These findings suggest that competition often occurs due to the modification of some plant traits crucial for the herbivores.

The great majority of plant species interact with a diverse suite of herbivores. Thus, plants are simultaneously and/or sequentially consumed by leaf-chewers, gall-makers, stem-borers, sap-feeders, invertebrate and vertebrate seed predators, mammal browsers, and other guilds of herbivores. This multispecific assemblage of co-occurring herbivores creates the opportunity for competitive interactions between highly dissimilar organisms (Davidson *et al.* 1984, 1985, Brown *et al.* 1986, Hochberg and Lawton 1990, Hunter 1992, Roche and Fritz 1997, Tscharntke 1997, Gómez and González-Megías 2002, González-Megías and Gómez 2003). Unfortunately, most experiments testing for the role of interspecific competition have been performed under the assumption that it occurs mainly between members of the same guild and among taxonomically, morphologically, and ecologically similar species (Stewart 1996, Denno *et al.* 2000). Thus there is a great ignorance about the functioning of food webs that involve both vertebrates and invertebrates, and in particular about the types of forces determining the abundance of phytophagous insects in these complex communities.

In this chapter, our main goal is to show that highly dissimilar herbivores can interact when sharing the same resource in a variety of ways. We first show that herbivores can decrease the availability of resources by affecting plant performance, abundance, and diversity. Next, we describe the different ways by which plants can phenotypically respond to herbivore damage. Using these two pieces of information, we explore the different mechanisms that underlie the interaction between ungulates and insects. We finish the chapter by describing an example of a system where ungulates interact with phytophagous insects by more than one mechanism.

Effects of herbivores on plant performance, populations, and communities

The main mechanism creating exploitative competition between herbivores is the strong impact that herbivores usually have on their resources. The reproduction and growth of many plant species is negatively affected by the severe damage inflicted by herbivores (Marquis 1992, Crawley 1997, 2000, Zamora *et al.* 1999). This detrimental effect of herbivory is evident irrespective of plant life history, since significant negative effects on plant performance have been reported for a diverse array of plants including annuals, perennial herbs, shrubs, and trees (Marquis 1992, Zamora *et al.* 1999 and references therein). Nevertheless, the existing data demonstrate that the severity of herbivore impact on plants strongly depends on several extrinsic and intrinsic factors, such as the availability of resources for the plants, the plant age, the plant compensatory ability, the type of herbivores or the kind of plant tissue consumed by the herbivores (Gómez and Zamora 2000a, 2000b, Hawkes and Sullivan 2001, Zamora *et al.* 2001, Hambäck and Beckerman 2003, Allcock and Hik 2004).

Herbivores can also decrease plant population densities, although in contrast to the copious literature concerning the effect of herbivores on individual plants, little empirical information is available on this topic (Carson and Root 2000, Maron and Simms 2001, Gómez 2005). Unfortunately, the effect of herbivores on plant population dynamics is commonly inferred from their effect on some demographic components like seed germination, seedling emergence, or juvenile recruitment, and this may give an inaccurate picture of the overall impact of herbivory (e.g., Louda and Potvin 1995, Juenger and Bergelson 2000, Maron and Gardner 2000, Maron *et al.* 2002).

Herbivores also affect the diversity of plant communities (Crawley 1983). In fact, some herbivores are considered keystone species because they regulate plant diversity by selective grazing (Hunter 1992). The net effect of herbivores on plant species diversity is controversial, since they enhance diversity in some cases while they decrease plant diversity in other cases (Zamora *et al.* 1999). They can also affect vegetation through such nontrophic mechanisms as trampling, burrowing, and soil compaction. These effects usually produce an increase in spatial heterogeneity and a change in plant architecture and vegetation structure.

Mammalian versus insect herbivory

Most studies considering the simultaneous activity of vertebrate and invertebrate herbivores have found that the former have a much stronger impact on plant performance than the latter (Hulme 1994, 1996, Palmisano and Fox 1997, Gómez and Zamora 2000a, 2000b, Sessions and Kelly 2001, Goheen et al. 2004). Furthermore, according to the scarce experimental evidence available, it seems that vertebrates also have a greater impact on plant populations and diversity than do invertebrates (Danell and Bergström 2002; but see Tscharntke 1997, Prittinen et al. 2003). Several reasons could explain why vertebrate herbivores have a higher impact than invertebrate herbivores, both in the short and long term, including: their greater body size, polyphagy, individual bite size, mobility, and tolerance to starvation (Danell and Bergström 2002).

Herbivore effect on plant phenotype

As a consequence of the two aforementioned main features common to most herbivore–plant interactions, nonlethality and plant phenotypic plasticity, plant phenotype is modified by herbivores in a variety of ways.

Induced resistance

One very common and well-studied plant response to herbivory is the so-called induced resistance response (Karban and Myers 1989, Agrawal 1999). As a consequence of the strong effect that herbivores have on plant fitness, plants have evolved a myriad of anti-herbivore resistance traits like chemical compounds, spines, thorns, etc. (Marquis 1992). However, most resistance traits are costly because resource allocation to these traits represents a trade-off with other activities, such as growth (Herms and Mattson 1992). An effective way to save the cost of defensive traits is to produce or activate them only when needed, usually following herbivore attack. In fact, most current theories assume that inducible defenses are cost-saving mechanisms (Karban and Baldwin 1997) and, that therefore they are a kind of adaptive phenotypic plasticity because they allow plants to maximize fitness in each environment (Cipollini 2004).

Induced plant responses to damage produced by both vertebrate and invertebrate herbivores are ubiquitous in nature (Karban and Baldwin 1997). Inducible traits are very diverse, ranging from phytochemical compounds (Agrawal 1999, 2000) to mechanical structures such as thorns, spines, and trichomes (Young and

Okello 1998, Gómez and Zamora 2002, Young *et al.* 2003), nutritional quality (Bi *et al.* 1997), and even some other more baroque traits such as volatiles that attract predators and parasitoids of herbivores (Takabayashi and Dicke 1996), and extrafloral nectaries that attract defensive ants (Agrawal and Rutter 1998, Huntzinger *et al.* 2004).

Tolerance to damage

Plants can compensate for herbivore damage (Strauss and Agrawal 1999) by increasing their net photosynthetic rate, relative growth rate, or branching. Alternatively, they can reallocate carbon from roots to shoots after damage. A consequence of compensation for damage is a change in the phenotypic value of some traits. Thus, compensating plants usually have delayed reproductive phenology (flowering and fruiting) and a longer reproductive period because they have to produce new tissue when co-occurring conspecifics are flowering. Morphological as well as life-history traits are affected. Thus, it is very common that plants compensating for damage produce smaller and thicker leaves than undamaged plants (Strauss and Agrawal 1999).

Nonadaptive phenotypic plasticity

Herbivores cause changes in plant traits not related at all to defense against herbivory. Many times changes in plant traits are due to a reallocation of resources made by damaged plants. For example, leaf damage decreases pollen production per flower in *Cucurbita texana* and *Raphanus raphanistrum* (Quesada *et al.* 1995, Strauss *et al.* 1996). Furthermore, the flowers of *Raphanus raphanistrum* individuals damaged by *Pieris rapae* larvae are smaller than the flowers of undamaged siblings (Strauss 1997). In addition, plant architecture is dramatically modified in many plant species as a consequence of herbivore activity. For example, many typically monopodial species grow to become shrub-shaped plants because ungulates repeatedly consume their leader shoots (Cuartas and García-González 1992, Zamora *et al.* 2001).

Changes in plant distribution pattern

Herbivores can also shape the habitat distribution and spatial structure of plant populations (Louda and Rodman 1996, Kleijn and Steinger 2002, Passos and Oliveira 2002, DeWalt *et al.* 2004). This effect occurs when the impact of herbivores varies spatially, between microhabitats. In some cases, a plant species escapes damage by growing beneath a protective mechanical barrier produced by other species, whereas in other cases damage is avoided by associating with unpalatable or less-preferred plants (Hjältén and Price 1997, WallisDeVries *et al.* 1999). By influencing spatial distribution patterns, herbivores can indirectly modify plant phenotype. For example, plants growing under low light conditions generally have larger leaves and longer and narrower stalks than plants growing in the sun.

Interactions between mammalian and insect herbivores

Mammalian herbivores, by having a strong impact on shared plant species, can have profound direct and indirect effects on insect populations (Hunter 1992, Tscharntke 1997). Unfortunately, due to the scarcity of studies, there is no consensus about the magnitude and sign of the effects of ungulates on insects. Several studies have shown that mammals have a negative effect on insects (Baines et al. 1994, Dennis et al. 1997, Tscharntke 1997, Strand and Merritt 1999, Suominen et al. 1999a, Kruess and Tscharntke 2002a, 2002b, González-Megías et al. 2004). However, other studies have shown that the presence of mammalian herbivores has a positive effect on many phytophagous insects (Danell and Hus-Danell 1985, Abensperg-Traun et al. 1996, Bestelmeyer and Wiens 1996, Elligsen et al. 1997, Martinsen et al. 1998, Seymour and Dean 1999, Suominen et al. 1999b, Goheen et al. 2004, González-Megías et al. 2004). Finally, some studies have even found that mammalian herbivores do not have any perceptible effect on invertebrates (Rambo and Faeth 1999). This absence of consensus about the effect of mammals on insects is surely the result of the scarcity of studies designed to explore the mechanisms underlying this interaction, since most studies are designed solely to determine the overall effect that the exclusion of mammals from a portion of the landscape has on arthropod communities.

Density-mediated indirect interactions

Based on the strong impact of ungulates on the availability of resources, DMII is the most often invoked type of competition between herbivorous mammals and phytophagous insects (Table 5.1). Thus, mammalian herbivores can have a detrimental effect on insects by decreasing the abundance of their shared host plant. Nevertheless, DMIIs can also arise even when herbivores do not affect plant abundance. Thus, when foraging on plants, some herbivores consume much of the vegetative (leaves) and reproductive (flowers and fruits) tissue, depressing the available food for other herbivores.

When herbivores differ in major traits, such as morphology or size, the outcome of the competition between them is usually asymmetrical (Davidson et al. 1985, Brown et al. 1986, Christensen and Whitham 1993, Lucas et al. 1998). For this reason, the DMIIs arising between mammals and insects are presumably highly asymmetrical, because most of the time mammals affect insect abundance but the converse rarely happens (Gómez and González-Megías 2002).

Table 5.1. *Different types of interactions occurring between ungulates and phytophagous insects sharing the same host plant*

Type of interaction	Mechanism	Lethality[a]
Indirect		
Density-mediated (DMII)	Decrease in the shared resource	No
Trait-mediated (TMII)	Modification of plant phenotype	No
Direct		
Density-mediated (DMI)	Insect ingestion	Yes
Trait-mediated (TMI)	Effect on insect behavior	No

Note:
[a]Lethality refers to the probability of the insects being killed by the ungulates.

Nevertheless, accurate information on this topic is very scarce, since factorial experiments that exclude each type of herbivore separately are very difficult to perform (Tscharntke 1997).

Trait-mediated indirect interactions

Browsing and grazing can damage phytophagous insects that do not coincide spatiotemporally with mammalian herbivory. This frequent but overlooked pattern is mostly produced by TMIIs (Table 5.1). When foraging, mammalian herbivores alter plant growth and architecture. This structural modification severely affects those insects dependent on complex structures, like some sap-feeders and Lepidoptera larvae (Tscharntke and Greiler 1995, Abensperg-Traun *et al.* 1996, Bestelmeyer and Wiens 1996, Dennis *et al.* 1997, Tscharntke 1997, Seymour and Dean 1999, Kruess and Tscharntke 2002a, 2002b). Mammalian herbivores also induce the production of defensive responses by plants that could negatively affect insects (Gómez and González-Megías 2002). Induced defenses can be specific or general, and this latter point is crucial to understanding the TMIIs between mammalian and insect herbivores. Some studies have demonstrated that many insect herbivores are able to discriminate between damaged and undamaged plants (Reznik 1991, Dolch and Tscharntke 2000). For example, experimental defoliation on alder *Alnus glutinosa* simulating damage by ungulates reduces consumption by the leaf beetle *Agelastica alni* (Dolch and Tscharntke 2000). Insects could also have a significant counter-effect on mammalian herbivores by inducing defensive responses in plants (Karban and Baldwin 1997 and references therein). In this case, the occurrence of TMIIs may change the usual asymmetrical outcome of the mammal–insect competition, since it allows the smaller organism to outcompete the larger ones.

Unfortunately, no study has been conducted yet to explore these potential indirect effects of insects on mammals.

Trait-mediated interactions could be positive in some cases, shifting the interaction from being competitive to facilitative or even mutualistic. For example, damage by mammalian herbivores can induce overcompensatory growth in plants and/or the rejuvenation of plant tissues. In the first case the outcome of ungulate damage is an increase in the amount of resource available for insects, whereas in the second case the consequence is the increased availability of plant tissue of higher quality (Kruess and Tscharntke 2002a). Nitrogen-rich plants are preferred by insects, which respond positively to nitrogen contents in plant tissues (Speight *et al.* 1999). In this way, Danell and Huss-Danell (1985) reported that *Betula pendula* browsed by moose, *Alces alces*, were preferred by some phytophagous insects (such as aphids, psyllids, leaf miners, and leaf-gallers) more than unbrowsed trees. Martinsen *et al.* (1998) similarly found that the leaf beetle *Chrysomela confluens* prefers the resprouts from *Populus fremontii* and *P. angustifolia* individuals damaged by the beaver *Castor canadensis* because they are richer in nitrogen.

Direct interactions

As a consequence of their huge size difference the competition between mammalian herbivores and insects is sometimes so asymmetrical that it can even turn into direct interference (Tscharntke 1997). This occurs, for example, when insects fall from plants as ungulates approach to feed upon them, a behavior that presumably reduces the time insects have available for feeding and ingestion. This kind of interaction mediated by a change in the behavior of one interacting species is the classical example of TMII reported for three-trophic systems, with the prey altering its behavior in the presence of the predator (Dill *et al.* 2003, Luttbeg *et al.* 2003, Werner and Peacor 2003). However, in this kind of herbivore two-trophic level interaction, the trait-mediated interaction (TMI) is direct rather than indirect, since one herbivore affects the other without the mediation of a third species (Table 5.1). Unfortunately, it has been overlooked, and consequently there is no information about its frequency and importance as a mechanism influencing insect performance and population density.

Moreover, a more extreme consequence of the size difference between mammals and insects takes place when the outcome of the interaction is lethal for the smaller organism because they are incidentally ingested by the bigger one (density-mediated interaction (DMI) in Table 5.1) (Tscharntke 1997, Lucas *et al.* 1998). Indeed, some studies have reported that browsing ungulates consume plant material that contains endophagous insect larvae (Zamora and Gómez 1993, Baines *et al.* 1994, Retamosa 1997, Gómez and González-Megías

2002). Red and fallow deer ingest seed-predators when consuming *Iberis contracta* flower heads in south west Spain (Retamosa 1997), whereas Spanish ibex in the alpine of the Sierra Nevada, south east Spain, consume gall-makers and weevil seed-predators when feeding on *Hormathophylla spinosa* (Zamora and Gómez 1993, Gómez and González-Megías 2002). This incidental predation may possibly amplify the harmful effect that the asymmetrical exploitative competition by mammalian herbivores has on insects (Baines et al. 1994). Despite incidental predation which could be a major mortality factor for many phytophagous insects, especially for those endophagous species living concealed within plant structures, not many studies have been designed to explore its effects. In addition, when incidentally consuming insects, ungulates affect multiple trophic levels, with insects being consumed irrespective of their trophic level ranging from phytophages to saprophages and from predators to parasitoids (Tscharntke 1997). In this case, mammalian herbivores can act as keystone species, producing cascading effects in invertebrate food webs that include direct and indirect interactions among herbivores which could modify the interactions between other minor herbivores and the host plant (Hunter 1992, Tscharntke 1997, Gómez and González-Megías 2002).

Evolutionary consequences of the TMIIs between mammalian and insect herbivores

Herbivores, by affecting plant relative fitness, can exert selective pressure on many plant traits, including flower number, plant size, phenology, constitutive concentration of phytochemical compounds, or mechanical defensive traits (Marquis 1992). As a prerequisite for this herbivore-mediated evolution to occur, it is necessary that herbivore preference and/or performance be determined by specific plant traits. For instance, by avoiding plants with higher concentrations of chemical compounds, herbivores are potentially selecting for these resistance traits in many plant species. Nevertheless, as shown in this chapter, most plant species are attacked by several to many antagonistic organisms. That is, the plants are most times under multispecific and complex selective pressures in which each interacting species can constitute a distinctive selective pressure.

In this context, it is easy to imagine that the selection imposed by a given species on a plant trait can be frequently modified by the action of one or several other co-occurring species on the same plant traits through TMIIs. The most probable outcome of this similarity in preferences is the occurrence of synergic selection pressures against the target trait. For example, by preferring plants with low concentrations of a given chemical compound, the two kinds of herbivores are selecting for plants with higher concentrations of that compound, which will evolve as a broad-spectrum defensive trait. In this case, a possibility for the occurrence of TMIIs exists, since the phenotypic modification of the plant in response to one herbivore affects the other herbivore.

Under these circumstances, TMIIs are conceptually similar to diffuse coevolution among multiple herbivores and the plant (Inouye and Stinchcombe 2001). Adaptations to avoid phytophagous insects can constrain the evolution of plant–ungulate interaction, and vice versa. Unfortunately, empirical information on the evolutionary outcomes of the TMIIs between herbivores sharing a host plant is still extremely scarce (Inouye and Stinchcombe 2001, Agrawal and van Zandt 2003 and references therein).

A case study: interaction among *Erysimum mediohispanicum*, ungulates, and phytophagous insects

Erysimum mediohispanicum (Cruciferae) is a biannual to perennial monocarpic herb found in many montane habitats of the Iberian Peninsula. Plants usually grow for 2–3 years as vegetative rosettes, and then die after producing one to eleven reproductive stalks which can display between a few and several hundred flowers. In the Sierra Nevada (south east Spain), reproductive individuals are fed upon by many different species of herbivores: (1) Several species of sap-feeders (mostly the bugs *Eurydema oleracea, E. fieberi,* and *E. ornata* Heteroptera: Pentatomidae) feed on the reproductive stalks, both during flowering and fruiting. (2) The stalks are bored into by the larvae of a weevil species (presumably *Lixus ochraceus* Coleoptera: Curculionidae), which consume the inner tissues. (3) The seeds are preyed upon before dispersal by another weevil species (presumably *Ceutorhynchus chlorophanus* Curculionidae), which develops inside the fruits, living on developing seeds. (4) Some floral buds do not open because they are galled by flies (presumably *Dasineura* sp. Diptera: Cecidomidae). (5) The entire stalks are consumed by Spanish ibex (*Capra pyrenaica* Bovidae), which consume flowers and green fruits by browsing on the reproductive stalks. Whereas 1 and 5 are free-living herbivores, moving among different plants easily, 2, 3, and 4 are endophagous herbivores tightly associated with a single plant during their development.

All these herbivores are abundant in the study site: Spanish ibex damaged 33% of the plants, whereas stem-borers damaged 30%, seed predators damaged 36%, sap-suckers damaged 45%, and gall-makers damaged 31% (data from 1997–2003, 150–300 plants yr^{-1}). The most detrimental effect on plant reproductive output and population dynamic was produced by ungulates. Damaged plants lost to ungulate herbivory on average 50.7% of their potential reproductive output, and in some years these losses increasing up to 75%. Ungulates also affected plant abundance and population dynamics, since after 7 years of ungulate exclusion the growth rate of the populations was 0.334 in ungulate-excluded plots but only 0.114 in control plots. Finally, ungulates also influenced the spatial distribution pattern of the plants, since they grow mostly under the canopy of co-occurring shrubs when ungulates were present, but colonized open sites promptly after the exclusion of ungulates (Gómez 2005).

Ungulates not only decrease plant populations, they also affect plant phenotype. First, ungulates have an immediate effect on some plant traits. After being damaged, over 70% of the plants compensate and regrow to flower again. However, compensation is not total, and damaged plants have shorter stalks and display fewer flowers than undamaged plants. Second, ungulates seem also to have a delayed effect on plant traits. Thus, after several years of ungulate exclusion, plants were taller and had a lower density of trichomes than plants exposed to ungulates. Finally, there is an additional indirect pathway by which ungulates can affect plant phenotype: they affect the spatial distribution of plants. This occurs because plant morphology varies among microhabitats. Plants had taller and narrower stalks when located under shrubs than in open sites because they overgrow shrubs before flowering.

Interaction between ungulates and co-occurring herbivorous insects

Ungulates outcompeted all co-occurring phytophagous insects, since the exclusion of ungulates provoked a significant increase in the abundance of seed predators, sap-feeders, gall-makers, and stem-borers (Fig. 5.1a). We presume that, as a consequence of their severe impact on both plant performance and abundance, ungulates affect the four taxa by means of DMIIs. Nevertheless, other mechanisms are also involved. Thus, we found that ungulates have a significant indirect effect on gall-makers, sap-feeders, and stem-borers, as suggested by the change in the insect attack rate when ungulates were present (Fig. 5.1b). This is strong evidence for the existence of TMIIs between these invertebrates and the ungulates (Inouye and Stinchcombe 2001). Sap-feeders prefer plants with many flowers and few trichomes, a preference pattern that supports the existence of TMIIs for these insects since these plant traits are modified by ungulates. However, stem-borers preferred plants with thicker stalks and gall-makers preferred shorter plants with fewer flowers.

There is an indirect pathway by which ungulates can affect stem-borers through TMIIs, the ungulate effect on the spatial distribution of plants. Plants are located under shrubs to protect against ungulate damage. However, plants under shrubs have narrow stalks, and therefore are not suitable hosts for stem-borers. When ungulates were excluded, plants started to colonize the open

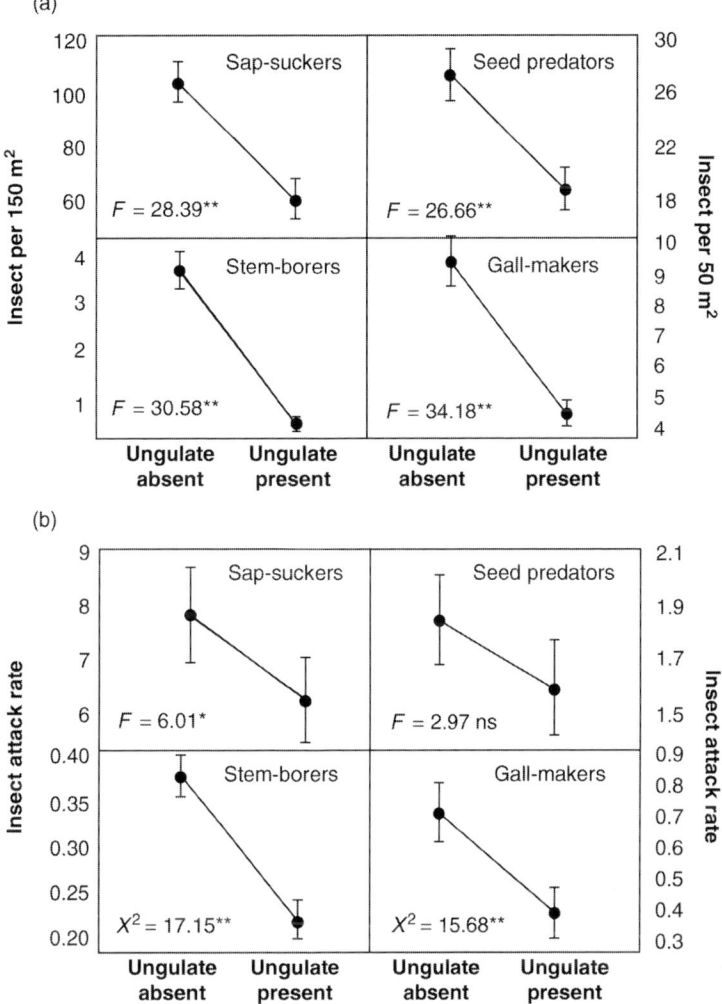

Figure 5.1. Difference between ungulate-excluded and control plots ($n = 3$ plots per treatment, ten 50 m² transects per plot, and over 7 years) in Sierra Nevada (south east Spain) in (a) the abundance and (b) the per capita interaction strength of four taxa of phytophagous insects. The effect of ungulate exclusion was tested by three-way mixed ANOVAs and log-linear analyses, including ungulate, year, and plot as main factors. * indicates $p < 0.01$; ** indicates $p < 0.0001$.

sites where they developed thicker stalks. As a consequence, the abundance of stem-borers increased in these areas. Interestingly, the spatial distribution of stem-borers also changed following ungulate exclusion. While in control plots the insects were more abundant in the plants located under shrubs, in ungulate-excluded plots the stem-borers started to colonize the plants located in open areas between shrubs. This microhabitat is the most risky for stem-borers

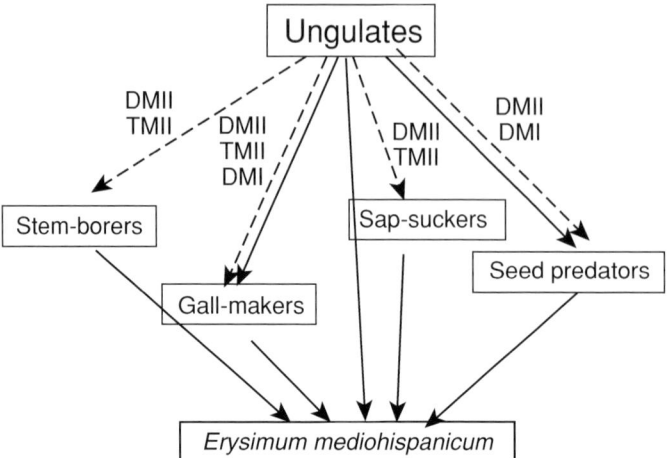

Figure 5.2. Diagram showing the hypothetical interactions that may occur among ungulates, four taxa of phytophagous insects, and their shared host plant, *Erysimum mediohispanicum* (Cruciferae) in the high mountains of the Sierra Nevada (south east Spain). Solid lines indicate direct interactions, broken lines indicate indirect interactions.

because it is the one where the plants suffer the most damage by ungulates. That means that when ungulates are present, stem-borers are located in the less-suitable microhabitat in relation to host plant morphology, presumably to escape from the interaction with ungulates.

Finally, ungulates affect seed predators and gall-makers directly by ingesting them during browsing. These two species are endophagous, and are unable to avoid ungulate effect by leaving the plant. They are usually located in the upper part of the plant, which is the part more likely to be ingested by ungulates. We estimated that ungulates consumed around 15–16% of the population of both species. This incidental predation may explain the increase in gall-makers above the expected from the increase in the resource alone when ungulates are excluded.

In brief, this study shows that ungulates can affect co-occurring phytophagous insects through several, nonexclusive mechanisms (Fig. 5.2).

Prospects for future research

The importance of interspecific competition as a force organizing phytophagous insect communities has been hotly disputed during the last three decades (Strong *et al.* 1984, Denno *et al.* 1995, Stewart 1996, Schoonhoven *et al.* 1998, Speight *et al.* 1999). In this chapter we have intended to show that mammalian herbivores may determine the abundance and diversity of phytophagous insects, because they can interact in diverse and sometimes unusual ways, as in the case of incidental predation. For this reason, ecologists should pay more attention to the interactions arising between these two groups of organisms. In this respect, we think that future research should consider the following proposals:

1. Most studies to date on mammal–insect interactions have been designed to determine the increase in abundance and/or diversity of insects after excluding mammalian herbivores from a given portion of the landscape. Although we think that these kinds of studies are necessary to understand the role of these mammalian herbivores as disturbance factors on the structure and dynamics of insect communities, they are phenomenological and thereby unable to discover the mechanisms underlying the observed interactions between vertebrate and invertebrate herbivores. We encourage experimental ecologists to develop mechanistic studies in order to accurately separate the effects that ungulates have on insects via the shared resource from those produced by alternative mostly nontrophic mechanisms.
2. Following this mechanistic approach, we also encourage ecologists to explore the relative roles that different types of interactions may play in the overall outcome of the interactions between ungulates and insects. In particular, it would be interesting to determine the proportional effects of DMII, TMII, interference competition, and direct predation in shaping herbivore populations.

Addressing these proposals will require innovative and sometimes complex experimental designs rooted in a deep knowledge of the natural history of the study systems. By doing this, we will surely discover that herbivore populations are affected by many more complex interactions than most current theories have assumed.

Acknowledgments

We are most grateful to Regino Zamora, Jose Antonio Hódar, Elena Baraza, and Francisco Sanchez Piñero for their encouragement and constant discussion on this topic. We also deeply thank Takayuki Ohgushi and Tim Craig for their invitation to participate in the symposium "Impact of trait-mediated indirect effects of plants on the structure of herbivorous insect communities" held at the 22nd International Conference of Entomology (Brisbane, Australia, 2004). An early version of this manuscript has been thoroughly reviewed by T. Ohgushi, Joe Bailey, and one anonymous reviewer. David Nesbitt improved the English. Our studies in ungulate–insect interactions have been partially supported by the Andalusian government grant (PAI) rnm-220, the University of Granada grant 2002-30P-3176 and the Ministerio de Ciencia y Tecnología grant BOS2003-01095.

References

Abensperg-Traun, M., G. T. Smith, G. W. Arnold, and D. E. Steven. 1996. The effects of habitat fragmentation and livestock grazing on animal communities in remnants of gimlet *Eucalyptus salubris* woodland in the Western Australian wheat belt. I. Arthropods. *Journal of Applied Ecology* **33**:1281–1301.

Abrams, P. A. 1995. Implications of dynamically variable traits for identifying, classifying and measuring direct and indirect effects in ecological communities. *American Naturalist* **146**:112–134.

Abrams, P. A., B. A. Menge, G. G. Mittelbach, D. A. Spiller, and P. Yodzis. 1996. The role of indirect effects in food webs, pp. 371–395 in G. Polis and K. Winemiller (eds.) *Food Webs: Dynamics and Structure*. New York: Chapman and Hall.

Agrawal, A. A. 1999. Induced plant defense: evolution of induction and adaptive phenotypic plasticity, pp. 251–268 in A. A. Agrawal, S. Tuzun, and E. Bent (eds.) *Inducible Plant Defenses against Pathogens and Herbivores*. St. Paul, MN: American Phytopathological Society Press.

Agrawal, A. A. 2000. Benefits and costs of induced plant defense for *Lepidium virginicum* (Brassicaceae). *Ecology* **81**:1804–1813.

Agrawal, A. A., and M. T. Rutter. 1998. Dynamic anti-herbivore defense in ant-plants: the role of induced responses. *Oikos* **83**:227–236.

Agrawal, A. A., and P. A. van Zandt. 2003. Ecological play in the coevolutionary theatre: genetic and environmental determinants of attack by a specialist weevil on milkweed. *Journal of Ecology* **91**:1049–1059.

Allcock, K. G., and D. S. Hik. 2004. Survival, growth, and escape from herbivory are determined by habitat and herbivore species for three Australian woodland plants. *Oecologia* **138**:231–241.

Baines, D., R. B. Sage, and M. M. Baines. 1994. The implications of red deer grazing to ground vegetation and invertebrate communities of Scottish native pinewoods. *Journal of Applied Ecology* **31**:776–783.

Bestelmeyer, R., and J. A. Wiens. 1996. The effect of land use on the structure of ground-foraging ant communities in the Argentine Chaco. *Ecological Applications* **6**:1225–1240.

Bi, J. L., J. B. Murphy, and G. W. Felton. 1997. Antinutritive and oxidative components as mechanisms of induced resistance in cotton to *Helicoverpa zea*. *Journal of Chemical Ecology* **23**:97–117.

Bolker, B., M. Holyoak, V. Krivan, L. Rowe, and O. Schmitz. 2003. Connecting theoretical and empirical studies of trait-mediated interactions. *Ecology* **84**:1101–1114.

Brown, J. H., D. W. Davidson, J. C. Munger, and R. S. Inouye. 1986. Experimental community ecology: the desert granivorous system, pp. 41–61 in J. Diamond and T. J. Case (eds.) *Community Ecology*. New York: Harper and Row.

Carson, W. P., and R. B. Root. 2000. Herbivory and plant species coexistence: community regulation by an outbreaking phytophagous insect. *Ecological Monographs* **70**:73–99.

Christensen, K. M., and T. G. Whitham. 1993. Impact of herbivores on competition between birds and mammals for pinyon pine seeds. *Ecology* **74**:2270–2278.

Cipollini, D. 2004. Stretching the limits of plasticity: can a plant defend against both competitors and herbivores? *Ecology* **85**:28–37.

Crawley, M. J. 1983. *Herbivory*. Oxford, UK: Blackwell Scientific Publications.

Crawley, M. J. 1997. Plant–herbivore dynamics, pp. 401–474 in M. J. Crawley (ed.) *Plant Ecology*, 2nd edn. Oxford, UK: Blackwell Scientific Publications.

Crawley, M. J. 2000. Seed predators and plant population dynamics, pp. 167–182 in M. Fenner (ed.) *Seeds: The Ecology of Regeneration in Plant Communities*, 2nd edn. Wallingford, UK: CAB International.

Cuartas, P., and R. García-González. 1992. *Quercus ilex* browses utilization by Caprini in Sierra de Cazorla and Segura (Spain). *Vegetatio* **99/100**:317–330

Danell, K., and R. Bergström. 2002. Mammalian herbivory in terrestrial environments, pp. 107–131 in C. M. Herrera and O. Pellmyr (eds.) *Plant–Animal Interactions: An Evolutionary Approach*. Oxford, UK: Blackwell Science.

Danell, K., and K. Hus-Danell. 1985. Feeding by insects and hares on birches earlier affected by moose browsing. *Oikos* **44**:75–81.

Davidson, D. W., R. S. Inouye, and J. H. Brown. 1984. Granivory in a desert ecosystem: experimental evidence for indirect facilitation of ants by rodents. *Ecology* **65**:1780–1786.

Davidson, D. W., D. A. Samson, and R. S. Inouye. 1985. Granivory in the Chihuahuan desert: interactions within and between trophic levels. *Ecology* **66**:486–502.

Dennis, P., M. R. Young, C. L. Howard, and I. J. Gordon. 1997. The response of epigeal beetles (Col.: Carabidae, Staphylinidae) to varied grazing regimes on upland *Narduus stricta* grasslands. *Journal of Applied Ecology* **34**:433–443.

Denno, R. F., M. S. McClure, and J. R. Ott. 1995. Interspecific interactions in phytophagous insects: competition reexamined and resurrected. *Annual Review of Entomology* **40**:297–331.

Denno, R. F., M. A. Peterson, C. Gratton, *et al.* 2000. Feeding-induced changes in plant quality mediate interspecific competition between sap-feeding herbivores. *Ecology* **81**:1814–1827.

DeWalt, S. J., J. S. Denslow, and K. Ickes. 2004. Natural-enemy release facilitates habitat expansion of the invasive tropical shrub *Clidemia hirta*. *Ecology* **85**:471–483.

Dill, L. M., M. R. Heithaus, and C. J. Walters. 2003. Behaviorally mediated indirect interactions in marine communities and their conservation implications. *Ecology* **84**:1151–1157.

Dolch, R., and T. Tscharntke. 2000. Defoliation of alders (*Alnus glutinosa*) affects herbivory by leaf beetles on undamaged neighbours. *Oecologia* **125**:504–511.

Elligsen, H., B. Beinlich, and H. Plachter. 1997. Effects of large-scale cattle grazing on populations of *Coenonympha glycerion* and *Lasiommata megera* (Lepidoptera: Satyridae). *Journal of Insect Conservation* **1**:13–23.

Faeth, S. H. 1992. Interspecific and intraspecific interactions via plant responses to folivory: an experimental field-test. *Ecology* **73**:1802–1813.

Faeth, S. H., and D. Wilson. 1997. Induced responses in trees: mediator of interactions among macro- and micro-herbivores? pp. 201–215 in A. C. Gange and V. K. Brown (eds.) *Multitrophic Interactions in Terrestrial Systems*. Oxford, UK: Blackwell Science.

Ferrenberg, S. M., and R. F. Denno. 2003. Competition as a factor underlying the abundance of an uncommon phytophagous insect, the salt-marsh planthopper *Delphacodes penedetecta*. *Ecological Entomology* **28**:58–66.

Fisher, A. E. I., S. E. Hartley, and M. Young. 2000. Direct and indirect competitive effects of foliage feeding guilds on the performance of the birch leaf-miner *Eriocrania*. *Journal of Animal Ecology* **69**:165–176.

Goheen, J. R., F. Keesing, B. F. Allan, D. L. Ogada, and R. S. Ostfeld. 2004. Net effects of large mammals on *Acacia* seedling survival in an African savanna. *Ecology* **85**:1555–1561.

Gómez, J. M. 2005. Ungulate effect on the performance, abundance and spatial structure of two montane herbs: a 7-yr experimental study. *Ecological Monographs* **75**:231–258.

Gómez, J. M., and A. González-Megías. 2002. Asymmetrical interactions between ungulates and phytophagous insects: being different matters. *Ecology* **83**:201–211.

Gómez, J. M., and R. Zamora. 2000a. Differential impact of vertebrate and invertebrate herbivores on *Hormathophylla spinosa* reproductive output. *Ecoscience* **7**:299–306.

Gómez, J. M., and R. Zamora. 2000b. Spatial variation in the selective scenarios of *Hormathophylla spinosa* (Cruciferae). *American Naturalist* **155**:657–668.

Gómez, J. M., and R. Zamora. 2002. Thorns as induced mechanical defense in a long-lived shrub (*Hormathophylla spinosa*, Cruciferae). *Ecology* **83**:885–890.

González-Megías, A., and J. M. Gómez. 2003. Consequences of removing a keystone herbivore for the abundance and diversity of arthropods associated with a cruciferous shrub. *Ecological Entomology* **28**:299–308.

González-Megías, A., J. M. Gómez, and F. Sanchez Piñero. 2004. Effects of ungulates on epigeal arthropods in Sierra Nevada National Park (southeast Spain). *Biodiversity and Conservation* **13**:733–752.

Hambäck, P. A., and A. P. Beckerman. 2003. Herbivory and plant resource competition: a review of two interacting interactions. *Oikos* **101**:26–37.

Hawkes, C. V., and J. J. Sullivan. 2001. The impact of herbivory on plants in different resource conditions: a meta-analysis. *Ecology* **82**:2045–2058.

Hjältén, J., and P. W. Price. 1997. Can plants gain protection from herbivory by association with unpalatable neighbours? A field experiment in a willow–sawfly system. *Oikos* **78**:317–322.

Herms, D. A., and W. J. Mattson. 1992. The dilemma of plants: to grow or defend. *Quarterly Review of Biology* **67**:283–335.

Hochberg, M. E., and J. H. Lawton. 1990. Competition between kingdoms. *Trends in Ecology and Evolution* **5**:367–371.

Hulme, P. E. 1994. Seedling herbivory in grassland: relative impact of vertebrate and invertebrate herbivores. *Journal of Ecology* **82**:873–880.

Hulme, P. E. 1996. Herbivores and the performance of grassland plants: a comparison of arthropod, mollusc, and rodent herbivory. *Journal of Ecology* **84**:43–51.

Hunter, M. D. 1992. Interactions within herbivore communities mediated by the host plant: the keystone herbivore concept, pp. 287–325 in M. D. Hunter, T. Ohgushi, and P. W. Price (eds.) *Effect of Resource Distribution on Plant–Animal Interactions*. San Diego, CA: Academic Press.

Huntzinger, M., R. Karban, T. P. Young, and T. M. Palmer. 2004. Relaxation of induced indirect defenses of acacias following exclusion of mammalian herbivores. *Ecology* **85**:609–614.

Inouye, B., and J. R. Stinchcombe. 2001. Relationships between ecological interaction modifications and diffuse coevolution: similarities, differences, and causal links. *Oikos* **95**:353–360.

Juenger, T., and J. Bergelson. 2000. Factors limiting rosette recruitment in scarlet gilia, *Ipomopsis aggregata*: seed and disturbance limitation. *Oecologia* **123**:358–363.

Karban, R., and I. T. Baldwin. 1997. *Induced Responses to Herbivory*. Chicago, IL: University of Chicago Press.

Karban, R., and J. H. Myers. 1989. Induced plant responses to herbivory. *Annual Review of Ecology and Systematics* **20**:331–348.

Kleijn, D., and T. Steiner. 2002. Contrasting effects of grazing and hay cutting on the spatial and genetic population structure of *Veratrum album*, an unpalatable, long-lived, clonal plant species. *Journal of Ecology* **90**:360–370.

Kruess, A., and T. Tscharntke. 2002a. Contrasting responses of plant and insect diversity to variation in grazing intensity. *Biological Conservation* **106**:293–302.

Kruess, A., and T. Tscharntke. 2002b. Grazing intensity and the diversity of grasshoppers, butterflies, and trap-nesting bees and wasps. *Conservation Biology* **16**:1570–1580.

Louda, S. M., and M. A. Potvin. 1995. Effect of inflorescence-feeding insects on the demography and lifetime fitness of a native plant. *Ecology* **76**:229–245.

Louda, S. M., and J. E. Rodman. 1996. Insect herbivory as a major factor in the shade distribution of a native crucifer (*Cardamine cordifolia* A. Gray, bittercress). *Journal of Ecology* **84**:229–237.

Lucas, E., D. Coderre, and J. Brodeur. 1998. Intraguild predation among aphid predators: characterization and influence of extraguild prey density. *Ecology* **79**:1084–1092.

Luttbeg, B., L. Rowe, and M. Mangel. 2003. Prey state and experimental design affect relative size of trait- and density-mediated indirect effects. *Ecology* **84**:1140–1150.

Maron, J. L., and S. N. Gardner. 2000. Consumer pressure, seed versus safe-site limitation, and plant population dynamics. *Oecologia* **124**:260–269.

Maron, J. L., and E. L. Simms. 2001. Rodent-limited establishment of bush lupine: field experiments on the cumulative effect of granivory. *Journal of Ecology* **89**:578–588.

Maron, J. L., J. K. Combs, and S. M. Louda. 2002. Convergent demographic effects of insect attack on related thistles in coastal vs. continental dunes. *Ecology* **83**:3382–3392.

Marquis, R. J. 1992. The selective impact of herbivores, pp. 301–325 in R. S. Fritz and E. L. Simms (eds.) *Plant Resistance to Herbivores and Pathogens: Ecology, Evolution and Genetics*. Chicago, IL: University of Chicago Press.

Martinsen, G. D., E. M. Driebe, and T. G. Whitham. 1998. Indirect interactions mediated by changing plant chemistry: beaver browsing benefits beetles. *Ecology* **79**:192–200.

Master, G. J., and V. K. Brown. 1997. Host-plant mediated interactions between spatially separated herbivores: effects on community structure, pp. 217–328 in A. C. Ganges and V. K. Brown (eds.) *Multitrophic Interactions in Terrestrial Systems*. Oxford, UK: Blackwell Science.

Palmisano, S., and L. R. Fox. 1997. Effects of mammal and insect herbivory on population dynamics of a native Californian thistle, *Cirsium occidentale*. *Oecologia* **111**:413–421.

Passos, L., and P. S. Oliveira. 2002. Ants affect the distribution and performance of seedlings of *Clusia criuva*, a primarily bird-dispersed rain forest tree. *Journal of Ecology* **90**:517–528.

Pimm, S. L. 2002. *Food Webs*, 2nd edn. Chicago, IL: University of Chicago Press.

Prittinen, K., J. Pusenius, K. Koivunoro, M. Rousi, and H. Roininen. 2003. Mortality in seedling populations of silver birch: genotype variation and herbivore effects. *Functional Ecology* **17**:658–663.

Quesada, M., K. Bollman, and A. G. Stephenson. 1995. Leaf damage decreases pollen production and hinders pollen performance in *Cucurbita texana*. *Ecology* **76**:437–443.

Raimondi, P. T., S. E. Forde, L. F. Delph, and C. M. Lively. 2000. Processes structuring communities: evidence for trait-mediated indirect effects through induced polymorphisms. *Oikos* **91**:353–361.

Rambo, J. L., and S. H. Faeth. 1999. Effect of vertebrate grazing on plant and insect community structure. *Conservation Biology* **13**:1047–1054.

Retamosa, E. C. 1997. Efecto de los herbívoros sobre *Iberis contracta* en el Parque Nacional de Doñana. Ph.D. dissertation, University of Córdoba, Spain.

Reznik, S. Y. 1991. The effects of feeding damage in ragweed *Ambrosia artemisiifolia* (Asteraceae) on populations of *Zygogramma suturalis* (Coleoptera, Chrysomelidae). *Oecologia* **88**:204–210.

Roche, B. M., and R. S. Fritz. 1997. Genetics of resistance of *Salix sericea* to a diverse community of herbivores. *Evolution* **51**:1490–1498.

Schmitz, O., V. Krivan, and O. Ovadia. 2004. Trophic cascades: the primacy of trait-mediated indirect interactions. *Ecology Letters* **7**:153–163.

Schoonhoven, L. M., T. Jermy, and J. J. A. van Loon. 1998. *Insect–Plant Biology: From Physiology to Evolution*. London: Chapman and Hall.

Sessions, L. A., and D. Kelly. 2001. Heterogeneity in vertebrate and invertebrate herbivory and its consequences for New Zealand mistletoes. *Austral Ecology* **26**:571–581.

Seymour, C. L., and W. R. J. Dean. 1999. Effects of heavy grazing on invertebrate assemblages in the Succulunt Karoo, South Africa. *Journal of Arid Environment* **43**:267–286.

Speight, M. R., M. D. Hunter, and A. D. Watt. 1999. *Ecology of Insects: Concepts and Applications*. Oxford, UK: Blackwell Scientific Publications.

Stewart, A. J. A. 1996. Interspecific competition reinstated as an important force structuring insect herbivore communities. *Trends in Ecology and Evolution* **11**:233–234.

Strand, M., and R. W. Merritt. 1999. Impact of livestock grazing activities on stream insect communities and the riverine environment. *American Entomologist* **45**:13–27.

Strauss, S. Y. 1997. Floral characters link herbivores, pollinators, and plant fitness. *Ecology* **78**:1640–1645.

Strauss, S. Y., and A. A. Agrawal. 1999. The ecology and evolution of plant tolerance to herbivory. *Trends in Ecology and Evolution* **14**:179–185.

Strauss, S. Y., J. K. Conner, and S. Rush. 1996. Foliar herbivory affects floral characters and plant attractiveness to pollinators: implications for male and female plant fitness. *American Naturalist* **147**:1098–1107.

Strong, D. R., J. H. Lawton, and T. R. E. Southwood. 1984. *Insects on Plants: Community Patterns and Mechanisms.* Oxford, UK: Blackwell Scientific Publications.

Suominen, O., K. Danell, and R. Bergström. 1999a. Moose, trees, and ground-living invertebrates: indirect interactions in Swedish pine forest. *Oikos* **84**:215–226.

Suominen, O., K. Danell, and J. P. Bryant. 1999b. Indirect effects of mammalian browsers on vegetation and ground-dwelling insects in an Alaskan floodplain. *Ecoscience* **6**:505–510.

Takabayashi, J., and M. Dicke. 1996. Plant–carnivore mutualisms through herbivore-induced carnivore attractants. *Trends in Plant Science* **1**:109–113.

Trussell, G. C., P. J. Ewanchuk, and M. D. Bertness. 2003. Trait-mediated effects in rocky intertidal food chains: predator risk cues alter prey feeding rates. *Ecology* **84**:629–640.

Tscharntke, T. 1997. Vertebrate effects on plant–invertebrate food webs, pp. 277–297 in A. C. Ganges and V. K. Brown (eds.) *Multitrophic Interactions in Terrestrial Systems.* Oxford, UK: Blackwell Science.

Tscharntke, T., and H. J. Greiler. 1995. Insect communities, grasses, and grasslands. *Annual Reviews of Entomology* **40**:535–558.

WallisDeVries, M. F., E. A. Laca, and M. W. Demment. 1999. The importance of scale patchiness for selectivity in grazing herbivores. *Oecologia* **121**:355–363.

Werner, E. E., and S. D. Peacor. 2003. A review of trait-mediated indirect interactions in ecological communities. *Ecology* **84**:1083–1100.

Wootton, J. T. 1994. The nature and consequences of indirect effects in ecological communities. *Annual Review of Ecology and Systematics* **25**:443–466.

Young, T. P., and B. D. Okello. 1998. Relaxation of an induced defense after exclusion of herbivores: spines on *Acacia drepanolobium. Oecologia* **115**:508–513.

Young, T. P., M. L. Stanton, and C. E. Christian. 2003. Effects of natural and simulated herbivory on spine lengths of *Acacia drepanolobium* in Kenya. *Oikos* **101**:171–179.

Zamora, R., and J. M. Gómez. 1993. Vertebrate herbivores as predators of insect herbivores: an asymmetrical interaction mediated by size differences. *Oikos* **66**:223–228.

Zamora, R., J. A. Hódar, and J. M. Gómez. 1999. Plant–herbivore interaction: beyond a binary vision, pp. 677–718 in F. Valladares and F. I. Pugnaire (eds.) *Handbook of Plant Functional Ecology.* New York: Marcel Dekker.

Zamora, R., J. M. Gómez, J. A. Hódar, J. Castro, and D. García. 2001. Effect of browsing by ungulates on sapling growth of Scots pine in a Mediterranean environment: consequences for forest regeneration. *Forest Ecology and Management* **144**:33–42.

6

Insect–mycorrhizal interactions: patterns, processes, and consequences

ALAN C. GANGE

Introduction

A wide variety of fungi form an intimate association with the roots of plants, and the word "mycorrhiza" is used to describe the overall structure formed by the union of these partners. There are seven different types of mycorrhiza, but the two that are most abundant in nature and of most importance ecologically, are the arbuscular and ectomycorrhiza. An excellent review of all aspects of mycorrhizal biology can be found in Smith and Read (1997). Arbuscular mycorrhizae are formed by about 150 different species of fungi within the Glomeromycota and are mostly associated with herbaceous plants. The fungus enters the roots of plants where hyphae grow intercellularly. In addition, unique highly branched structures are formed within the cells of plant roots, called arbuscules. These are thought to be sites of nutrient exchange. Arbuscular mycorrhizal (AM) fungi occur in all ecosystems of the world and associate with the roots of about 70% of all vascular plants (Hodge 2000). Ectomycorrhizal (ECM) fungi generally associate with woody plants. They are formed by about 5000 species of fungi, mainly from the Basidiomycotina, with some representatives from the Ascomycotina also. The fungus forms a sheath over the root tips and there is some intercellular, but no intracellular, growth of the hyphae. Ectomycorrhizae also have worldwide distributions, and although they associate with only about 3% of seed plants, their global importance is huge, as they associate with important timber and natural forest trees.

Ecological Communities: Plant Mediation in Indirect Interaction Webs, ed. Takayuki Ohgushi, Timothy P. Craig, and Peter W. Price. Published by Cambridge University Press. © Cambridge University Press 2007.

Essentially, arbuscular and ectomycorrhizae function in the same way. The association is regarded as mutualistic, because there is a bidirectional transfer of nutrients between plant host and its fungal partners. Carbon compounds are passed from plant to fungus and in return there is a transfer of mineral nutrients, principally nitrate and phosphate, although many other micronutrients are probably transferred too (Smith and Read 1997).

Insects are the most numerous animals on earth and about half of the 900 000 estimated species feed on plants (Schoonhoven et al. 1998). It is safe to say that every mycorrhizal plant in the world suffers attack from at least one species of insect and, in most cases, many species. Insects require a variety of nutrients from plants for conversion into their own biomass, and might be regarded as competitors with a mycorrhiza for carbon. One might therefore hypothesize that insects may affect the mycorrhizal symbiosis detrimentally, through a process of carbon limitation within the host plant. My first aim is to examine this proposal and provide examples of the effects of insect herbivory on the mycorrhiza. Meanwhile, there are many potential ways in which insects may be affected by mycorrhizal presence, because the association has a wide range of morphological and physiological effects on plants. All of these effects have the potential to alter the growth and reproduction of insects and some of them have been the subject of experimental work. The majority of experiments involving these organisms have been laboratory-based, involving just one or two insect and mycorrhizal species, and have been reviewed by Gehring and Whitham (1994, 2002), Gange and Bower (1997), and Gange and Brown (2002a). Therefore, my second aim is to review briefly these studies and attempt to provide some explanations for the patterns that are beginning to emerge.

Indirect interactions between mycorrhizae and insects are a perfect example of plant-trait-mediated effects in communities. Usually, the fungi and the insects are spatially and temporally separated, but they are linked through their dependence on a common host plant. Their interactions may be of major importance in the determination of fungal, plant, and insect community structure. My third aim is thus to discuss the consequences of the interactions in community ecology terms. Much of this will be speculative, because there are many gaps in our knowledge. I hope that this review will stimulate research into this fascinating, but complex, indirect interaction web between insects, plants, and fungi.

Effects of insect herbivory on mycorrhizae

The first review of this subject (Gehring and Whitham 1994) listed the effects of herbivory on 37 different plant species. However, the vast majority of these were subject to either manual leaf removal or grazing by vertebrate herbivores. Subsequent to this paper, most studies examining effects of foliar herbivory by insects on mycorrhizae have concentrated on ECM fungi and are listed by Gehring and Whitham (2002). There are also a number of recent experiments that have studied manual defoliation or vertebrate grazing effects on mycorrhizae and the results from these provide an interesting contrast to those of the insect experiments.

In general, insect herbivory has been found to reduce ECM colonization of tree roots. In some cases (e.g., Gehring and Whitham 1991), ECM levels in trees not subject to herbivory may be 40–50% higher. Furthermore, detrimental effects on the mycorrhiza are persistent, with Del Vecchio et al. (1993), Gehring and Whitham (1995), and Gehring et al. (1997) finding that several years of insect removal were necessary before mycorrhizal levels recovered.

Due to the nature of the plant growth patterns, one would expect any effects of herbivory on arbuscular mycorrhizae in herbaceous plants to be seen on much shorter timescales. Gange et al. (2002a) found that foliar herbivory reduced AM colonization of *Plantago lanceolata* after about 15 weeks of growth. The reduction was also large, being about 40%. The most rapid case of insect herbivory affecting a mycorrhizal association was reported by Wamberg et al. (2003). These authors studied the effects of leaf herbivory by the beetle *Sitona lineatus* on annual pea (*Pisum sativum*) plants. Their results are important because they differ depending on the age of the plant and give a clue to the mechanism involved in herbivore–mycorrhizal interactions. Pea plants that were 25 days old and had been subjected to 10 days of herbivory (from day 15 to day 25 in the life of the plant) had AM colonization levels 45% higher than those that were not eaten. However, 39-day-old plants that had had 16 days of herbivory had AM levels that were 22% lower than unattacked plants.

The results of Wamberg et al. (2003) are one of only two examples in which the mycorrhizal colonization of a plant has been shown to increase in response to insect herbivory. The only other is the work of Currie (2004) in which the effects of root herbivory by an insect (*Tipula paludosa*) on mycorrhizal colonization of *Agrostis capillaris* was examined. The Currie study is unique because it took a different approach to that of all the aboveground experiments, where the insect was allowed to feed on a plant that had already been colonized by mycorrhizae. By allowing *T. paludosa* larvae to feed on uncolonized roots, Currie was able to determine the effect of herbivory on subsequent colonization by the mycorrhiza. The results were quite striking: plants that had suffered recent herbivory, or which continued to be attacked, had considerably higher levels of AM colonization than those that had not been eaten (Fig. 6.1). Currie (2004) suggests that positive effects of herbivores on AM colonization occurred because feeding

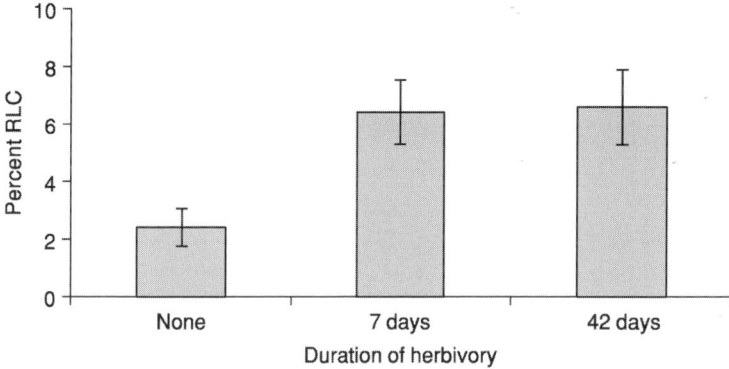

Figure 6.1. The effect of root herbivory by *Tipula paludosa* larvae on subsequent mycorrhizal colonization of the grass *Agrostis capillaris*. Larvae were added to plants and after 7 days were removed from one treatment. At this stage, all plants, including no herbivory controls, were inoculated with AM fungi and the plants grown for a further 42 days. Any duration of herbivore attack increased the percentage root length colonized (RLC). Data from Currie (2004).

caused an increase in root exudation. Indeed, it is possible that temporarily enhanced root exudation was also the explanation for the positive effects seen by Wamberg *et al.* (2003). An increase in root exudation in response to foliage removal has been reported (Holland *et al.* 1996), and root damage can increase the quantity and alter the composition of root exudates (Rovira 1969). Mycorrhizal spore germination and hyphal growth are known to be increased by root exudates, so it is possible that under moderate levels of foliar or root herbivory, root exudation is increased, providing colonization signals for a mycorrhiza and hence leading to enhanced colonization.

Curiously, the experiments showing a positive effect of insect herbivory on mycorrhizae have provided similar results to some studies of clipping and vertebrate grazing (Gehring and Whitham 1994, Eom *et al.* 2001). Wamberg *et al.* (2003) argue that herbivore-induced transfer of carbon below ground early in the life of the plant caused the increase in AM colonization and this suggestion urgently needs testing by experiment. The hypothesis proposed was that early in the nutrient acquisition phase of plants, herbivory causes carbon to be translocated to the mycorrhiza and other rhizosphere microorganisms, to aid in the acquisition of further nutrients to support growth. Certainly, root removal by insects causes a transfer of carbon to the belowground tissues, as structural repair is often rapid (Brown and Gange 1990). If this carbon was intercepted by a mycorrhiza then it could enable greater colonization to take place.

It is well known that mycorrhizal colonization can increase the photosynthetic capacity of a plant (Smith and Read 1997). However, the extra carbon fixed is allocated to the mycorrhiza, not the plant itself (Staddon et al. 1999). Furthermore, some authors have argued that plants may overcompensate in their response to insect attack by an increase in photosynthetic rate and the production of extra biomass. The idea of overcompensation is one that has received much attention and debate in the literature (Crawley 1987); however, a feature of these arguments is that no one has considered whether the plant in question was mycorrhizal or not. If herbivore pressure is low and the plant is mycorrhizal, then the elevation of photosynthesis may be sufficient to sustain the overcompensatory response for a time. However, as the level of herbivory increases, a limit must soon be reached, beyond which increase in photosynthesis is no longer possible, meaning that the carbon supply to the mycorrhiza is impaired and, eventually, colonization reduced. The simple way in which to test this idea is to examine the responses of mycorrhizae to different degrees of defoliation. Surprisingly few studies have been performed on this topic, but when they have, be it with insects and ECM (Kolb et al. 1999) or insects and AM fungi (Gange et al. 2002a) they have shown that high levels of defoliation reduce colonization, but low levels of attack have less of an effect. One could argue that a similar response was found by Wamberg et al. (2003), as at the second harvest (i.e., higher amount of leaf damage), AM colonization was reduced by herbivory.

Gehring and Whitham (2002) have also suggested that carbon limitation is the main reason why herbivory seems to reduce colonization levels. They link herbivory to the prevailing nutrient conditions, since Gehring and Whitham (1995) found that clipping of pinyon pine shoots reduced ECM colonization in a low nutrient and water site, but not in a site where water and nutrients were plentiful. Gehring and Whitham (2002) provide a curvilinear model that describes how mycorrhizal colonization declines progressively as herbivore pressure increases. They also argue that the decline is steeper in stressful environments because the herbivore-induced reduction in belowground carbon allocation occurs more quickly in stressful areas. I believe there are two problems with this model: a line depicting a continuous decline does not explain why there are positive responses of mycorrhizae to herbivory, nor does it explain those studies in which plants have been subject to foliage removal and no response is found in terms of percentage mycorrhizal colonization. Examples of the latter situation are provided by the clipping studies of Saikkonen et al. (1999), Cullings et al. (2001), and Kuikka et al. (2003).

A feature of the three latter studies is that in every case, defoliation caused a significant change in the mycorrhizal community inhabiting the roots, this being measured by morphotypes (Saikkonen et al. 1999), polymerase chain reaction (PCR) analyses (Cullings et al. 2001), or sporocarp production (Kuikka et al. 2003). Furthermore, Gehring and Whitham (2002) also provide data from a molecular analysis of the fungal communities inhabiting the roots of the scale-resistant and susceptible pinyon pines (above) and show that these are quite different in their composition. A consistent theme emerging from these studies is that herbivory can alter the species composition of the mycorrhizal fungi in the roots of a plant. However, all of the work has so far been conducted with ECM fungi and the experiment needs to be repeated with AM fungi. We know that plants are colonized by some or many AM species at any one time and that these can vary during a season (Helgason et al. 1999). However, we have no idea if insect herbivores can affect the composition of these mycorrhizal communities.

In all of the three ECM studies, the authors suggest that different fungal species have different carbon demands and that competition between the fungal species causes changes in the community when defoliation occurs. Saikkonen et al. (1999) provide a useful model of fungal growth rates in relation to available carbon and explain how loss of photosynthate to herbivores should give a competitive advantage to those species with the lowest carbon demands (grazing-tolerant mycorrhizae: Eom et al. 2001). Thus, total ECM levels do not seem to change in response to herbivory, as one fungus may "take over" root space vacated by another. If this is true, then it would suggest the overall percentage colonization level is not adequate as a response variable when examining the effects of herbivores on mycorrhizae. Instead, we need to think in terms of species composition, hence I propose a redefinition of the Gehring and Whitham (2002) mycorrhizal response model, in terms of fungal species composition (Fig. 6.2).

In this model, at very low or no herbivore grazing, mycorrhizal species composition is limited by competition for carbon, and a relatively low number of species coexist. Root exudation is minimal. As herbivory increases, a stimulation of photosynthesis can lead to an increase in translocation of carbon below ground and root exudation. The latter may act as colonization signals while the former

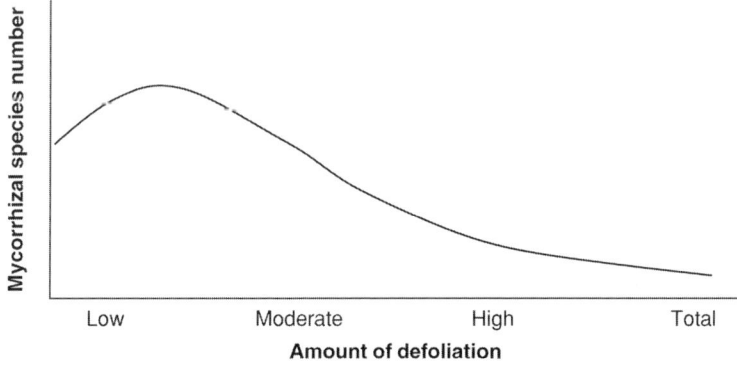

Figure 6.2. Predicted relationship between the intensity of insect herbivory and the diversity of mycorrhizal fungi colonizing a root system.

provides resources for colonizing species. As resources (carbon) become temporarily more abundant, so a greater diversity of species may coexist. If a single species of fungus is present, such as in the Wamberg et al. (2003) study, then greater fungal growth is possible, due to the enhanced carbon supply. However, as herbivory increases still further, the transfer of carbon below ground is impaired and those fungal species with high carbon demands drop out, leading to a reduction in diversity. Eventually, at extremely high levels of herbivory, very few mycorrhizal species can persist, due to the extreme limitation on carbon. At present, we do not know the definitions of "low," "moderate," and "high" for any particular mycorrhiza–plant insect system and it is likely that a family of curves really exist, depending on the host plant and its mycorrhizal affinity. For example, for a strongly mycotrophic plant, the increase in diversity is probably small and transient, while for weakly mycotrophic plants, the starting point would be lower, but the complete response is probably closer to a horizontal line (no change in diversity). I acknowledge that no empirical study has provided support for this model, but this is largely because authors have tended to compare the situations of "high" vs. "low" on the graph. What we now need to do is to consider mycorrhizal responses along a continuum of herbivory. This is much easier said than done, as performing analyses of mycorrhizal communities is not straightforward. However, the development of molecular techniques is opening up exciting opportunities in which the species composition of mycorrhizal fungi in roots, and as extraradical hyphae in the soil, may be examined.

Effects of mycorrhizae on insect herbivores

One of the commonest outcomes of a plant being mycorrhizal is an increase in size, relative to nonmycorrhizal conspecifics (Smith and Read 1997). Insects often use visual cues to orientate towards their host plant and there are many examples where insect host location is positively related to plant architecture (Schoonhoven et al. 1998). One might therefore expect that positive effects on plant size might result in increased host acceptance, with the converse also being true. Linked to this is the plant vigor hypothesis of Price (1991), which states that insect herbivores perform best on the most vigorous plants, although it has also been argued that herbivore performance may be negatively related to plant vigor (White 1984). Although being of fundamental importance in insect ecology, host acceptance has hardly featured in any study of insect–plant–mycorrhizal interactions, since most experiments have involved presenting captive insects with leaf material from mycorrhizal and nonmycorrhizal plants.

Gange and West (1994) reduced AM colonization of *P. lanceolata* in a field experiment and compared insect attack rates on mycorrhizal and nonmycorrhizal plants. Mycorrhizal plants were considerably larger at five successive harvests, yet at four of these, insect attack rates were highest on the nonmycorrhizal (i.e., smallest) plants. A similar lack of response was found by Gange and Nice (1997) in which mycorrhizal and nonmycorrhizal thistle (*Cirsium arvense*) plants were exposed to ovipositing females of the gall-forming fly *Urophora cardui*. Although nonmycorrhizal plants were smaller than mycorrhizal individuals, this did not have any effect on the proportion of plants that were galled or the average gall number per plant. Meanwhile, Gange *et al.* (2003) experimentally colonised plants of oxeye daisy (*Leucanthemum vulgare*) with all possible combinations of three AM species and exposed these to ovipositing adults of the leaf mining fly *Chromatomyia syngenesiae*. Many (but not all) combinations of mycorrhizal species increased plant size, but none of these increased fly attack rate, and some of them decreased it. Finally, Manninen *et al.* (1998) could find no evidence that ECM colonization of *Pinus sylvestris* affected egg-laying of the bug *Lygus rugulipennis*. Therefore, although it is an attractive hypothesis, there is no evidence that mycorrhizal colonization of plants affects insect host plant choice.

However, there is evidence that some insects perform better on more vigorous mycorrhizal plants and these are listed by Gehring and Whitham (2002). To date, there is no good explanation for these data and no study has managed to link mycorrhizal-induced changes in host plant quality, and thus insect nutrition, to changes in insect growth. Indeed, when authors have sought for nutritional changes in plants, they have either failed to find any or these have been reductions, rather than increases (Rabin and Pacovsky 1985, Gange and West 1994, Borowicz 1997, Goverde *et al.* 2000). Positive responses of aphids to mycorrhizae have been explained by the fact that AM fungi are known to increase the size of the vascular bundle in some plants (Krishna *et al.* 1981), thus increasing the chance of successful phloem location. Given that host plant resistance is often caused by a failure of the aphid to locate the phloem (Tjallingii and Esch 1993) this is a plausible hypothesis that needs to be tested.

However, enhanced vascular penetration cannot explain the results of Borowicz (1997) or Goverde *et al.* (2000) and in these studies, mycorrhizal improvement in diet quality seems the most likely explanation. Borowicz (1997) found that beetle larval performance was maximized on plants that had intermediate growth rates, and Gange *et al.* (1999) also found evidence for the fact that aphid growth rates were maximized on plants showing intermediate levels of stress. Insect responses to variation in food quality, such as total nitrogen or protein content, are known to follow a similar pattern of nonlinearity (Schoonhoven *et al.* 1998), implying some response to food quality. Borowicz (1997) makes the comment that the beetle larvae probably responded to some unmeasured plant quality parameter (phosphorus, soluble carbohydrate, and soluble protein were measured) and this illustrates the difficulty of identifying which parameter is responsible. However, plant sterol content is one example that may be worth exploring. Insects lack the ability to synthesize sterols that are required in lipid structures, hormone precursors, and regulators of developmental processes (Behmer and Nes 2003). Goverde *et al.* (2000) found that adult butterfly lipid content was increased by larval feeding on plants colonized by one AM species and AM fungi are known to synthesize sterols in vivo (Fontaine *et al.* 2001).

Cases of AM colonization reducing insect performance have been noted in a variety of herbivores (Gehring and Whitham 2002). Since that paper, Vicari *et al.* (2002) and Wamberg *et al.* (2003) have also found similar effects. In most cases, a mycorrhizal-induced reduction in food quality effect was ruled out (e.g., Rabin and Pacovsky 1985, Gange and West 1994, Gange 2001); however, Gange and Nice (1997) did suggest this was responsible for impaired growth of the gall fly *U. cardui* feeding on *C. arvense*. The larvae feed mainly on the rich nutritive tissue, which appears in the gall relatively late in its development. Mycorrhizae reduced total plant nitrogen and it was suggested that this delayed the appearance of nutritive tissue, hence reducing larval growth.

The most likely explanation is one of a mycorrhizal modification of plant defense, though few studies have established the causal link between defense alteration and herbivore performance. An exception is that of Gange and West (1994) where it was found that concentrations of two known insect feeding deterrents, the iridoid glycosides aucubin and catalpol, were elevated in mycorrhizal plants. Although AM colonization reduced many other foliage food-quality parameters, such as soluble-neutral sugars, starch, and total nitrogen, the food assimilation efficiency of insects did not differ on mycorrhizal and nonmycorrhizal plants. Therefore, it does seem that the increases in plant defense were responsible for the reduced growth of insect. In this study (and in the others, above), mycorrhizae increased the ratio of carbon to nitrogen in the plant mainly through a reduction in nitrogen and it was proposed that the excess carbon was made available to defense in the synthesis of the two glycosides. The carbon–nutrient balance hypothesis of plant defense allocation has been criticized recently (Hamilton *et al.* 2001), because it fails to predict defense allocation in many cases. Hamilton *et al.* (2001) suggest that adaptation or evolution-based cost-defense theories such as the optimal defense hypothesis (Rhoades 1979) should be pursued. However, this argument rests on the fact that insect attack in time or on particular plant parts is predictable. No mention of mycorrhizae is made in the Hamilton *et al.* (2001) paper, and the presence of the association will serve to confound such theories. This is because insects seem to ignore whether a plant is mycorrhizal or not when locating their hosts (described above), but when feeding upon it, their growth may be reduced, increased, or unaffected. Indeed,

Gange and West (1994) have shown that two insects feeding on a mycorrhizal host at the same time can have opposite reactions to the mycorrhiza. Therefore, the amount of damage that a plant receives may become less predictable, if a mycorrhiza is present. I suggest that the future development of plant defense hypotheses includes a consideration of the mycorrhizal status of the plant. Further evidence for defense modification comes from the fact that when specialist and generalist insects are reared on a mycorrhizal plant, growth of the former is increased while that of the latter decreased (Gange et al. 2002b). Indeed, these authors argue that mycorrhizae alter host plant chemistry to the benefit of specialist and detriment of generalist insects, leading in evolutionary time to the fact that most insects are specialist.

Studies with root-feeding insects are consistent in that the association (AM or ECM) reduces the incidence of attack (Gange et al. 1994, Halldorsson et al. 2000, Gange 2001). The explanation given is again in terms of plant chemistry. This is because there is little evidence that root herbivores are limited by root nutritional quality (Brown and Gange 1990), because roots are such a poor food source. However, mycorrhizal colonization can increase the amounts of many chemicals within roots, all of which have known activity against pathogenic fungi and nematodes (Strack et al. 2003) and certain activity against insects. Of particular importance may be the fact that AM colonization can alter gene expression leading to jasmonate biosynthesis in roots (Hause et al. 2002) which is known to be important in the defense response of plants to phytophagous insects (Thaler et al. 2002). It is also known that a number of aboveground defensive chemicals are synthesized or presynthesized in the roots and transported to foliar parts. There is a need to incorporate plant defense below ground into future multitrophic studies, and van der Putten (2003) expands upon this concept in more detail. It is a distinct possibility that mycorrhizae may alter foliar chemistry to the detriment of insects, through altered gene expression and the induction of chemicals that have a variety of activities. Such molecular "cross-talk" is well known for plant pathogen–insect interactions (Paul et al. 2000) and would be a most fruitful area of research for mycorrhizal fungi too.

However, it is critical that future studies use combinations of host plant, insects, and fungi that are known to occur in the field. In many studies investigating the interactions between mycorrhizae and Collembola (Gange 2000, see below), the results may be doubted because convenient combinations of fungi and insects were used, with no evidence that these naturally coexist. Similar problems exist with some of the laboratory experiments involving insects and AM fungi and one must question the relevance of these artificial combinations to community ecology. The root-feeding insect–mycorrhizal experiments provide good evidence that different mycorrhizal species have different effects on insect growth (Gange 1996, 2001). In both of these studies, AM colonization by either of two fungal species reduced growth of the black vine weevil (*Otiorhynchus sulcatus*) while colonization by two species had no effect. Furthermore, different combinations of AM species had varying effects on the proportion of leaves attacked by a leaf-mining fly (*Chromatomyia syngenesiae*) (Gange et al. 2003). To date there has been no experiment involving naturally occurring combinations of ECM fungi, even though we know that root systems are colonized by a variety of species at once (e.g., Cullings et al. 2001).

It is clear that certain combinations of AM species may differ in their effects on insect herbivores. However, in a few special cases the situation may be even more complex as some plant families are able to form both AM and ECM associations simultaneously (known as dual mycorrhizal plants). Examples are the family Salicaceae and the economically important genus *Eucalyptus*. There is only one published comparison of the simultaneous effects of AM and ECM on an insect (*Chaitophorus populicola*) (Gehring and Whitham 2002). Here, aphid numbers were much lower on AM plants, while those on ECM were much higher than controls. Unfortunately, there was no factorial treatment of dual colonization. In a recent experiment, Gange et al. (2005) have grown sapling trees of *Eucalyptus urophylla* in a field location near Guangzhou, south China. Seedling trees were given an inoculum of AM (*Glomus caledonium*), ECM (*Laccaria laccata*), both or neither fungi and grown for 18 weeks, by which time the saplings were nearly 2 m tall. At the final harvest, the percentage of leaves that had been attacked by different species of insect were recorded on each sapling. Data for leaf chewing by adults of *Anomala cupripes* (Scarabaeidae) (a pest of *Eucalyptus* in China) are given in Fig. 6.3. ECM colonization significantly reduced attack,

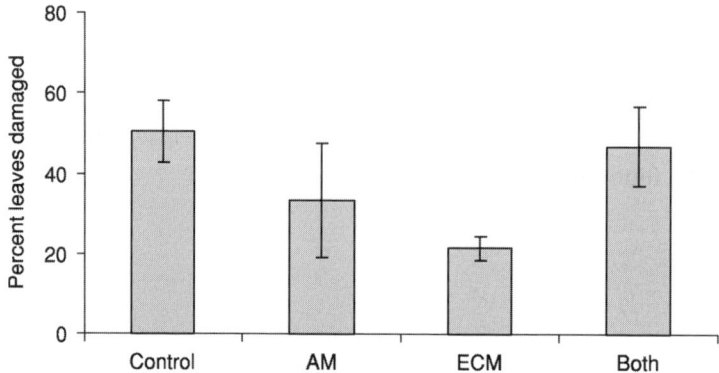

Figure 6.3. The amount of leaf-feeding by the beetle *Anomala cupripes* on saplings of field-grown *Eucalyptus urophylla* and colonized by either AM or ECM fungi, both types of mycorrhiza or none. (A. C. Gange et al. 2005.)

while AM fungi had no effect. However, there was a significant interaction between the fungi, as the dual colonization treatment had damage levels equal to those of the uncolonized controls. These two experiments have provided data in marked contrast to other mycorrhiza–insect experiments and suggest that further studies in this area would be worthwhile, if only to understand the complex phenomenon of the dual association (Lodge 2000).

Building the interaction web

In the light of the studies presented in the preceding two sections, we are now in a position to construct a basic indirect interaction web between mycorrhizae, insects, and plants (Fig. 6.4). It must be emphasized that this diagram is extremely simplistic, as each of the boxes (with the exception of that labeled "Plant") could contain one or many species. In the latter case, not all species may react in the same way to the presence of other insects or mycorrhizae. However, a basic summary of this web is that mycorrhizae seem to have negative effects on root-feeding insects and negative or positive effects on foliar-feeding insects. Foliar herbivores usually reduce mycorrhizal abundance, while root herbivores seem to enhance colonization.

The web in Fig. 6.4 is also simplistic because it just considers bi- or trilateral interactions. To date, virtually all of the controlled experiments investigating the effect of herbivory on the mycorrhiza or vice versa have involved simple combinations of one to three fungi and one insect. None of these experiments takes into account the fact that a plant may be simultaneously attacked by a number of other herbivores, fungi, or diseases, while the insect may also be subject to predation and parasitism. Only very recently have ecologists begun to include other organisms in their insect–mycorrhizal studies.

The foliage of many grasses is infected by asymptomatic fungal endophytes, which often (but not always) confer resistance to insect herbivores (Saikkonen *et al.* 1998). Endophytes can reduce the incidence of mycorrhizal colonization (Muller 2003) and so have the potential to alter the mycorrhiza–insect interaction. Vicari *et al.* (2002) found that a number of plant-mediated interactions occur between the endophyte, mycorrhiza, and an insect. When endophytes were absent, AM colonization reduced the growth of a moth (*Phlogophora meticulosa*), but if endophytes were present, the mycorrhiza had no effect (Fig. 6.5). In a second experiment involving host choice, the mycorrhizal effect again disappeared in the presence of the endophyte, though here the endophyte-induced reduction in host choice remained, whether or not a plant was mycorrhizal. The mechanism by which these two fungal types interact to affect insect herbivores is unknown, but critical, if we are to understand the consequences for plant and insect

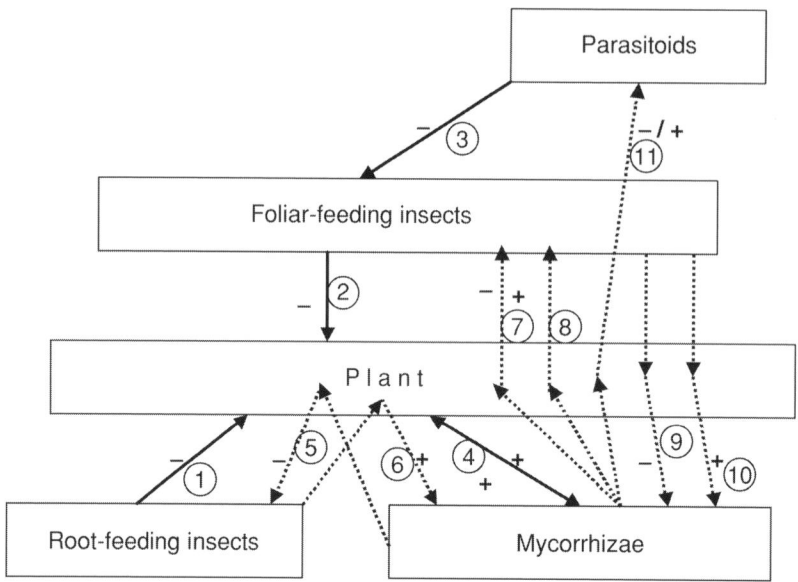

Figure 6.4. The indirect interaction web involving phytophagous insects and mycorrhizae. There are four (one positive (+), three negative (−)) direct interactions, indicated by solid lines. These are the detrimental effects of rhizophagous insects on plants (1), foliar-feeding insects on plants (2), and parasites on insects (3). The fourth direct interaction is double-headed as the mycorrhizal association is assumed to be mutualistic (4).

Mycorrhizae can alter plant defensive chemistry, leading to negative indirect interactions on root-feeding (5) and foliar-feeding insects (7). Root-feeding insects may (temporarily, at least) increase root exudation or the transfer of carbon below ground, leading to a positive indirect effect on the mycorrhiza (6). Mycorrhizae can also have a positive effect on foliar-feeders, by enhancing their nutrition (8). Foliar insects may decrease mycorrhizal levels (9) through competition for carbon or (temporarily, at least) increase these levels through an induced transfer of carbon below ground (10). Finally, mycorrhizae can increase or decrease plant size and lead to an indirect effect on higher trophic levels, by increasing or reducing parasitoid searching efficiency (11). In the indirect interaction web, there are 11 interactions, indicated by dotted lines, of which at least five are positive.

population dynamics. Endophytes generally reduce insect herbivore growth through the production of toxic alkaloids, which are nitrogen-based defenses. As mycorrhizae may often reduce the total nitrogen content of plants (see above), it may be that the endophyte effect is mitigated by the lack of nitrogen in a plant. Given the ubiquitous nature of mycorrhizae and endophytes in grasses, this is an important, but neglected area of research and future experiments should consider variations in soil nutrient availability, particularly nitrogen and phosphorus.

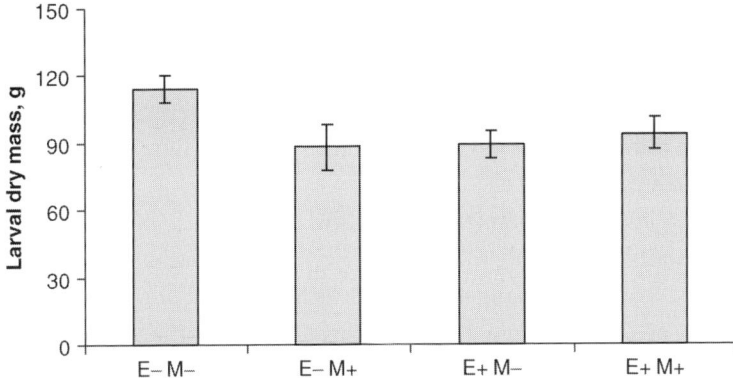

Figure 6.5. Larval dry masses of the moth *Phlogophora meticulosa*, when fed plants of *Lolium perenne* that were infected; by either endophytes or mycorrhizae. E−, endophyte-free; E+, endophyte infected; M−, mycorrhizal-free; M+, mycorrhizal colonized. Redrawn from Vicari et al. (2002).

Apart from endophytes, mycophagous Collembola are the most often cited regulators of the mycorrhizal symbiosis. Their interactions with AM fungi have been reviewed by Gange (2000), in which it was shown that many previous experiments purporting to show that Collembola reduce mycorrhizal functioning are flawed. In fact, Collembola prefer to feed on nonmycorrhizal fungi in the rhizosphere and in so doing allow increased levels of mycorrhizal colonization. If this is so, then their activities might, through changes in colonization levels, increase or decrease foliar-feeding insect performance. In an experiment without mycorrhizae, Collembola were found to decrease reproduction of the aphid *Myzus persicae* (Scheu et al. 1999) though the mechanism was not clear. It is therefore certain that these primitive insects also play a role in determining the outcome of insect–mycorrhizal interactions and need to be included in multitrophic experiments in the future.

Another important component of the soil fauna are earthworms and their presence has been shown to increase (Scheu et al. 1999) or decrease aphid reproduction (Wurst et al. 2004). Only one study has examined the combined effects of worms and AM fungi on insect performance (Wurst et al. 2004). Mycorrhizae had no effect on development time of *M. persicae*, but when worms and fungi were present together development time was significantly shortened. Although not measured, this could result in reduced insect size, since aphids that develop fastest are usually small (Dixon 1997).

There are many plant antagonists in the soil and all have been the subject of studies involving mycorrhizae. Deleterious effects of plant-feeding nematodes are generally (but not always) reduced by mycorrhizae, while good evidence exists for the mycorrhizal protection of plants against pathogenic fungi (Borowicz 2001). However, to date, neither nematodes nor pathogens have featured in insect–mycorrhizal experiments. There are clearly many experiments that need to be done if we are to understand fully the indirect interaction web in which insects and mycorrhizae are placed.

As noted above, mycorrhizae reduce the performance of root-feeding insects and it is well known that root removal can have a variety of effects on plant physiology, leading to changes in the performance of foliar-feeding insects. Only the study of Gange (2001) has considered whether the mycorrhizal-induced reduction of root-feeding may have consequences for a foliar-feeding insect on the same host. It was found that in the absence of mycorrhizae, adult black vine weevils consumed more leaf material from plants that had suffered root-feeding by their larvae. However, this effect disappeared completely if either of two mycorrhizae were present singly. This preliminary study shows that the positive effect of root-feeding on foliar insects may be eliminated by mycorrhizae and so future insect–plant–insect experiments need to take mycorrhizae into account.

Foliar-feeding insects are subject to many so-called "top–down" forces of predation and parasitism, which can regulate their populations (Schoonhoven et al. 1998). The only experiment that has so far considered whether mycorrhizal-induced changes in foliar herbivore performance may have consequences for a higher trophic level organism is that of Gange et al. (2003). These authors studied parasitism of C. syngenesiae when feeding on Leucanthemum vulgare which had been colonized by three different species of AM fungus, in all possible combinations. Some combinations of mycorrhizae increased parasitism rates, while others decreased it. Parasitism did not seem to be related to phytophage density and the conclusion was that mycorrhizal effects on host plant size were the cause. This is interesting, given the lack of such effects on herbivore choice (above), but larger plants seemed to have lower rates of parasitism, due to reduced search efficiency, and vice versa. Reviews of multitrophic interactions highlight the paucity of studies involving higher trophic levels (e.g., van der Putten et al. 2001) and many more are needed, as parasites and predators are likely to be important components of the indirect interaction web.

It is therefore clear that the web depicted in Fig. 6.4 needs to be expanded to include (at least) endophytes, pathogenic fungi, other soil fauna, and insect predators. Our knowledge of the influence of each of these factors on insect–mycorrhizal interactions exists currently as single studies, hence I feel it would be too speculative to include them in Fig. 6.4. Nevertheless, this must be the main goal for the future and I hope that many more experiments will be conducted, to enable a more complete interaction web to be constructed.

Consequences for communities

A perennial problem in ecology is whether interaction effects seen in simple laboratory experiments scale up to detectable effects at the community level. In terms of insect–mycorrhizal interactions, it is almost too early to tell. There are no studies of foliar insect–mycorrhiza interactions in a community context and only a few with root-feeding insects. As plants in the field are colonized by several mycorrhizal species at least, and combinations of fungi seem to have no effect on root-feeding insects, it might be thought that community interactions would be of little consequence. However, Gange and Brown (2002b) found that this was not so, in an experiment that manipulated levels of AM fungi and soil insects during the first 5 years of succession. They found that AM fungi tended to increase plant species richness, while insects decreased it. There were many statistical interactions between the treatments, because the mycorrhizal effect was most clearly seen when insects were excluded. This suggests that insects disrupt the association, perhaps by their feeding or burrowing activities. Meanwhile, the effect of insects was most clearly seen when fungi were present. We would not expect to see such an event if the deleterious effect of single mycorrhizae (e.g., Gange 2001) was repeated, but we would if the null effect of multiple mycorrhizae occurred. The mechanism of community structuring is thought to be one of differential seedling establishment, especially of perennial forbs (Gange and Brown 2002b) yet there have been no controlled studies of rhizophagous insects, mycorrhizae, and seedling plants. Furthermore, the outcome of any interaction will depend on whether a plant is first colonized by a mycorrhiza or eaten by a root-feeding insect, and we do not know the sequence probabilities of these events. Gange and Brown (2002b) and Brown and Gange (2002) provide models of different sequences that need to be tested.

Perhaps the most important community consequence is to be found in the spatial distribution of insects, plants, and fungi. In Hart and Klironomos (2002) there is an excellent graphical depiction of the heterogeneous distributions of four AM fungal species in soil of a 50 m × 50 m field. By overlaying the four distributions, one can easily see that any plant in this field could be uncolonized or colonized by one, two, three, or four species of mycorrhiza, in any combination. Therefore, if one then proceeded to rear an insect on these plants, a large variation in growth rate might be found, depending on whether a particular fungal combination increased, decreased, or had no effect on plant growth. Such variation in insect growth rates is widely reported and often attributed to abiotic factors or variation in plant defense (Schoonhoven *et al.* 1998). However, it is likely that mycorrhizae contribute significantly to this variation and field experiments need to relate mycorrhizal occurrence to insect growth rates.

Of course, one must also ask why mycorrhizae are so heterogeneously distributed. As has been shown, there is good evidence that insect herbivory can reduce the abundance and alter the species composition of mycorrhizae in field soil, especially with ECM fungi. Assuming that each mycorrhizal species has a different carbon demand, then a patchy distribution of species will result from insect herbivory occurring on certain plants. Patchy distributions of insect attack are common and, by varying the species composition of fungi in field soil, may contribute to subsequent variation in local plant species richness, through differential seedling establishment (van der Heijden et al. 1998).

Phytophagous insects are known to have significant effects on the structure of plant communities and plant succession (Brown and Gange 2002). Arbuscular mycorrhizae might mitigate these effects, through enhancing the tolerance (capacity to regrow after herbivory) of certain plants to attack. By elevating photosynthetic rate, and/or nutrient acquisition, tolerance could occur in situations where defense investment is costly (van der Putten 2003). To date, there is no evidence that mycorrhizae affect tolerance (Borowicz 1997), but it is a mechanism that deserves attention.

Whether defense investment or tolerance occurs is likely to be determined in part by environmental conditions, such as light and nutrient availability. Jones and Last (1991) provide predictions for the effects of ECM on insects, related to abiotic conditions. It is highly likely that they can be applied to AM fungi too and experiments are needed to test them, so that community-level effects of mycorrhizae on insects can be understood.

Further variation will be introduced to the structure of the fungal, plant, and insect community through the indirect interaction web outlined in this chapter and summarized in Fig. 6.4. However, until we know the relative strengths and chronology of the links in this figure, much of our interpretation will be speculative. The important point is that the interactions between insects and mycorrhizae are likely to vary in space and time and be influenced by many other organisms. The end result is the heterogeneous distributions of organisms that we see in natural communities. Can we understand the nature of this heterogeneity? I believe that we can, and are now at an exciting stage of discovery, where advanced molecular techniques can be integrated with ecological studies, so that community ecology becomes a more predictable science.

Acknowledgments

I would like to thank the British Ecological Society, the British Entomological and Natural History Society, the European Union (through programmes CONSIDER and

BIORHIZ), the Natural Environment Research Council, the Nuffield Foundation and the University of London (Central Research Fund) for supporting my experimental work on insects and mycorrhizae. My past and present research colleagues David Aplin, Erica Bower, Valerie Brown, Amanda Currie, Philip Murray, Penny Stagg, Lena Ward, James Wearn, and Helen West deserve special thanks for their ideas and support.

References

Behmer, S.T., and W.D. Nes. 2003. Insect sterol nutrition and physiology: a global overview. *Advances in Insect Physiology* **31**:1–72.

Borowicz, V.A. 1997. A fungal root symbiont modifies plant resistance to an insect herbivore. *Oecologia* **112**:534–542.

Borowicz, V.A. 2001. Do arbuscular mycorrhizal fungi alter plant–pathogen relations? *Ecology* **82**:3057–3068.

Brown, V.K., and A.C. Gange. 1990. Insect herbivory below ground. *Advances in Ecological Research* **20**:1–58.

Brown, V.K., and A.C. Gange. 2002. Tritrophic below- and above-ground interactions in succession, pp. 197–222 in T. Tscharntke and B.A. Hawkins (eds.) *Multitrophic Level Interactions*. Cambridge, UK: Cambridge University Press.

Crawley, M.J. 1987. Benevolent herbivores? *Trends in Ecology and Evolution* **2**:167–168.

Cullings, K.W., D.R. Vogler, V.T. Parker, and S. Makhija. 2001. Defoliation effects on the ectomycorrhizal community of a mixed *Pinus contorta/Picea engelmannii* stand in Yellowstone Park. *Oecologia* **127**:533–539.

Currie, A.F. 2004. Interactions between root-feeding insects and arbuscular mycorrhizal fungi. Ph.D. dissertation, University of London, UK.

Del Vecchio, T.A., C.A. Gehring, N.S. Cobb, and T.G. Whitham. 1993. Negative effects of scale insect herbivory on the ectomycorrhizae of juvenile pinyon pine. *Ecology* **74**:2297–2302.

Dixon, A.F.G. 1997. *Aphid Ecology: An Optimization Approach*. Glasgow, UK: Blackie.

Eom, A.-H., G.W.T. Wilson, and D.C. Hartnett. 2001. Effects of ungulate grazers on arbuscular mycorrhizal symbiosis and fungal community structure in tallgrass prairie. *Mycologia* **93**:233–242.

Fontaine, J., A. Grandmougin-Ferjani, M.A. Hartmann, and M. Sancholle. 2001. Sterol biosynthesis by the arbuscular mycorrhizal fungus *Glomus intraradices*. *Lipids* **36**:1357–1363.

Gange, A.C. 1996. Reduction in vine weevil larval growth by mycorrhizal fungi. *Mitteilungen aus der Biologischen Bundesanstalt für Land- und Forstwirtschaft* **316**:56–60.

Gange, A.C. 2000. Arbuscular mycorrhizal fungi, Collembola and plant growth. *Trends in Ecology and Evolution* **15**:369–372.

Gange, A.C. 2001. Species-specific responses of a root- and shoot-feeding insect to arbuscular mycorrhizal colonization of its host plant. *New Phytologist* **150**:611–618.

Gange, A.C., and E. Bower. 1997. Interactions between insects and mycorrhizal fungi, pp. 115–132 in A.C. Gange and V.K. Brown (eds.) *Multitrophic Interactions in Terrestrial Systems*. Oxford, UK: Blackwell Science.

Gange, A. C., and V. K. Brown. 2002a. Actions and interactions of soil invertebrates in affecting the structure of plant communities, pp. 321–344 in M. G. A. van der Heijden and I. R. Sanders (eds.) *Mycorrhizal Ecology*. Berlin, Germany: Springer-Verlag.

Gange, A. C., and V. K. Brown. 2002b. Soil food web components affect plant community structure during early succession. *Ecological Research* **17**:217–222.

Gange, A. C., and H. E. Nice. 1997. Performance of the thistle gall fly, *Urophora cardui*, in relation to host plant nitrogen and mycorrhizal colonization. *New Phytologist* **137**:335–343.

Gange, A. C., and H. M. West. 1994. Interactions between arbuscular-mycorrhizal fungi and foliar-feeding insects in *Plantago lanceolata* L. *New Phytologist* **128**:79–87.

Gange, A. C., V. K. Brown, and G. S. Sinclair. 1994. Reduction of black vine weevil growth by vesicular–arbuscular mycorrhizal fungi. *Entomologia Experimentalis et Applicata* **70**:115–119.

Gange, A. C., E. Bower, and V. K. Brown. 1999. Positive effects of an arbuscular mycorrhizal fungus on aphid life history traits. *Oecologia* **120**:123–131.

Gange, A. C., E. Bower, and V. K. Brown. 2002a. Differential effects of insect herbivory on arbuscular mycorrhizal colonization. *Oecologia* **131**:103–112.

Gange, A. C., P. G. Stagg, and L. K. Ward. 2002b. Arbuscular mycorrhizal fungi affect phytophagous insect specialism. *Ecology Letters* **5**:11–15.

Gange, A. C., V. K. Brown, and D. M. Aplin. 2003. Multitrophic links between arbuscular mycorrhizal fungi and insect parasitoids. *Ecology Letters* **6**:1051–1055.

Gange, A. C., D. R. J. Gane, Y. L. Chen, and M. Q. Gong. 2005. Dual colonization of *Eucalyptus urophylla* S. T. Blake by arbuscular and ectomycorrhizal fungi affects levels of insect herbivore attack. *Agricultural and Forest Entomology* **7**:253–263.

Gehring, C. A., and T. G. Whitham. 1991. Herbivore-driven mycorrhizal mutualism in insect-susceptible pinyon pine. *Nature* **353**:556–557.

Gehring, C. A., and T. G. Whitham. 1994. Interactions between aboveground herbivores and the mycorrhizal mutualists of plants. *Trends in Ecology and Evolution* **9**:251–255.

Gehring, C. A., and T. G. Whitham. 1995. Duration of herbivore removal and environmental stress affect the ectomycorrhizae of pinyon pines. *Ecology* **76**:2118–2123.

Gehring, C. A., and T. G. Whitham. 2002. Mycorrhizae-herbivore interactions: population and community consequences, pp. 295–320 in M. G. A. van der Heijden and I. R. Sanders (eds.) *Mycorrhizal Ecology*. Berlin, Germany: Springer-Verlag.

Gehring, C. A., N. S. Cobb, and T. G. Whitham. 1997. Three-way interactions among ectomycorrhizal mutualists, scale insects, and resistant and susceptible pinyon pines. *American Naturalist* **149**:824–841.

Goverde, M., M. G. A. van der Heijden, A. Wiemken, I. R. Sanders, and A. Erhardt. 2000. Arbuscular mycorrhizal fungi influence life history traits of a lepidopteran herbivore. *Oecologia* **125**:362–369.

Halldorsson, G., H. Sverrisson, G. G. Eyjolfsdottir, and E. S. Oddsdottir. 2000. Ectomycorrhizae reduce damage to Russian larch by *Otiorhynchus* larvae. *Scandinavian Journal of Forest Research* **15**:354–358.

Hamilton, J. G., A. R. Zangerl, E. H. DeLucia, and M. R. Berenbaum. 2001. The carbon-nutrient balance hypothesis: its rise and fall. *Ecology Letters* **4**:86–95.

Hart, M., and J. N. Klironomos. 2002. Diversity of arbuscular mycorrhizal fungi and ecosystem functioning, pp. 225–242 in M. G. A. van der Heijden and I. R. Sanders (eds.) *Mycorrhizal Ecology*. Berlin, Germany: Springer-Verlag.

Hause, B., W. Maier, O. Miersch, R. Kramell, and D. Strack. 2002. Induction of jasmonate biosynthesis in arbuscular mycorrhizal barley roots. *Plant Physiology* **130**:1213–1220.

Helgason, T., A. H. Fitter, and J. P. W. Young. 1999. Molecular diversity of arbuscular mycorrhizal fungi colonising *Hyacinthoides non-scripta* (bluebell) in a seminatural woodland. *Molecular Ecology* **8**:659–666.

Hodge, A. 2000. Microbial ecology of the arbuscular mycorrhiza. *FEMS Microbiology Ecology* **32**:91–96.

Holland, J. N., C. Weixin, and D. A. Crossley. 1996. Herbivore-induced changes in plant carbon allocation: assessment of below-ground C fluxes using carbon-14. *Oecologia* **107**:87–94.

Jones, C. G., and F. T. Last. 1991. Ectomycorrhizae and trees: implications for aboveground herbivory, pp. 65–103 in P. Barbosa, V. A. Krischik, and C. G. Jones (eds.) *Microbial Mediation of Plant–Herbivore Interactions*. Chichester, UK: John Wiley.

Kolb, T. E., K. A. Dodds, and K. M. Clancy. 1999. Effect of western spruce budworm defoliation on the physiology and growth of potted Douglas-fir seedlings. *Forest Science* **45**:280–291.

Krishna, K. R., H. M. Suresh, J. Syamsunder, and D. J. Bagyaraj. 1981. Changes in the leaves of finger millet due to VA mycorrhizal infection. *New Phytologist* **87**:717–722.

Kuikka, K., E. Härma, A. Markkola, *et al.* 2003. Severe defoliation of Scots pine reduces reproductive investment by ectomycorrhizal symbionts. *Ecology* **84**:2051–2061.

Lodge, D. J. 2000. Ecto- or arbuscular mycorrhizae: which are best? *New Phytologist* **146**:353–354.

Manninen, A.-M., T. Holopainen, and J. K. Holopainen. 1998. Susceptibility of ectomycorrhizal and non-mycorrhizal Scots pine (*Pinus sylvestris*) seedlings to a generalist insect herbivore, *Lygus rugulipennis* at two nitrogen availability levels. *New Phytologist* **140**:55–63.

Muller, J. 2003. Artificial infection by endophytes affects growth and mycorrhizal colonization of *Lolium perenne*. *Functional Plant Biology* **30**:419–424.

Paul, N. D., P. E. Hatcher, and J. E. Taylor. 2000. Coping with multiple enemies: an integration of molecular and ecological perspectives. *Trends in Plant Science* **5**:220–225.

Price, P. W. 1991. The plant vigor hypothesis and herbivore attack. *Oikos* **62**:244–251.

Rabin, L. B., and R. S. Pacovsky. 1985. Reduced larva growth of two lepidoptera (Noctuidae) on excised leaves of soybean infected with a mycorrhizal fungus. *Journal of Economic Entomology* **78**:1358–1363.

Rhoades, D. F. 1979. Evolution of plant chemical defense against herbivores, pp. 3–54 in G. A. Rosenthal and D. H. Janzen (eds.) *Herbivores: Their Interaction with Secondary Plant Metabolites*. New York: Academic Press.

Rovira, A. D. 1969. Plant root exudates. *Botanical Review* **35**:35–57.

Saikkonen, K., S. H. Faeth, M. Helander, and T. J. Sullivan. 1998. Fungal endophytes: a continuum of interactions with host plants. *Annual Review of Ecology and Systematics* **29**:319–343.

Saikkonen, K., U. Ahonen-Jonnarth, A. Markkola, *et al.* 1999. Defoliation and mycorrhizal symbiosis: a functional balance between carbon sources and below-ground sinks. *Ecology Letters* **2**:19–26.

Scheu, S., A. Theenhaus, and T. H. Jones. 1999. Links between the detrivore and the herbivore system: effects of earthworms and Collembola on plant growth and aphid development. *Oecologia* **119**:541–551.

Schoonhoven, L. M., T. Jermy, and J. J. A. van Loon. 1998. *Insect–Plant Biology: From Physiology to Evolution.* London: Chapman and Hall.

Smith, S. E., and D. J. Read. 1997. *Mycorrhizal Symbiosis.* San Diego, CA: Academic Press.

Staddon, P. L., A. H. Fitter, and D. Robinson. 1999. Effects of mycorrhizal colonization and elevated atmospheric carbon dioxide on carbon fixation and below-ground carbon partitioning in *Plantago lanceolata*. *Journal of Experimental Botany* **50**:853–860.

Strack, D., T. Fester, B. Hause, W. Schliemann, and M. H. Walter. 2003. Arbuscular mycorrhiza: biological, chemical, and molecular aspects. *Journal of Chemical Ecology* **29**:1955–1979.

Thaler, J. S., M. A. Farag, P. W. Pare, and M. Dicke. 2002. Jasmonate-deficient plants have reduced direct and indirect defences against insect herbivores. *Ecology Letters* **5**:764–774.

Tjallingii, W. F., and T. H. Esch. 1993. Fine-structure of aphid stylet routes in plant tissues in correlation with EPG signals. *Physiological Entomology* **18**:317–328.

van der Heijden, M. G. A., T. Boller, A. Wiemken, and I. R. Sanders. 1998. Different arbuscular mycorrhizal fungal species are potential determinants of plant community structure. *Ecology* **79**:2082–2091.

van der Putten, W. H. 2003. Plant defense belowground and spatiotemporal processes in natural vegetation. *Ecology* **84**:2269–2280.

van der Putten, W. H., L. E. M. Vet, J. A. Harvey, and F. L. Wäckers. 2001. Linking above- and belowground multitrophic interactions of plants, herbivores, pathogens, and their antagonists. *Trends in Ecology and Evolution* **16**:547–554.

Vicari, M., P. E. Hatcher, and P. G. Ayres. 2002. Combined effect of foliar and mycorrhizal endophytes on an insect herbivore. *Ecology* **83**:2452–2464.

Wamberg, C., S. Christensen, and I. Jakobsen. 2003. Interaction between foliar-feeding insects, mycorrhizal fungi, and rhizosphere protozoa on pea plants. *Pedobiologia* **47**:281–287.

White, T. C. R. 1984. The abundance of herbivores in relation to the availability of nitrogen in stressed food plants. *Oecologia* **63**:90–105.

Wurst, S., D. Dugassa-Gobena, R. Langel, M. Bonkowski, and S. Scheu. 2004. Combined effects of earthworms and vesicular–arbuscular mycorrhizae on plant and aphid performance. *New Phytologist* **163**:169–176.

Part III PLANT-MEDIATED INDIRECT EFFECTS IN MULTITROPHIC SYSTEMS

7

Plant-mediated interactions between below- and aboveground processes: decomposition, herbivory, parasitism, and pollination

KATJA POVEDA, INGOLF STEFFAN-DEWENTER, STEFAN SCHEU, AND TEJA TSCHARNTKE

Introduction

All terrestrial ecosystems are constituted of a belowground and an aboveground subsystem. These subsystems depend on each other, since above the ground primary producers are the main source of organic carbon for the system, whereas below the ground soil organisms are responsible for the breakdown and recycling of organic matter and the mineralization of the nutrients therein (Scheu and Setälä 2002, Wardle 2002, Porazinska et al. 2003). To understand community and ecosystem-level processes it is necessary to study the interactions within and between these subsystems. However, most ecologists have investigated belowground and aboveground communities separately, leaving the "between subsystems" interactions unstudied (but see Wardle 1999, 2002, Bonkowski et al. 2001, Masters et al. 2001, van der Putten et al. 2001, Brown and Gange 2002). Moreover, studies of different types of plant–animal interactions, such as pollination, herbivory, and seed dispersal, have traditionally progressed separately, focusing on just one kind of interaction and ignoring the possible interference with others (Herrera et al. 2002). This leads to an oversimplification of our understanding of plant–animal interactions, as most plants interact simultaneously with a broad spectrum of animals. Aboveground

Ecological Communities: Plant Mediation in Indirect Interaction Webs, ed. Takayuki Ohgushi, Timothy P. Craig, and Peter W. Price. Published by Cambridge University Press. © Cambridge University Press 2007.

communities are affected by both direct and indirect effects of soil organisms on plants. Soil biota exert direct effects on plants by feeding on roots and forming antagonistic or mutualistic relationships with their host plants (Wardle et al. 2004). Such direct interactions with plants influence not only the performance of the host plants themselves, but also that of the herbivores and their predators. With regard to indirect pathways, feeding activities in the detrital food web stimulate nutrient turnover, plant nutrient acquisition, and plant performance and thereby should indirectly influence the aboveground animal community (Scheu 2001, Wardle et al. 2004).

The aim of this chapter is to evaluate plant-mediated interactions between aboveground and belowground processes. First, we review patterns of how belowground processes, such as decomposition and root herbivory, are linked to the aboveground biota. Second, we discuss in more detail the way in which decomposers and root herbivores alone and in combination cause aboveground responses. Finally, we stress that decomposition and root herbivory are only one part of the diverse plant–animal and plant–fungi interactions linking aboveground and belowground subsystems.

The belowground biota

Soil organisms are densely packed: underneath one footprint of forest soil there may be billions of protozoa, hundreds of thousands of nematodes, thousands of Collembola and mites, and a large number of isopods, spiders, beetles, and other invertebrates (Scheu and Setälä 2002). The soil biota has been divided into four main groups depending on size: (1) the microflora, consisting mainly of bacteria and fungi with body width less than 50 µm; (2) the microfauna, composed primarily of nematodes and protozoa with body widths of less than 0.1 mm; (3) the mesofauna, including microarthropods such as mites and springtails as well as enchytraeids, with a body width between 0.1 and 2.0 mm; and (4) the macrofauna, with body widths greater than 2 mm, which is composed of earthworms, termites, millipedes, and other arthropods that live in and above the soil (Lavelle and Spain 2001). The soil biota can also be divided into two main groups depending on the way they interact with plants. The first group are the root-associated organisms, i.e., root herbivores and their consumers, which influence the plant directly, for example, by feeding on roots. The second is the decomposer community, which breaks down dead plant material and indirectly regulates plant growth and community composition by determining the supply of available soil nutrients (Scheu and Setälä 2002, Wardle et al. 2004).

Effects of root herbivores

Animals feeding on belowground plant parts include both vertebrates, mainly rodents, and invertebrates, such as herbivorous insects and nematodes. The role of nematode herbivory has been reviewed earlier (Stanton 1988, Mortimer *et al.* 1999) and it will not be considered in this chapter. The main functions of the root systems are anchorage and the acquisition of water and nutrients in order to support rapid growth and reproduction of the plant. The loss of roots to herbivores affects the vegetative growth of the plant not only through diminished nutrient and water uptake, but also through the loss of storage tissue (Mortimer *et al.* 1999).

Effects on plant growth

Root herbivory reduces plant growth and increases plant mortality (Brown and Gange 1990, Wardle 2002). For example, a study of Nötzold *et al.* (1998) on the effects of the weevil *Hylobius transvittatus* on purple loosestrife (*Lytrum salicaria*) showed that root herbivory reduced plant height in the first year and plant biomass in the second year. A review of the effect of mechanical root pruning showed increases in growth of root tissue and a reduction in shoot growth for many plant species (Andersen 1987). Mortimer *et al.* (1999) stated that differences in responses of plants to belowground herbivory can be related to characteristics of their life histories, such as levels of stored resources. Long-lived clonal species appear to be able to respond to herbivory by compensatory regrowth, whilst annual and monocarpic species are usually more susceptible and respond with reduced growth or fecundity. It has also been noted that damage to belowground plant parts leads to altered source–sink relationships within the plant, with compensatory root growth occurring at the expense of shoot growth (Mortimer *et al.* 1999).

The effects of root herbivores may be altered by belowground interactions in the soil food web. For example, ghost moth (*Hepialus californicus*) larvae are known to attack bush lupine (*Lupinus arboreus*), occasionally eradicating entire stands (Maron 2001). In the same way as above the ground, root damage by herbivores may be counteracted by predators, as is the case in the ghost moth being attacked by nematodes (*Heterorhabditis marelatus*) (Strong *et al.* 1999, Preisser 2003).

Effects on floral traits and pollination

In animal-pollinated plants fitness may be influenced by floral traits that function as advertisement and reward for pollinators. In order to attain outcrossing, plants with flowers offer an extraordinary range of attractants to increase pollinator visitation. Floral advertisements include olfactory cues, short-range tactile cues, and visual cues, such as size, shape, and color of inflorescences and flowers. Floral rewards include nectar and pollen that are highly attractive for flower visitors (Dafni 1992, Pellmyr 2002). Floral advertisement and seed set can be affected by a range of organisms interacting with the plant. Leaf and floral herbivory reduce pollinator visitation in damaged plants through changes in floral traits (Lehtilä and Strauss 1997, 1999, Strauss 1997, Mothershead and Marquis 2000). Effects of aboveground herbivory on pollination have been considered in detail (see Bronstein *et al.* Chapter 4 this volume), and belowground herbivores have also been shown to affect floral traits. For example, root herbivory by the weevil *Hylobius transvittatus* on purple loosestrife (*Lytrum salicaria*) delays the flowering period and decreases the number of flowers and the size of the inflorescences (Nötzold *et al.* 1998), although it is unknown if this translates into effects on plant fitness or visitation by pollinators. In addition, reduction in root herbivory resulted in larger flower heads and an earlier onset of flowering in thistles (*Cirsium palustre*) (Masters *et al.* 2001). In contrast to these findings, the annual herb *Sinapis arvensis* (Brassicaceae) attacked by root herbivores did not suffer from changes in floral traits, but had an increased number of flower visitors in comparison to control plants (Poveda *et al.* 2003), suggesting that root herbivory may somehow enhance the attractiveness of plants to their pollinators.

Effects on aboveground herbivores and their parasitism

A series of studies have shown that root herbivory increases the susceptibility of plants to the attack of aboveground herbivores. For example, Gange and Brown (1989) assessed the effects of the chafer larva *Phyllopertha horticola* (Coleoptera: Scarabaeidae) on the performance of the black bean aphid (*Aphis fabae*), mediated via a common annual host plant, *Capsella bursa-pastoris*. Root feeding caused an increase in the weight, growth rate, fecundity, and adult longevity of the aphid. They also reported a positive effect of root herbivory on the growth and performance of a foliar feeder, but this effect was mitigated at high soil moisture levels.

Masters (1995b) tested the effect of insect root herbivory on aphid performance under field conditions and in an experiment with controlled environmental conditions. In both experiments, root feeding by insects affected the performance of foliar-feeding aphids beneficially. Plants subjected to belowground insect herbivory in the field supported greater numbers of aphids, and root feeding in the laboratory increased adult aphid weight and growth rate, thereby increasing fecundity. Similarly, Masters *et al.* (2001) showed that the abundance of tephritid flies, which induce galls in the flower heads of *Cirsium palustre*, increased as a result of root herbivory. Seed predation and parasitism also increased, the latter indicating an indirect effect of root herbivory on higher trophic levels.

Reviewing studies on the effect of belowground herbivores on aboveground herbivores Bezemer et al. (2002) documented that seven out of eight studies showed a positive effect of root herbivory on aboveground herbivory. Masters et al. (1993) put forward a conceptual model explaining the positive effects of root herbivores on aboveground herbivores (termed "stress response hypothesis" by Bezemer et al. 2002). They suggested that root feeding limits the ability of the plant to take up water and nutrients, and leads to a reduction in the relative water content of the foliage, increasing levels of soluble nitrogen (especially amino acids) and carbohydrates. The higher-quality food resource leads to increased insect growth, fecundity, and population size of foliar-feeding insects. Bezemer et al. (2002) proposed as an alternative the "defense induction hypothesis." It predicts that root herbivores detrimentally affect aboveground insect performance through the induction of secondary plant compounds in the foliage. Supporting this hypothesis they showed that root herbivory by wireworms (Agriotes lineatus) induced cotton plants (Gossypium herbaceum) to increase the concentration of terpenoids in the leaves, resulting in a reduced performance of Spodoptera exigua larvae on these plants.

Root and leaf feeders can also interact via the host plant in a plus–minus fashion, when belowground herbivory facilitates aboveground herbivory, but aboveground herbivory inhibits belowground herbivory. Moran and Whitham (1990) reported that root-feeding aphids (Pemphigus batae) affected neither their host plant Chenopodium album nor an aboveground leaf-galling aphid (Hayhurstia atriplicis). In contrast, aboveground aphids reduced overall plant biomass by more than 50%, seed set by 60%, and the number of belowground aphids by 91%. Similarly, root feeding by chafer larvae (Phyllopertha horticola) increased the fecundity of a leaf miner (Chromatamyia syngensiae, Diptera: Agromyzidae), whereas leaf feeding decreased the growth rate of the belowground insect herbivore (Masters and Brown 1992, Masters 1995a). Subsequent laboratory experiments were performed to test the effects of chafer larvae on different foliar-feeding guilds, namely the leaf chewer Mamestra brassicae (Lepidoptera: Noctuidae), the phloem feeder Myzus persicae (Homoptera: Aphididae), and the leaf miner Chromatamyia syngensiae (Masters and Brown 1997). In each case, the growth rate of the root feeders decreased when foliar feeders were present. On the other hand, root herbivory significantly increased the performance of phloem-feeding aphids and leaf miners but there was no significant effect on any performance-related parameter of the leaf chewer. In this case, root herbivory reduced the consumption rate of leaf material. However, caterpillars still maintained their growth rate and developed normally.

In summary, aboveground responses appear to depend on the level of root herbivory, with low levels but not high levels being compensated for by the plant. Root herbivory presumably increases aboveground nutrient quality via soluble amino acids, but so far the mechanism is not clear. Depending on the plant species, this effect may be masked by the induction of secondary compounds. Further experimentation is necessary to generalize the importance of each of these mechanisms.

Effects of decomposers

The primary consumers within the decomposer food web, bacteria and fungi, are directly responsible for most of the mineralization of nutrients in the soil, and are therefore the primary biotic regulators of nutrient supply for plants. This is due to their unique capacity to directly break down complex carbohydrates and to mineralize the nutrients contained therein (Wardle 2002). The nutrient mineralization process driven by the soil microflora is in turn influenced by the soil food webs in which micro- and mesofauna, including microarthropods, nematodes, and protozoa, feed upon the microflora and on each other (Ruess et al. 2004), releasing nutrients that are locked up in the bacterial biomass (Bonkowski 2004). The importance of soil fauna is considerable given their influence on the growth and activity of the microflora (Wardle et al. 2004). The largest scale of the decomposer subsystem includes earthworms and termites and is characterized by their ability to build physical structures that create habitats for smaller organisms and to function as litter transformers. These large, structure-forming invertebrates process a major part of the detritus available and function as ecosystem engineers (Jones et al. 1994). Earthworms have been shown to transform soil systems from moder (medium humified humus) to mull (well-humified organic matter) type humus due to physical action, i.e., by engineering (Bohlen et al. 2004). Given that decomposers are responsible for the breakdown of organic matter and the release and cycling of nutrients (Haimi and Einbork 1992, Wardle 2002), they could be expected to stimulate plant growth and herbivore performance.

Effects on plant growth

The activity of decomposers often results in increased plant growth and plant nitrogen content (Scheu and Parkinson 1994, Bonkowski et al. 2000, 2001, Wardle 2002). Scheu (2003) reviewed the response of plants to the presence of earthworms. He found that in 79% of all studies, shoot biomass of plants significantly increased in the presence of earthworms, in 9% it declined, and in 12% no significant effect was found. Root biomass increased in 50% of the cases and decreased in 38%. For example, the presence of the earthworm *Dendrobaena octaedra* enhanced the shoot biomass of the grass *Agropyron trachycaulum* and increased the shoot-to-root ratio during early plant growth (Scheu and Parkinson 1994). Spain *et al.* (1992) showed that transfer of ^{15}N from microbial biomass to plants was enhanced by the addition of earthworms and that ^{15}N incorporated into both microbial biomass and earthworms served as a source of nutrients to plants. On the other hand, Newington *et al.* (2004) showed that earthworms increase soil nitrate and foliar nitrogen concentrations in *Veronica persica* and *Cardamine hirsuta*, but that there was no associated increase in plant biomass. More details on earthworm–plant interactions are given by Scheu (2003) and Brown *et al.* (2004).

In contrast to this scenario of decomposer effects, Scheu *et al.* (1999) found collembolans (*Heteromurus nitidus* and *Onychiurus scotarius*) to cause a reduction in plant biomass of *Poa annua* mainly in roots, while plant tissue nitrogen concentration was increased.

Generally, results from microcosm studies suggest that larger soil fauna can have strong positive effects on plant growth and nutrient acquisition, presumably through promoting microbial activity and therefore nutrient mineralization (e.g., Setälä and Huhta 1991, Haimi *et al.* 1992, Alphei *et al.* 1996, Bardgett *et al.* 1997, 1998). An important mechanism responsible for these processes is known as the "microbial loop" in the soil (Clarholm 1985, Coleman 1994, Moore *et al.* 2003). It is triggered by the release of root exudates from plants that increase bacterial growth in the rhizosphere. Plant nutrients may be sequestered during microbial growth and remain locked up in bacterial biomass. Grazing by decomposer invertebrates remobilizes these nutrients, making them available for plant uptake. Due to the relatively small differences in the C:N ratios between decomposers and bacterial prey and the relatively low assimilation efficiency of the decomposers, only a small percentage of the consumed nitrogen is used for biomass production. The excess nitrogen is excreted as ammonia and hence is readily available for plant roots (see Bonkowski 2004, Scheu *et al.* 2004).

Effects on herbivores

As mentioned above, earthworms enhance nitrogen uptake from litter and soil into the plant (Wurst *et al.* 2003, 2004a) leading to an enhanced nutrient concentration in plant tissue (Alphei *et al.* 1996, Callaham and Hendrix 1998, Schmidt and Curry 1999, Bonkowski *et al.* 2001). Since herbivore performance is known to depend strongly on plant tissue nitrogen concentration (White 1993), the effects of earthworms on plant growth likely propagate into the herbivore community.

Effects of earthworms on aboveground herbivores likely vary with soil type, litter distribution, and plant species. Earthworm presence has been shown to increase the reproduction of aphids (*Myzus persicae*) on *Poa annua* and *T. repens*

(Scheu et al. 1999) and *Cardamine hirsuta* (Wurst and Jones 2003). However, earthworms increased aphid reproduction on *T. repens* only when the litter was concentrated in patches in the soil and not when litter was mixed homogeneously into the soil (Wurst et al. 2003). Also, Bonkowski et al. (2001) report no changes in aphid reproduction on wheat in the presence of earthworms. Wurst et al. (2003) even found aphid reproduction on *Plantago lanceolata* to be reduced in the presence of earthworms. The reduction of aphid reproduction in the presence of earthworms was associated with changes in the phytosterol content in leaves of *P. lanceolata*, indicating that not only belowground herbivores but also decomposers increase plant defense against herbivores (Wurst et al. 2004b). Newington et al. (2004) investigated the effect of earthworms on the development of *Mamestra brassicae* larvae on *Cardamine hirsuta* and *Veronica persica* plants. They found no effect of earthworms on the biomass of the larvae. However, in feeding trials the consumption rate of *V. persica* foliage by *M. brassicae* was higher when plants were grown in the presence of earthworms. Highest larval mortality occurred in microcosms without earthworms, suggesting that studying only larval biomass gives a misleading picture.

Although decomposers generally affect plant growth and herbivore development beneficially, this appears to be not always true; in fact, they may affect herbivores detrimentally. Secondary compounds in plants induced by decomposers have been ignored so far, but may significantly affect plant–herbivore interactions (Wurst et al. 2004a). Effects of decomposers on higher trophic levels and plant–pollinator interactions are unknown and need to be addressed in order to complete our understanding on the interrelationships between decomposers and aboveground food webs.

Multitrophic belowground–aboveground interactions: a case study

The interactions between above- and belowground organisms have only recently become a major field of study. Many of these studies focus on the effects of decomposers on plant growth (Scheu and Parkinson 1994, Bonkowski et al. 2000) and on aboveground plant–herbivore interactions (Scheu et al. 1999, Bonkowski et al. 2001), as well as on the effects of belowground and aboveground herbivores on plants and their natural enemies (Brown and Gange 1989, Gange and Brown 1989, Moran and Whitham 1990, Masters and Brown 1992, Masters et al. 1993, 2001, Nötzold et al. 1998). However, there are several limitations in the way that the effects of belowground organisms on aboveground organisms have been studied:

1. Most experiments have been done in microcosms.
2. Plants were not naturally colonized by herbivores.

3. Studies have been performed on plant–herbivore interactions, but not on plant–pollinator interactions.
4. None of the studies linked the effects of the decomposer and the root-feeding fauna on aboveground processes.

We established an experiment to address some of these deficiencies. We examined the single and combined effects of root herbivores and decomposers on plant growth, floral traits, flower visitation, and herbivore–parasitoid interactions of *Sinapis arvensis*. By exposing plants to root herbivores (five wireworms, larvae of the click beetle *Agriotes* sp.) and decomposers (two earthworms of the species *Octolasion tyrtaeum*) in a full factorial design, we investigated how direct and indirect belowground plant–animal interactions affect plant performance and the insect community associated aboveground. The decomposer density chosen (50 earthworms m^{-2}), is in the lower range of densities in the field (9–239 earthworms m^{-2}: Pizl 1999). Wireworms are patchily distributed in soil and its densities may vary from zero up to several hundreds per square meter (Poveda *et al.* 2003). Plants were transferred to a fallow field, and pots were buried into the soil to simulate field conditions. In a first experiment, plants were exposed to natural herbivore colonization and pollination, whereas in a second experiment, plants were covered with a gauze tent to exclude effects of other organisms such as herbivores, seed predators, and/or pollinators on plant growth and floral traits.

In the first experiment total plant biomass (Fig. 7.1a) was reduced when root herbivores were present, but decomposers counteracted this negative effect when they co-ocurred with root herbivores (Table 7.1). This suggests that decomposers play an important role in counteracting detrimental effects of root herbivores on plant growth and reproduction. The marginally significant interaction between root herbivores and decomposers on the production of seeds per plant (Table 7.1) suggests that the presence of earthworms mitigated the negative effects of root herbivores (Fig. 7.1b). Root herbivores and decomposers affected the total number of aphids on the plants (Table 7.1). In treatments with only earthworms or with only root herbivores, the number of aphids was higher than control plants (Fig. 7.1c). In contrast, aphid numbers were not significantly different from the control when both organisms were present. The enhanced numbers of aphids on plants with earthworms or root herbivores suggest an increased quality or quantity of assimilates transferred in the phloem. Nitrogen availability is one of the main factors limiting herbivore development (White 1993). Both decomposers and root herbivores likely increased nitrogen availability to aphids. By stimulating nitrogen mineralization, earthworms enhanced plant nitrogen uptake and increased nitrogen concentration in plant tissues

156 K. Poveda *et al.*

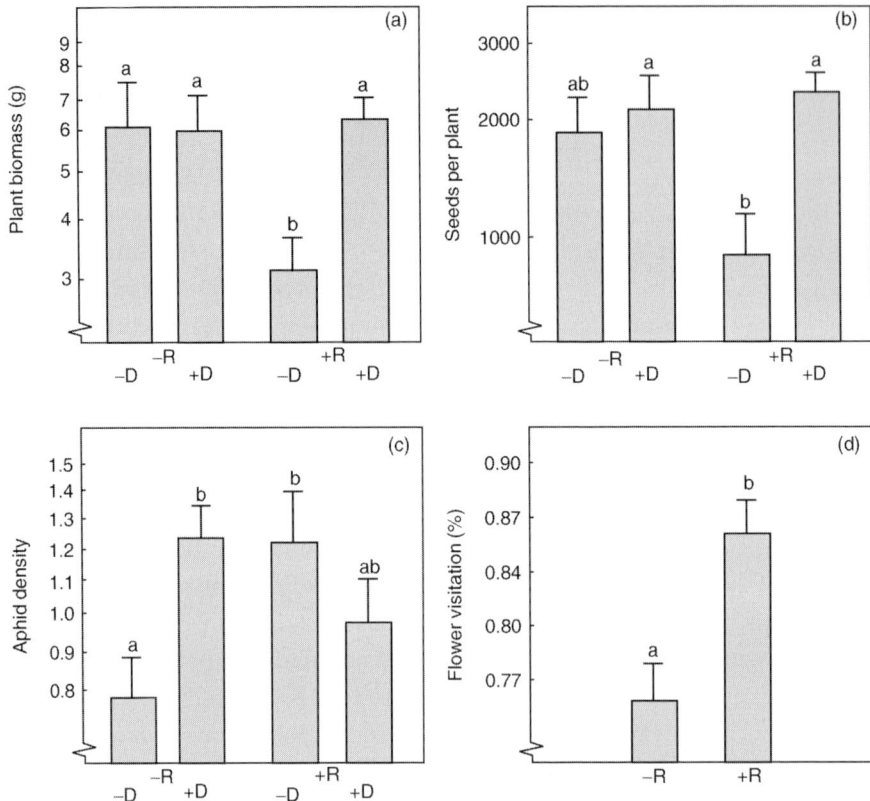

Figure 7.1. Effects of decomposers (D) and root herbivores (R) on (a) plant biomass, (b) number of seeds per plant, (c) number of aphids on a plant, and (d) flower visitation rate (mean ± 1 standard error). Treatments with different letters are significantly different (Tukey's honestly significant difference test, $p <0.05$). Note the log scale of y-axis.

(see above). By feeding on roots, wireworms likely decreased the water uptake by the plant and consequently increased the nitrogen concentration in the phloem (Gange and Brown 1989, Masters and Brown 1992, Masters *et al.* 1993, 2001). There was no effect of herbivores or decomposers on the rate of parasitism of the aphids (Table 7.1). The number of parasitoids was only affected by the number of aphids per plant (simple correlation: $r = 0.74$, $n = 40$, $p <0.001$). Flower visitation rate was higher in plants with root herbivores (Table 7.1, Fig. 7.1d); the most abundant flower visitor was the honeybee, *Apis mellifera*, comprising 61% of the visits. This suggests that root herbivory increases plant attractiveness to flower visitors. In a previous study, lower quantities of root herbivores (two wireworms of the genus *Agriotes*) also enhanced the flower visitation rate (Poveda *et al.* 2003).

Table 7.1. *ANOVA table of F and p values (in parentheses) on the effect of root herbivores (R) and decomposers (D) on plant biomass, seed production, flowering period, aphid abundance, parasitation rate, and number of visited flowers per flower of* Sinapis arvensis (n = 40)

	Plant biomass	Seeds/ plant	Flowering period	Aphid abundance	Parasitism rate	Visited flowers (%)
R	3.39 (n.s.)	2.11 (n.s.)	0.55 (n.s.)	0.59 (n.s.)	0.89 (n.s.)	8.92 (0.003)
D	4.19 (0.047)	6.88 (0.012)	4.19 (0.047)	0.81 (n.s.)	0.77 (n.s.)	0.37 (n.s.)
R D	4.72 (0.036)	3.81 (0.058)	0.40 (n.s.)	7.56 (0.009)	0.55 (n.s.)	1.75 (n.s.)

Source: Data are from the first experiment.

Table 7.2. *ANOVA table of F and p values (in parentheses) on the effect of root herbivores (R) and decomposers (D) on plant biomass, seed production, anther length, nectar production, and petal size of* Sinapis arvensis (n = 24)

	Plant biomass	Seeds/plant	Anther length	Nectar	Petal size
R	0.34 (n.s.)	0.03 (n.s.)	2.25 (n.s.)	0.0006 (n.s.)	1.83 (n.s.)
D	8.30 (0.009)	6.47 (0.02)	1.05 (n.s.)	0.09 (n.s.)	0.31 (n.s.)
R D	2.86 (n.s.)	0.51 (n.s.)	1.37 (n.s.)	0.002 (n.s.)	0.06 (n.s.)

Source: Data are from the second experiment.

In the present experiment, the density of wireworms was more than twice as high as in our previous study (five instead of two wireworms per pot), but effects on flower visitation were similar. This suggests that root herbivores, even at relatively high densities, may stimulate the attractiveness of flowers to flower visitors.

In the second experiment we measured floral traits, such as the amount of nectar, pollen production, and size of the petals, because these parameters may reveal the mechanisms underlying the enhanced attractiveness of flowering plants in the herbivore treatment. In contrast to our expectations, none of the parameters of floral traits measured was affected by the presence of root herbivores or decomposers (Table 7.2). Potentially, changes in nectar concentration may have been responsible for the observed changes in flower visitation. Masters et al. (1993) shows that water stress caused by root herbivory on plants led to the accumulation of soluble amino acids and carbohydrates in the foliage. It is possible that the carbohydrate concentration of nectar also increased in the presence of root herbivores, thereby augmenting the attractiveness to flower visitors, but this was not measured in our study.

In both experiments, decomposers enhanced plant biomass and fruit and seed set, but did not affect floral traits. This suggests that decomposers increase the plants' fitness. It is challenging to investigate the evolutionary forces that resulted in the observed variation of responses of *S. arvensis* to a complex of plant-associated invertebrate guilds, and to relate these to other plant species of different functional groups and with different life histories. Such studies may lead to a more detailed understanding of how belowground and aboveground plant–animal interactions shaped the evolution of plant traits.

Conclusions and future research

Belowground communities affect not only plant growth but also aboveground processes, including plant–herbivore–parasitoid interactions and even plant–pollinator interactions. Effects of indirect interactions mediated through the decomposer subsystem presumably are as important as direct interactions caused by root herbivory, leading in some cases to the same aboveground response. The mechanisms responsible for these effects are still poorly known, and further research is needed to understand the physiological processes in the plants. Our case studies shed some light on the effects of decomposers and root herbivores on aboveground plant–animal interactions (Fig. 7.2), but integrating these effects into the whole soil food web remains to be done. The links between belowground and aboveground biotic interactions are much more complex than what we have presented here. Future research on the mechanisms responsible for aboveground–belowground interrelationships is essential for a better understanding of the ecology and evolution of plant–animal interactions. Results of the studies presented suggest that the effects are highly dependent on the plant species involved. Field studies are needed to understand more clearly the role of belowground organisms on aboveground biotic interactions in the context of natural plant communities. Feedbacks between the aboveground and the belowground systems should also be taken into account, since it is known that the amount of carbon translocated into roots and into the rhizosphere may significantly increase if plants are subjected to aboveground herbivory. This may strongly influence the rhizosphere food web (Mikola *et al.* 2001, Bardgett and Wardle 2003), which in turn likely affects aboveground interactions.

Until recently the role of soil biota in modifying the attractiveness of plants to herbivores and pollinators has been largely ignored. Particularly in natural habitats where plant species interact with each other, soil biota may play a key role in modifying the insect community associated with plants. Future work needs to focus upon more complex systems consisting of plants and soil biota in

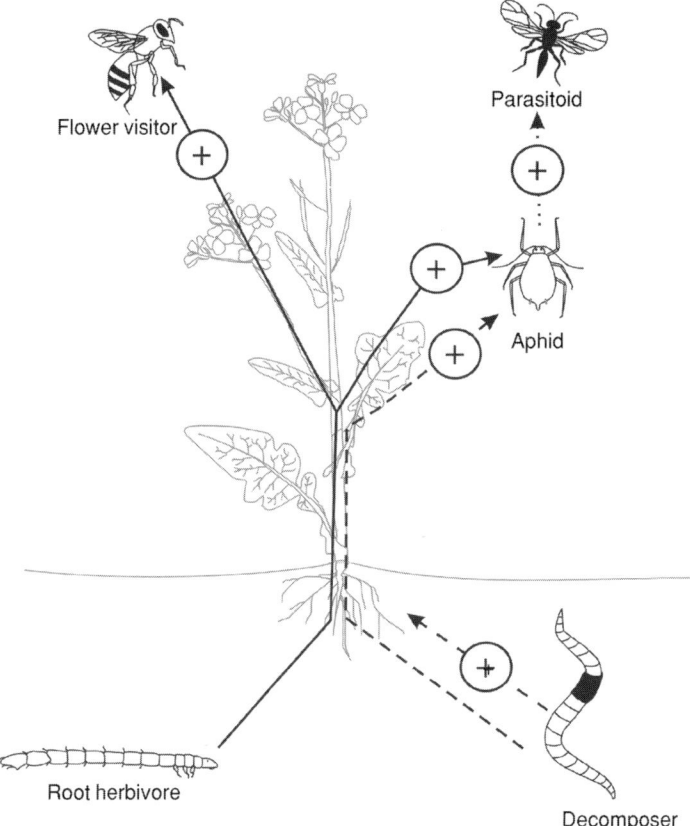

Figure 7.2. Ways in which root herbivores and decomposers affect flower visitors, aphids, and their parasitoids through *Sinapis arvensis* plants. Feeding activities by root herbivores (continuous line) cause an indirect positive effect on aphids and flower visitors. Decomposers (dashed line) indirectly increase plant growth and the number of aphids. An increase in the number of aphids caused an indirect increase in the number of parasitoids (dotted line).

different spatial and temporal contexts to better understand the functioning of terrestrial systems.

Acknowledgments

The authors thank the German National Academic Foundation (Studienstiftung des Deuschen Volkes) and the Universidad Nacional de Colombia for financial support, and Janet Lensing, Erika Garcia, Judith Bronstein, and an anonymous reviewer for comments on early drafts of the manuscript. Our special thanks go to Gabriela Poveda and Jorge Jácome for invaluable help with field work.

References

Alphei, J., M. Bonkowski, and S. Scheu. 1996. Protozoa, Nematoda, and Lumbricidae in the rhizosphere of *Hordelymus europaeus* (Poaceae): faunal interactions, response to microorganisms and effects on plant growth. *Oecologia* **106**:111–126.

Andersen, D. 1987. Below-ground herbivory in natural communities: a review emphasizing fossorial animals. *Quarterly Review of Biology* **62**:261–286.

Bardgett, R. D., and D. A. Wardle. 2003. Herbivore mediated linkages between aboveground and belowground communities. *Ecology* **84**:2258–2268.

Bardgett, R. D., D. K. Leemans, R. Cook, and P. Hobbs. 1997. Temporal and spatial variation in soil biota of grazed and ungrazed hill grassland in Snowdonia, North Wales. *Soil Biology and Biochemistry* **29**:1286–1294.

Bardgett, R. D., S. Keiller, R. Cook, and A. S. Gilburn. 1998. Dynamic interactions between soil animals and microorganisms in upland grassland soils amended with sheep dung: a microcosm experiment. *Soil Biology and Biochemistry* **30**:531–539.

Bezemer, T. M., R. Wagenaar, N. M. van Dam, and F. L. Wäckers. 2002. Interactions between root and shoot feeding insects are mediated by primary and secondary plant compounds. *Proceedings of the Section Experimental and Applied Entomology of the Netherlands Entomological Society (NEV)* **13**:117–121.

Bohlen, P., S. Scheu, C. Hale, *et al.* 2004. Non-native invasive earthworms as agents of change in northern temperate forests. *Frontiers in Ecology and the Environment* **2**:427–435.

Bonkowski, M. 2004. Protozoa and plant growth: the microbial loop in soil revisited. *New Phytologist* **162**:617–631.

Bonkowski, M., W. Cheng, B. Griffiths, J. Alphei, and S. Scheu. 2000. Microbial-faunal interactions in the rhizosphere and effects on plant growth. *European Journal of Soil Biology* **36**:135–147.

Bonkowski, M., I. Geoghegann, A. Nicholas, E. Birch, and B. Griffiths. 2001. Effects of soil decomposer invertebrates (protozoa and earthworms) on above-ground phytophagous insects (cereal aphid) mediated through changes in the host plant. *Oikos* **95**:441–450.

Brown, G. G., C. A. Edwards, and L. Brussaard. 2004. How earthworms affect plant growth: burrowing into mechanisms, pp. 13–49 in C. A. Edwards (ed.) *Earthworm Ecology*, 2nd edn. Boca Raton, FL: CRC Press.

Brown, V. K., and A. C. Gange. 1989. Herbivory by soil-dwelling insects depresses plant species richness. *Functional Ecology* **3**:667–671.

Brown, V. K., and A. C. Gange. 1990. Insect herbivory below ground. *Advances in Ecological Research* **20**:1–58.

Brown, V. K., and A. C. Gange. 2002. Tritrophic below- and above-ground interactions in succession, pp. 197–222 in T. Tscharntke and B. Hawkins (eds.) *Multitrophic Level Interactions*. Cambridge, UK: Cambridge University Press.

Callaham, M. A., and P. E. Hendrix. 1998. Impact of earthworms (Diplocardia, Megascolecidae) on cycling and uptake of nitrogen in coastal plain forest soils from northwest Florida, USA. *Applied Soil Ecology* **9**:233–239.

Clarholm, M. 1985. Possible roles for roots, bacteria, protozoa and fungi in supplying nitrogen to plants, pp. 355–365 in A. H. Fitter (ed.) *Ecological Interactions in Soils.* Oxford, UK: Blackwell Science.

Coleman, D. C. 1994. The microbial loop as used in terrestrial soil ecology. *Microbial Ecology* **28**:245–250.

Dafni, A. 1992. *Pollination Ecology: A Practical Approach.* Oxford, UK: Oxford University Press.

Gange, A. C., and V. K. Brown. 1989. Effects of root herbivory by an insect on a foliar feeding species, mediated through changes in the host plant. *Oecologia* **81**:38–42.

Haimi, J., and M. Einbork. 1992. Effects of endogeic earthworms on soil processes and plant growth in coniferous forest soil. *Biology and Fertility of Soils* **13**:6–10.

Haimi, J., V. Huhta, and M. Boucelham. 1992. Growth increase of birch seedlings under the influence of earthworms: a laboratory study. *Soil Biology and Biogeochemistry* **24**:1525–1528.

Herrera, C. M., M. Medrano, P. J. Rey, *et al.* 2002. Interaction of pollinators and herbivores on plant fitness suggests a pathway for correlated evolution of mutualism- and antagonism-related traits. *Proceedings of the National Academy of Sciences of the USA* **99**:16823–16828.

Jones, C. G., J. H. Lawton, and M. Shachak. 1994. Organisms as ecosystem engineers. *Oikos* **69**:373–386.

Lavelle, P., and A. V. Spain. 2001. *Soil Ecology.* Dordrecht, The Netherlands: Kluwer Academic Publishers.

Lehtilä, K., and S. Strauss. 1997. Leaf damage by herbivores affects attractiveness to pollinators in wild radish, *Raphanus raphanistrum*. *Oecologia* **111**:396–403.

Lehtilä, K., and S. Strauss. 1999. Effects of foliar herbivory on male and female reproductive traits of wild radish, *Raphanus raphanistrum*. *Ecology* **80**:116–124.

Maron, J. L. 2001. Intraspecific competition and subterranean herbivory: individual and interactive effects on bush lupine. *Oikos* **92**:178–186.

Masters, G. J. 1995a. The effect of herbivore density on host plant mediated interactions between two insects. *Ecological Research* **10**:125–133.

Masters, G. J. 1995b. The impact of root herbivory on aphid performance: field and laboratory evidence. *Acta Oecologica* **16**:135–142.

Masters, G. J., and V. K. Brown. 1992. Plant mediated interactions between two spatially separated insects. *Functional Ecology* **6**:175–179.

Masters, G. J., and V. K. Brown. 1997. Host-plant mediated interactions between spatially separated herbivores: effects on community structure, pp. 217–237 in A. C. Gange and V. K. Brown (eds.) *Multitrophic Interactions in Terrestrial Systems.* Oxford, UK: Blackwell Science.

Masters, G. J., V. K. Brown, and A. C. Gange. 1993. Plant mediated interactions between above- and below-ground insect herbivores. *Oikos* **66**:148–151.

Masters, G. J., H. Jones, and M. Rogers. 2001. Host-plant mediated effects of root herbivory on insect seed predators and their parasitoids. *Oecologia* **127**:246–250.

Mikola, J., G. W. Yeates, G. M. Barker, D. A. Wardle, and K. I. Bonner. 2001. Effects of defoliation intensity on soil food web properties in an experimental grassland community. *Oikos* **92**:333–343.

Moore, J. C., K. McCann, H. Setälä, and P. C. De Ruiter. 2003. Top-down is bottom-up: does predation in the rhizosphere regulate aboveground dynamics? *Ecology* **84**:846–857.

Moran, N. A., and T. G. Whitham. 1990. Interspecific competition between root-feeding and leaf-galling aphids mediated by host-plant resistance. *Ecology* **71**:1050–1058.

Mortimer, S. R., W. van der Putten, and V. K. Brown. 1999. Insect and nematode herbivory below ground: interactions and role in vegetation succession, pp. 205–238 in H. Olff, V. K. Brown, and R. H. Drent (eds.) *Herbivores: Between Plants and Predators*. Oxford, UK: Blackwell Science.

Mothershead, K., and R. Marquis. 2000. Fitness impacts of herbivory through indirect effects on plant–pollinator interactions in *Oenothera macrocarpa*. *Ecology* **81**:30–40.

Newington, J. E., H. Setälä, T. M. Bezemer, and T. H. Jones. 2004. Potential effects of earthworms on leaf-chewer performance. *Functional Ecology* **18**:746–751.

Nötzold, R., B. Blossey, and E. Newton. 1998. The influence of below ground herbivory and plant competition on growth and biomass allocation of purple loosestrife. *Oecologia* **113**:82–93.

Pellmyr, O. 2002. Pollination by animals, pp. 157–184 in C. M. Herrera and O. Pellmyr (eds.) *Plant–Animal Interactions: An Evolutionary Approach*. Oxford, UK: Blackwell Science.

Porazinska, D. L., R. D. Bardgett, M. B. Blaauw, et al. 2003. Relationships at the aboveground–belowground interface: plants, soil biota and soil processes. *Ecological Monographs* **73**:377–395.

Poveda, K., I. Steffan-Dewenter, S. Scheu, and T. Tscharntke. 2003. Effects of below- and above-ground herbivores on plant growth, flower visitation and seed set. *Oecologia* **135**:601–605.

Preisser, E. L. 2003. Field evidence for a rapidly cascading underground food web. *Ecology* **84**:869–874.

Ruess, L., M. Häggblom, R. Langel, and S. Scheu. 2004. Nitrogen isotope ratios and fatty acid composition as indicators of animal diets in belowground systems. *Oecologia* **139**:336–346.

Scheu, S. 2001. Plants and generalist predators as links between the belowground and aboveground system. *Basic and Applied Ecology* **2**:3–13.

Scheu, S.. 2003. Effects of earthworms on plant growth: patterns and perspectives. *Pedobiologia* **47**:846–856.

Scheu, S., and D. Parkinson. 1994. Effects of invasion of an aspen forest (Canada) by *Dendrobaena octaedra* (Lumbricidae) on plant growth. *Ecology* **75**:2348–2361.

Scheu, S., and H. Setälä. 2002. Multitrophic interactions in decomposer food-webs, pp. 223–264 in T. Tscharntke and B. Hawkins (eds.) *Multitrophic Level Interactions*. Cambridge, UK: Cambridge University Press.

Scheu, S., A. Theenhaus, and T. H. Jones. 1999. Links between the detritivore and the herbivore system: effects of earthworms and Collembola on plant growth and aphid development. *Oecologia* **119**:541–551.

Scheu, S., L. Ruess, and M. Bonkowski. 2004. Interactions between micro-organisms and soil micro- and mesofauna, pp. 253–278 in F. Buscot and A. Varma (eds.) *Microorganisms in Soils: Roles in Genesis and Function*. New York: Springer-Verlag.

Schmidt, O., and J. P. Curry. 1999. Effects of earthworms on biomass production, nitrogen allocation and nitrogen transfer in wheat–clover intercropping model systems. *Plant and Soil* **214**:187–198.

Setälä, H., and V. Huhta. 1991. Soil fauna increase *Betula pendula* growth: laboratory experiments with coniferous forest floor. *Ecology* **76**:1844–1851.

Spain, A. V., P. Lavelle, and A. Mariotti. 1992. Stimulation of plant growth by tropical earthworms. *Soil Biology and Biochemistry* **26**:1179–1184.

Stanton, N. L. 1988. The underground in grasslands. *Annual Review of Ecology and Systematics* **19**:573–589.

Strauss, S. 1997. Floral characters link herbivores, pollinators, and plant fitness. *Ecology* **78**:1640–1645.

Strong, D. R., A. V. Whipple, A. L. Child, and B. Dennis. 1999. Model selection for a subterranean trophic cascade: root feeding caterpillars and entomopathogenic nematodes. *Ecology* **80**:2750–2761.

van der Putten, W., L. E. M. Vet, J. Harvey, and F. Wäckers. 2001. Linking above- and belowground multitrophic interactions of plants, herbivores, pathogens, and their antagonists. *Trends in Ecology and Evolution* **16**:547–554.

Wardle, D. 1999. How soil food webs make plants grow. *Trends in Ecology and Evolution* **14**:418–420.

Wardle, D. 2002. *Communities and Ecosystems: Linking the Aboveground and Belowground Components*. Princeton, NJ: Princeton University Press.

Wardle, D. A., R. D. Bardgett, J. N. Klironomos, et al. 2004. Ecological linkages between aboveground and belowground biota. *Science* **304**:1629–1633.

White, T. C. R. 1993. *The Inadequate Environment: Nitrogen and the Abundance of Animals*. Berlin, Germany: Springer-Verlag.

Wurst, S., and T. H. Jones. 2003. Indirect effects of earthworms (*Aporrectodea caliginosa*) on an above-ground tritrophic interaction. *Pedobiologia* **47**:91–97.

Wurst, S., R. Langel, A. Reineking, M. Bonkowski, and S. Scheu. 2003. Effects of earthworms and organic litter distribution on plant performance and aphid reproduction. *Oecologia* **137**:90–96.

Wurst, S., D. Dugassa-Gobena, and S. Scheu. 2004a. Earthworms and litter distribution affect plant defence chemistry. *Journal of Chemical Ecology* **30**:691–701.

Wurst, S., D. Dugassa-Gobena, R. Langel, M. Bonkowski, and S. Scheu. 2004b. Combined effects of earthworms and vesicular-arbuscular mycorrhizae on plant and aphid performance. *New Phytologist* **163**:169–176.

8

Bottom–up cascades induced by fungal endophytes in multitrophic systems

ENRIQUE J. CHANETON AND MARINA OMACINI

Introduction

Current views on ecological communities have been shaped by conceptual and empirical progress in two related themes. First, the last two decades have seen a shift from the traditional emphasis on pair-wise species interactions towards a broader, multispecies approach, which highlights the impact of *indirect interactions* on community dynamics (Wootton 1994, Abrams *et al.* 1996, Gange and Brown 1997, Schmitz 1998, Werner and Peacor 2003). Second, there is increasing recognition that resources and productivity at the base of the food web (*bottom–up* forces) as well as predation from higher-level consumers (*top–down* forces) are both important in community structuring (Oksanen *et al.* 1981, Hunter and Price 1992, Leibold *et al.* 1997, Polis 1999, Meserve *et al.* 2003). Nevertheless, some issues remain to be addressed before a comprehensive understanding of community regulation can be achieved. One aspect that needs attention is the role of microbial symbionts in trophic dynamics (Hunter and Price 1992, Polis and Strong 1996). Ecosystems would probably look very different without such "hidden players," yet we are only beginning to uncover the influence of plant and animal symbionts in multitrophic systems (Omacini *et al.* 2001, Gange *et al.* 2003, Oliver *et al.* 2003).

In this chapter, we focus on fungal endophytes as potential elicitors of bottom–up indirect effects transmitted to upper trophic levels through a modification of host plant traits. Endophytic fungi grow systemically within the aerial

Ecological Communities: Plant Mediation in Indirect Interaction Webs, ed. Takayuki Ohgushi, Timothy P. Craig, and Peter W. Price. Published by Cambridge University Press. © Cambridge University Press 2007.

tissues of many grass species (Clay 1990, Clay and Schardl 2002). These symbionts have been regarded as "defensive mutualists" or "acquired defenses" (Clay 1988, Cheplick and Clay 1988; cf. Faeth 2002), because of their purported role in plant protection from herbivores and pathogens (Clay 1990, 1991, Breen 1994, Popay and Bonos 2005). We consider the indirect effects of grass endophytes on phytophagous insects and their natural enemies to illustrate two general points: (1) the potential impact of cryptic symbionts on multitrophic interactions, and (2) the pervasive influence of biotic and abiotic heterogeneity in shaping consumer responses to plant-trait-mediated indirect effects. In this way we hope to contribute to a broader appreciation of ecological interactions involving microbial symbionts.

We begin with an overview of current ideas on indirect interactions and food web regulation, which provides the theoretical context for studying endophyte–grass symbioses from a multitrophic perspective. Second, we integrate endophytic fungi into a simple model of "bottom-up cascades" that combines density- and trait-mediated interactions. Third, we discuss endophyte mediation of plant–insect relations and review the evidence for endophyte effects on higher trophic levels transmitted by changes in plant quality and herbivore traits. We end by discussing our recent work showing how fungal endophytes can indirectly alter the structure of multitrophic insect assemblages.

Conceptual background

Ecological communities are structured by complex interaction networks involving organisms of varied sizes and trophic functions. An emergent property of this complexity is the occurrence of *indirect* interactions (or effects), that arise when the effect of one species on another depends on the presence of a third species (Wootton 1994, Abrams et al. 1996). By contrast, *direct* interactions occur between two species even if they are in isolation. The significance of indirect effects was initially recognized through field experiments yielding unexpected responses to species manipulations (e.g., Brown et al. 1986). It is now clear that indirect effects are common and can have community-wide impacts comparable to those of direct effects (Wootton 1994, Abrams et al. 1996, Werner and Peacor 2003).

Two general types of indirect interaction have been distinguished regarding the mechanism whereby intermediary species transmit effects between donor and receiver species (Abrams et al. 1996). Indirect effects are often transmitted by changes in population densities of intermediary species; such "density-mediated" indirect interactions (DMII) typically propagate through chains of pair-wise consumer–resource relations (also termed "interaction chains": Wootton 1994). Alternatively, indirect effects may arise when the intermediary species alters its phenotype in the presence of other species, and thus changes its per capita effect on the receiver species. These "trait-mediated" indirect interactions (TMII; also "interaction modifications": Wootton 1994) can be transmitted via morphological, physiological, biochemical, or behavioral changes, even if the abundance of the intermediary remains unaltered (Werner and Peacor 2003). Whereas the outcome of DMII can be predicted from knowledge of pair-wise direct effects between intervening species, TMII are essentially unpredictable from species pairs, since trait modifications are contingent on the particular species present (Wootton 1994, Werner and Peacor 2003). Indirect effects mediated by trait plasticity may be relatively rapid compared to density-mediated effects requiring changes in demographic rates to arise (Brown et al. 1986, Abrams 1995). Therefore, the distinction between DMII and TMII has implications for predicting community responses to species loss, invasions, and environmental change (Wootton 1994).

One type of indirect effect that has attracted much attention in recent studies of terrestrial communities are the so-called "trophic cascades" (Polis 1999, Schmitz et al. 2000, Halaj and Wise 2001, Dyer and Coley 2002). Trophic cascades occur when the effects of top carnivores on middle-level consumers (primary carnivores and/or herbivores) produce significant changes in plant damage, biomass, density, or diversity. Such top–down indirect effects are often transmitted by density changes of intermediary consumers (Schmitz et al. 2000, Halaj and Wise 2001), but can also be mediated by trait modifications associated with predation risk (Schmitz et al. 2004), or through a combination of density- and trait-based interaction linkages (Schmitz 1998, Werner and Peacor 2003). Evidence largely from arthropod webs indicates, however, that not all top–down effects lead to observable changes in plant abundance. Cascades initiated by top carnivores may be attenuated as they pass through reticulate middle-consumer webs, and thus may "trickle down" to affect herbivore assemblages but not their food plants (Polis and Strong 1996, Schmitz et al. 2000, Halaj and Wise 2001). Further, because plant species greatly differ in vulnerability to herbivores, carnivore-induced variation in herbivory patterns may not always translate into changes in plant biomass (Leibold et al. 1997, Polis 1999).

Together with the potential importance of top–down control, ecological theory allows for a prominent role of productivity and other bottom–up factors in multitrophic dynamics (Hunter and Price 1992, Abrams 1993, Leibold et al. 1997, Polis 1999, Oksanen and Oksanen 2000, Price 2002). This perspective is embodied in the "bottom–up cascade" concept, according to which substantial changes in the amount or quality of resources at the base of the food web should propagate upwards altering the abundance of higher-level consumers and, eventually, their impact on intermediary consumers. Note that this definition accommodates both density- and trait-mediated indirect effects, as we shall illustrate below (see section "Mechanisms for endophyte-induced bottom–up cascades"). Dynamic models based on linear food chains (Oksanen et al. 1981, Oksanen and Oksanen 2000) predict increased abundance of alternate trophic levels in response to energy/resource inputs. In contrast, models incorporating various forms of within-trophic-level heterogeneity (Hunter and Price 1992) generally predict positive effects of increased basal resources on all upper trophic levels, although effects can vary depending on food web topology and species composition (Abrams 1993, Leibold et al. 1997).

The potential for top–down feedbacks generated by greater energy transfer from the bottom up is a key element of trophic regulation models (Oksanen et al. 1981, Hunter and Price 1992, Leibold et al. 1997). Several studies on plant-arthropod systems have manipulated both resources and carnivores to examine interactions between bottom–up and top–down forces (e.g., Fraser and Grime 1998, Dyer and Letourneau 1999, Forkner and Hunter 2000, Denno et al. 2002, Moon and Stiling 2002). This body of evidence indicates not only that plant productivity may affect the abundance of higher-level consumers, but also that bottom–up factors set the stage for top–down forces (Hunter and Price 1992). Moreover, the strength of top–down effects may vary with changes in either growth or quality of basal plants (Forkner and Hunter 2000). On the other hand, experimental studies have shown that the propagation of cascading effects induced by predator exclusion or nutrient addition depends on the identity of intermediary herbivore species (Denno et al. 2002, Moon and Stiling 2002).

Both theory and data suggest that bottom–up cascades initiated by environmentally modified plant traits play a significant role in terrestrial communities (Price 2002). Heterogeneity in plant quality and architecture may, indeed, reflect the influence of abiotic and biotic sources of phenotypic plasticity. It is well known that plants respond to changing abiotic conditions by adjusting their morphology, physiology, and biochemistry (Callaway et al. 2003). Plant tissue quality may also reflect changes in secondary chemistry induced by interactions with herbivores and pathogens (Karban and Baldwin 1997, Agrawal et al. 1999). The consequences of such variation in plant quality for insect herbivores and their natural enemies have been well researched (Price et al. 1980, Price 1991, Karban and Baldwin 1997, Tollrian and Harvell 1999, Ohgushi Chapter 10 this volume). In addition, most plants harbor different kinds of microbial symbionts, including shoot and root mutualists or antagonists, which may strongly influence plant phenotypic expression and plant–herbivore relations. While plant symbionts have been studied for over a century (Boucher et al. 1982), their role in multitrophic systems has yet to be fully discovered. Systemic endophytes that protect their host grasses from insect herbivores via mycotoxin production

(Bush et al. 1997) provide a useful system to examine plant-trait-mediated effects of symbiotic microbes in multitrophic assemblages.

Endophyte–grass symbiosis

The term "endophyte" refers to microorganisms that colonize and live inside the tissues of living plants without causing visible signs of disease. Endophytic fungi have been found in algae, mosses, and ferns, as well as in conifers and angiosperms including grasses, palms, and a diversity of shrubs and trees (Saikkonen et al. 1998). For the purpose of this chapter, we consider fungal endophytes in the tribe Balansieae (Clavicipitaceae, Ascomycota), which form lifelong symbiotic associations with C_3 grasses in the subfamily Pooideae by growing systemically in the aerial organs of the host plant (Clay 1990, Clay and Schardl 2002). Other endophytic associations with non-grass hosts (Saikkonen et al. 1998) will not be discussed here, since little is known of their interactions with higher-level consumers (Faeth and Bultman 2002).

Grass endophytes comprise a diverse group of closely related fungal species with contrasting life histories. A variety of asexual forms (*Neotyphodium* spp.) have radiated from members of the genus *Epichloë*, most commonly through interspecific hybridization (Moon et al. 2000, Clay and Shardl 2002). The fungus mode of reproduction varies with the type of endophyte–grass symbiosis, being partly correlated with the nature of the interaction, which spans a continuum from antagonism to mutualism (Saikkonen et al. 1998, Clay and Schardl 2002). Antagonistic, sexually reproducing endophytes form fruiting bodies (stroma) on flowering tillers and spread contagiously by ascospores, while suppressing host seed set. Unlike their *Epichloë* ancestors, *Neotyphodium* species produce asymptomatic infections, which are vertically transmitted by hyphae growing into the host seeds. Because of obligate asexual transmission, the long-term survival of these endophytes is largely determined by the fitness of the host plant (Clay and Schardl 2002, Saikkonen et al. 2004).

Seed-borne, *Neotyphodium* endophytes occur in many turf and forage grasses as well as in wild, native grass species (Saikkonen et al. 1998, Clay and Schardl 2002, Faeth 2002). In the field, endophyte-infected grasses have been found to increase in abundance over successional time at the expense of uninfected conspecifics and other co-occurring plant species (Clay 1990, Clay and Holah 1999). Asexual endophytes rely entirely on their host for nutrition as well as reproduction. The hyphae colonize intercellular spaces of leaves and stems obtaining carbohydrates and amino acids from the apoplast. Potential plant benefits from such endophyte infections are manifold and include increased vigor and competitive ability (Clay 1994), greater tolerance to abiotic stress (Malinowski and Belesky 2000), and herbivore resistance due to accumulation of fungal alkaloids (Clay 1991, Dahlman et al. 1991, Bush et al. 1997). These endophyte–grass symbioses have generally been regarded as mutualistic (Clay 1988, 1990, Breen 1994, Clay and Schardl 2002). However, most evidence to date is based on cultivars of two widespread forage grasses, *Festuca arundinacea* (tall fescue) and *Lolium perenne* (perennial ryegrass), and on several generalist insect pests (Clay 1991), whereas little is known about the nature of endophyte associations with native grasses (Faeth 2002, Saikkonen et al. 2004). Indeed, recent work on the grass *F. arizonica* (Arizona fescue) suggests that positive endophyte effects on host plant fitness may not be universal, with *Neotyphodium* becoming parasitic under certain conditions (Faeth and Sullivan 2003, Faeth et al. 2004).

Mechanisms for endophyte-induced bottom–up cascades

We now turn to the integration of fungal endophytes into a generalized multitrophic system. We start by considering a simple three-trophic-level chain to illustrate the types of indirect effects that may arise in the presence of endophytes (Fig. 8.1). Note, however, that this rationale may be readily extended to cover more complex food web topologies. In the classic three-level food chain, plant productivity and quality determine the energy available to herbivores and, ultimately, to secondary consumers. Consumption pressure from top carnivores may in turn decrease herbivory and trigger trophic cascades benefiting plants. Bottom–up and top–down cascades of this kind (Fig. 8.1, right-hand dashed arrows) have already been discussed here.

In addition, plants may trigger bottom–up effects by directly modifying key traits of insect herbivores and natural enemies (Price et al. 1980, Müller and Godfray 1999). In Fig. 8.1 TMIIs are depicted as dotted arrows pointing toward the intermediary agent whose phenotype is being modified (after Werner and Peacor 2003). For example, changes in aggregative behavior and physiological condition of insect herbivores may reflect, respectively, the relative abundance and nutritional quality of food plants and will impact on enemy attack rates (Price 1991, 2002). Plants may also affect natural enemies through the effect of canopy architecture on prey location by predatory arthropods and parasitoids (Clark and Messina 1998, Denno et al. 2002). Furthermore, emission of volatile compounds by wounded plants has been found to attract natural enemies, reinforcing top–down control on herbivorous insects (Kessler and Baldwin 2001).

The presence of endophytic grasses may, in theory, generate indirect effects that will reverberate through the food web. Clay (1994) first pointed out the potential impact of endophytes on higher-level consumers. Endophyte effects may cascade up through trait or density changes in host grass species (Fig. 8.1).

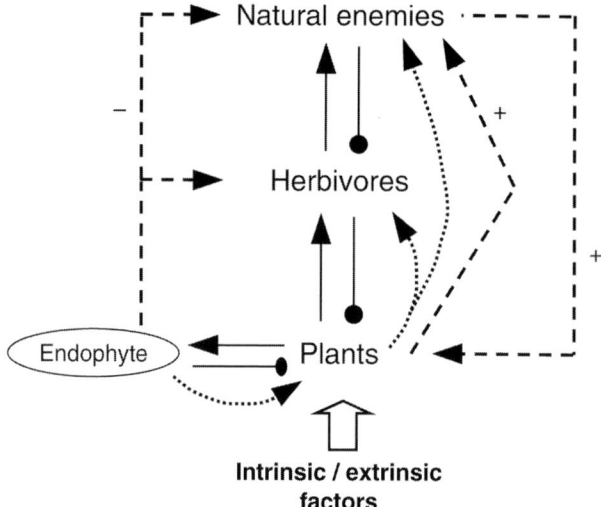

Figure 8.1. Pathways for bottom–up cascades induced by fungal endophytes in multitrophic insect assemblages. Natural enemies include predators and parasites. Direct interactions are depicted by solid lines indicating the direction of energy flow (arrow-head) and negative effects from consumption (bullet-head). Indirect interactions comprise density-mediated (dashed lines) and trait-mediated (dotted lines) effects. The curved dotted arrows point toward the species whose phenotypic traits are modified by the interaction.

Systemic endophytes are energetically maintained by plant photosynthates. A mutualistic or parasitic association will then be realized depending on whether beneficial modifications in host traits, and ultimately fitness, override negative effects from the fungus exploitation of plant-acquired resources. Infected grasses exhibit phenotypic differences relative to their uninfected conspecifics that can alter the levels of herbivory they experience. For example, endophyte-infected plants may show architectural changes associated with higher tillering (Clay 1990). Endophyte infection may also alter tissue quality to herbivorous insects by enhancing plant uptake of water and mineral nutrients (Malinowsky and Belesky 2000). However, predicting how such morphological and physiological changes will affect herbivory may not be straightforward (Bultman and Bell 2003).

The best-known allelochemical modification in endophytic grasses is the accumulation of bioprotective metabolites, in particular endophyte-produced ergot, diterpene, peramine, and pyrrolizidine (loline) alkaloids. The presence of alkaloids in host tissues may act as a potent feeding deterrent, and was found to decrease reproductive performance in several insect taxa (Dahlman et al. 1991,

Bush et al. 1997). Additionally, insects feeding on endophyte-infected plants may show impaired growth, together with behavioral or biochemical changes, which can influence interactions with natural enemies (Faeth and Bultman 2002, Bultman et al. 2003). Few data exist on endophyte-induced changes in plant volatile emissions (Yue et al. 2001), and it remains to be seen whether they are relevant to insect herbivores and/or their natural enemies.

Endophyte-induced resistance to herbivory is not limited to aerial insects but involves root feeders as well (Breen 1994, Popay et al. 2003). Fungal alkaloids have been found in roots of infected grasses (Bush et al. 1997). Where present, endophyte effects on soil-dwelling herbivores may propagate to upper trophic levels of belowground food chains. We further speculate that insects embedded in detritus-based consumer webs may be susceptible to bottom–up control from the litter deposited by endophyte-infected grasses (Omacini et al. 2005). Endophyte-induced changes in grass litter quality and persistence could indirectly alter the structure of detritivore communities (see Davidson and Potter 1995, Bernard et al. 1997).

We have so far considered the pathways for plant-trait-mediated effects of endophytes on primary and secondary consumers. Such TMII may interact with plant density changes to affect herbivorous insects and associated enemies. The tendency of endophytic grasses to become dominant in cultivated and semi-natural communities (Clay 1990, Clay and Schardl 2002), and the strong dependence of insects on food plant abundance and canopy structure (Price 2002), offer ample scope for endophyte-driven DMII. Clay (1994) posited that enhanced productivity from endophyte infection will be accompanied by reduced energy transfer and consumer abundance at higher trophic levels (Fig. 8.1: negative, bottom–up DMII). Even if plant productivity remains constant, competitive displacement of co-occurring plant species by endophytic grasses could lead to greater dominance by specialized, endophyte-resistant herbivores, and hence may alter insect diversity (Clay 1994, S. Faeth personal communication). Yet observed community-level effects of fungal endophytes might well depend on whether infected grasses hold a substantial fraction of the total plant biomass (Omacini et al. 2005).

Endophyte infection interacts with both intrinsic and extrinsic factors in determining the host plant phenotype (Fig. 8.1), and this variation will strongly affect the way endophytes mediate interactions between host plants and their insect herbivores. Effects of systemic endophytes on plant growth and biochemistry depend not only on the species involved, but also on the fungal strain and host plant genotype. Alkaloid profiles of infected grasses depend chiefly on the endophyte species (Bush et al. 1997). Moreover, extrinsic factors such as resource availability and grazing history may alter individual traits of endophyte-infected plants including tillering, biomass production, and tissue nitrogen and alkaloid levels (Malinowsky and Belesky 2000, Faeth *et al.* 2002, Bultman *et al.* 2004). In the following sections we briefly discuss endophyte interactions with herbivorous insects (for extensive discussions, see Clay 1991, Breen 1994, Faeth 2002, Popay and Bonos 2005), and then review evidence for indirect endophyte effects on upper trophic levels. Further, we consider biotic and abiotic sources of heterogeneity that may influence the role of endophyte–grass symbioses in multitrophic systems.

Insect herbivory on endophytic grasses

Many studies have been conducted to determine the significance of fungal endophytes in the modulation of grass–insect herbivore relations. Endophytes were found to adversely affect phytophagous insects from different feeding guilds, including leaf chewers, sap-suckers, stem-borers, leaf miners, root feeders, and seed predators (Clay 1988, Dahlman *et al.* 1991, Breen 1994, Prestige and Ball 1997, Popay and Bonos 2005). Studies have shown that endophyte presence can reduce insect feeding and deter oviposition on host grasses. Also, controlled feeding trials revealed that endophyte infection may alter individual growth, survival, and developmental rates at different stages of the insect life cycle (Clay 1990, 1991, Dahlman *et al.* 1991). Bioassays using parthenogenetic (aphid) insects have shown reduced reproductive rates on endophyte-infected plants (e.g., Eichenseer and Dahlman 1992, Bultman *et al.* 2004). Field evidence also indicates that plots sown with endophyte-infected grasses often experience reduced herbivory and support lower insect densities, compared to plots sown with uninfected conspecifics (Prestige and Ball 1997, Omacini *et al.* 2001, Popay *et al.* 2003).

While most studies on agronomic grasses have reported negative endophyte impacts on insect herbivores, various lines of evidence call into question the generality of this symbiosis as an anti-herbivore defense (Saikkonen *et al.* 1998, 2004, Faeth 2002, Bultman and Bell 2003). The nature and strength of endophyte mediation of plant–insect interactions depends in complex ways on the specific grass–endophyte association, fungal strain, insect species, and environmental conditions. An inspection of bioassays using the same herbivore species on different grass–endophyte systems shows that the outcome may change with both plant and fungal genotypes (Clay 1991, Dahlman *et al.* 1991, Clement *et al.* 1997). Likewise, work on single grass–fungus associations revealed insect-species-specific variation in herbivore feeding and growth responses to endophyte presence (Dahlman *et al.* 1991, Bultman and Bell 2003). To some extent, this variability would reflect insect feeding modes and alkaloid concentrations in the different plant parts on which the herbivores feed (Breen 1994, Popay and Bonos 2005).

Extrinsic environmental forces may play a critical influence in the expression of herbivore resistance in endophytic grasses. Factors such as water stress, soil fertility, and shoot damage are known to alter alkaloid levels in host plant tissues (Clay 1991, Breen 1994, Bultman et al. 2004). We notice, however, that interactive effects of plant resources and endophyte infection on insect herbivory have been addressed in only a few experiments. In two recent studies, Bultman and co-workers investigated effects of abiotic stress and simulated grazing on resistance of endophyte-infected and uninfected tall fescue (F. arundinacea) to insect herbivory. Prior drought increased alkaloid levels and decreased caterpillar performance in the presence of Neotyphodium (Bultman and Bell 2003). Clipping was also found to increase loline alkaloids in infected plants, resulting in lower susceptibility to aphids following damage (Bultman et al. 2004). The latter suggests that endophytes may provide a mechanism of inducible resistance to insect herbivory, as well as conferring constitutive defense against certain herbivores.

In summary, despite considerable heterogeneity in herbivore responses to endophyte–grass symbioses, effects of endophytic fungi on insect herbivores are expected to be widespread. More work is clearly needed on native grass–insect interactions, where specialist insect herbivores may have found their way to circumvent toxicity from fungal alkaloids (Faeth 2002, Saikkonen et al. 2004). Even if herbivory levels and insect densities are not affected by endophyte infection, changes in herbivore performance or behavior may still impact on natural enemies (Faeth and Bultman 2002).

Endophyte interactions with insect natural enemies

Empirical evidence for endophyte effects on higher trophic levels is relatively scarce and limited to a handful of systems. All published studies showing multitrophic responses to endophytes refer to insect host–parasite or pathogen interactions. This pattern reflects the interest in studying insect pests on agronomically important grasses. It is nonetheless remarkable that evidence exists for endophyte effects on natural enemies attacking insects with disparate feeding styles (leaf chewers, stem-borers, sap-suckers, and root feeders).

Endophyte effects on natural enemies may occur through altered population densities or individual traits of herbivores (Fig. 8.1: DMII vs. TMII). We have documented strong numerical responses to endophyte infection in an aphid–parasitoid assemblage (Omacini et al. 2001, see section "Endophytes in multitrophic interaction webs" below). Unfortunately, we are not aware of any other experiment in which changes in herbivore densities had been implicated in the indirect interaction between endophytes and higher-level consumers. In perennial ryegrass (*L. perenne*) pastures in New Zealand, decreased densities of the Argentine stem weevil (*Listronotus bonariensis*) were inversely correlated with increasing frequencies of endophyte-infected plants (Prestige and Ball 1997). A later study showed that weevil parasitism by the braconid wasp *Microctonus hyperodae* decreased with peramine alkaloid concentration in ryegrass plants (Goldson *et al.* 2000). There was no relationship between weevil density and peramine levels across pastures, and therefore the authors suggested that lower parasitism could result from reduced parasitoid searching/attack efficiency in highly infected plots.

Several studies were designed to examine individual-level mechanisms that might explain endophyte modification of host–parasite interactions. Endophytes may elicit either positive or negative indirect effects on natural enemies via altered herbivore traits (Faeth and Bultman 2002). As noted for TMII in general, the sign of the indirect effect on the receiving species is contingent on the identity of the players in the system and may not be predicted simply from their trophic positions (Wootton 1994, Werner and Peacor 2003) (notice that no sign was attached to dotted arrows in Fig. 8.1). We found that most studies so far show reduced susceptibility to natural enemies for herbivorous insects feeding on *Neotyphodium*-infected plant material. Grewal *et al.* (1995) reported an interesting exception to this pattern. The authors found that susceptibility of the root scarab *Popillia japonica* to entomopathogenic nematodes increased when fed endophyte-infected fescue species. Results suggested that ergot alkaloids could reduce beetle vigor making them more vulnerable to nematode parasitism.

Negative effects of endophytes on third-level consumers have been found in above- and belowground grazing chains. Bultman *et al.* (1997) studied endophyte-mediated interactions between the fall armyworm, *Spodoptera frugiperda*, and *Euplectrus* ectoparasitoids under laboratory conditions. Feeding fall armyworm larvae with endophyte-infected tall fescue decreased pupal mass of two parasitoid species. Presence of loline alkaloids in artificial diets of caterpillars also reduced survival for one parasitoid species. Much attention has been given to effects of *N. lolii* on interactions between the Argentine stem weevil and its parasitoid *M. hyperodae* (Goldson *et al.* 2000). Endophyte infection of perennial ryegrass decreased parasitoid larval development and survival (Barker and Addison 1996, Bultman *et al.* 2003). Parasitoid development was also retarded in weevils fed artificial diets with diterpene alkaloids (Barker and Addison 1996). Bultman *et al.* (2003) observed that *N. lolii* strains differing in their alkaloid profiles had similarly reduced effects on weevil feeding and survival, but differed in their negative impacts on parasitoid performance. Parasitoid survival was negatively related with the presence of ergovaline in the weevil diet. In addition, in cage experiments, weevil behavior was altered on endophyte-infected plants so that they spent less time in positions exposed to parasitoid attack (Gerard 2000).

Richmond et al. (2004) found that fall armyworm larvae fed *N. lolii*-infected perennial ryegrass suffered reduced mortality when exposed to the parasitic nematode *Steinernema carpocapsae*. Larvae fed infected plants had lower body mass than those on an endophyte-free grass diet suggesting that host vigor could not account for the reduced susceptibility to nematodes. By injecting the nematode's symbiotic bacteria (*Xenorhabdus nematophila*) into the larvae, the authors were able to link the lower nematode virulence with a reduction in the pathogenicity of its bacterial symbiont inside the hemocoel of insects fed infected ryegrass. Indirect effects of *N. lolii* endophytes on a similar five-species interaction web were reported by Kunkel et al. (2004). Endophyte infection of perennial ryegrass reduced the susceptibility of black cutworms (*Agrotis ipsilon*) to the nematode *S. carpocapsae* (Kunkel and Grewal 2003). Insect larvae hosted similar nematode densities when fed endophyte-free or infected plants, however they exhibited greater mortality on the endophyte-free diet. Detailed work revealed that the endophyte alkaloid ergocristine decreased nematode virulence and suppressed the nematode symbiotic bacteria (Kunkel et al. 2004).

The responses of root-feeding insects to endophyte infection have been less frequently reported and, in general, effects appear to be more variable (Breen 1994, Popay and Bonos 2005). It is therefore not surprising that tests of endophyte indirect effects on natural enemies of root insect pests had yielded contradictory results. For example, susceptibility of Japanese beetles (*P. japonica*) and other white grub species to soil pathogens (nematodes and bacteria) has been found to either increase (Grewal et al. 1995) or remain unchanged (Walston et al. 2001, Koppenhöfer and Fuzy 2003) when insects were fed endophyte-infected plants. Other arthropod consumers of grazing or detritus soil webs (collembolans, oribatid mites, etc.) showed inconsistent and temporally variable responses to endophyte infection in field-sown pastures (Davidson and Potter 1995, Bernard et al. 1997). The critical role of soil microarthropods in organic matter cycling suggests that endophyte-driven belowground cascades deserve further experimental consideration (Omacini et al. 2005).

Endophytes in multitrophic interaction webs

The work reviewed above reveals two major shortcomings in our present knowledge of endophyte-driven multitrophic interactions. Most studies have involved tall fescue and perennial ryegrass endophytes, and usually looked at three-species food chains in laboratory or glasshouse conditions. We now discuss our recent work showing the impacts of a Neotyphodium–grass association on a species-diverse assemblage of aphid herbivores and their hymenopteran parasitoids in a field setting (Omacini et al. 2001). Based on previous evidence for agronomic grasses, we predicted that bottom-up control from endophytes would propagate upwards *reducing* both herbivore densities and the strength of top-down forces (parasitism) on aphids.

Interactions between aphids and parasitoids can be readily quantified by laboratory incubation of parasitized aphids or "mummies" (Müller et al. 1999). Aphid primary parasitoids lay eggs inside the aphid body. The developing larvae kills the host aphid, which turns into a mummy in which the parasitoid pupates. Primary parasitoids are attacked by two types of secondary parasitoid (Müller and Godfray 1999). Hyperparasitoids oviposit inside primary parasitoids within a living aphid; mummy parasitoids attack primary parasitoids after the host aphid was mummified. Because secondary parasitoids consume their host before emergence, only one adult parasitoid (primary or secondary) emerges from each mummy. Müller and Godfray (1999) review indirect interactions in aphid–parasitoid systems.

We established replicated plots of endophyte-infected and uninfected *Lolium multiflorum* (Italian ryegrass) monocultures, and exposed them to natural colonization by insect herbivores and parasitoids (for details, see Omacini et al. 2001). We used Italian ryegrass populations from pampean old fields, which were known to be highly infected (95%) by a *Neotyphodium* endophyte that morphologically resembles *N. occultans* (Moon et al. 2000). Forty ryegrass monocultures were grown from seed in large containers filled with natural grassland soil. Each plot was sowed with endophyte-infected (E+) or endophyte-free (E−) seeds and maintained by hand weeding. E− seeds were obtained by treating infected seeds with a systemic fungicide; microscopic examination (Moon et al. 2000) showed endophyte infection to be zero in fungicide-treated seeds. We measured densities of living and parasitized aphids in ten plots per treatment and collected aphid mummies from all monocultures. Parasitoids were individually reared and identified upon emergence. Two aphid species attacked by seven different parasitoids were found in the grass plots (Fig. 8.2). There were three species of primary parasitoids, three mummy parasitoids, and one hyperparasitoid. Most secondary parasitism (95%) was due to mummy parasitoids (Omacini et al. 2001).

Endophyte removal enhanced aphid density ($t_{18} = 2.42$, $p < 0.03$) and parasitoid activity. The density of mummified aphids increased eight-fold in E− plots relative to that in E+ plots ($t_{18} = 4.27$, $p < 0.001$). Aphid species showed contrasting responses to endophyte presence (Fig. 8.3a). *Rhopalosiphum padi* was strongly reduced by endophyte infection ($t_{18} = 3.0$, $p < 0.01$), whereas *Metopolophium festucae* was virtually unaffected. Previous work on *R. padi* (Eichenseer and

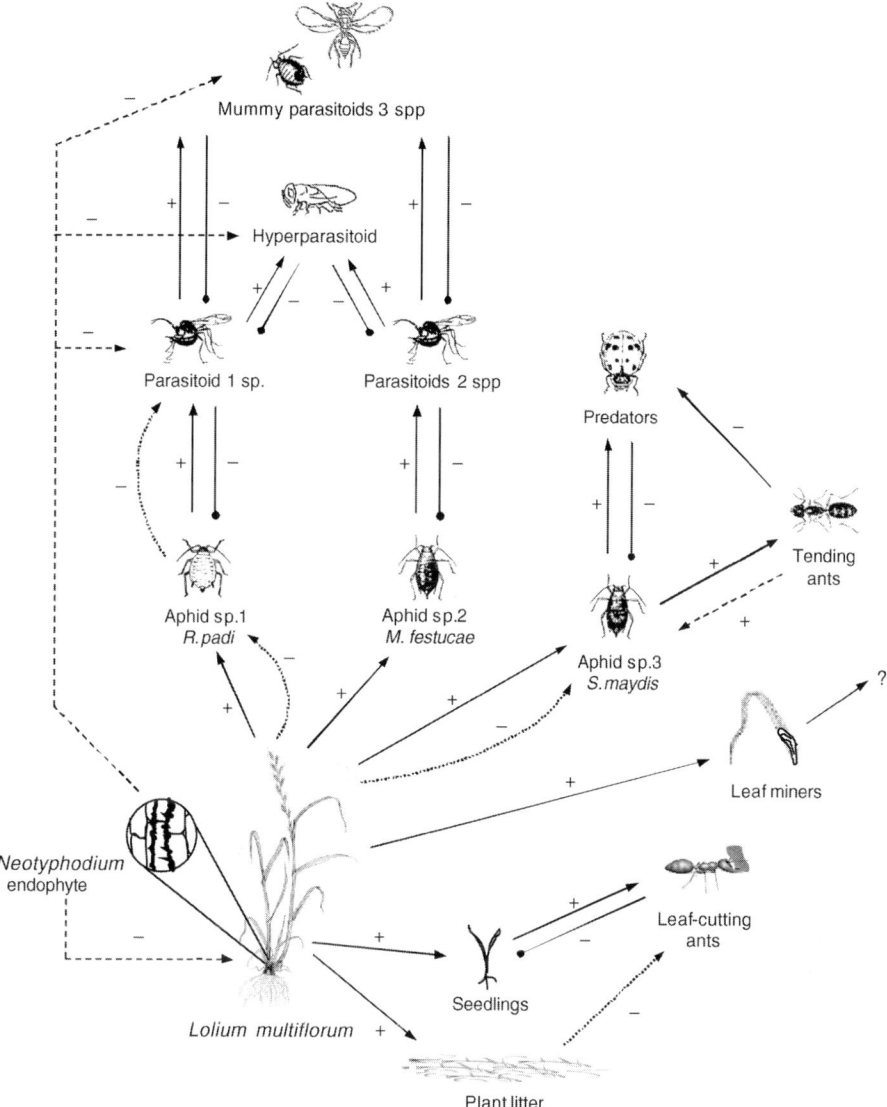

Figure 8.2. Interaction web for insect consumers on endophyte-infected Italian ryegrass. Direct and indirect interactions are shown by different line styles (as in Fig. 8.1). Indirect effects were revealed through manipulation of seed-borne endophytes in grass monocultures exposed to natural colonization by herbivorous insects and associated enemies and mutualists. Top-down cascades were not explicitly tested and therefore were omitted. The aphid–parasitoid web was drawn from data in Omacini *et al.* (2001). *M. festucae, Metopolophium festucae; R. padi, Rhopalosiphum padi; S. maydis, Sipha maydis.*

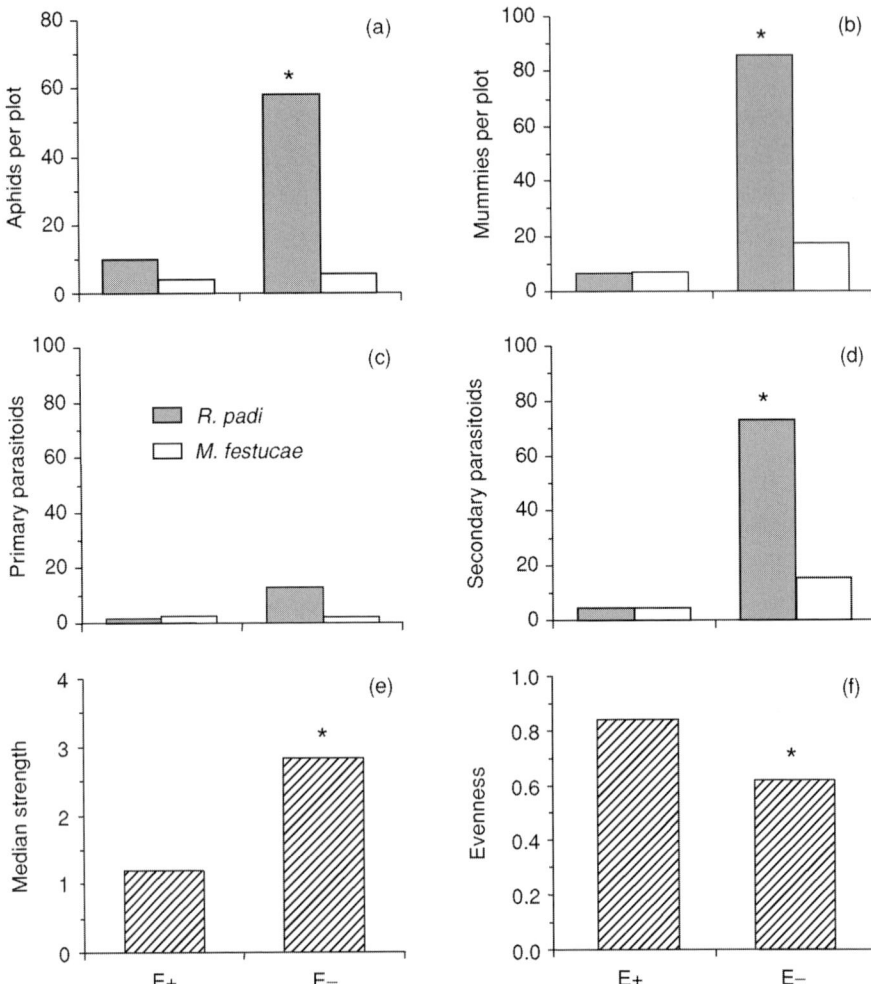

Figure 8.3. Indirect endophyte effects on the structure of an aphid–parasitoid food web. Top panels show mean densities of (a) living and (b) parasitized aphids in 0.25 m^2 plots of *Lolium multiflorum*, with (E+) and without (E−) endophytic fungi ($n = 10$). Middle panels show the actual numbers of (c) primary and (d) secondary parasitoids in response to endophyte presence. Bottom panels depict differences for two descriptors of food web structure, (e) the median strength of all pair-wise aphid–parasitoid interactions, and (f) the evenness of those interaction strengths. Panels (a–d) show effects on two aphids: *Rhopalosiphum padi* and *Metopolophium festucae*. Asterisks denote significant effects ($p < 0.05$) of endophyte removal on the corresponding variables. (After Omacini *et al.* 2001.)

Dahlman 1992) suggests that low aphid densities on E+ plants reflected reduced reproduction, rather than low immigration. As a result, endophyte effects on parasitized aphid densities were accounted for by the greater number of *R. padi* mummies in E− plots ($t_{18} = 2.91$, $p < 0.01$) (Fig. 8.3b). That parasitized aphids were more abundant in E− plots may seem unsurprising given the likely aggregative response of parasitoids to increased host numbers. However, aphids feeding on E− grasses experienced proportionally greater parasitism rates (0.63) than those on E+ plots (0.31) ($p < 0.025$, randomization test). This suggests that endophyte effects on aphid−parasitoid interactions were driven both by reduced aphid densities and by altered parasitoid per-capita attack rates.

The success of primary and secondary parasitoids was estimated using emergence rates in the laboratory. Secondary/primary parasitoid ratios were higher in endophyte-removal plots (E+ = 2.1 vs. E− = 6.5, χ^2 test: $p < 0.05$), indicating that primary parasitoids were relatively more successful in the presence of endophytes. Higher parasitism on aphids in E− plots did not translate into more emerging primary parasitoids, but resulted in higher secondary parasitism, especially on *R. padi* (Fig. 8.3c, d). This pattern prompts two broad conclusions. First, endophyte effects appeared to be transmitted from aphids to top parasitoids by changes in primary parasitoid density and quality as hosts to mummy parasitoids. Not only were aphid mummies less common on E+ grass plots but they were less susceptible to secondary parasitism. The reduced quality of primary parasitoids on E+ monocultures was supported by the smaller body size of secondary parasitoids emerging from *R. padi* in those plots (ANCOVA, endophyte effect: $p < 0.025$, mummy size covariate: $p < 0.10$). Second, endophyte removal induced a bottom-up cascade that increased the strength of top−down forces, thus having unequal impacts on adjacent trophic levels. Whereas aphids and mummy parasitoids were positively affected in E− plots, primary parasitoids were not (Fig. 8.3). This finding is consistent with Oksanen et al.'s (1981) trophic exploitation hypothesis, even though this model does not allow for within-trophic level diversity, nor for trait-mediated interactions (Abrams 1993).

The aphid−parasitoid web was structured as two quasi-linear food chains connected at the top by a guild of mummy parasitoids, while primary parasitoids were specific to each aphid host (Omacini et al. 2001) (Fig. 8.2). To quantify overall endophyte effects on food web structure, we estimated the strength of pair-wise aphid−parasitoid interactions from the mummies taken from E+ and E− plants (Müller et al. 1999, Omacini et al. 2001). Endophyte removal increased the median strength of parasitic interactions (Fig. 8.3e), an effect associated with the dominance of *R. padi* on E− plots (Fig. 8.3a, b). Parasitoid links with *M. festucae* became weaker, and therefore the evenness of interaction strengths was lower in food webs assembled on E− plots (Fig. 8.3f). These results support the view that species heterogeneity may strongly influence the propagation of bottom−up cascades to higher trophic levels (Hunter and Price 1992, Abrams 1993).

The multitrophic impact of endophytes occurred in the absence of changes in aerial plant biomass ($p = 0.97$, t test); only subtle differences in plant traits were found between treatments. Tiller densities were slightly higher in E+ plots ($p < 0.05$), which resulted in mean tiller mass being higher in E− plots. Endophyte-free plants also showed reduced nitrogen content (E+ = 1.02% vs. E− = 0.89%, $p < 0.01$). We think it unlikely that these changes were responsible for the altered food web structure observed in E+ monocultures. Rather, we believe that endophyte effects on insect webs were mediated by changes in other aspects of plant quality. Grass aphids, particularly *R. padi*, can be highly sensitive to loline alkaloids accumulated in E+ plants (Siegel et al. 1990, Bultman and Bell 2003). In addition, parasitoid development and survival may be affected by lolines in the herbivore host diet (Bultman et al. 1997, 2003). Lolines have been found in infected Italian ryegrass at total concentrations of 49–63 $\mu g\, g^{-1}$ dry mass, but were undetectable in E− plants (M. Omacini and L. Bush unpublished data). However, we still do not know the extent to which different alkaloids may influence aphid–parasitoid food webs.

Effects of *L. multiflorum* endophytes were not restricted to aphid–parasitoid interactions (Fig. 8.2). We found that the frequency of tillers colonized by leaf miners (Agromizidae) was also decreased by endophyte infection (Omacini et al. 2001). This suggests that endophyte effects may cascade up through various trophic pathways. In a similar experiment we found no difference in natural densities of another aphid species, *Sipha maydis*, between ryegrass monocultures with and without endophyte infection. However, the number of ants (*Camponotus* sp.) attending aphid colonies at the peak of the season was reduced on E+ plots (Fig. 8.4). While the mechanism for this indirect effect is unknown, we speculate that the lower ant attendance of colonies attached to E+ plants might reflect a change in aphid honeydew quality and/or productivity. Our observations indicate that tending ants were effective at driving ladybird beetles away from aphid colonies, which suggests that top–down effects could be stronger for *S. maydis* on E+ plants. We believe this is the first record of an indirect, non-density-mediated interaction between endophytic fungi and an insect herbivore mutualist (see Wimp and Whitham Chapter 12 this volume).

Finally, in an experiment testing for the impact of litter accumulation on Italian ryegrass recruitment the frequency of seedlings damaged by leaf-cutting ants was lower in patches covered by litter produced by E+ plants than in patches with litter from E− plants (authors' unpublished data). This form of interaction modification (Wootton 1994) appeared to be mediated by a change in microhabitat conditions for the ants, and was independent of whether seedlings had emerged from endophyte infected or uninfected seeds. More work is needed to determine the mechanism for this indirect effect and how it might feedback on the host plant population dynamics (Fig. 8.2). Overall, these experimental results demonstrate the potential for grass endophytes to affect interactions

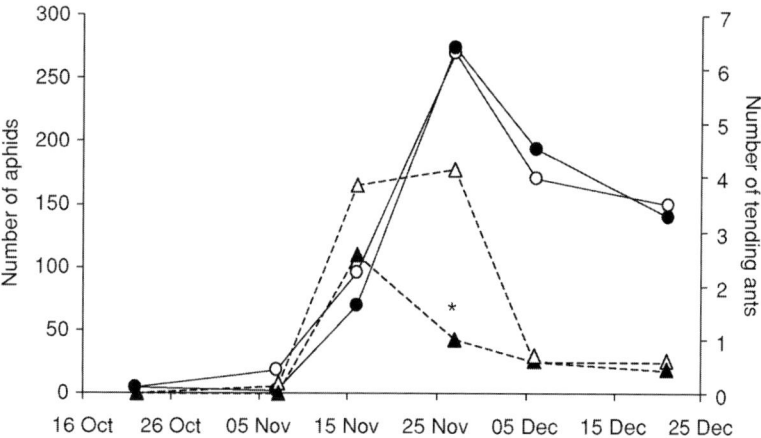

Figure 8.4. Trait-mediated endophyte effects on an aphid–ant mutualism. Data show mean number of aphids (*Sipha maydis*, solid lines) and tending ants (dashed lines) recorded on endophyte-infected (filled symbols) and uninfected (empty symbols) *Lolium multiflorum* monocultures. The asterisk denotes a significant endophyte effect for ant-tending activity on aphid colonies ($t = 2.2$, $p = 0.037$, d.f. $= 26$). No significant difference in aphid density was detected (authors' unpublished data).

involving various insect consumers in a multispecies context. Future studies should examine how endophyte modification of host plant traits alter interaction webs under more realistic scenarios, especially where endophytic grasses occur as mosaics of E+ and E− plants within species-rich communities (Faeth 2002, Omacini *et al.* 2005), and also where environmental variation may act to attenuate indirect effects on higher trophic levels.

Conclusions and perspectives

Fungal endophytes are cryptic symbionts that are capable of "engineering" the phenotype of their grass hosts in a variety of ways. Increasing evidence shows that manipulation of plant traits by endophytes can indirectly affect consumers at various trophic levels. While still limited in scope, data suggest that endophyte-induced TMII are most often negative on the receiver consumer species. The accumulation of fungal alkaloids in host tissues appears to play a critical role in the initiation of bottom–up cascades from endophyte infection. Moreover, sequestration of bioprotective alkaloids by insect herbivores might be involved in the transmission of indirect effects to natural enemies. Possible trophic impacts from other traits of endophytic grasses have been overlooked but should be given more attention. Theory suggests that plant-trait-mediated indirect effects of endophytes could rapidly propagate through interaction webs as they trigger density reactions of insect species with different trophic functions. However, current evidence for endophyte-induced DMII comes from just one aphid–parasitoid system, and field studies testing for endophyte indirect effects on community structure and diversity at relevant scales are virtually nonexistent (Omacini et al. 2005).

We have shown that biotic heterogeneity influences the occurrence of bottom–up cascades induced by endophytes even in the simplest food chains. This variability is generated by genotypic variation among endophyte–grass symbiota, and also by interspecific differences within the herbivore and carnivore trophic levels. Admittedly, the consideration of such horizontal heterogeneity in a multitrophic context opens a "Pandora's box" of potential endophyte-driven indirect effects. For instance, what kinds of consumer-mediated interactions arise between endophytic grasses and nonendophytic plant neighbors? Furthermore, how do bottom–up forces such as resource enrichment alter endophyte interactions with higher trophic levels? Environmental conditions modulate endophyte–grass relations and, therefore, we expect abiotic resources to affect the nature and strength of endophyte-induced cascades.

Finally, we do not know what are the consequences of endophyte-induced cascades for host plant growth and fitness (Faeth and Bultman 2002). Taken together, the studies reviewed here suggest that host benefits from endophyte-based defenses against foliar and stem herbivores may be counterbalanced by the reduced performance of natural enemies. To begin solving this puzzle, we need experiments that manipulate both endophyte presence and top consumers in realistic settings, allowing feedbacks from top–down forces to be exposed. Simply assuming bottom–up control in the presence of endophytes would be an oversimplification, which has been implicit in the experiments done to date. That is, we still know very little about the relative strength of bottom–up and top–down forces in consumer webs sustained by endophytic grasses. This issue has not only theoretical but also applied implications, especially regarding the use of biological control agents in agricultural systems where endophyte infections affect both forage and livestock production.

Acknowledgments

We thank I. Miranda, M. Rabadan, and P. Tognetti for technical assistance, and C. Müller for her invaluable contribution to the aphid–parasitoid study. D. Srivastava, N. Mazía, S. Faeth, and one anonymous reviewer provided useful comments and criticisms on the manuscript. This work was supported by Consejo Nacional de Investigaciones Científicas y Técnicas de Argentina, Agencia Nacional de Promoción Científica y Tecnológica (FONCYT), and the University of Buenos Aires.

References

Abrams, P. A. 1993. Effect of increased productivity on the abundances of trophic levels. *American Naturalist* **141**:351–371.

Abrams, P. A. 1995. Implications of dynamically variable traits for identifying, classifying, and measuring direct and indirect effects in ecological communities. *American Naturalist* **146**:112–134.

Abrams, P. A., B. A. Menge, G. G. Mittelbach, D. A. Spiller, and P. Yodzis. 1996. The role of indirect effects in food webs, pp. 371–395 in G. Polis and K. O. Winemiller (eds.) *Food Webs: Integration of Patterns and Dynamics.* New York: Chapman and Hall.

Agrawal, A. A., S. Tuzun, and E. Bent (eds.) 1999. *Induced Plant Defenses against Pathogens and Herbivores.* St. Paul, MN: American Phytopathological Society Press.

Barker, G. M., and P. J. Addison. 1996. Influence of clavicipitaceous endophyte infection in ryegrass on development of the parasitoid *Microctonus hyperodae* Loan (Hymenoptera: Braconidae) in *Listronotus bonariensis* (Kuschel) (Coleoptera: Curculionidae). *Biological Control* **7**:281–287.

Bernard, E. C., K. D. Gwinn, C. D. Pless, and C. D. Williver. 1997. Soil invertebrate species diversity and abundance in endophyte-infected tall fescue pastures, pp. 125–135 in C. W. Bacon and N. S. Hill (eds.) Neotyphodium *Grass Interactions.* New York: Plenum Press.

Boucher, D. H., S. James, and K. H. Keeler. 1982. The ecology of mutualism. *Annual Review of Ecology and Systematics* **13**:315–347.

Breen, J. P. 1994. Acremonium endophyte interactions with enhanced plant resistance to insects. *Annual Review of Entomology* **39**:401–423.

Brown, J. H., D. W. Davidson, J. C. Munger, and R. S. Inouye. 1986. Experimental community ecology: the desert granivore system, pp. 41–61 in J. Diamond and T. J. Case (eds.) *Community Ecology.* New York: Harper and Row.

Bultman, T. L., and G. D. Bell. 2003. Interaction between fungal endophytes and environmental stressors influences plant resistance to insects. *Oikos* **103**:182–190.

Bultman, T. L., K. L. Borowicz, R. M. Schneble, T. A. Coudron, and L. P. Bush. 1997. Effect of a fungal endophyte on the growth and survival of two *Euplectrus* parasitoids. *Oikos* **78**:170–176.

Bultman, T. L., M. R. McNeill, and S. L. Goldson. 2003. Isolate-dependent impacts of fungal endophytes in a multitrophic interaction. *Oikos* **102**:491–496.

Bultman, T. L., G. Bell, and W. D. Martin. 2004. A fungal endophyte mediates reversal of wound-induced resistance and constrains tolerance in a grass. *Ecology* **85**:679–685.

Bush, L. P., H. H. Wilkinson, and C. L. Schardl. 1997. Bioprotective alkaloids of grass-fungal endophyte symbioses. *Plant Physiology* **114**:1–7.

Callaway, R. M., S. C. Pennings, and C. L. Richards. 2003. Phenotypic plasticity and interactions among plants. *Ecology* **84**:1115–1128.

Cheplick, G. P., and K. Clay. 1988. Acquired chemical defences in grasses: the role of fungal endophytes. *Oikos* **52**:309–318.

Clark, T. L., and F. Messina. 1998. Plant architecture and the foraging success of ladybird beetles attacking the Russian wheat aphid. *Entomologia Experimentalis et Applicata* **86**:153–161.

Clay, K. 1988. Fungal endophytes of grasses: a defensive mutualism between plants and fungi. *Ecology* **69**:10–16.

Clay, K. 1990. Fungal endophytes of grasses. *Annual Review of Ecology and Systematics* **21**:275–297.

Clay, K. 1991. Fungal endophytes, grasses, and herbivores, pp. 199–226 in P. Barbosa, V. A. Krischik, and C. G. Jones (eds.) *Microbial Mediation of Plant–Herbivore Interactions*. New York: John Wiley.

Clay, K. 1994. The potential role of endophytes in ecosystems, pp. 73–86 in C. W. Bacon and J. F. White, Jr. (eds.) *Biotechnology of Endophytic Fungi of Grasses*. Boca Raton, FL: CRC Press.

Clay, K., and J. Holah. 1999. Fungal endophyte symbiosis and plant diversity in successional fields. *Science* **285**:1742–1744.

Clay, K., and C. Schardl. 2002. Evolutionary and ecological consequences of endophyte symbiosis with grasses. *American Naturalist* **160**:S99–S127.

Clement, S. L., A. Wilson, D. Lester, and C. Davitt. 1997. Fungal endophytes of wild barley and their effects on *Diuraphis noxia* population development. *Entomologia Experimentalis et Applicata* **82**:275–281.

Dahlman, D. L., H. Eichenseer, and M. R. Siegel. 1991. Chemical perspectives on endophyte–grass interactions and their implications to insect herbivory, pp. 227–252 in P. Barbosa, V. A. Krischik, and C. G. Jones (eds.) *Microbial Mediation of Plant–Herbivore Interactions*. New York: John Wiley.

Davidson, A. W., and D. A. Potter. 1995. Response of plant-feeding, predatory, and soil-inhabiting invertebrates to *Acremonium* endophyte and nitrogen fertilization in tall fescue turf. *Journal of Economical Entomology* **88**:367–379.

Denno, R. F., C. Gratton, M. A. Peterson, *et al.* 2002. Bottom-up forces mediate natural-enemy impact in a phytophagous insect community. *Ecology* **83**:1433–1442.

Dyer, L. A., and P. D. Coley. 2002. Tritrophic interactions among host plants, herbivores, and natural enemies, pp. 67–88 in T. Tscharntke and B. A. Hawkins (eds.) *Multitrophic Level Interactions*. Cambridge, UK: Cambridge University Press.

Dyer, L. A., and D. K. Letourneau. 1999. Relative strengths of top-down and bottom-up forces in a tropical forest community. *Oecologia* **119**:265–274.

Eichenseer, H., and D. L. Dahlman. 1992. Antibiotic and deterrent qualities of endophyte-infected tall fescue to two aphid species (Homoptera: Aphididae). *Environmental Entomology* **21**:1046–1051.

Faeth, S. H. 2002. Are endophytic fungi defensive plant mutualists? *Oikos* **98**:25–36.

Faeth, S. H., and T. L. Bultman. 2002. Endophytic fungi and interactions among host plants, herbivores, and natural enemies, pp. 89–123 in T. Tscharntke and B. A. Hawkins (eds.) *Multitrophic Level Interactions*. Cambridge, UK: Cambridge University Press.

Faeth, S. H., and T. J. Sullivan. 2003. Mutualistic asexual endophytes in a native grass are usually parasitic. *American Naturalist* **161**:310–325.

Faeth, S. H., L. P. Bush, and T. J. Sullivan. 2002. Peramine alkaloid variation in *Neotyphodium*-infected Arizona fescue: effects of endophyte and host genotype and environment. *Journal of Chemical Ecology* **28**:1511–1526.

Faeth, S. H., M. Helander, and K. Saikkonen. 2004. Asexual *Neotyphodium* endophytes in a native grass reduce competitive abilities. *Ecology Letters* **7**:304–313.

Forkner, R. E., and M. D. Hunter. 2000. What goes up must come down? Nutrient addition and predation pressure on oak herbivores. *Ecology* **81**:1588–1600.

Fraser, L. H., and J. P. Grime. 1998. Top-down control and its effects on the biomass and composition of three grasses at high and low soil fertility in outdoor microcosms. *Oecologia* **113**:239–246.

Gange, A. C., and V. K. Brown (eds.) 1997. *Multitrophic Interactions in Terrestrial Systems*. Oxford, UK: Blackwell Science.

Gange, A. C., V. K. Brown, and D. M. Aplin. 2003. Multitrophic links between arbuscular mycorrhizal fungi and insect parasitoids. *Ecology Letters* **6**:1051–1055.

Gerard, P. 2000. Ryegrass endophyte infection affects argentine stem weevil adult behaviour and susceptibility to parasitism. *New Zealand Plant Protection* **53**:406–409.

Goldson, S. L., J. R. Proffitt, L. R. Fletcher, and D. B. Baird. 2000. Multitrophic interaction between the ryegrass *Lolium perenne*, its endophyte *Neotyphodium lolii*, the weevil pest *Listronotus bonariensis*, and its parasitoid *Microctonus hyperodae*. *New Zealand Journal of Agricultural Research* **43**:227–233.

Grewal, S., P. Grewal, and R. Gaugler. 1995. Endophytes of fescue grasses enhance susceptibility of *Popillia japonica* larvae to an entomophagous nematode. *Entomologia Experimentalis et Applicata* **74**:219–224.

Halaj, J., and D. H. Wise. 2001. Terrestrial trophic cascades: how much do they trickle? *American Naturalist* **157**:262–281.

Hunter, M. D., and P. W. Price. 1992. Playing chutes and ladders: heterogeneity and the relative roles of bottom-up and top-down forces in natural communities. *Ecology* **73**:724–732.

Karban, R., and I. T. Baldwin. 1997. *Induced Responses to Herbivory*. Chicago, IL: University of Chicago Press.

Kessler, A., and I. T. Baldwin. 2001. Defensive function of herbivore-induced plant volatile emissions in nature. *Science* **291**:2141–2144.

Koppenhöffer, A., and E. Fuzy. 2003. Effects of turfgrass endophytes (Clavicipitaceae: Ascomycetes) on white grub (Coleoptera: Scarabaeidae) control by the entomopathogenic nematode *Heterorhabditis bacteriophora* (Rhabditida: Heterorhabditae). *Environmental Entomology* **32**:392–396.

Kunkel, B. A., and P. S. Grewal. 2003. Endophyte infection in perennial ryegrass reduces the susceptibility of black cutworm to an entomopathogenic nematode. *Entomologia Experimentalis et Applicata* **107**:95–104.

Kunkel, B. A., P. S. Grewal, and M. F. Quigley. 2004. A mechanism of acquired resistance against an entomopathogenic nematode by *Agrotis ipsilon* feeding on perennial ryegrass harboring a fungal endophyte. *Biological Control* **29**:100–108.

Leibold, M. A., J. M. Chase, J. B. Shurin, and A. L. Downing. 1997. Species turnover and the regulation of trophic structure. *Annual Review of Ecology and Systematics* **28**:467–494.

Malinowski, D. P., and D. P. Belesky. 2000. Adaptations of endophyte-infected cool-season grasses to environmental stresses: mechanisms of drought and mineral stress tolerance. *Crop Science* **40**:923–940.

Meserve, P. L., D. A. Kelt, W. B. Milstead, and J. R. Gutierrez. 2003. Thirteen years of shifting top-down and bottom-up control. *BioScience* **53**:633–646.

Moon, D. C., and P. Stiling. 2002. The influence of species identity and herbivore feeding mode on top-down and bottom-up effects in a salt marsh system. *Oecologia* **133**:243–253.

Moon, C. D., B. Scott, C. L. Schardl, and M. J. Christensen. 2000. Evolutionary origins of *Epichloë* endophytes from annual ryegrasses. *Mycologia* **92**:1103–1118.

Müller, C. B., and H. C. J. Godfray. 1999. Indirect interactions in aphid-parasitoid communities. *Researches on Population Ecology* **41**:93–106.

Müller, C. B., I. C. T. Adriaanse, R. Belshaw, and H. C. J. Godfray. 1999. The structure of an aphid-parasitoid community. *Journal of Animal Ecology* **68**:346–370.

Oksanen, L., and T. Oksanen. 2000. The logic and realism of the hypothesis of exploitation ecosystems. *American Naturalist* **155**:703–723.

Oksanen, L., S. D. Fretwell, J. Arruda, and P. Niemelä. 1981. Exploitation ecosystems in gradients of primary productivity. *American Naturalist* **118**:240–261.

Oliver, K. M., J. A. Russell, N. A. Moran, and M. S. Hunter. 2003. Facultative bacterial symbionts in aphids confer resistance to parasitic wasps. *Proceedings of the National Academy of Sciences of the USA* **100**:1803–1807.

Omacini, M., E. J. Chaneton, C. M. Ghersa, and C. B. Müller. 2001. Symbiotic fungal endophytes control insect host-parasite interaction webs. *Nature* **409**:78–81.

Omacini, M., E. J. Chaneton, and C. M. Ghersa. 2005. A hierarchical framework for understanding the ecosystem consequences of endophyte-grass symbioses, pp. 141–162 in C. Roberts, C. P. West, and D. Spiers (eds.) Neotyphodium *in Cool-Season Grasses: Current Research and Applications*. Ames, IA: Blackwell.

Polis, G. A. 1999. Why are parts of the world green? Multiple factors control productivity and the distribution of biomass. *Oikos* **86**:3–15.

Polis, G. A., and D. R. Strong. 1996. Food web complexity and community dynamics. *American Naturalist* **147**:813–846.

Popay, A. J., and S. A. Bonos. 2005. Biotic responses in endophytic grasses, pp. 163–186 in C. Roberts, C. P. West, and D. Spiers (eds.) Neotyphodium *in Cool-Season Grasses: Current Research and Applications* Ames, IA: Blackwell.

Popay, A. J., R. J. Townsend, and L. R. Fletcher. 2003. The effect of endophyte (*Neotyphodium uncinatum*) in meadow fescue on grass grub larvae. *New Zealand Plant Protection* **56**:123–128.

Price, P. W. 1991. The plant vigor hypothesis and herbivore attack. *Oikos* **62**:244–251.

Price, P. W. 2002. Resource-driven terrestrial interaction webs. *Ecological Research* **17**:241–247.

Price, P. W., C. E. Bouton, P. Gross, *et al.* 1980. Interaction among three trophic levels: influence of plants on interactions between insect herbivores and natural enemies. *Annual Review of Ecology and Systematics* **11**:41–65.

Prestige, R. A., and O. J. P. Ball. 1997. A catch 22: the utilization of endophytic fungi for pest management, pp. 171–192 in A. C. Gange and V. K. Brown (eds.) *Multitrophic Interactions in Terrestrial Systems*. Oxford, UK: Blackwell Science.

Richmond, D. S., B. A. Kunkel, N. Somasekhar, and P. S. Grewal. 2004. Top-down and bottom-up regulation of herbivores: *Spodoptera frugiperda* turns tables on endophyte-mediated plant defence and virulence of an entomopathogenic nematode. *Ecological Entomology* **29**:353–360.

Saikkonen, K., S. H. Faeth, M. Helander, and T. J. Sullivan. 1998. Fungal endophytes: a continuum of interactions with host plants. *Annual Review of Ecology and Systematics* **29**:319–343.

Saikkonen, K., P. Wäli, M. Helander, and S. H. Faeth. 2004. Evolution of endophyte–plant symbioses. *Trends in Plant Science* **9**:275–280.

Schmitz, O. J. 1998. Direct and indirect effects of predation and predation risk in old-field interaction webs. *American Naturalist* **151**:327–342.

Schmitz, O. J., P. A. Hambäck, and A. P. Beckerman. 2000. Trophic cascades in terrestrial systems: a review of the effects of carnivore removal on plants. *American Naturalist* **155**:141–153.

Schmitz, O. J., V. Krivan, and O. Ovadia. 2004. Trophic cascades: the primacy of trait-mediated indirect interactions. *Ecology Letters* **7**:153–163.

Siegel, M. R., G. C. M. Latch, L. P. Bush, *et al.* 1990. Fungal endophyte-infected grasses: alkaloid accumulation and aphid response. *Journal of Chemical Ecology* **16**:3301–3315.

Tollrian, R., and C. D. Harvell (eds.) 1999. *The Ecology and Evolution of Inducible Defenses*. Princeton, NJ: Princeton University Press.

Walston, A., D. Held, N. Mason, and D. Potter. 2001. Absence of interaction between endophytic perennial ryegrass and susceptibility of Japanese beetle (Coleoptera: Scarabaeidae) grubs to *Paenibacillus popilliae* Dutky. *Journal of Entomological Science* **36**:105–108.

Werner, E. E., and S. D. Peacor. 2003. A review of trait-mediated indirect interactions in ecological communities. *Ecology* **84**:1083–1100.

Wootton, J. T. 1994. The nature and consequences of indirect effects in ecological communities. *Annual Review of Ecology and Systematics* **25**:443–466.

Yue, Q., C. Wang, T. J. Gianfagna, and W. A. Meyer. 2001. Volatile compounds of endophyte-free and infected tall fescue (*Festuca arundinacea* Schreb.). *Phytochemistry* **58**:935–941.

9

Ecology meets plant physiology: herbivore-induced plant responses and their indirect effects on arthropod communities

MAURICE W. SABELIS, JUNJI TAKABAYASHI, ARNE JANSSEN, MERIJN R. KANT, MICHIEL VAN WIJK, BEATA SZNAJDER, NAYANIE S. ARATCHIGE, IZABELA LESNA, BELEN BELLIURE, AND ROBERT C. SCHUURINK

Introduction

Herbivory by arthropods induces a wealth of changes in the primary and secondary chemistry of plants (Karban and Baldwin 1997, Constabel 1999, Agrawal et al. 1999, Kessler and Baldwin 2002). These chemical changes in turn do not only affect the inducer, but also other herbivore species attacking the induced plant (Denno et al. 1995, Denno and Kaplan Chapter 2 this volume). This effect of one herbivore species on other herbivores is called "indirect," because it can only arise via the plant as an intermediate organism (Wootton 1994). Moreover, it is called trait-mediated, because the immediate effect of herbivory is an induced change in plant quality, not in plant quantity (Werner and Peacor 2003, Schmitz et al. 2004).

The herbivore-induced state of plants may influence the community of arthopods that live on them. When the induced plant allocates much of its energy in compensatory growth or defense specifically aimed at the inducer, other herbivore species may profit from the increased nutritional quality or weakened defense of the plant, thereby giving rise to interspecific aggregations of herbivores on individual plants (Denno et al. 1995). If, however, the induced plant

Ecological Communities: Plant Mediation in Indirect Interaction Webs, ed. Takayuki Ohgushi, Timothy P. Craig, and Peter W. Price. Published by Cambridge University Press. © Cambridge University Press 2007.

mounts a sufficiently generalized defense, the plant becomes "vaccinated" against attack by other herbivores, leading to species-poor communities of herbivorous arthropods on the plant (Karban and Baldwin 1997). Much the same reasoning applies to herbivore genotypes within a single species. Thus, in theory, herbivore-induced changes in the chemistry of plants may be a key to understanding the role of within-host plant interactions among herbivore genotypes and species in structuring herbivore communities on plants.

Such plant-mediated, indirect effects on herbivore communities may also cascade up to higher trophic levels. A herbivore-induced change in plant chemistry may not only alter the composition and density of herbivore species, but also the type and abundance of enemies that visit the plant. Thus, trait-mediated, indirect effects on herbivore communities may translate into density-mediated effects at higher trophic levels. However, herbivorous arthropods can also induce chemical processes in the plant that affect higher trophic levels. These trait-mediated, indirect effects arise when plants undergo herbivore-induced changes that promote the effectiveness of the herbivore's enemies (Price et al. 1980). For example, herbivory may induce the plant to release volatile chemicals that betray the presence of herbivores to their enemies, or induce the plant to produce food, such as extrafloral nectar, that promote predator survival or even reproduction (e.g., Sabelis et al. 1999a, 1999b, 1999c, 2002, 2005). Because the plant does not act directly against the second trophic level, but indirectly via the third, this is termed indirect plant defense (Price et al. 1980). Thus, plants may actively influence the trophic structure of arthropod communities living on them and they may manipulate the food web such that it acts like an "external immune system."

In this chapter we highlight some major advances in our understanding of how herbivore-induced chemical changes in plants modify community structure of arthropods that live on those plants. These advances have emerged from targeted manipulation of the plant defense chemistry either through (1) manual application of compounds that alter the defense chemistry in a specific way (e.g., phytohormones) or through (2) genetic transformation of plants in order to silence or overexpress specific parts of the defense chemistry. These new ways to manipulate the plant response to herbivores allowed analytic experiments to assess the role of the plant in mediating interactions between species at the second trophic level and with species at the third trophic level. Here, we discuss whether the plant's response to herbivory results in community changes of arthropods at the second and third trophic level and whether these changes result in a net benefit to the individual plant. This is done with an open eye for the – mutually nonexclusive – alternative in which it is not the plant but the herbivore that manipulates other trophic levels.

Plant-mediated changes at the second trophic level

There exists an extensive literature on how plant responses induced by one herbivore species affect the performance of other herbivores on the same plant (Denno et al. 1995, Karban and Baldwin 1997, Denno and Kaplan Chapter 2 this volume). A recent example is the study by Van Zandt and Agrawal (2004) on seven herbivorous insects on common milkweed (*Asclepias syriaca*). Early-season herbivory by one of three herbivore species (Monarch larvae, stem-feeding weevils and leaf beetle larvae) had an impact on the herbivore community structure 2 months later, but the effect strongly depended on which of the three species was the early colonizer. The underlying mechanisms are not yet clear, however. Competition for food cannot be distinguished from indirect effects via changes in plant traits, let alone whether the latter result from changes in nutritional state of the plant, the plant's (internal) immunity, or its external immunity due to protection from enemies of the herbivores.

Plant response to one herbivore influences other herbivores

To prove that the effect on other herbivores does not result from competition for food or from direct interference among the herbivores, the plant has to be triggered to produce the induced response without being fed upon. Induction in the absence of herbivores can be achieved by applying chemical elicitors of plant defense. Of particular interest are isolated/purified herbivore-derived elicitors or herbivore saliva/regurgitate containing such elicitors. For example, to show that corn leaf aphids (*Rhopalosiphum maidis*) respond to plant volatiles induced by caterpillars of the cotton leafworm (*Spodoptera littoralis*), Bernasconi et al. (1998) treated mechanically damaged corn plants with caterpillar regurgitate, which induces the plants to produce volatiles (Turlings et al. 1993b). They found that aphids avoid regurgitate-induced plants and prefer clean plants, both under laboratory conditions and in the field. By inducing the plant with herbivore regurgitate instead of the herbivore itself, effects of the presence of herbivore-specific odors or their feces were ruled out as a possible cue for avoidance. According to Bernasconi et al. (1998) the corn leaf aphids avoided regurgitate-treated plants because the induced plant-borne volatiles indicate (1) that the plant has started to produce toxic compounds in response to damage, (2) that potential competitors are present on the plant, or (3) that the plant is attractive to natural enemies. Moreover, they suggest that aphids avoid damaged plants because one of the induced plant volatiles (i.e., (*E*)-β-farnesene) is identical to an aphid alarm pheromone. This is not necessarily the case, however, because the same volatile compound is part of odor from undamaged hop plants that are known to be attractive to aphids (Campbell et al. 1993). Hence, it seems more likely that the response of the aphids to this volatile depends on context rather than on the particular compound itself.

Another approach to avoid herbivore damage, yet trigger induced response is to manually treat plants with plant-specific elicitors that activate and regulate their defense (e.g., phytohormones). For example, the application of various kinds of plant-specific elicitors (or their analogs) to tomato (*Lycopersicon esculentum*) plants in the field resulted in less damage by several, but not all, herbivorous arthropods and plant pathogens (Inbar et al. 1998). Another example is the application of jasmonic acid (JA), a product of the octadecanoid pathway, but also a phytohormone stimulating this pathway and ultimately plant defense against herbivorous arthropods (Farmer and Ryan 1992, McConn et al. 1997, Li et al. 2002, 2004). Jasmonic acid mediates most – but not all – plant gene expression that is generated by feeding of caterpillars of the small white cabbage butterfly (*Pieris rapae*) on wall cress (*Arabidopsis thaliana*) (Reymond et al. 2004) and by feeding of two-spotted spider mites (*Tetranychus urticae*) on tomato (Ament et al. 2004). The accumulation of JA leads to local and systemic production of plant toxins (e.g., nicotine in wild tobacco (*Nicotiana attenuata*): Baldwin 1998), defense proteins (e.g., digestion inhibitors acting on proteinases) or the release of volatile chemicals, similar to those induced by herbivorous arthropods (e.g., two-spotted spider-mites in tomato: Ament et al. 2004). Treating plants with JA or exposing them to its volatile methylated form (MeJA) elicits defenses very similar to those triggered by herbivory. When the roots are exposed to MeJA, wild tobacco plants in the field (i.e., juniper habitats in south east Nevada and south west Utah, USA) suffer less damage and produce more seeds, provided herbivore attack rates were moderate, i.e., not too high so as to overwhelm plant defenses (Baldwin 1998). At very low herbivore attack rates, seed production of JA-treated plants was lower than that of control plants, suggesting that JA-mediated plant defense comes at a cost and requires tuning to actual attack (Baldwin 1998). Application of exogenous JA to tomato plants in the field caused accumulation of defensive proteins (proteinase inhibitors, polyphenol oxidase) and reduced the abundance of herbivores in three feeding guilds: parenchyme feeders (thrips), leaf chewers (caterpillars, flea beetles) and phloem sap feeders (aphids) (Thaler et al. 2001). Moreover, two-choice tests with flea beetles showed an increased tendency to visit the untreated plants when JA-treated plants were the alternative. This is exactly what is expected because the control plants are the less defended hosts. The flea beetles may have tasted and/or smelled the difference, but this was not tested by Thaler et al. (2001).

A problem with the application of exogenous JA is that it is hard to control and mimicks the plant-tissue-specific, endogenous concentration during actual herbivory. Ideally, one would like to manipulate the genes involved in JA biosynthesis or genes involved in defense downstream of JA. Kessler *et al.* (2004) developed such an approach by knocking out genes involved in JA biosynthesis. They transformed wild tobacco to silence three genes involved in the oxylipin–octadecanoid signaling pathway, one coding for lipoxygenase (LOX3, a core enzyme of this pathway), one for hydroperoxide lyase (essential for production of green leaf volatiles, such as hexenal), and one for allene oxide synthase (essential for production of JA). Pairs of transformed plant lines and wild-type plants were planted along a transect in a field near Santa Clara in Utah. After 2 weeks, the overall herbivore damage was increased on the plant line with the silenced LOX3 gene, relative to the other plant lines. Leaf-chewing caterpillars of sphingid hornworms reached a larger body mass and – most surprisingly – two new herbivore species appeared to cause major plant damage: a leafhopper and a cucumber beetle, two generalist herbivores that otherwise do not visit wild tobacco plants, but other herbs growing in the same field. Two-choice tests showed that these novel herbivores even preferred to feed on the plant line transformed to silence the LOX3 gene. Possibly, differential emission of volatile chemicals between the plant lines may have helped these herbivores to make a choice, but this was not tested. All in all, the experiments by Kessler *et al.* (2004) show that downregulation of direct and indirect defenses has consequences for the amount of damage a plant incurs and for the structure of the herbivore community. It remains to be elucidated to what extent the effects of silencing the LOX3 gene are due to reduced internal immunity or lower external immunity resulting from the low emission of herbivore-induced plant volatiles (e.g., *cis*-α-bergamotene).

Multiple types of attackers, multiple types of plant responses

Besides the octadecanoid route, there are several other signaling pathways in plants involved in establishing and fine-tuning a diversity of defense responses. One of the best known pathways is the shikimate pathway that leads to the production of salicylic acid (SA). SA is a phytohormone that regulates local and systemic expression of defense products such as pathogenesis-related (PR) proteins. SA facilitates the functioning of a protein called NPR1 (nonexpressor of PR genes 1) that binds to a transcription factor to activate SA-responsive nuclear PR genes (Pieterse and van Loon 1999, Spoel *et al.* 2003). In this way, SA regulates systemic acquired resistance (SAR) (Schneider *et al.* 1996), which acts mainly against biotrophic, necrotizing pathogens. SA and JA are both involved in the production of herbivore-induced plant volatiles (HIPV) in Lima bean (*Phaseolus lunatus*) and wall cress (*Arabidopsis thaliana*) (Ozawa *et al.* 2000, Van Poecke and Dicke 2002, Ament *et al.* 2004) and, possibly, there is an SA-mediated, but NPR1-independent pathway leading to HIPV (Van Poecke and Dicke 2003). SA can inhibit the octadecanoid signaling pathway and thereby JA biosynthesis, as well as JA-mediated defenses (Doares *et al.* 1995, Felton *et al.* 1999, Pieterse and van Loon 1999, 2004, Felton and Korth 2000). Especially when applied simultaneously and in high concentrations, defense against bacterial pathogens (*Pseudomonas syringae* pv. *tomato* DC3000) is unaltered, whereas defense against herbivores decreases due to downregulation of JA-controlled, defensive proteins (proteinase inhibitors, polyphenol oxidase) that would otherwise be upregulated by JA alone (Thaler *et al.* 1999, 2001, 2002a, Thaler 2002).

This raises the question how well plants can cope with a diversity of attackers. The trade-off between SA and JA implicates that plants that are first attacked by pathogens may become vulnerable to herbivores, thereby changing the community of attackers on these plants. For example, tobacco plants infected with tobacco mosaic virus displayed decreased ability to induce JA in response to mechanical wounding and increased susceptibility to a herbivore, the tobacco hornworm (*Manduca sexta*) (Preston *et al.* 1999). Another example is that saliva of the corn earworm (*Helicoverpa zea*) contains glucose oxidase which suppresses the JA-dependent defense response in plants, but simultaneously enhances plant protection against pathogens (Musser *et al.* 2005a, 2005b). The vulnerability of plant defense due to the SA–JA trade-off is also illustrated by the fact that some virulent *Pseudomonas* bacteria produce coronatine which activates JA-mediated defenses and thereby shuts off SA-mediated defenses that would otherwise harm the bacteria (Zhao *et al.* 2003). Wall cress (*Arabidopsis thaliana*) infected with a non-coronatine-producing *Pseudomonas* strain is indeed more susceptible to herbivory by the cabbage looper (*Trichoplusia ni*) than when infected by a coronatine-producing *Pseudomonas* strain (Cui *et al.* 2005). Similarly, tomato deficient in the induced JA response has increased resistance to virulent *Pseudomonas*, but increased susceptibility to spider mites (*Tetranychus urticae*) (Li *et al.* 2002).

Why does SA inhibit the JA pathway? Although the causal mechanism is beginning to be elucidated (Pieterse and van Loon 2004), the selective advantage is not immediately obvious. Possibly, the plant gains by prioritizing defense against pathogens because they propagate faster, they can be stopped only by direct defense, and there is a suite of muggers in the microbial world waiting for opportunities where a plant's defenses are in some way compromised. If plants prioritize defense against pathogens, however, opportunities emerge for herbivores that transmit plant pathogens, because the pathogens cause the plant to shut off the JA-mediated defense response and this benefits the herbivore vectoring the pathogen. In fact, there is evidence for this. Western flower thrips (*Frankliniella occidentalis*) thrive on pepper plants that are thrips-inoculated or mechanically inoculated with tomato-spotted wilt virus, but not on unattacked plants and they perform poorly on plants previously under attack of virus-free thrips (Belliure et al. 2005). Under natural conditions this may select for plants with reduced interaction between signaling pathways leading to well-defendedness against both pathogens and herbivores, unless there is a trade-off in the energy spent in different modes of defense. We therefore hypothesize that co-regulation of signaling pathways in wild plants reflects fine-tuning to the risks imposed by many different plant attackers present in their environment and that – if constrained by a trade-off in energy allocation to different defenses – the first attacker may benefit the second and they may even join forces (e.g., herbivores vectoring plant disease, which in turn benefits herbivores).

There is also evidence for JA antagonizing the SA-mediated pathway (Kunkel and Brooks 2002, Rojo et al. 2003). Multiple plant defense pathways that modulate each other have probably evolved to allow plants to fine-tune their defense responses, such that they deploy the appropriate combination of defenses against specific pathogens and herbivores. Mutually antagonistic interactions, such as between SA- and JA-mediated pathways, may ensure that certain defenses are prioritized over others and that inappropriate defenses are not activated. However, they could also create increased vulnerability to particular pathogens or herbivores. For example, *Pseudomonas syringae* produces the phytotoxin coronatine, which structurally and functionally resembles MeJA, and utilizes it to activate the JA signaling pathway, thereby interfering with the induction of SA-dependent signaling (Kloek et al. 2001, Zhao et al. 2003, Li et al. 2004, Cui et al. 2005). This could inhibit or delay defenses, thereby providing the pathogen with a window of opportunity during which it can colonize host tissue. Moreover, mutually antagonistic, SA- and JA-mediated pathways may be exploited by herbivores by vectoring pathogens that switch off anti-herbivore defenses in the plant.

Not only may herbivores and pathogens join forces, but plants may also join forces with their attackers in the competitive battle against neighboring plants. This possibility was recognized by Van Dam et al. (2000) and Baldwin (2001) who wondered why wild tobacco plants suppress JA by an ethylene burst following attack by young larvae of the tobacco hornworm (*Manduca sexta*), yet mount the nicotine defense when attacked by old larvae. They noted that a single fourth or fifth instar larva can consume an entire tobacco plant, whereas first and second instars consume a negligible part of the total leaf mass. Therefore, these authors hypothesized that the plant gains by being mild to young larvae and by mounting a vigorous nicotine defense against old larvae, if this would motivate them to move from the focal plant to neighboring plants. The old larvae would then consume the neighbors, thus reducing competition of the focal plant with them. This interesting hypothesis on herbivores as weapons in plant competition is not yet tested, however.

Plant-mediated changes at the third trophic level

Herbivory also causes plants to respond in ways that affect community structure at the third trophic level. It induces plants to produce more extrafloral nectar that can serve as alternative food for natural enemies of herbivores (e.g., Sabelis et al. 2005). It may also induce the plant to release specific volatile chemicals (e.g., Dicke and Sabelis 1988, Dicke et al. 1990a, 1990b, Turlings et al. 1990, 1995, Takabayashi and Dicke 1996) that arrest and/or attract natural enemies of the herbivores. There are several lines of evidence for an active role of the plant in releasing HIPV: (1) whereas many natural enemies are attracted to plants harboring herbivores as prey, herbivorous arthropods and some of the products they deposit on the plant (e.g., feces, silk) can be eliminated as the source of volatiles found in blends of HIPV (e.g., Sabelis et al. 1984), (2) the compounds identified in blends of HIPV have been reported to occur in plants (e.g., Dicke et al. 1990a, Turlings et al. 1990, Scutareanu et al. 1997, Du et al. 1998), (3) pathways for biosynthesis of compounds in blends of HIPV exist in the plants and are shown to be inducible by herbivory (e.g., Bouwmeester et al. 1999, Paré et al. 1999), (4) blends of HIPV do not only emanate from the leaf under attack by the herbivore, but are also systemically induced in the plant (Turlings and Tumlinson 1992, Dicke 1994, Guerrieri et al. 1999), (5) application of the phytohormone jasmonic acid to wild-type plants or defense-signaling mutants (e.g., JA) leads to induction of volatiles similar to HIPV (Hopke et al. 1994, Ozawa et al. 2000, Thaler et al. 2002a, Gols et al. 2003, Van Poecke and Dicke 2003, Ament et al. 2004), (6) herbivore-induced gene expression patterns in plants are similar to those mediated by jasmonates (e.g., Ament et al. 2004, Kant et al. 2004, Mercke et al. 2004, Reymond et al. 2004), (7) elicitors of HIPV synthesis in the plant have been found in the regurgitate/saliva of herbivores (N-(17-hydroxylinolenoyl)-L-glutamine or volicitin: Alborn et al. 1997, 2000; β-glucosidase: Hopke et al. 1994, Mattiacci et al. 1995). Evidence for herbivore-induced production of extrafloral nectar is much less comprehensive, but it is known that this response can be elicited by JA treatment of the ant-attended plant macaranga (*Macaranga tanarius*) (Heil et al. 2001). A phylogenetic study of wattles (*Acacia* spp.) showed that inducibility of extrafloral nectar is likely to be the ancestral state, whereas constitutive extrafloral nectar production is derived (Heil et al. 2004). Thus, a constitutive trait involved in indirect plant defense has evolved from an inducible defense, presumably to establish a more permanent food-for-protection mutualism with ants. Constitutive production is commonplace for another mode of indirect plant defense, the provisioning of shelter (so called domatia). It is known to exist in plants from the Eocene (O'Dowd et al. 1991). Herbivore-induced shelter formation is to be expected, but as yet not demonstrated experimentally.

Enemies of herbivores respond to HIPV

For plants to benefit from the herbivore-induced production of extrafloral nectar or plant volatiles, there should be adequate foraging responses from natural enemies of the herbivores. Whereas there is a sizeable literature on behavioral responses in various laboratory or small-scale greenhouse settings (Sabelis and Dicke 1985, Turlings et al. 1995, Dicke et al. 1998, Janssen et al. 1998, Sabelis et al. 1999a), evidence from greenhouse crops, agricultural fields, or natural environments is limited. The first demonstration of the impact of HIPV in the field came from research on biocontrol of psyllids in pear (*Pyrus communis*) orchards in the Netherlands (Drukker et al. 1995). Pear trees suffer from only a few plant pests, the most important of which are leaf suckers (*Psylla pyri* and *P. pyricola*). These psyllids are absent in spring (March–May), emerge in the course of June and may become numerous from July to September. In spring and early summer the predators thought to be most effective against the psyllids (i.e., heteropteran bugs such as *Anthocoris nemorum*, *A. nemoralis*, and various *Orius* spp.) feed on aphids on other trees (e.g., alder (*Alnus*), hawthorn (*Crataegus*)). By the time psyllid populations increase in July, adult predators begin to immigrate into the pear orchard in June–July. The predators then increase in numbers and their local population sizes show strong positive correlations with those of the psyllids (Scutareanu et al. 1999). It was hypothesized that the immigration of adult predators was triggered by HIPV from *Psylla*-infested pear trees. This was assessed in the field by sampling trees next to predator-proof cages with trees harboring *Psylla* populations of various size. Densities of predatory bugs on trees next to the cages increased with the density of psyllids in the cages. Since covering the cage harboring an infested pear tree by a plastic sheet led to a sudden drop in the density of predatory bugs on trees next to the cages and removal of the sheet was quickly followed by a build-up in predator numbers on these trees, there was support for HIPV-triggered immigration of predators (Drukker et al. 1995). This was further substantiated by identification of headspace volatiles from *Psylla*-infested leaves (methyl salicylate and (E,E)-α-farnesene) and by testing the behavioral response to *Psylla*-infested vs. clean leaves and to single HIPV compounds in a Y-tube olfactometer (Scutareanu et al. 1997).

To demonstrate that such responses of natural enemies to HIPV affect the abundance of herbivores on a plant, Thaler (1999a, 1999b) positioned cups with caterpillars of the beet armyworm (*Spodoptera exigua*) and artificial diet near tomato plants to rule out induction of direct defenses and treated the tomato plants with exogenous JA, to induce the release of many – but not all – volatile chemicals present in odor blends from herbivore-induced plants (Hopke et al. 1994, Thaler et al. 2002a, Gols et al. 2003, Ozawa et al. 2004). Near these JA-treated plants, Thaler found significantly higher parasitism of caterpillars (by the endoparasitic wasp *Hyposoter exiguae*), than near control plants. Thaler's experiment shows in an elegant way that – in absence of direct defense – plants can reduce herbivore numbers by releasing odors that attract parasitoids. Similarly, Birkett et al. (2000) used the plant volatile (Z)-jasmone (qualitatively different, but biosynthetically related to JA) to show its repellence toward various species of aphids and its ability to induce attractiveness to seven-spot ladybirds and aphid parasitoids.

One may wonder which of all the various effects of HIPV have been most decisive in generating a selective advantage to the plant expressing such a response to herbivory. Possibly, the original function was in regulation of heat and oxidative stress and/or in direct defense based on toxicity to pathogens (Holopainen 2004). Once plants evolved to attract natural enemies of herbivores by HIPV, repellency of HIPV to herbivores may have evolved as a by-product, because it pays them to avoid the herbivore's enemies attracted to the HIPV-releasing plant. Moreover, less effective natural enemies of herbivores (i.e., from the plant's perspective) may evolve attraction to HIPV, once plants evolved HIPV to attract effective natural enemies. Attraction of many species of parasitic wasps to HIPV may represent such a by-product. Unlike true predators that instantaneously kill their prey and thereby free the plant of attackers, koinobiont (as opposed to idiobiont) parasitoids require their host to live and feed to enable parasitoid development. For some species this causes the host to increase its feeding activity and the time spent feeding, resulting in an increase of plant damage (e.g., Sabelis and de Jong 1988, Van der Meijden and Klinkhamer 2000, Janssen et al. 2002). The role of parasitic wasps in plant protection was tested in the greenhouse by Van Loon et al. (2000) for wall cress (*Arabidopsis thaliana*) with larvae of the small white cabbage butterfly (*Pieris rapae*), as herbivores, and the solitary endoparasitoid *Cotesia rubecula* as the enemy of the herbivore. They found that seed production of intact *A. thaliana* plants, plants that were mechanically damaged, and plants that were fed upon by parasitized caterpillars, did not differ significantly from each other, but was significantly higher than seed production of plants fed upon by unparasitized caterpillars. Similar results were obtained by Fritzsche-Hoballah and Turlings (2001) for maize (*Zea mays*) plants with larvae of the corn leafworm (*Spodoptera littoralis*) as herbivores and the endoparasitoids *Cotesia marginiventris* and *Campoletis sonorensis*, as the herbivore's enemies. They found reduced feeding and weight gain in the parasitized corn leafworm larvae. As a result, young maize plants attacked by a single parasitized larva suffered much less feeding damage and produced about 30% more seed than plants that were attacked by an unparasitized larva. Such reproductive benefits to the plant can thus accrue from herbivore suppression by parasitic wasps alone. The exceptions to this rule are still to be shown (Van der Meijden et al. 2000, Janssen et al. 2002).

To date, the only study demonstrating a role of HIPV under natural conditions was carried out by Kessler and Baldwin (2001). They quantified volatile emissions from a postfire annual, wild tobacco (*Nicotiana attenuata*), growing in natural populations in Utah during attack by three species of leaf-feeding herbivores: tobacco hornworm larvae (*Manduca quinquemaculata* or *M. sexta*), mirid bugs (*Diciphus minimus*), and chrysomelid flea beetles (*Epitrix hirtipennis*). By applying lanolin paste with a chemical compound to the stem of each plant, they mimicked the release of each of five commonly emitted volatiles and they mimicked the release of HIPV blends by treating the plant with MeJA. The effect of these volatiles on natural enemies was quantified by assessing the rate of predation on *M. sexta* eggs that were glued to the underside of a leaf of each test plant. By using a nonfeeding stage of the herbivore and by mimicking the production of HIPV blends or components thereof, Kessler and Baldwin (2001) were able to eliminate the confounding effect of direct plant defense on herbivore abundance. Blends of volatiles emanating from MeJA-treated plants and three individual volatile compounds ((Z)-3-hexen-1-ol, linalool, and *cis*-α-bergamotene) increased egg predation rates by a heteropteran predator, *Geocoris pallens*. Linalool and the MeJA-mediated blend decreased oviposition rates of tobacco hornworms on treated plants. Thus, the release of JA-induced or herbivore-induced volatiles acts a double-edged sword: it causes predator attraction and herbivore avoidance. In this way, a wild tobacco plant reduced the number of herbivores by more than 90%, proving that indirect plant defenses can be effective in nature.

Coordination of defense responses

Effective plant defense requires coordination of direct and indirect defenses such that they do not interfere or even act synergistically. A nice example comes from studies by Baldwin (2001), who wondered why the nicotine defense of wild tobacco is downregulated by an ethylene burst following attack by young tobacco hornworm larvae. He formulated three – as yet untested – hypotheses to explain this. First, the hornworm larvae may suppress the plant's nicotine response or they may reduce their food intake and therefore run less risk to alert the plant's nicotine response. Second, the tobacco plant may improve the impact of its indirect plant defenses by reducing possibilities for hornworm larvae to sequester nicotine as a defense against predators and parasitoids, and by slowing down hornworm growth (through production of digestion inhibitors), thereby causing prolonged exposure of hornworm larvae to their predators. In this way, the plant saves fitness costs from investment in nicotine defense and gains more protection from indirect defense (Kahl et al. 2000). Third, the plant may suppress the nicotine response until the hornworm larvae reach a size (i.e., fourth or fifth instar) where they impose a serious death risk to the tobacco plant and then switch to produce nicotine to make these large larvae move to their neighboring competitors (Van Dam et al. 2000). These three hypotheses – herbivore stealth, indirect defense optimization, and herbivores as allies in plant competition – do not necessarily exclude each other. If the plant first chooses to optimize indirect defense against young hornworm larvae, it may switch to direct defense by the time the hornworm larvae are large enough, thereby deterring the larvae and imposing immediate and present danger to neighboring (i.e., competing) plants. This defense scenario is not likely to be general because the best scenario depends on the effectivity of alternative defenses. There are examples of plants switching on direct defenses in response to feeding by caterpillars even though this reduces the suitability of the caterpillars as hosts for parasitic wasps (e.g., Havill and Raffa 2000, Thaler 2002).

Another example of plants coordinating different defenses stems from work on tulip bulbs (Lesna et al. 2004). When attacked by the rust mite *Aceria tulipae*, bulbs become attractive to the predatory mite *Neoseiulus cucumeris* (Aratchige et al. 2004). The rust mites on the outside can be easily cleared from the bulb by these predatory mites, but thanks to their minute size the rust mites can also move in between the scales of the bulb via the so-called "nose" of the bulb. Between the scales, the rust mites would be in a refuge from predation because the predatory mites are just too large to enter the bulb. However, in response to rust mite attack, the bulb starts producing ethylene and this causes the distance between bulb scales in the nose to widen just enough to allow the predatory mites to enter. Without this response, the bulbs are eaten from within by the rust mites, as is demonstrated by treating the bulbs with an ethylene inhibitor (1-methylcyclopropene). Thus, the bulb coordinates the release of HIPV with an ethylene-mediated change in bulb structure, providing predatory mites access to the bulb's interior.

Specificity of HIPV

One of the striking features of HIPV as a means to attract natural enemies is that there are usually several compounds involved and that these mixtures of volatiles contain a great deal of specific information. In an olfactometer study by Sabelis and van der Baan (1983) predatory mites (*Phytoseiulus persimilis* and *Metaseiulus occidentalis*) known to control spider mites (*Tetranychus urticae* and *T. viennensis*) were shown to prefer volatiles from Lima bean leaves and apple leaves infested by these *Tetranychus* mites when clean leaves were the alternative, but they did not respond to apple leaves infested by non-prey, such as fruit-tree red spider mites (*Panonychus ulmi*) and apple rust mites (*Aculus schlechtendahli*). Similarly, predatory mites (*Amblyseius andersoni* and *Neoseiulus finlandicus*) known to control fruit-tree red spider mites and apple rust mites were shown to respond to apple leaves with these herbivorous mites, but not to Lima bean leaves or apple leaves infested with *Tetranychus* mites. This differential response makes sense because *Tetranychus* mites use silken webs as an effective defense against *A. andersoni* and *N. finlandicus*, but it is much less effective against *P. persimilis* and *M. occidentalis* (Sabelis and van der Baan 1983, Sabelis and Bakker 1992). Although a role of prey-produced odors in this attraction cannot be fully excluded, evidence so far points at odor components that are of plant origin (Dicke et al. 1990a, 1990b, Takabayashi et al. 1991, Dicke 1994, Thaler et al. 2002a, Gols et al. 2003). This would imply that the plant provides information to predators that is sufficiently specific to allow discrimination by olfaction. This hypothesis was confirmed by De Moraes et al. (1998) in a study on olfactory responses of the parasitoid *Cardiochiles nigriceps* to three species of host plants (tobacco, maize, and cotton) that were attacked by two closely related herbivore species, the tobacco budworm (*Heliothis virescens*) and the corn earworm (*Helicoverpa zea*). These authors found that tobacco, cotton, and maize plants each released distinct volatile blends in response to damage by the two herbivore species, and that the parasitoid exploits these differences in odor blends to distinguish infestation by its host, *H. virescens*, from that by the non-host *H. zea*. Most convincingly, these preferences persisted when the leaves with the herbivores had been removed and, thus, only systemically produced odors were offered. Another comprehensive study showing herbivore specificity of HIPV was done by studying the choice of a braconid parasitoid (*Aphidius ervi*) for broad bean plants (*Vicia faba*) infested by either of two aphid species, one suitable host (the pea aphid, *Acyrthosiphum pisum*) and one non-host (the black bean aphid, *Aphis fabae*) (Du et al. 1998, Guerrieri et al. 1999). Air entrainment extracts of volatiles collected from a broad bean plant infested by the non-host aphid or from an uninfested broad bean plant elicited few oriented flights and landing responses by female parasitoids. These extracts were significantly less attractive than extracts collected from a broad bean plant infested by the host aphid, indicating the specificity of HIPV from the same plant species attacked by different aphid species. A similar example exists for whiteflies (*Trialeurodes vaporariorum*). They induce four de novo emitted volatiles in beans and three of those elicit oriented flight and landing of *Encarsia formosa*, when offered in pure form, most effective being a mixture of (Z)-3-hexen-1-ol and 3-octanone (Birkett et al. 2003).

Learning to cope with variability in HIPV

Because the same species of herbivore can induce different plant species to produce different odors and different species of herbivores can induce the same plant species to produce different odors, predators that use their olfactory senses to find their preferred prey face a difficult task. The number of chemicals in blends of HIPV is large, but limited. Thus, blends of HIPV may differ qualitatively, but they often share components and hence differ quantitatively. There is evidence for a genetic component of olfactory responses in predatory mites (Margolies et al. 1997, Maeda et al. 1999, 2001, Jia et al. 2002) and parasitoids (Gu and Dorn 2000, Wang and Dorn 2003, Wang et al. 2004), but it is hard to see how fixed responses suffice in a world full of variation in odors. The ability to associate odors with rewards is well established for parasitoids (Papaj and Lewis 1993, Roitberg et al. 1993, Turlings et al. 1993a, Vet et al. 1995) and this is now also shown for HIPV as a cue associated with the occurrence of hosts for the parasitoids (Vet and Dicke 1992, Geervliet et al. 1998a, 1998b, Storeck et al. 2000).

Predatory arthropods were much less intensively investigated for their ability to learn, but it was known that their olfactory responses may depend on conditions during rearing (Takabayashi and Dicke 1992, Krips et al. 1999, De Boer and Dicke 2004). That predatory arthropods can also learn to associate HIPV with herbivores as profitable prey is a new finding (Drukker et al. 2000a, 2000b). This has led to the following hypothesis on the role of fixed and plastic responses to HIPV (Sabelis et al. 1999a, Drukker et al. 2000a, 2000b). Predatory arthropods often grow up on a plant with herbivores as prey. On this host plant, they experience HIPV in presence of abundant prey, resulting in a positive association. However, when prey densities decline to levels where they are not sufficiently abundant to the predator, they will experience HIPV in absence of abundant prey. The herbivore-infested plants continue to produce HIPV, even when herbivores have declined in numbers. This causes the predators to build up a negative association with HIPV, triggering dispersal away from the herbivore-poor plant. Although not yet proven for predators (but see Papaj et al. (1994) for evidence in parasitic wasps), it is likely that this negative association will quickly fade away with time spent in dispersal. Then, they will switch to rely on the initial positive association (Papaj et al. 1994). Upon arrival on a new plant with prey the starved, predatory arthropods start to feed on this prey. While experiencing this reward the animals learn that contextual cues such as HIPV are associated with the presence of their newly found prey. To stay in those areas of the plant where the prey density is highest, the arthropods may sample the prey densities on the plant and upon depleting a patch they may search locally to find other prey patches on the same plant. Predatory arthropods switch from local searching behavior to dispersal behavior (e.g., aerial) when their prey decline to levels where they are not sufficiently abundant to the predator. To find their next meal, the predators rely on both their memory and their innate preferences. It is conceivable that during this phase the predators are attracted to the HIPV of non-prey-infested plants and it is expected that the predators after a local exploration disperse again until they find yet another plant with herbivores as prey. Some predators may also be able to learn from the encounters with non-prey and use this experience to avoid future mistakes and enhance their ability to differentiate between HIPV associated with non-prey and HIPV associated with prey.

A number of cases have been reported where predatory arthropods and parasitoids respond to HIPV from plants with non-prey or non-host herbivores (Agrawal and Colfer 2000, Shimoda and Dicke 2000, Thaler 2002, Dicke et al. 2003, Van Poecke and Dicke 2003, Janssen and Van Wijk 2005). Shimoda and Dicke (2000) collected females of the predatory mite *P. persimilis* from a culture that was fed spider mites (*T. urticae*) on bean and starved them for 1 hour or 1 day. Predators that were starved for 1 hour responded positively to HIPV from bean leaves infested by a non-prey herbivore (the beet armyworm, *Spodoptera exigua*), but females that were starved for 1 day did not. These authors argue that 1-hour starvation corresponds to a real-life situation where the predator recently lost track of prey-associated cues, and thus must be close to its prey (and thus at little risk of visiting leaves with non-prey). For that case, the authors hypothesize that the predator responds to a subset of volatile chemicals from the full HIPV blend, in particular those shared between blends from bean infested with prey (spider mites) and bean infested with non-prey (*S. exigua*). One day of starvation would then represent a situation in which the original site where they were fed with prey is too distant. For that case, the authors hypothesize that the predators avoid the risk of visiting a plant with non-prey. Thus, the predators may only initially (1 hour) generalize between the learned blend associated with prey (spider mites) on bean and the blend associated with non-prey (beet armyworm) on bean. After 24 hours, all predators equally prefer a blend associated with their prey (spider mites). Strikingly, 1-hour starved predators appear not to generalize between blends when they were collected from a culture with prey (spider mites) on cucumber and offered the blend associated with non-prey (beet armyworm) on bean.

In another series of experiments with the same arthropods, De Boer et al. (2004) showed that *P. persimilis* predators fed with spider mites on bean plants respond positively to HIPV from bean plants infested by a suffciently high number of spider mites, when HIPV from other plants infested by beet armyworm larvae were the alternative. Chemical analysis of the different odor blends from plants infested by either one or the other herbivore species showed mainly quantitative differences in composition, suggesting that the predatory mites discriminate between ratios of volatile chemicals in the HIPV blends (De Boer et al. 2004). Thus, the predatory mite *P. persimilis* can locate plants with prey in a mixture of HIPV from plants infested by prey herbivores and plants with non-prey herbivores. De Boer et al. (2005) also found that nonrewarding experiences (e.g., starvation in the presence of HIPV from *S. exigua*-infested bean plants) have much less effect on olfactory choice between HIPV from *S. exigua*-infested and *T. urticae*-infested bean plants than rewarding experiences (feeding on *T. urticae* in the presence of HIPV from *T. urticae*-infested bean plants), whereas providing these rewarding and nonrewarding experiences shortly after each other had the strongest effect.

Janssen and colleagues investigated innate and acquired responses of three species of predatory mites (*Phytoseiulus persimilis*, *Iphiseius degenerans*, *Neoseiulus cucumeris*) to HIPV produced by cucumber plants with prey (the spider mite *T. urticae* for *P. persimilis* and the western flower thrips *Frankliniella occidentalis* for the other two predator species) and non-prey (thrips for *P. persimilis* and spider mites for the other two predator species) (Takabayashi et al. 2006). Although all three species discriminated between plants with prey and plants with non-prey, this resulted from quite different processes. One species (*N. cucumeris*) showed innate attraction to volatiles from plants with prey, but not to volatiles from plants with non-prey. A second species (*I. degenerans*) showed an innate aversion to volatiles from plants with non-prey and was not attracted to plants with prey. A third species (*P. persimilis*) was attracted to plants with prey and non-prey, but preferred volatiles from plants with prey when offered a choice. When this last species was given an experience of 1 day with non-prey in presence of HIPV induced by non-prey, it was still attracted to these volatiles, but an experience of 2 days resulted in loss of such attraction. Thus, predatory arthropods are equipped with a remarkable set of innate responses. To cope with the variability in HIPV, they are able to associate HIPV with the presence of prey. It is far less clear how long these memories persist and what factors may contribute to loss or retention. Negative experiences such as starvation or the presence of non-prey in combination with HIPV can further fine-tune the response of the predatory arthropods to HIPV. A key question for the future remains how predators quickly assess that an initially attractive plant is actually infested with non-prey and how, and at what timescale, predators decide to leave.

Although learning is a relatively fast process compared to the rate of population dynamical change, it does take time and – more importantly – it does require a certain frequency of similar experiences, which takes even more time. This may yield a head-start advantage to herbivores colonizing a host plant already occupied by a herbivore that does not have the same natural enemies. Plants that are simultaneously infested by two species of herbivores are unlikely to produce blends similar to those infested by one of the two herbivore species. Hence, the natural enemies of the herbivores have to learn the association between their victim and the HIPV released from plants with prey and non-prey herbivores. This learning process has not yet been studied in depth, but an advantage to the herbivore of joining other herbivore species on a plant has been demonstrated. Shiojiri et al. (2002) studied how host plant (cabbage) infestation by two herbivore species (the diamondback moth, *Plutella xylostella*, and the small cabbage white butterfly, *Pieris rapae*) influenced mortality by specialist parasitoids (*Cotesia plutellae* and *C. glomerata*). They found that diamondback moth larvae suffered much lower parasitization risk on plants also infested by cabbage white larvae. Moreover, they found that female diamondback moths preferred to oviposit on plants infested by cabbage white larvae, rather than on uninfested or diamondback-moth-infested plants that were offered as an alternative. Such preferences for host plants colonized by other herbivore species may evolve if such plants represent enemy-scarce space. Such herbivore refuges may have an important impact on population dynamics, as they are shown to promote the persistence of otherwise unstable parasitoid–host systems (Hassell 1978, Vos et al. 2001).

Do entomopathogens and nonarthropod insectivores use HIPV as a signal?

Whereas predatory arthropods (phytoseiid mites and heteropteran bugs) and parasitoid wasps have been best explored for their responses to HIPV, other classes of natural enemies have received little attention. Some major advances have been achieved recently with respect to entomopathogens. Van Tol et al. (2001) were the first to show that indirect plant defenses also operate underground. They found that the roots of a coniferous plant (*Thuja occidentalis*) release chemicals upon attack by vine weevil larvae (*Otiorhynchus sulcatus*) and that these chemicals attract nematodes that parasitize weevils (*Heterorhabditis megidis*). Hountondji et al. (2005) were the first to show that volatiles emanating from cassava plants infested by green mites (*Mononychellus tanajoa*) trigger production of conidia, the infectious stage, in different isolates of a mite-pathogenic fungus (*Neozygites tanajoae*), whereas volatiles from clean plants suppress conidiation. These opposing effects make sense in that the entomopathogenic fungus tunes the release of conidia, the stage most vulnerable to environmental conditions, to herbivore-induced plant chemicals that signal the presence of hosts. Apart from entomopathogens, insectivorous birds and mammals are also expected to exploit HIPV. Recently, Mantyla et al. (2004) showed that foliar damage by sawfly larvae on mountain birch (*Betula pubescens*) led passerine birds (the willow warbler, *Phylloscopus trochilus*) to prefer intact (nondamaged) branches from trees with sawfly larvae to intact branches from control trees. The authors offer two hypotheses on the underlying mechanism: (1) olfactory perception of HIPV from sawfly-infested birch, (2) perception of UV reflected by herbivore-inducible leaf compounds, such as surface flavonoids, which have UV spectral maxima well within the range of UV perception in birds.

Plant-mediated interactions in food webs: who manipulates whom?

To understand the ecological impact of herbivore-induced plant defenses, it is necessary to elucidate the responses of all species that are loosely or tightly, directly or indirectly linked by interactions in the food web (Dicke 1999, Sabelis *et al.* 1999a, 1999b, 1999c, 2002). In this chapter, we reviewed how novel insights and techniques from plant physiology can be used to gain insight in interactions at the community level. Conversely, plant physiology, as a field of science, probably also benefited from ecological insight into the way plants cope with multiple attackers (e.g., Paul *et al.* 2000) and how they protect themselves against herbivores by coordinating direct and indirect modes of defense (e.g., Baldwin 2001).

Ecologists will continue to provide insight into plant defense scenarios by assessing the fitness effects on plants and the organisms on plants alike. The interactions that have shaped communities and plant defense scenarios can only be identified by careful analysis of the overlap and conflict in interests of plants and the arthropods that live on them. Several areas of novel enquiry stand out. The first arises from the fact that, once plants mount indirect defenses in response to herbivory, they have no control over which organisms will reap the benefits (Sabelis *et al.* 1999a, 1999b, 1999c, 2002, 2005). For example, plants emitting HIPV due to herbivory by one species may become conspicuous to other herbivore species. If these other species avoid this plant, thus avoiding natural enemies that they share with the herbivore on the plant, the plant gains. However, if they settle on that plant because they have different enemies and even gain a refuge from their own enemies, the plant may be worse off. This is illustrated by experiments with western flower thrips (*Frankliniella occidentalis*) and two-spotted spider mites (*Tetranychus urticae*) on cucumber. Spider mites avoid thrips-infested plants (Pallini *et al.* 1997), but western flower thrips do not avoid spider-mite infested plants (Pallini *et al.* 1999). Western flower thrips even profit from joining spider mites on a plant because they can feed on spider mite eggs and they use the silken webs produced by spider mites as a refuge from predatory mites that are otherwise effective against thrips alone, but not against spider mites due to protection from the silken web they spin (Pallini *et al.* 1998). Another example of herbivores reaping the benefits relates to plant-provided foods (Sabelis *et al.* 2005). Western flower thrips and the predatory mite *Iphiseius degenerans* can both feed on plant-provided foods. Admittedly, the alternative food here is pollen, which is not shown to be inducible. However, this study illustrates some points that may well apply to inducible plant foods, such as extrafloral nectar. Van Rijn et al. (2002) found that a patchy supply of the plant-provided alternative foods allowed the predatory mites to monopolize its consumption, because western flower thrips either refrained from visiting the alternative food patches due to the presence of predatory mites or they were eaten by the predatory mites on the alternative food patches. The consumption of alternative foods boosted the numerical response of the predatory mites and this in turn caused the predation pressure on western flower thrips to increase. Models predicted that western flower thrips would have better access to the plant-provided foods, thereby boosting their reproduction, if predatory mites would not aggregate around plant-provided food patches or if the plant-provided food was not patchily distributed. This would cause the plant to receive more damage from herbivory than it would incur if it did not provide alternative foods.

The idea that herbivores may reap the benefits of indirect plant defenses is fundamentally important for the way we interpret phenomena that do not conveniently fit in the scenario of plants and enemies of herbivores being allies. For example, there are several reports on natural enemy responses to volatiles induced by non-prey herbivores (Agrawal and Colfer 2000, Shimoda and Dicke 2000, Thaler et al. 2002a, Dicke et al. 2003, Van Poecke and Dicke 2003, De Boer et al. 2004). Not only may these enemy responses be overruled by learning to avoid associations between non-prey and HIPV (De Boer 2004), but there should also be room for the idea that herbivores manipulate the plant to induce the release of volatiles attracting the enemies of other herbivore species that would act as competitors of the inducer.

A second area of novel inquiry comes from the idea that natural selection may favor herbivores and pathogens that do not alarm the plant's defenses. This is particularly innovative if it concerns indirect plant defenses against herbivores. For example, after its formation in the regurgitate, volicitine, an elicitor of HIPV, is decomposed by midgut enzymes of the tobacco budworm (*Heliothis virescens*) more than in the corn earworm (*Helicoverpa zea*) (Mori et al. 2001). To what extent these differences matter to the induction of volatiles in the plant is still to be determined. However, it might indicate a counter-adaptation of the herbivore to reduce production of induced volatiles by the plant.

A third area of novel inquiry comes from the idea that antagonistic species may join forces in their battle against the defending plant. For example, herbivores may benefit plant pathogens by transmitting them to new hosts and the plant pathogens may benefit the herbivores by inducing plant defense against plant pathogens at the expense of defense against the herbivores (Belliure et al. 2005). Another example is that plants first allow herbivores to grow by shutting off direct defenses and then later force herbivores to move to their neighbor competitor when they reach a size where they would impose a serious threat to any nonalerted plant (Van Dam et al. 2000).

A final area of novel inquiry is based on the idea that the interests of plants and the natural enemies of herbivores do not fully overlap and that plants compete for protection from natural enemies of herbivores. The plant gains by acquiring protection from "bodyguards" even in advance of attack by herbivores, whereas the "bodyguards" expect food in return. This creates a conflict of interest between the plant and the herbivore's enemies. This conflict has profound consequences for the signals plants are expected to produce, which can be understood in nonmathematical terms, as follows. In a population of plants that produce signals only after herbivore attack there is room for invasion of mutants that produce the same signal even when they are not under attack. In fact, these dishonest signals can be seen as a form of "mimicry," the advantage of which will persist until the mutants have become so abundant that mutant plants that provide new "honest" information about the presence of herbivores will be preferred by the herbivore's enemies. This preference will arise either as a result of associative learning of the new information by the enemies or as a result of selection for enemy genotypes that respond to this information. Once the honest signalers dominate, there will be room again for invasion of a mutant signaler that employs mimicry and provides dishonest information. In theory, this process of frequency-dependent change gives rise to cycles in the composition of signal traits in the plant population under a broad range of conditions (Van Baalen and Jansen 2003, M. Van Baalen and M. W. Sabelis unpublished data). Thus, it is quite possible that the chemical languages in plant–predator communication are not stable, but that new dialects develop again and again in space and time. Whether such dynamical changes in the genetics of plant signals and enemy responses occur and how these will influence the long-term dynamics of tritrophic systems of plants, herbivorous arthropods, and their enemies, are major questions for future research.

References

Agrawal, A. A., and R. G. Colfer. 2000. Consequences of thrips-infested plants for attraction of conspecifics and parasitoids. *Ecological Entomology* 25:493–496.

Agrawal, A. A., S. Tuzun, and E. Bent (eds.) 1999. *Induced Plant Defenses against Pathogens and Herbivores*. St. Paul, MN: American Phytopathological Society Press.

Alborn, H. T., T. C. J. Turlings, T. H. Jones, G. Stenhagen, J. H. Loughrin, and J. H. Tumlinson. 1997. An elicitor of plant volatiles from beet armyworm oral secretion. *Science* **276**:945–949.

Alborn, H. T., T. H. Jones, G. S. Stenhagen, and J. H. Tumlinson. 2000. Identification and synthesis of volicitin and related components from beet armyworm oral secretions. *Journal of Chemical Ecology* **26**:203–220.

Ament, K., M. R. Kant, M. W. Sabelis, M. A. Haring, and R. C. Schuurink. 2004. Jasmonic acid is a key regulator of spider mite-induced volatile terpenoid and methyl salicylate emission in tomato. *Plant Physiology* **135**:1–13.

Aratchige, N. S., I. Lesna, and M. W. Sabelis. 2004. Below-ground plant parts emit herbivore-induced volatiles: olfactory responses of a predatory mite to tulip bulbs infested by rust mite. *Experimental and Applied Acarology* **33**:21–30.

Baldwin, I. T. 1998. Jasmonate-induced responses are costly but benefit plants under attack in natural populations. *Proceedings of the National Academy of Sciences of the USA* **95**:8113–8118.

Baldwin, I. T. 2001. An ecologically motivated analysis of plant–herbivore interactions in native tobacco. *Plant Physiology* **127**:1449–1458.

Belliure, B., A. Janssen, P. C. Maris, D. Peters, and M. W. Sabelis. 2005. Herbivore arthropods benefit from vectoring plant viruses. *Ecology Letters* **8**:70–79.

Bernasconi, M. L., T. C. J. Turlings, L. Ambrosetti, P. Bassetti, and S. Dorn. 1998. Herbivore-induced emissions of maize volatiles repel the corn leaf aphid, *Rhopalosiphum maidis*. *Entomologia Experimentalis et Applicata* **87**:133–142.

Birkett, M. A., C. A. M. Campbell, K. Chamberlain, et al. 2000. New roles for cis-jasmone as an insect semiochemical and in plant defense. *Proceedings of the National Academy of Sciences of the USA* **97**:9329–9334.

Birkett, M. A., K. Chamberlain, E. Guerrieri, et al. 2003. Volatiles from whitefly-infested plants elicit a host-locating response in the parasitoid, *Encarsia formosa*. *Journal of Chemical Ecology* **29**:1589–1600.

Bouwmeester, H. J., F. W. A. Verstappen, M. A. Posthumus, and M. Dicke. 1999. Spider mite-induced (3S)-(E)-nerolidol synthase activity in cucumber and lima bean: the first dedicated step in acyclic C11-homoterpene biosynthesis. *Plant Physiology* **121**:173–180.

Campbell, C. A. M., J. Pettersson, J. A. Pickett, L. J. Wadhams, and C. M. Woodcock. 1993. Spring migration of damson-hop aphid, *Phorodon humuli* (Homoptera, Aphididae), and summer host plant-derived semiochemicals released on feeding. *Journal of Chemical Ecology* **81**:1569–1576.

Constabel, C. P. 1999. A survey of herbivore-inducible defensive proteins and phytochemicals, pp. 137–166 in A. A. Agrawal, S. Tuzun, and E. Bent (eds.) *Induced Plant Defenses against Pathogens and Herbivores*. St. Paul, MN: American Phytopathological Society Press.

Cui, J., A. K. Bahrami, E. G. Pringle, et al. 2005. *Pseudomonas syringae* manipulates systemic plant defenses against pathogens and herbivores. *Proceedings of the National Academy of Sciences of the USA* **102**:1791–1796.

De Boer, J. G. 2004. Bugs in odour space: how predatory mites respond to variation in herbivore-induced plant volatiles. Ph.D. dissertation, Wageningen University, the Netherlands.

De Boer, J. G., and M. Dicke. 2004. Experience with methyl salicylate affects behavioural responses of a predatory mite to blends of herbivore-induced plant volatiles. *Entomologia Experimentalis et Applicata* **110**:181–189.

De Boer, J. G., M. A. Posthumus, and M. Dicke. 2004. Identification of volatiles that are used in discrimination between plants infested with prey or nonprey herbivores by a predatory mite. *Journal of Chemical Ecology* **30**:2215–2230.

De Boer J. G., T. A. L. Snoeren, and M. Dicke. 2005. Predatory mites learn to discriminate between plant volatiles induced by prey and nonprey herbivores. *Animal Behaviour* **69**:869–879.

De Moraes, C. M., W. J. Lewis, P. W. Paré, H. T. Alborn, and J. H. Tumlinson. 1998. Herbivore-infested plants selectively attract parasitoids. *Nature* **393**:570–573.

Denno, R. F., M. S. McClure, and J. R. Ott. 1995. Interspecific interactions in phytophagous insects: competition revisited and resurrected. *Annual Review of Entomology* **40**:297–331.

Dicke, M. 1994. Local and systemic production of volatile herbivore-induced terpenoids: their role in plant–carnivore mutualism. *Journal of Plant Physiology* **143**:465–472.

Dicke, M. 1999. Are herbivore-induced plant volatiles reliable indicators of herbivore identity to foraging carnivorous arthropods? *Entomologia Experimentalis et Applicata* **91**:131–142.

Dicke, M., and M. W. Sabelis. 1988. How plants obtain predatory mites as bodyguards. *Netherlands Journal of Zoology* **38**:148–165.

Dicke, M., T. A. van Beek, M. A. Posthumus, et al. 1990a. Isolation and identification of volatile kairomone that affects acarine predator–prey interactions: involvement of host plant in its production. *Journal of Chemical Ecology* **16**:381–396.

Dicke, M., M. W. Sabelis, J. Takabayashi, J. Bruin, and M. A. Posthumus. 1990b. Plant strategies of manipulating predator–prey interactions through allelochemicals: prospects for application in pest control. *Journal of Chemical Ecology* **16**:3091–3118.

Dicke, M., J. Takabayashi, M. A. Posthumus, C. Schütte, and O. E. Krips. 1998. Plant–phytoseiid interactions mediated by herbivore-induced plant volatiles: variation in production of cues and in responses of predatory mites. *Experimental and Applied Acarology* **22**:311–333.

Dicke, M., J. G. de Boer, M. Hofte, and M. C. Rocha-Granados. 2003. Mixed blends of herbivore-induced plant volatiles and foraging success of carnivorous arthropods. *Oikos* **101**:38–48.

Doares, S. H., T. Syrovets, E. W. Weiler, and C. A. Ryan. 1995. Oligogalacturonides and chitosan activate plant defensive genes through the octadecanoid pathway. *Proceedings of the National Academy of Sciences of the USA* **92**:4095–4098.

Drukker, B., P. Scutareanu, and M. W. Sabelis. 1995. Do anthocorid predators respond to synomones from *Psylla*-infested pear trees under field conditions? *Entomologia Experimentalis et Applicata* **77**:193–203.

Drukker, B., J. Bruin, G. Jacobs, A. Kroon, and M. W. Sabelis. 2000a. How predatory mites learn to cope with variability in volatile plant signals in the environment of their herbivorous prey. *Experimental and Applied Acarology* **24**:881–895.

Drukker, B., J. Bruin, and M.W. Sabelis. 2000b. Anthocorid predators learn to associate herbivore-induced plant volatiles with presence or absence of prey. *Physiological Entomology* **25**:260–265.

Du, Y.J., G.M. Poppy, W. Powell, *et al.* 1998. Identification of semiochemicals released during aphid feeding that attract parasitoid *Aphidius ervi*. *Journal of Chemical Ecology* **24**:1355–1368.

Farmer, E.E., and C.A. Ryan. 1992. Octadecanoid precursors of jasmonic acid activate the synthesis of wound-inducible proteinase inhibitors. *Plant Cell* **4**:129–134.

Felton, G., and K. Korth. 2000. Trade-offs between pathogen and herbivore resistance. *Current Opinion in Plant Biology* **3**:309–314.

Felton, G.W., K.I. Korth, J.L. Bi, *et al.* 1999. Inverse relationship between systemic resistance of plants to microorganisms and to insect herbivory. *Current Biology* **9**:317–320.

Fritzsche-Hoballah, M.E., and T.C.J. Turlings. 2001. Experimental evidence that plants under caterpillar attack may benefit from attracting parasitoids. *Evolutionary Ecology Research* **3**:533–565.

Geervliet, J.B.F., A.I. Vreugdenhil, M. Dicke, and L.E.M. Vet. 1998a. Learning to discriminate between infochemicals from different plant-host complexes by the parasitoids *Cotesia glomerata* and *C. rubecula*. *Entomologia Experimentalis et Applicata* **86**:241–252.

Geervliet, J.B.F., S. Ariens, M. Dicke, and L.E.M. Vet. 1998b. Long-distance assessment of patch profitability through volatile infochemicals by the parasitoids *Cotesia glomerata* and *C. rubecula* (Hymenoptera: Braconidae). *Biological Control* **11**:113–121.

Gols, R., M. Roosjen, H. Dijkman, and M. Dicke. 2003. Induction of direct and indirect plant responses by jasmonic acid, low spider mite densities, or a combination of jasmonic acid treatment and spider mite infestation. *Journal of Chemical Ecology* **29**:2651–2666.

Gu, H., and S. Dorn. 2000. Genetic variation in behavioral response to herbivore-infested plants in the parasitic wasp *Cotesia glomerata* (L.) (Hymenoptera: Braconidae). *Journal of Insect Behavior* **13**:141–156.

Guerrieri, E., G.M. Poppy, W. Powell, E. Tremblay, and F. Pennacchio. 1999. Induction and systemic release of herbivore-induced plant volatiles mediating in-flight orientation of *Aphidius ervi*. *Journal of Chemical Ecology* **25**:1247–1261.

Hassell, M.P. 1978. *The Dynamics of Arthropod Predator–Prey Systems*. Princeton, NJ: Princeton University Press.

Havill, N.P., and K.F. Raffa. 2000. Compound effects of induced plant responses on insect herbivores and parasitoids: implications for tritrophic interactions. *Ecological Entomology* **25**:171–179.

Heil, M., T. Koch, A. Hilpert, *et al.* 2001. Extrafloral nectar production of the ant-associated plant, *Macaranga tanarius*, is an induced, indirect, defensive response elicited by jasmonic acid. *Proceedings of the National Academy of Sciences of the USA* **98**:1083–1088.

Heil, M., S. Greiner, H. Meimberg, *et al.* 2004. Evolutionary change from induced to constitutive expression of an indirect plant resistance. *Nature* **430**:205–208.

Holopainen, J. K. 2004. Multiple functions of inducible plant volatiles. *Trends in Plant Science* **9**:529–533.

Hopke, J., J. Donath, S. Blechert, and W. Boland. 1994. Herbivore-induced volatiles: the emission of acyclic homoterpenes from leaves of *Phaseolus lunatus* and *Zea mays* can be triggered by a beta-glucosidase and jasmonic acid. *FEBS Letters* **352**:146–150.

Hountondji, F. C. C., M. W. Sabelis, R. Hanna, and A. Janssen. 2005. Do herbivore-induced plant volatiles trigger sporulation in an entomopathogenic fungus? A study on cassava green mite and *Neozygites tanajoae*. *Journal of Chemical Ecology* **31**:1003–1021.

Inbar, M., H. Doostdar, R. M. Sonoda, G. L. Leibee, and R. T. Mayer. 1998. Elicitors of plant defensive systems reduce insect densities and disease incidence. *Journal of Chemical Ecology* **24**:135–149.

Janssen, A., A. Pallini, M. Venzon, and M. W. Sabelis. 1998. Behaviour and food web interactions among plant inhabiting mites and thrips. *Experimental and Applied Acarology* **22**:497–521.

Janssen, A., M. W. Sabelis, and J. Bruin. 2002. Evolution of herbivore-induced plant volatiles. *Oikos* **97**:134–138.

Jia, F., D. C. Margolies, J. E. Boyer, and R. E. Charlton. 2002. Genetic variation in foraging traits among inbred lines of a predatory mite. *Heredity* **89**:371–379.

Kahl, J., D. H. Siemens, R. J. Aerts, *et al*. 2000. Herbivore-induced ethylene suppresses a direct defense but not a putative indirect defense against an adapted herbivore. *Planta* **210**:336–342.

Kant, M. R., K. Ament, M. W. Sabelis, M. Haring, and R. Schuurink. 2004. Differential timing of spider mite-induced direct and indirect-defenses in tomato plants. *Plant Physiology* **135**:483–495.

Karban, R., and I. T. Baldwin. 1997. *Induced Responses to Herbivory*. Chicago, IL: University of Chicago Press.

Kessler, A., and I. T. Baldwin. 2001. Defensive function of herbivore-induced volatiles in nature. *Science* **291**:2141–2144.

Kessler, A., and I. T. Baldwin. 2002. Plant responses to insect herbivory: the emerging molecular analysis. *Annual Review of Plant Biology* **53**:299–328.

Kessler, A., R. Halitschke, and I. T. Baldwin. 2004. Silencing the jasmonate cascade: induced plant defenses and insect populations. *Science* **305**:665–668.

Kloek, A. P., M. L. Verbsky, S. B. Sharma, *et al*. 2001. Resistance to *Pseudomonas syringae* conferred by an *Arabidopsis thaliana* coronatine-insensitive (coi1) mutation occurs through two distinct mechanisms. *Plant Journal* **26**:509–522.

Krips, O. E., P. E. L. Willems, R. Gols, M. A. Posthumus, and M. Dicke. 1999. The response of *Phytoseiulus persimilis* to spider mite-induced volatiles from *Gerbera*: influence of starvation and experience. *Journal of Chemical Ecology* **25**:2623–2641.

Kunkel, B. N., and D. M. Brooks. 2002. Cross-talk between signaling pathways in pathogen defense. *Current Opinion in Plant Biology* **5**:325–331.

Lesna, I., C. G. M. Conijn, and M. W. Sabelis. 2004. From biological control to biological insight: rust-mite induced change in bulb morphology, a new mode of indirect plant defence? *Phytophaga* **14**:1–7.

Li, C. Y., M. M. Williams, Y. T. Loh, G. I. Lee, and G. A. Howe. 2002. Resistance of cultivated tomato to cell content-feeding herbivores is regulated by the octadecanoid-signaling pathway. *Plant Physiology* **130**:494–503.

Li, L., Y. F. Zhao, B. C. McCaig, et al. 2004. The tomato homolog of CORONATINE-INSENSITIVE1 is required for the maternal control of seed maturation, jasmonate-signaled defense responses, and glandular trichome development. *Plant Cell* **16**:126–143.

Maeda, T., J. Takabayashi, S. Yano, and A. Takafuji. 1999. Response of the predatory mite, *Amblyseius womersleyi* (Acari: Phytoseiidae), toward herbivore-induced plant volatiles: variation in response between two local populations. *Applied Entomology and Zoology* **34**:449–454.

Maeda, T., J. Takabayashi, S. Yano, and A. Takafuji. 2001. Variation in the olfactory response of predatory mite, *Amblyseius womersleyi* (Acari: Phytoseiidae), of 13 populations to herbivore-induced plant volatiles. *Experimental and Applied Acarology* **25**:55–64.

Mantyla, E., T. Klemola, and E. Haukioja. 2004. Attraction of willow warblers to sawfly-damaged mountain birches: novel function of inducible plant defences? *Ecology Letters* **7**:915–918.

Margolies, D. C., M. W. Sabelis, and J. E. Boyer. 1997. Response of a phytoseiid predator to herbivore-induced plant volatiles: selection on attraction and effect of prey exploitation. *Journal of Insect Behavior* **10**:695–709.

Mattiacci, L., M. Dicke, and M. A. Posthumus. 1995. β-Glucosidase: an elicitor of herbivore-induced plant odor that attracts hostsearching parasitic wasps. *Proceedings of the National Academy of Sciences of the USA* **92**:2036–2040.

McConn, M., R. A. Creellman, E. Bell, J. E. Mullet, and J. Browse. 1997. Jasmonate is essential for insect defense in *Arabidopsis*. *Proceedings of the National Academy of Sciences of the USA* **93**:5473–5477.

Mercke, P., I. F. Kappers, F. W. A. Verstappen, et al. 2004. Combined transcript and metabolite analysis reveals genes involved in spider mite induced volatile formation in cucumber plants. *Plant Physiology* **135**:2012–2024.

Mori, N., H. T. Alborn, P. E. A. Teal, and J. H. Tumlinson. 2001. Enzymatic decomposition of elicitors of plant volatiles in *Heliothis virescens* and *Helicoverpa zea*. *Journal of Insect Physiology* **47**:749–757.

Musser, R. O., D. F. Cipollini, S. M. Hum-Musser, et al. 2005a. Evidence that the caterpillar salivary enzyme glucose oxidase provides herbivore offense in solanaceous plants. *Archives of Insect Biochemistry and Physiology* **58**:128–137.

Musser, R. O., H. S. Kwon, S. A. Williams, et al. 2005b. Evidence that caterpillar labial saliva suppresses infectivity of potential bacterial pathogens. *Archives of Insect Biochemistry and Physiology* **58**:138–144.

O'Dowd, D. J., C. F. R. Brew, D. C. Christophel, and R. A. Norton. 1991. Mite–plant associations from the Eocene of Southern Australia. *Science* **252**:99–101.

Ozawa, R., G. Arimura, J. Takabayashi, T. Shimoda, and T. Nishioka. 2000. Involvement of jasmonate- and salicylate-related signaling pathway for the production of specific herbivore-induced volatiles in plants. *Plant Cell Physiology* **41**:391–398.

Ozawa, R., K. Shiojiri, M. W. Sabelis, *et al.* 2004. Corn plants treated with jasmonic acid attract more specialist parasitoids, thereby increasing parasitization of the common armyworm. *Journal of Chemical Ecology* **30**:1797–1808.

Pallini, A., A. Janssen, and M. W. Sabelis. 1997. Odour-mediated responses of phytophagous mites to conspecific and heterospecific competitors. *Oecologia* **110**:179–185.

Pallini, A., A. Janssen, and M. W. Sabelis. 1998. Predators induce interspecific herbivore competition for food in refuge space. *Ecology Letters* **1**:171–176.

Pallini, A., A. Janssen, and M. W. Sabelis. 1999. Do western flower thrips avoid plants infested with spider mites? Interactions between potential competitors, pp. 375–380 in J. Bruin, L. P. S. van der Geest, and M. W. Sabelis (eds.) *Ecology and Evolution of the Acari*. Dordrecht, The Netherlands: Kluwer Academic Publishers.

Papaj, D. R., and A. C. Lewis. 1993. *Insect Learning: Ecology and Evolutionary Perspectives*. New York: Chapman and Hall.

Papaj, D. R., H. Snellen, K. Swaans, and L. E. M. Vet. 1994. Unrewarding experiences and their effect on foraging in the parasitic wasp *Leptopilina heterotoma* (Hymenoptera, Eucoilidae). *Journal of Insect Behavior* **7**:465–481.

Paré, P. W., W. J. Lewis, and J. H. Tumlinson. 1999. Induced plant volatiles: biochemistry and effects on parasitoids, pp. 167–180 in A. A. Agrawal, S. Tuzun, and E. Bent (eds.) *Induced Defenses against Pathogens and Herbivores*. St. Paul, MN: American Phytopathological Society Press.

Paul, N. D., P. E. Hatcher, and J. E. Taylor. 2000. Coping with multiple enemies: an integration of molecular and ecological perspectives. *Trends in Plant Science* **5**:221–225.

Pieterse, C. M. J., and L. C. van Loon. 1999. Salicylic acid-independent plant defence pathways. *Trends in Plant Science* **4**:52–58.

Pieterse, C. M. J., and L. C. van Loon. 2004. NPR1: the spider in the web of induced resistance signaling pathways. *Current Opinion in Plant Biology* **7**:456–464.

Preston, C. A., C. Lewandowski, A. J. Enyedi, and I. T. Baldwin. 1999. Tobacco mosaic virus inoculation inhibits wound-induced jasmonic acid-mediated responses within but not between plants. *Planta* **209**:87–95.

Price, P. W., C. E. Bouton, P. Gross, *et al.* 1980. Interactions among three trophic levels: influence of plants on interactions between insect herbivores and natural enemies. *Annual Review of Ecology and Systematics* **11**:41–65.

Reymond, P., N. Bodenhausen, R. M. P. van Poecke, *et al.* 2004. A conserved transcriptional pattern in response to a specialist and a generalist herbivore. *Plant Cell* **16**:3132–3147.

Rojo, E., R. Solano, and J. J. Sanchez-Serrano. 2003. Interactions between signaling compounds involved in plant defense. *Journal of Plant Growth Regulation* **22**:82–98.

Roitberg, B., M. L. Reid, and C. Li. 1993. Choosing hosts and mates: the value of learning, pp. 174–194 in D. R. Papaj and A. C. Lewis (eds.) *Insect Learning: Ecology and Evolutionary Perspectives*. New York: Chapman and Hall.

Sabelis, M. W., and F. M. Bakker. 1992. How predatory mites cope with the web of their tetranychid prey: a functional view on dorsal chaetotaxy in the Phytoseiidae. *Experimental and Applied Acarology* 16:203–225.

Sabelis, M. W., and M. C. M. de Jong. 1988. Should all plants recruit bodyguards? Conditions for a polymorphic ESS of synomone production in plants. *Oikos* 53:247–252.

Sabelis, M. W., and M. Dicke. 1985. Long-range dispersal and searching behaviour, pp. 141–160 in W. Helle and M. W. Sabelis (eds.) *Spider Mites: Their Biology, Natural Enemies and Control*. Amsterdam, The Netherlands: Elsevier.

Sabelis, M. W., and H. E. van der Baan. 1983. Location of distant spider mite colonies by phytoseiid predators: demonstration of specific kairomones emitted by *Tetranychus urticae* and *Panonychus ulmi*. *Entomologia Experimentalis et Applicata* 33:303–314.

Sabelis, M. W., B. P. Afman, and P. J. Slim. 1984. Location of distant spider-mite colonies by *Phytoseiulus persimilis*: localization and extraction of a kairomone, pp. 431–440 in D. A. Griffiths and C. E. Bowman (eds.) *Acarology VI*, vol. 1. New York: Halsted Press.

Sabelis, M. W., A. Janssen, A. Pallini, *et al.* 1999a. Behavioural responses of predatory and herbivorous arthropods to induced plant volatiles: from evolutionary ecology to agricultural applications, pp. 269–296 in A. A. Agrawal, S. Tuzun, and E. Bent (eds.) *Induced Plant Defenses against Pathogens and Herbivores*. St. Paul, MN: American Phytopathological Society Press.

Sabelis, M. W., M. van Baalen, F. M. Bakker, *et al.* 1999b. The evolution of direct and indirect plant defence against herbivorous arthropods, pp. 109–166 in H. Olff, V. K. Brown, and R. H. Drent (eds.) *Herbivores: Between Plants and Predators*. Oxford, UK: Blackwell Science.

Sabelis, M. W., A. Janssen, J. Bruin, *et al.* 1999c. Interactions between arthropod predators and plants: a conspiracy against herbivorous arthropods? pp. 207–230 in J. Bruin, L. P. S. van der Geest, and M. W. Sabelis (eds.) *Ecology and Evolution of the Acari*. Dordrecht, The Netherlands: Kluwer Academic Publishers.

Sabelis, M. W., M. van Baalen, B. Pels, M. Egas, and A. Janssen. 2002. Evolution of exploitation and defence in plant–herbivore–predator interactions, pp. 297–321 in U. Dieckmann, J. A. J. Metz, M. W. Sabelis, and K. Sigmund (eds.) *The Adaptive Dynamics of Infectious Diseases: In Pursuit of Virulence Management*. Cambridge, UK: Cambridge University Press.

Sabelis, M. W., P. C. J. van Rijn, and A. Janssen. 2005. Fitness consequences of food-for-protection strategies in plants, pp. 109–134 in F. L. Wäckers, P. C. J. van Rijn, and J. Bruin (eds.) *Plant-Provided Food and Herbivore–Carnivore Interactions*. Cambridge, UK: Cambridge University Press.

Schmitz, O. J., V. Krivan, and O. Ovadia. 2004. Trophic cascades: the primacy of trait-mediated indirect interactions. *Ecology Letters* 7:153–163.

Schneider, M., P. Schweitzer, P. Meuwly, and J. P. Metraux. 1996. Systemic acquired resistance in plants. *International Review of Cytology* **168**:303–339.

Scutareanu, P., B. Drukker, J. Bruin, M. A. Posthumus, and M. W. Sabelis. 1997. Isolation and identification of volatile synomones involved in the interaction between *Psylla*-infested pear trees and two anthocorid predators. *Journal of Chemical Ecology* **23**:2241–2260.

Scutareanu, P., R. Lingeman, B. Drukker, and M. W. Sabelis. 1999. Cross-correlation analysis of fluctuations in local populations of pear psyllids and anthocorid bugs. *Ecological Entomology* **24**:1–9.

Shimoda, T., and M. Dicke. 2000. Attraction of a predator to chemical information related to nonprey: when can it be adaptive? *Behavioral Ecology* **11**:606–613.

Shiojiri, K., J. Takabayashi, S. Yano, and A. Takafuji. 2002. Oviposition preferences of herbivores are affected by tritrophic interaction webs. *Ecology Letters* **5**:186–192.

Spoel, S. H., A. Koornneef, S. M. C. Claessens, *et al.* 2003. NPR1 modulates cross-talk between salicylate- and jasmonate-dependent defense pathways through a novel function in the cytosol. *Plant Cell* **15**:760–770.

Storeck, A., G. M. Poppy, H. F. van Emden, and W. Powell. 2000. The role of plant chemical cues in determining host preference in the generalist aphid parasitoid *Aphidius colemani*. *Entomologia Experimentalis et Applicata* **97**:41–46.

Takabayashi, J., and M. Dicke. 1992. Response of predatory mites with different rearing histories to volatiles of uninfested plants. *Entomologia Experimentalis et Applicata* **64**:187–193.

Takabayashi, J., and M. Dicke, 1996. Plant–carnivore mutualism through herbivore-induced carnivore attractants. *Trends in Plant Science* **1**:109–113.

Takabayashi, J., M. Dicke, and M. A. Posthumus. 1991. Variation in composition of predator-attracting allelochemicals emitted by herbivore-infested plants: relative influence of plant and herbivore. *Chemoecology* **2**:1–6.

Takabayashi, J., M. W. Sabelis, A. Janssen, K. Shiojiri, and M. van Wijk. 2006. Can plants betray the presence of multiple herbivore species to predators and parasitoids? The role of learning in phytochemical networks. *Ecological Research* **21**:3–8.

Thaler, J. S. 1999a. Jasmonic acid mediated interactions between plants, herbivores, parasitoids, and pathogens: a review of field experiments in tomato, pp. 319–334 in A. A. Agrawal, S. Tuzun, and E. Bent (eds.) *Induced Plant Defenses against Pathogens and Herbivores*. St. Paul, MN: American Phytopathological Society Press.

Thaler, J. S. 1999b. Jasmonate-inducible plant defenses cause increased parasitism of herbivores. *Nature* **399**:686–688.

Thaler, J. S., 2002. Effect of jasmonate-induced plant responses on the natural enemies of herbivores. *Journal of Animal Ecology* **71**:141–150.

Thaler J. S., A. L. Fidantsef, S. S. Duffey, and R. M. Bostock. 1999. Trade-offs in plant defense against pathogens and herbivores: a field demonstration of chemical elicitors of induced resistance. *Journal of Chemical Ecology* **25**:1597–1609.

Thaler, J. S., M. J. Stout, R. Karban, and S. S. Duffey. 2001. Jasmonate-mediated induced plant resistance affects a community of herbivores. *Ecological Entomology* **26**:312–324.

Thaler, J. S., M. A. Farag, P. W. Paré, and M. Dicke. 2002a. Jasmonate-deficient plants have reduced direct and indirect defenses against herbivores. *Ecology Letters* **5**:764–774.

Thaler, J. S., R. Karban, D. E. Ullman, K. Boege, and R. M. Bostock. 2002b. Cross-talk between jasmonate and salicylate plant defense pathways: effects on several plant parasites. *Oecologia* **131**:227–235.

Turlings, T. C. J., and J. H. Tumlinson. 1992. Systemic release of chemical signals by herbivore-injured corn. *Proceedings of the National Academy of Sciences of the USA* **89**:8399–8402.

Turlings, T. C. J., J. H. Tumlinson, and W. J. Lewis. 1990. Exploitation of herbivore-induced plant odors by host seeking parasitic wasps. *Science* **250**:1251–1253.

Turlings, T. C. J., F. Wäckers, L. E. M. Vet, W. J. Lewis, and J. H. Tumlinson. 1993a. Learning of host-finding cues by hymenopterous parasitoids, pp. 51–78 in D. R. Papaj and A. C. Lewis (eds.) *Insect Learning: Ecology and Evolutionary Perspectives*. New York: Chapman and Hall.

Turlings, T. C. J., P. J. McCall, H. T. Alborn, and J. H. Tumlinson. 1993b. An elicitor in caterpillar oral secretions that induces corn seedlings to emit chemical signals attractive to parasitic wasps. *Journal of Chemical Ecology* **19**:411–425.

Turlings, T. C. J., J. H. Loughrin, P. J. McCall, *et al.* 1995. How caterpillar-damaged plants protect themselves by attracting parasitic wasps. *Proceedings of the National Academy of Sciences of the USA* **92**:4169–4174.

Van Baalen, M., and V. A. A. Jansen. 2003. Common language or Tower of Babel? On the evolutionary dynamics of signals and their meanings. *Proceedings of the Royal Society of London Series B* **270**:69–76.

Van Dam, N. M., K. Hadwich, and I. T. Baldwin. 2000. Induced responses in *Nicotiana attenuata* affect behavior and growth of the specialist herbivore *Manduca sexta*. *Oecologia* **122**:371–379.

Van der Meijden, E., and P. G. L. Klinkhamer. 2000. Conflicting interests of plants and the natural enemies of herbivores. *Oikos* **89**:202–208.

Van Loon, J. J. A., J. G. de Boer, and M. Dicke. 2000. Parasitoid-plant mutualism: parasitoid attack of herbivore increases plant reproduction. *Entomologia Experimentalis et Applicata* **97**:219–227.

Van Poecke, R. M. P., and M. Dicke. 2002. Induced parasitoid attraction by *Arabidopsis thaliana*: involvement of the octadecanoid and the salicylic acid pathway. *Journal of Experimental Botany* **53**:1793–1799.

Van Poecke, R. M. P., and M. Dicke. 2003. Signal transduction downstream of salicylic and jasmonic acid in herbivory-induced parasitoid attraction by *Arabidopsis* is independent of JAR1 and NPR1. *Plant Cell and Environment* **26**:1541–1548.

Van Rijn, P. C. J., Y. M. van Houten, and M. W. Sabelis. 2002. How plants benefit from providing food to predators when it is also edible to herbivores. *Ecology* **83**:2664–2679.

Van Tol, R. W. H. M., A. T. C. van der Sommen, M. I. C. Boff, *et al.* 2001. Plants protect their roots by alerting the enemies of grubs. *Ecology Letters* **4**:292–294.

Van Zandt, P. A., and A. A. Agrawal. 2004. Community-wide impacts of herbivore-induced plant responses in milkweed (*Asclepias syriaca*). *Ecology* 85:2616–2629.

Vet, L. E. M., and M. Dicke, M. 1992. Ecology of infochemical use by natural enemies in a tritrophic context. *Annual Review of Ecology and Systematics* 37:141–172.

Vet, L. E. M., W. J. Lewis, and R. T. Cardé. 1995. Parasitoid foraging and learning, pp. 65–101 in R. T. Cardé and W. J. Bell (eds.) *Chemical Ecology of Insects*, vol. 2. New York: Chapman and Hall.

Vos, M., S. M. Berrocal, F. Karamaouna, L. Hemerik, and L. E. M. Vet. 2001. Plant-mediated indirect effects and the persistence of parasitoid–herbivore communities. *Ecology Letters* 4:38–45.

Wang, Q., and S. Dorn. 2003. Selection on olfactory response to semiochemicals from a host-plant complex in a parasitic wasp. *Heredity* 91:430–435.

Wang, Q., H. Gu, and S. Dorn. 2004. Genetic relationship between olfactory response and fitness in the *Cotesia glomerata* (L.). *Heredity* 92:579–584.

Werner, E. E., and S. D. Peacor. 2003. A review of trait-mediated indirect interactions in ecological communities. *Ecology* 84:1083–1100.

Wootton, J. T. 1994. The nature and consequences of indirect effects in ecological communities. *Annual Review of Ecology and Systematics* 25:443–466.

Zhao, Y. F., R. Thilmony, C. L. Bender, *et al.* 2003. Virulence systems of *Pseudomonas syringae* pv. tomato promote bacterial speck disease in tomato by targeting the jasmonate signaling pathway. *Plant Journal* 36:485–499.

Part IV PLANT-MEDIATED INDIRECT EFFECTS ON COMMUNITIES AND BIODIVERSITY

10

Nontrophic, indirect interaction webs of herbivorous insects

TAKAYUKI OHGUSHI

Introduction

Since ecological communities are structured by species interactions, an understanding of trophic interactions has been a central issue in ecology (Paine 1980, Hunter and Price 1992, Polis and Winemiller 1996, Borer *et al.* 2002, Berlow *et al.* 2004). In terrestrial systems predation has a lethal effect on prey. On the other hand, mature plants are rarely killed by herbivores, but herbivory can subsequently change plant traits such as allelochemistry, cell structure and growth, physiology, morphology, and phenology. There is increasing appreciation that such herbivore-induced responses of plants are common and widespread in natural and agricultural systems (Karban and Baldwin 1997). Since plant-based arthropod communities consist of many herbivores interacting with a host plant, the herbivore-induced plant responses have the potential to greatly influence the formation of arthropod community by indirectly linking major herbivore species via nontrophic linkages. In this context, I use the term "nontrophic" to refer to "without *direct* feeding relationship" in this chapter. Since a primary focus of previous studies of trophic dynamics has been on determining how the relative abundance of biomass or energy produced by one trophic level is transferred to another (Oksanen *et al.* 1981, Cyr and Pace 1993, Leibold *et al.* 1997), community consequences of ubiquitous nontrophic and indirect interactions in plant–herbivore systems have long been overlooked.

This chapter addresses the importance of herbivore-induced plant responses as a creator of nontrophic interactions among herbivores and indirect interactions

Ecological Communities: Plant Mediation in Indirect Interaction Webs, ed. Takayuki Ohgushi, Timothy P. Craig, and Peter W. Price. Published by Cambridge University Press. © Cambridge University Press 2007.

involving herbivores and their host plants, and incorporates nontrophic and indirect links into plant-based terrestrial food webs. I will illustrate how induced plant responses act as mediators in linking multiple plant–insect interactions and increasing direct and indirect interactions involved in insect herbivore communities. I mainly deal with plant–herbivorous insect systems, because insect herbivores exhibit diverse feeding relationships with plants that produce many well-understood induced plant responses.

Herbivore-induced plants responses

Plants respond to herbivore damage over spatial scales ranging from single leaves to whole trees and over temporal scales ranging from minutes to evolutionary time. The induced plant responses are major outcomes of nonlethal effects of herbivores on terrestrial plants, contrasting with herbivores that do not exhibit such induced responses following predation. Since previous reviews on induced plant responses to herbivory have intensively surveyed examples of induced plant resistance (Karban and Myers 1989, Karban and Baldwin 1997, Nykänen and Koricheva 2004), it is not my intention to add to this extensive literature, but instead to emphasize how plant traits change in response to herbivory. Table 10.1 summarizes the responses of plants following herbivory, which can affect the performance and abundance of associated herbivores.

Induced defense

It is well known that some plants induce the production of defensive chemicals when they are attacked by herbivores (Haukioja and Neuvonen 1987). Herbivore-induced defense chemicals, such as alkaloids, phenolics, and terpenes, are widespread among a range of plant species (Karban and Baldwin 1997). The induced defense chemicals inhibit herbivore growth and development, decrease herbivore population growth, and thus subsequently reduce herbivore pressure on the plant (Agrawal 1998). Note that many sources of environmental stress can produce variation in induced secondary metabolites. Hence, the induced plant defenses show enormous intra- and interspecific variation in response to variation in light or soil nutrition, and in the timing of herbivory (Coley *et al.* 1985, Bazzaz *et al.* 1987).

Herbivory also induces changes in the physical characteristics of plants. When attacked by herbivores induced physical defenses include increases in thorns (Milewski *et al.* 1991, Gómez and Zamora 2002), spines (Bazely *et al.* 1991, Myers and Bazely 1991), and trichomes (Pullin and Gilbert 1989, Dalin and Björkman 2003). Leaf herbivory sometimes induces premature leaf abscission, which is considered a plant defense against immobile insects such as aphids,

Table 10.1. *Plant responses following herbivory that can affect performance and abundance of herbivorous insects*

Plant trait	Response	Reference
Foliar	Increased secondary compounds	Karban and Baldwin 1997, Constabel, 1999, Nykänen and Koricheva 2004
	Increased trichomes, thorns, and spines	Pullin and Gilbert 1989, Milewski et al. 1991, Myers and Bazely 1991, Gömez and Zamora 2002, Dalin and Björkman 2003
	Increased premature abscission	Williams and Whitham 1986, Preszler and Price 1993
	Emission of volatile compounds	Vet and Dicke 1992, Dicke and Vet 1999, Turlings et al. 2002
	Decreased nitrogen	Tuomi et al. 1984, Leather 1993, Denno et al. 2000
	Increased nitrogen	Danell and Huss-Danell 1985, Gange and Brown 1989, Nakamura et al. 2003
Floral	Decreased flower numbers	Hendrix and Trapp 1989, Mauricio et al. 1993, Lehtilä and Strauss 1997
	Decreased pollen production	Quesada et al. 1995, Strauss et al. 1996, Krupnick et al. 1999
	Decreased pollen performance	Quesada et al. 1995, Mutikainen and Delph 1996
	Decreased nectar production	Lehtilä and Strauss 1997, Krupnick et al. 1999
Structure	Increased branching	Tscharntke 1989, Strauss 1991b, Pilson 1992, Nakamura et al. 2003
	Enhanced shoot growth	Pilson 1992, Roininen et al. 1997, Nozawa and Ohgushi 2002
	Increased structural complexity	Whitham and Mopper 1985

leaf rollers, leaf miners, and leaf gallers (Williams and Whitham 1986, Preszler and Price 1993).

The major herbivore-induced chemicals such as tannin and alkaloids are nonvolatile, but there is also increasing evidence that damage by herbivorous insects increases the emission of volatile chemicals enormously (Vet and Dicke 1992, Dicke and Vet 1999). Some of the chemicals disseminated from the damaged plants merely result from mechanical disruption of plant cells and thus are nonspecific for the herbivore. However, other plant volatiles that are released upon damage are specific indicators of herbivore identity (Vet and Dicke 1992). Some plants even emit novel volatiles after herbivory (Turlings et al. 2002).

Nutrients in damaged plants

Herbivory can alter the nutrient status of plants that are eaten. As the nitrogen content of plant tissues is a major determinant of food quality for herbivorous insects and mammals (Scriber and Slansky 1981), nutritional changes following herbivory have great impacts on herbivore performance. Defoliation by herbivorous insects or mammals often decreases foliage quality (Myers and Post 1981, Tuomi et al. 1984, Denno et al. 2000). In contrast, herbivory sometimes improves nutrient status of plants in terms of nitrogen and water content (Williams and Myers 1984, Danell and Huss-Danell 1985, Gange and Brown 1989). Changes in nitrogen levels following herbivory are dependent on the feeding habits of insect herbivores (Masters and Brown 1997). In addition to foliar herbivory, leaf manipulation by herbivorous insects such as leaf rollers, leaf tiers, leaf gallers, and leaf folders alter the nutrient status of leaves manipulated (Fukui et al. 2002).

Enhanced growth following herbivory

Herbivores alter plant architecture in diverse ways that range from superficial modifications to transformations of the entire plant form (Mopper et al. 1991, Whitham et al. 1991). Terrestrial plants have the ability to regrow in response to mechanical damage by herbivores (Strauss and Agrawal 1999). Meristem destruction due to herbivory usually results in the development of dormant buds adjacent or basal to the site of mechanical damage. Lateral bud development is stimulated by apical meristem damage, which increases plant structural complexity. Even low levels of herbivory can stimulate regrowth of plants attacked (Whitham and Mopper 1985). In addition to insect feeding, insect oviposition behavior that causes mechanical damage to plant tissue can subsequently induce shoot development (Nozawa and Ohgushi 2002). Several factors, such as light, water, nitrogen, competition with neighboring plants, growth form, C/N balance, plant genotype, and type of damage alter outcomes of the compensatory responses in plants (Maschinski and Whitham 1989, Whitham et al. 1991, Strauss and Agrawal 1999).

Effects of leaf herbivory on floral traits

Herbivory not only alters the subsequent quality and structure of vegetative tissues, but also the reproductive characteristics of damaged plants. There has been growing evidence that leaf herbivory changes the quantity and quality of floral traits, which are of crucial importance in pollinator service (Strauss 1997). Foliar leaf damage in early season often decreases flower number (Mauricio et al. 1993), flower size (Strauss et al. 1996), pollen production (Quesada et al. 1995, Strauss et al. 1996), and pollen performance

(Quesada *et al.* 1995, Mutikainen and Delph 1996), thereby subsequently influencing pollination success in damaged plants.

Habitat construction by ecosystem engineers

Another common form of herbivore-induced changes in plant structure is generated by ecosystem engineers. Ecosystem engineers are defined as organisms that directly or indirectly control the availability of resources to other species, by causing physical state changes in biotic or abiotic materials (Jones *et al.* 1994). This contrasts with the trait mediation by insects discussed above where herbivory induces responses in the plant itself. Ecosystem engineering commonly occurs on terrestrial plants, because common insect herbivore guilds, such as gall-makers, leaf rollers, stem-borers, and leaf miners, create new habitats for other herbivores and/or their natural enemies (Cappuccino 1993, Larsson *et al.* 1997, Martinsen *et al.* 2000, Lill and Marquis 2003, Nakamura and Ohgushi 2003, Kagata and Ohgushi 2004). In particular, shelter building is a common lifestyle of the microlepidoptera, some weevils, sawflies, and even grasshoppers. It occurs on many plants, ranging from trees and shrubs to herbs and even ferns. Recent studies have suggested that insect ecosystem engineers have the potential to greatly affect other arthropods on the host plant by providing shelters against natural enemies and adverse microclimates, and by improving food resources (Larsson *et al.* 1997, Martinsen *et al.* 2000, Fukui *et al.* 2002).

Insect herbivores as creators of interaction linkages

Plant-mediated indirect interactions among herbivorous insects

Since organisms interact not only directly but also indirectly with other organisms in nature, indirect effects as an important force forming ecological communities have received increasing attention (Strauss 1991a, Holt and Lawton 1994, Wootton 1994, Menge 1995, Abrams *et al.* 1996, Masters and Brown 1997, Ohgushi 1997). Indirect effects occur when the impacts of one species on another are mediated by the presence and biological activities of one or more intermediate species. There are two types of indirect effects: density-mediated and trait-mediated indirect effects. Density-mediated indirect effects are transmitted through changes in density by numerical responses of intervening species, while trait-mediated indirect effects are transmitted through changes in traits such as behavior, morphology, and the life histories of intervening species. Density-mediated indirect effects, which include trophic cascades (Pace *et al.* 1999, Polis *et al.* 2000) and apparent competition (Holt and Lawton 1994, Bonsall and Hassell 1997, Morris *et al.* 2004), have been well studied both in theory and in practice.

Although the potential importance of trait-mediated indirect effects in ecological communities has been widely accepted (Abrams *et al.* 1996, Werner and Peacor 2003, Schmitz *et al.* 2004, Ohgushi 2005), they have been studied much less frequently than density-mediated indirect effects. This may be because the detection of effects through trait changes requires well-designed experimental studies to identify the mechanisms of trait alteration. More recently, trait-mediated indirect effects have begun to receive attention in prey–predator systems when prey avoid predation risk through behavioral changes (Bolker *et al.* 2003, Dill *et al.* 2003, Werner and Peacor 2003, Schmitz *et al.* 2004) and in plant communities when herbivory alters the outcomes of competitive interactions among plants (Callaway *et al.* 2003).

Because terrestrial plants are usually not killed when they are attacked by herbivores, but instead subsequently alter various traits that provide food and habitat resources to other herbivores, these herbivore-induced responses of plants will undoubtedly be a mechanistic basis of trait-mediated indirect effects. There is increasing evidence that one insect herbivore indirectly affects the performance and/or abundance of another insect through changes in plant traits (Faeth 1988, Damman 1993, Denno *et al.* 1995, Masters and Brown 1997). In reviewing 83 cases of such plant-mediated indirect interactions among herbivores that cover a broad spectrum of plant–herbivore systems, Ohgushi (2005) pointed out the following important features of plant-mediated indirect effects. First, herbivore-induced indirect effects through trait change in plants are common and widespread in many plant–herbivore systems. Second, substantial indirect interactions caused by herbivore-induced changes frequently occur among temporally separated, spatially separated, and taxonomically distinct species. Third, herbivores sharing the same host plant can benefit each other, because herbivory often enhances resource availability through improved nutritional quality and/or increased biomass of plants due to compensatory regrowth.

Despite the fact that herbivore-induced changes in plant traits are extremely widespread in nature, trait-mediated indirect interactions have received little attention in plant–herbivore systems. This is because the traditional view of within-trophic-level interactions has been largely biased towards the importance of competitive interactions for resources (Connell 1983, Schoener 1983), and it was concluded that interactions among herbivorous insects are unlikely because of a lack of evidence of plant depletion (Lawton and Strong 1981, Strong *et al.* 1984). However, the indirect interactions mediated by plant traits commonly occur at low levels of herbivory that are most unlikely to cause the defoliation of host plants (Ohgushi 2005). Hence, the traditional view resulted in the underestimation of the ubiquity of indirect interactions among herbivorous insects through changes in plant traits.

Interaction linkage through herbivore-induced indirect effects

In terrestrial systems, there are a number of single food chains of herbivorous insects on individual plant species. These plant-based food chains are interconnected with each other, making up a network of interacting species. Nevertheless, previous studies on plant–insect interactions have often focused on a single interaction, as if it were independent of other interactions. To understand how a community is structurally organized by interacting species, it is necessary to know how interactions are connected to each other. In this context, plant-mediated indirect effects have the potential to link multiple interactions in plant-based insect communities. Werner and Peacor (2003) stressed that ecological communities are replete with trait-mediated indirect effects that arise from phenotypic plasticity, and that these effects are quantitatively important to community dynamics. Only recently have ecologists begun to appreciate the importance of such linkage among species interactions in communities (Strauss 1997, Jones *et al.* 1998, Wardle 2002, Agrawal and Van Zandt 2003, Stanton 2003). As I mentioned earlier, leaf herbivory can modify plant–pollinator interactions through changes in floral traits (see also Bronstein *et al.* Chapter 4 this volume). Also, aboveground herbivory can affect belowground herbivores and microfauna through changes in root carbon allocation, root exudation, root biomass, and morphology (Bardgett *et al.* 1998, Van der Putten *et al.* 2001). All of these studies have emphasized the importance of plant-mediated indirect effects in forming interaction linkages within a community.

How are plant–insect interactions linked to each other?

Case studies

As examples of the indirect interaction linkage through herbivore-induced indirect effects discussed above, I will illustrate how multiple plant–insect interactions are connected to each other on the willows *Salix miyabeana* and *S. eriocarpa*, and the goldenrod *Solidago altissima*.

Example 1: Interaction linkage on the willow Salix miyabeana

The willow *Salix miyabeana* is distributed in the northern part of Japan, and it is a common woody plant growing along riversides. In flood plains of the Ishikari River in Hokkaido, northern Japan, the spittlebug *Aphrophora pectoralis*, the leaf rollers consisting of caterpillars of 23 lepidopteran species, and the leaf beetle *Plagiodera versicolora* are the most common insects that feed on *S. miyabeana* (Ishihara *et al.* 1999, Craig and Ohgushi 2002, Nozawa and Ohgushi 2002, Nakamura and Ohgushi 2003). The spittlebug is a specialist herbivore on the genus *Salix*. In autumn, adult females lay eggs into the distal part of current shoots of willow plants. Although attacked shoots died within 1 week because of mechanical damage, this damage enhanced shoot growth in the following year, resulting in longer shoots with a greater number of leaves. This enhanced shoot growth, in turn, increased the density of leaf rollers in the following spring, probably because of an increased number of new leaves that provide suitable materials for caterpillars to make leaf shelters. After the caterpillars left leaf shelters for adult emergence, the majority of leaf shelters were colonized by other herbivorous insects. Among them the aphid *Chaitophorus saliniger* colonized >75% of leaf rolls when vacant leaf shelters were most available. The aphid is highly specialized for colonizing leaf rolls, and it was rarely observed outside leaf rolls in the study area. Once the aphids established themselves within the leaf rolls, the aphid colonies were tended by three species of ants: *Camponotus japonicus*, *Lasius hayashi*, and *Myrmica jessensis*, all of which harvested aphid honeydew. A field experiment using artificial leaf rolls revealed that the number of aphids and associated ants significantly increased with increasing number of artificial leaf rolls (Nakamura and Ohgushi 2003). The increased number of tending ants, in turn, negatively affected the larval survival of the leaf beetle *P. versicolora*. A larval transfer experiment showed that the presence of leaf rolls significantly reduced daily survival rate of the leaf beetle larvae by 60%, and that the number of ants on shoots with leaf rolls was significantly greater than that on shoots lacking leaf rolls (Nakamura and Ohgushi 2003).

The interaction linkage of herbivorous insects on *S. miyabeana* clearly demonstrated that indirect effects play an important role in connecting multiple plant–herbivore interactions (Fig. 10.1). There are three mediators linking interactions between plants and herbivores involved in the system: enhanced shoot growth in response to spittlebug oviposition, leaf shelters built by leaf-rolling caterpillars, and ants tending aphid colonies within leaf shelters. The enhanced shoot growth by compensatory response links spittlebugs and leaf-rolling caterpillars (interaction 7); the leaf shelters connect leaf-rolling caterpillars with aphids (interaction 8); the leaf-rolling caterpillars interact with ants through increased aphid densities (interaction 9); and the increased tending by ants links aphids and leaf beetles (interaction 10). Note that the ant-mediated indirect interaction between aphids and leaf beetles is maintained by the provision of leaf shelters built by leaf rolling caterpillars. Hence, the interaction linkage of willow and herbivorous insects is chiefly caused by herbivore-induced changes in plant traits and by ecosystem engineering. In addition, ant-mediated indirect effects, based on the provision of leaf rolls by the caterpillars to the aphids, suggests that multitrophic interactions caused by plant-mediated indirect effects are also important in connecting interactions on the willow.

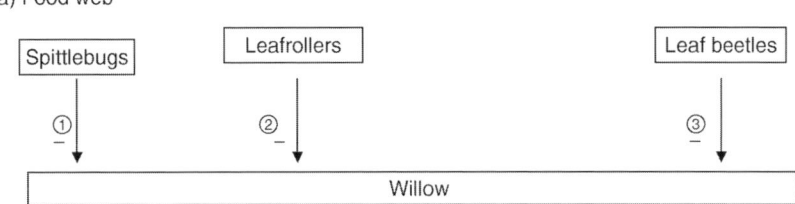
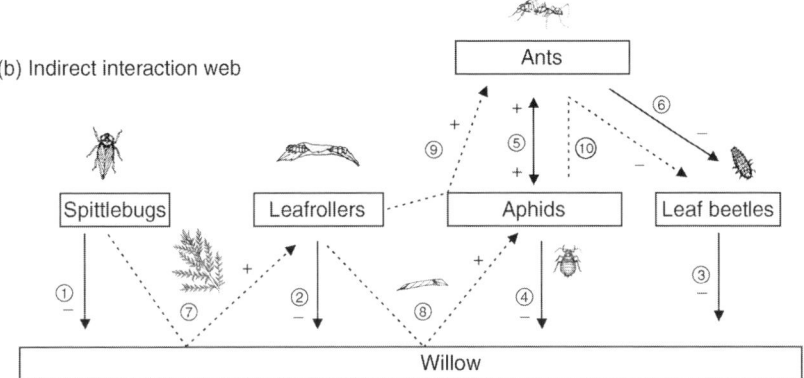

Figure 10.1. The interaction linkages among herbivorous insects on the willow *Salix miyabeana* in (a) the food web and (b) the indirect interaction web. (Adapted from Ohgushi 2005.)

Example 2: Interaction linkage on the willow *Salix eriocarpa*

We revealed another interaction linkage (Fig. 10.2), which is propagated through gall initiation by the gall midge *Rhabdophaga rigidae* on *Salix eriocarpa* in central Japan (Nakamura et al. 2003). The gall midge is a common insect herbivore on *S. eriocarpa*, and initiates stem galls on the apical regions of current-year shoots in mid-May. In response to gall initiation, *S. eriocarpa* rapidly developed vigorous lateral shoots, resulting in a secondary leaf flush. The colonization rate of the aphid *Aphis farinosa*, a common sap-feeder on *S. eriocarpa* on 1-year shoots, was four times greater on galled shoots than on ungalled shoots. Similarly, numbers of adults of two leaf beetles, *P. versicolora* and *Smaragdina semiaurantiaca*, were 3–10 times greater on galled than on ungalled shoots. The increased densities of aphids and leaf beetles were not only due to increased numbers of newly emerged shoots and leaves but also to improved leaf and stem quality for the herbivorous insects. Nitrogen and water contents were significantly increased but toughness was decreased in apical stems and upper leaves of galled shoots (Nakamura et al. 2003). Furthermore, we frequently observed that

(a) Food web

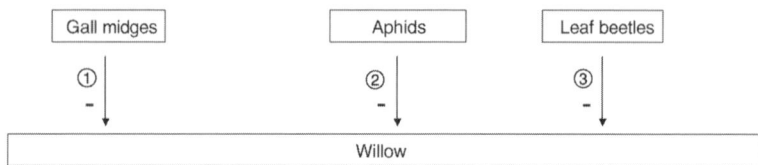

(b) Indirect interaction web

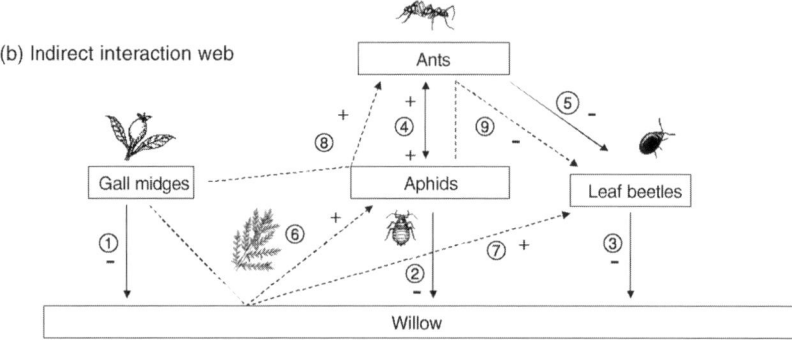

Figure 10.2. The interaction linkages among herbivorous insects on the willow *Salix eriocarpa* in (a) the food web and (b) the indirect interaction web.

increased ant tending in aphid colonies for honeydew resulted in the removal of leaf beetles from adjacent shoots.

In this case, there are two mediators linking interactions: enhanced shoot growth that resulted in increased plant quality from the development of lateral shoots for three herbivore species in response to gall initiation, and increased ant tending that resulted in increased aphid density. The interaction linkage is made up by the following positive and negative indirect interactions (Fig. 10.2b). The compensating growth links gall midges and aphids (interaction 6), and gall midges and two leaf beetle species (interaction 7), because of increased resource availability. Gall midges indirectly increased the number of ants by increasing aphid colonization (interaction 8). Ants connect aphids with leaf beetles through their removal behavior (interaction 9).

Example 3: Interaction linkage on the goldenrod *Solidago altissima*

Interaction linkages of multiple plant–insect interactions should occur not only on woody plants but also on herbaceous plants. Our recent work has revealed another example of the interaction linkages of herbivorous insects on

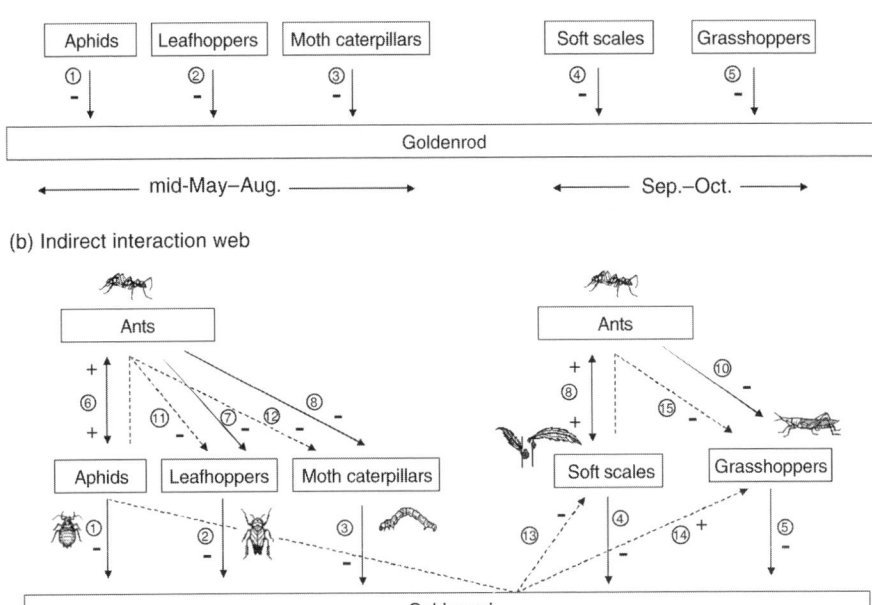

Figure 10.3. The interaction linkages among herbivorous insects on the goldenrod *Solidago altissima* in (a) the food web and (b) the indirect interaction web.

the goldenrod *Solidago altissima* (Fig. 10.3) (Y. Ando and T. Ohgushi unpublished data). *Solidago altissima* is a species from the USA introduced approximately 100 years ago, and it has been widely spread over Japan. It has become one of the most common weeds in disturbed fields. The common insect herbivores that feed on the goldenrod are aphids, leafhoppers, and leaf-chewing geometrid moth caterpillars in spring, and soft scales and grasshoppers in autumn. The aphid *Uroleucon nigrotuberculatum* is the most common species in early season, and it was also introduced from the USA, in the 1990s. It occurs from mid-May to early August. We frequently observed that the aphid colonies were tended by the native ant *Formica japonica* for honeydew (interaction 6). An aphid removal experiment resulted in an eight times greater abundance of the leafhopper *Nephotettix cincticeps* and of the geometrid moth caterpillars. This large increase in herbivore abundance was due to a decreased number of ants on aphid-free plants. Thus, the aphid had indirect, negative impacts on the herbivorous insects by removal behavior of tending ants (interactions 11 and 12). On the other hand, aphid feeding changed neither nitrogen nor water contents in leaves in spring.

It should be noted that the impacts of aphid colonization were evidently transmitted to the soft scale *Parasaissetia nigra* and to the grasshopper *Atractomorpha lata* in late September when the aphids were absent. We found that aphid infestation in the spring increased leaf nitrogen and stimulated lateral shoots in the autumn. The aphid removal experiment revealed that there was a four times greater increase of soft scales on aphid-free plants than on aphid-exposed plants. This is probably because decreased phloem quality in aphid-exposed plants reduced colonization of the soft scales (interaction 13). As the soft scales are also tended by *F. japonica* (interaction 9), the number of ants was significantly greater on aphid-free plants than on aphid-exposed plants, because the scale insects increased on aphid-free plants. Conversely, feeding damage by the grasshopper was twice as great on aphid-exposed plants as on aphid-free plants. The grasshopper probably increased because of the decreased impact of ants tending the scale insects and increased leaf nitrogen of aphid-exposed plants (interactions 14 and 15). Thus, the aphids early in the season negatively affected the soft scale by decreasing food quality but positively influenced the grasshopper through increased leaf quality and decreased ants tending on the soft scales late in the season.

In this case, there are two mediators linking interactions: ants tending aphids and scales for honeydew, and changes in leaf and sap quality in autumn due to the spring aphid infestation. The interaction linkage of herbivorous insects on the goldenrod involves following indirect interactions: the interactions between the aphids, and the leafhoppers, and the geometrid moth caterpillars through ant removal (interactions 11 and 12). On the other hand, plant-mediated indirect effects are apparent in autumn when the aphids disappear: the indirect interaction between the aphids, and the soft scales through changes in sap quality (interaction 13), and the indirect interaction between the aphids and the grasshoppers through increased leaf nitrogen (interaction 14). The ants also indirectly connect the soft scales with the grasshoppers through their removal behavior (interaction 15). Hence, the interaction linkage of goldenrod and herbivorous insects is chiefly caused by herbivore-induced changes in plants and ant-mediated indirect effects.

Herbivore-induced plant changes as mediators generating interaction linkages

The examples of interaction linkages of herbivorous insects on the two willow species and the goldenrod indicate that herbivore-induced changes in plants, such as enhanced shoot growth, changes in plant quality, and creation of habitats by ecosystem engineers, play an important role in connecting multiple interactions in plant-based food webs. Ant-mediated indirect effects are also an important mediator of links to other herbivorous insects and ant-attended homopterans, such as aphids and scales. Since plant traits that respond to insect herbivory have a large impact on colonization success and the abundance of ant-attending homopterans (Cushman 1991, Dixon 1998, Nakamura *et al.* 2003, Wimp and Whitham Chapter 12 this volume), the effects of ant-mediated indirect effects would be largely dependent on the herbivore-induced plant changes (Nakamura and Ohgushi 2003). Thus, plant-mediated indirect effects caused by herbivory and/or plant use by herbivorous insects structure insect communities in these multitrophic systems. Previous studies have demonstrated that such herbivore-induced changes in plant traits have the potential to generate indirect interactions among insect herbivores (see Ohgushi 2005 and references therein). For example, early-season herbivory alters the quality and quantity of host plants, which subsequently affects performance and/or population growth rate of herbivorous insects late in the season in negative (Harrison and Karban 1986, Leather 1993, Denno *et al.* 2000) or positive ways (Williams and Myers 1984, Hunter 1987). Compensatory growth in response to herbivory generally enhances performance and density of insect herbivores that use newly emerged plant tissues, because of increased resource availability (Strauss 1991b, Pilson 1992). Likewise, ecosystem engineering such as leaf folding provides a mechanistic basis for indirect interactions among insect herbivores (Cappuccino 1993, Damman 1993, Martinsen *et al.* 2000). Furthermore, recent studies have shown that insect ecosystem engineers have the potential to greatly affect species richness and the relative abundance of arthropods on the same host plant (Martinsen *et al.* 2000, Lill and Marquis 2003). It has been widely accepted that ants are keystone species that have a great impact on insect herbivores and thus insect community structure (Fowler and MacGarvin 1985). In this context, ant–aphid mutualism greatly affects other herbivorous insects and/or their natural enemies through the removal behavior of ants attending aphid colonies, resulting in alterations of species richness and relative abundance of insect herbivores (Floate and Whitham 1994, Wimp and Whitham 2001). Furthermore, the examples of indirect interaction linkages clearly demonstrated the important positive interactions involved in insect herbivore communities. This is because herbivory often enhances resource availability through improved nutritional quality and/or increased biomass of plants due to compensatory regrowth. In addition, ecosystem engineers mostly benefit secondary users that colonize newly constructed habitats later in the season (see also Marquis and Lill Chapter 11 this volume). Positive interactions are ubiquitous in many ecological communities (Bruno *et al.* 2003), but the beneficial interactions within the same trophic level have been largely ignored, because of the traditional view of community ecologists that emphasized negative interspecific competition as the interaction of primary importance between organisms on the same trophic level.

Nontrophic, indirect interaction webs

Incorporating nontrophic and indirect links into food web structure

We have long used food webs to depict interaction linkages in ecological communities (Pimm 1982, Polis and Winemiller 1996). As the food web concept is based on energy/matter flow among trophic levels, traditional food webs do not include nontrophic (within-trophic-level) links. The primary focus on predator–prey interactions in much of the research to date on food webs has meant that the potential significance of nontrophic interactions has been largely ignored in the simple food webs. It should be also noted that species whose primary impact on the community is not primarily trophic such as ecosystem engineers can be keystone species (Power *et al.* 1996). Thus, plant-based food webs that ignore nontrophic indirect links are not an adequate tool for understanding the structural organization of insect communities that are largely dependent on plant-mediated indirect effects. There is increasing recognition that nontrophic links are important in the dynamics of ecological communities (Jones *et al.* 1997, Werner and Peacor 2003, Berlow *et al.* 2004), and thus the inclusion of nontrophic and indirect links in models of food web dynamics is an essential new direction for future ecological research. To capture species interactions more realistically in an ecological community, I introduced the term "indirect interaction web" that includes nontrophic and indirect links, which represents the linkage of multiple plant–herbivore interactions (Ohgushi 2005). The indirect interaction webs are an alternative that incorporates nontrophic and indirect links into components of food webs based on direct feeding links.

How do indirect interaction webs illustrate interactions absent in food webs?

The structure of indirect interaction webs that include nontrophic indirect links on terrestrial plants differs greatly from traditional food webs, and reveals the types of interactions and interaction diversity in ecological communities (Figs. 10.1–10.3). In Example 1 the food web approach detected three independent insect-plant interactions (Fig. 10.1a): the interactions consisting of spittlebugs, leaf rollers, and leaf beetles that feed on *S. miyabeana* (interactions 1, 2, and 3). It is unlikely that interspecific competition occurs among the herbivorous insects for three reasons. First, the willow had a low level of leaf herbivory (<20% leaf consumption) throughout the season, indicating that interspecific competition among leaf chewers rarely occurs. We also found a positive correlation between spittlebugs and aphid density that is not consistent with a competitive interaction. Second, direct competition is expected to take place when two species utilize the same part of a plant at the same time, and these herbivorous insects are more or less separated temporally and in their feeding niches.

Third, spittlebug nymphs and leaf beetle larvae are mobile so that they can avoid plant tissues that have been severely damaged by other herbivores. In the interaction web the following indirect interactions were added (Fig. 10.1b): the positive interaction between spittlebugs and leaf rollers through enhanced shoot growth (interaction 7), the positive interaction between leaf rollers and aphids through leaf shelters (interaction 8), the positive interaction between leaf rollers and three ant species through aphid colonies within leaf shelters (interaction 9), and the negative interaction between aphids and leaf beetles mediated by ants tending aphids for honeydew (interaction 10). Since the aphids were included in this web when leaf shelters were available, three direct interactions were newly established: the negative interaction between aphids and willow (interaction 4), the positive interaction between aphids and three species of ants (interaction 5), and the negative interaction between ants and leaf beetles (interaction 6). Thus, the indirect interaction web revealed six direct and four indirect interactions including four positive interactions. On the other hand, the food web approach encompassed only three negative, direct interactions (Fig. 10.4, top).

In Example 2 three independent interactions were detected in the food web (Fig. 10.2a): the interactions consisting of gall midges, aphids, and leaf beetles feeding on the willow *S. eriocarpa* (interactions 1, 2, and 3). These herbivorous insects are unlikely to compete with each other directly, because they have different feeding niches, being stem gallers, sap-suckers, and leaf chewers: they are members of different guilds. In the indirect interaction web the following four indirect interactions were added (Fig. 10.2b): the positive interaction between gall midges and aphids through increased quantity and quality of food resource (interaction 6), the positive interaction between gall midges and two leaf beetle species through increased quantity and quality of the food resource (interaction 7), the positive interaction between gall midges and ants through increased density of the aphids (interaction 8), and the negative interaction between aphids and leaf beetles through tending ants for aphid honeydew (interaction 9). Furthermore, two direct interactions were established: the positive interaction between the aphids and the ants (interaction 4), and the negative interaction between the ants and the leaf beetles (interaction 5). Thus, the indirect interaction web revealed five direct and four indirect interactions, including four positive interactions, contrasting to three negative, direct interactions in the food web (Fig. 10.4, middle).

In Example 3 the food web detected six independent plant–insect interactions on the goldenrod, *S. altissima* (Fig. 10.3a): the feeding interactions consisting of aphids, leafhoppers, and leaf-chewing moth caterpillars in spring (interactions 1, 2, and 3), and soft scales and grasshoppers in autumn (interactions 4 and 5). As

236 Takayuki Ohgushi

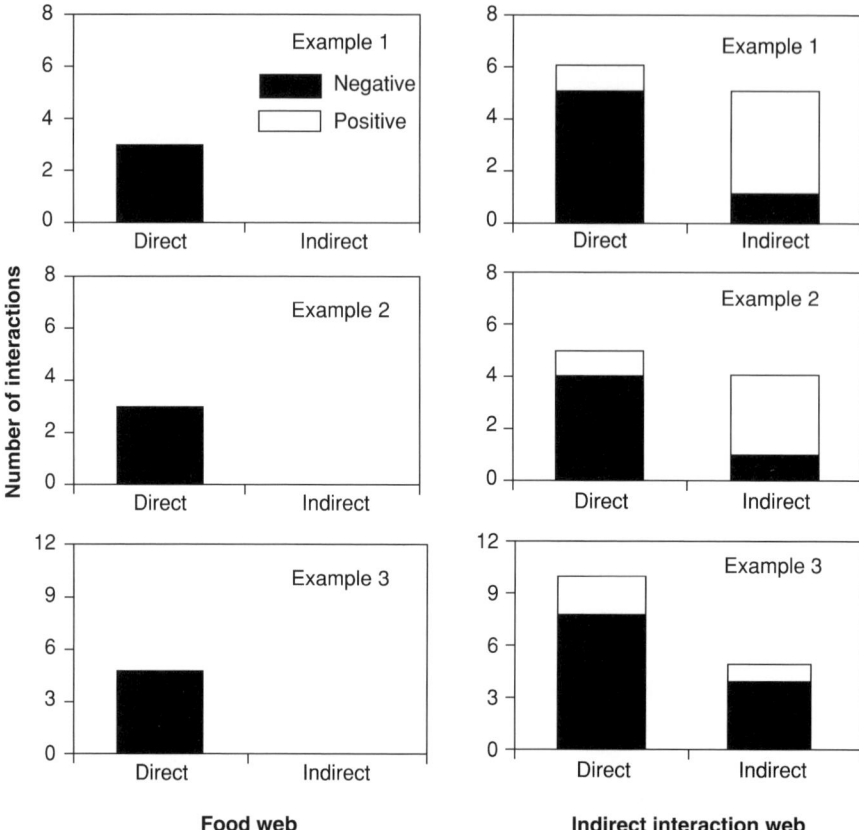

Figure 10.4. The numbers of negative and positive interactions identified in food webs and indirect interaction webs on *Salix miyabeana* (Example 1), *Salix eriocarpa* (Example 2), and on *Solidago altissima* (Example 3).

the spring feeders are temporally separated from autumn feeders in occurrence, there are no direct interactions for food resource between them. There is the potential for direct interactions if resource depletion occurred. In the indirect interaction web the following indirect interactions were detected (Fig. 10.3b): the negative interaction between aphids and leafhoppers through tending ants for aphid honeydew (interaction 11), the negative interaction between aphids and moth caterpillars through tending ants (interaction 12) in spring, the negative interaction between aphids and soft scales through change in sap quality (interaction 13), the positive interaction between aphids and grasshoppers through increased leaf nitrogen (interaction 14), and the negative interaction between soft scales and grasshoppers tending ants for aphid honeydew (interaction 15) in autumn. In addition, five direct interactions were newly established: the positive interaction between aphids and ants (interaction 6), the negative interaction between ants and leafhoppers (interaction 7), and the negative interaction between ants and moth caterpillar (interaction 8) in spring, the positive interaction between soft scales and ants (interaction 9), and the negative interaction between ants and grasshoppers (interaction 10) in autumn. Thus, the indirect interaction web revealed 10 direct and five indirect interactions, including three positive interactions, while the food web approach encompassed only five negative, direct interactions (Fig. 10.4, bottom).

Plant-based indirect interaction webs in terrestrial systems

Comparisons between food webs and indirect interaction webs of the three case studies clearly illustrated that we have long overlooked important components linking interactions in insect herbivore communities on terrestrial plants. This is because traditional food webs focus exclusively on links that are negative and direct in depicting the structure of interactions among species. It is widely appreciated that multitrophic interactions are important forces in structuring terrestrial communities (Gange and Brown 1997, Olff *et al.* 1999, Tscharntke and Hawkins 2002). The primary focus of studies in this field has been on vertical direct and indirect interactions through more than three trophic levels, while the role of horizontal interactions within a trophic level, which I have emphasized in this chapter, has been less explored. Thus, the inclusion of horizontal, indirect interactions connecting multiple trophic interactions should enrich the theory of multitrophic dynamics in ecological communities.

Since plant responses following herbivory are common and widespread in a diverse array of terrestrial plants including annuals, perennial herbs, shrubs, and trees (Karban and Baldwin 1997), I predict that interaction linkages among herbivorous insects generated by herbivore-induced plant changes will frequently occur in terrestrial systems. However, interspecific interactions among herbivorous insects have long been discounted, because the traditional view of insect ecology has argued that interspecific competition is too weak or infrequent to be a major structuring force in herbivorous insect communities (Lawton and Strong 1981, Strong *et al.* 1984). A well-known assumption of the arguments against the importance of interspecific interactions among insect herbivores is that herbivores rarely deplete green plants and thus herbivore populations are not limited by bottom-up forces (Hairston *et al.* 1960). The initial failure to test this assumption resulted in the ubiquitous plant-mediated indirect interactions among herbivores being overlooked. Unlike direct interspecific competition, plant-mediated indirect interactions can occur at low levels of herbivory for several reasons. First, the plant defense theory predicts that a rapid induction of defense should occur before herbivory reaches the high levels that cause plant mortality (Karban and Baldwin 1997). Second, plants that are overexploited by insect outbreaks are incapable of compensating regrowth for damaged tissues. Third, habitats previously created by ecosystem engineers are difficult to maintain at high levels of herbivory. Hence, heavy herbivory can decrease plant-mediated indirect interactions among herbivorous insects. A lack of visible depletion of green plants, therefore, does not mean that interspecific interactions among insect herbivores rarely occur, but rather it strongly increases the likelihood of indirect interactions among herbivores mediated by changes in plant traits. This underestimation of interspecific interactions also stems from the assumption that within-trophic-level interactions should be most prevalent among closely related species within guilds or among species that utilize the same part of a resource at the same time. In contrast to this previous view, more recent reviews have demonstrated that substantial indirect interactions occur frequently among temporally separated, spatially separated, and even taxonomically distinct herbivore species (Damman 1993, Denno et al. 1995, Ohgushi 2005). For example, based on 193 pair-wise interactions between herbivorous insects, Denno et al. (1995) found that interspecific competition occurs far more frequently than was once thought, and that over half of the cases of interspecific competition among mandibulate herbivores involved delayed, plant-mediated competition, in which previous feeding by one species induced either nutritional or allelochemical changes in the plant that adversely affected the performance of another species feeding later in the season (see also Denno and Kaplan Chapter 2 this volume).

Trait-mediated indirect interactions are more prevalent in terrestrial than in aquatic systems. In terrestrial systems, the average consumption rate by herbivores varies from 4% to 18% of aboveground plant biomass (Polis 1999), while in aquatic systems herbivore consumption often exceeds 50% of primary production. Indeed, in pelagic systems in which primary production is mainly by phytoplankton, the primary producers are killed by predation leaving an absence of organisms that can retain induced responses. In other words, the low levels of herbivory in terrestrial systems indicate that herbivore-induced changes in plant traits are most likely to occur, resulting in the prevalence of important nontrophic, indirect links.

Studies of indirect interaction webs will provide a better understanding of interaction biodiversity of terrestrial arthropods and plants, which are the groups that provide the majority of biodiversity. Biodiversity consists both of the number of species and the interactions among species (Thompson 1996, Price 2002), and long-term maintenance of biodiversity requires conserving both species and interaction diversity (Vázquez and Simberloff 2003). However, conservation research has primarily focused on species diversity, and it has largely ignored interaction diversity. It is crucially important that a clearer understanding of how all kinds of species interactions organize ecological communities in order to develop strategies to conserve biodiversity. Recent studies support the hypothesis that increasing biodiversity results in increasing stability in plant communities (Tilman 1999), and this trend may extend to herbivore communities. Indirect interactions, interspecific facilitation, and weak interactions, which have rarely been included in traditional food webs, may play an important role in promoting community stability and resistance, and ecosystem functioning (McCann *et al.* 1998, Cardinale *et al.* 2002, Neutel *et al.* 2002, Worm and Duffy 2003). Indirect interaction webs include more of the critical direct and indirect interactions that structure plant-based arthropod communities. Therefore the indirect interaction web approach will offer important insights into how interaction networks structure ecological communities, and this knowledge will aid efforts to conserve interaction biodiversity in nature.

Acknowledgments

I thank Peter Price, Tim Craig, Chris Sacchi, and Os Schmitz for invaluable comments on this manuscript. My work was partly supported by the Ministry of Education, Culture, Sports, Science and Technology Grant-in-Aid for Creative Basic Research (09NP1501), Scientific Research (A-15207003), and Biodiversity Research of the 21st Century COE Program (A14).

References

Abrams, P. A., B. A. Menge, G. G. Mittelbach, D. A. Spiller, and P. Yodzis. 1996. The role of indirect effects in food webs, pp. 371–395 in G. A. Polis and K. O. Winemiller (eds.) *Food Webs: Integration of Patterns and Dynamics.* New York: Chapman and Hall.

Agrawal, A. A. 1998. Induced responses to herbivory and increased plant performance. *Science* **279**:1201–1202.

Agrawal, A. A., and P. A. Van Zandt. 2003. Ecological play in the coevolutionary theatre: genetic and environmental determinants of attack by a specialist weevil on milkweed. *Journal of Ecology* **91**:1049–1059.

Bardgett, R. D., D. A. Wardle, and G. W. Yeates. 1998. Linking above-ground and below-ground interactions: how plant responses to foliar herbivory influence soil organisms. *Soil Biology and Biochemistry* **30**:1867–1878.

Bazely, D. R., J. H. Myers, and K. B. da Silva. 1991. The response of numbers of bramble prickles to herbivory and depressed resource availability. *Oikos* **61**:327–336.

Bazzaz, F. A., N. R. Chiariello, P. D. Coley, and L. F. Pitelka. 1987. Allocating resources to reproduction and defense. *BioScience* **37**:58–67.

Berlow, E. L., A. M. Neutel, J. E. Cohen, *et al.* 2004. Interaction strengths in food webs: issues and opportunities. *Journal of Animal Ecology* **73**:585–598.

Bolker, B., M. Holyoak, V. Krivan, L. Rowe, and O. Schmitz. 2003. Connecting theoretical and empirical studies of trait-mediated interactions. *Ecology* **84**:1101–1114.

Bonsall, M. B., and M. P. Hassell. 1997. Apparent competition structures ecological assemblages. *Nature* **388**:371–373.

Borer, E. T., K. Anderson, C. A. Blanchette, et al. 2002. Topological approaches to food web analyses: a few modifications may improve our insights. *Oikos* **99**:397–401.

Bruno, J. F., J. J. Stachowicz, and M. D. Bertness. 2003. Inclusion of facilitation into ecological theory. *Trends in Ecology and Evolution* **18**:119–125.

Callaway, R. M., S. C. Pennings, and C. L. Richards. 2003. Phenotypic plasticity and interactions among plants. *Ecology* **84**:1115–1128.

Cappuccino, N. 1993. Mutual use of leaf-shelters by lepidopteran larvae on paper birch. *Ecological Entomology* **18**:287–292.

Cardinale, B. J., M. A. Palmer, and S. L. Collins. 2002. Species diversity enhances ecosystem functioning through interspecific facilitation. *Nature* **415**:426–429.

Coley, P. D., J. P. Bryant, and F. S. I. Chapin. 1985. Resource availability and plant antiherbivore defense. *Science* **230**:895–899.

Connell, J. H. 1983. On the prevalence and relative importance of interspecific competition: evidence from field experiments. *American Naturalist* **122**:661–696.

Constabel, C. P. 1999. A survey of herbivore-inducible defensive proteins and phytochemicals, pp. 137–166 in A. A. Agrawal, S. Tuzun, and E. Bent (eds.) *Induced Plant Defenses against Pathogens and Herbivores*. St. Paul, MN: American Phytopathological Society Press.

Craig, T. P., and T. Ohgushi. 2002. Preference and performance are correlated in the spittlebug *Aphrophora pectoralis* on four species of willow. *Ecological Entomology* **27**:529–540.

Cushman, J. H. 1991. Host-plant mediation of insect mutualisms: variable outcomes in herbivore–ant interactions. *Oikos* **61**:138–144.

Cyr, H., and M. L. Pace. 1993. Magnitude and patterns of herbivory in aquatic and terrestrial ecosystems. *Nature* **361**:148–150.

Dalin, P., and C. Björkman. 2003. Adult beetle grazing induces willow trichome defense against subsequent larval feeding. *Oecologia* **134**:112–118.

Damman, H. 1993. Patterns of interaction among herbivore species, pp. 132–169 in N. E. Stamp and T. M. Casey (eds.) *Caterpillars: Ecological and Evolutionary Constraints on Foraging*. New York: Chapman and Hall.

Danell, K., and K. Huss-Danell. 1985. Feeding by insects and hares on birches earlier affected by moose browsing. *Oikos* **44**:75–81.

Denno, R. F., M. S. McClure, and J. R. Ott. 1995. Interspecific interactions in phytophagous insects: competition reexamined and resurrected. *Annual Review of Entomology* **40**:297–331.

Denno, R. F., M. A. Peterson, C. Gratton, et al. 2000. Feeding-induced changes in plant quality mediate interspecific competition between sap-feeding herbivores. *Ecology* **81**:1814–1827.

Dicke, M., and L. E. M. Vet. 1999. Plant–carnivore interactions: consequences for plant, herbivore and carnivore, pp. 483–520 in H. Olff, V. K. Brown, and R. H. Drent (eds.) *Herbivores: Between Plants and Predators*. Oxford, UK: Blackwell Science.

Dill, L. M., M. R. Heithaus, and C. J. Walters. 2003. Behaviorally mediated indirect interactions in marine communities and their conservation implications. *Ecology* **84**:1151–1157.

Dixon, A. F. G. 1998. *Aphid Ecology*, 2nd edn. London: Chapman and Hall.

Faeth, S. H. 1988. Plant-mediated interactions between seasonal herbivores: enough for evolution or coevolution? pp. 391–414 in K. C. Spencer (ed.) *Chemical Mediation of Coevolution*. New York: Academic Press.

Floate, K. D., and T. G. Whitham. 1994. Aphid–ant interaction reduces chrysomelid herbivory in a cottonwood hybrid zone. *Oecologia* **97**:215–221.

Fowler, S. V., and M. MacGarvin. 1985. The impact of hairy wood ants, *Formica lugubris*, on the guild structure of herbivorous insects on birch, *Betula pubescens*. *Journal of Animal Ecology* **54**:847–855.

Fukui, A., M. Murakami, K. Konno, M. Nakamua, and T. Ohgushi. 2002. A leaf-rolling caterpillar improves leaf quality. *Entomological Science* **5**:263–266.

Gange, A. C., and V. K. Brown. 1989. Effects of root herbivory by an insect on a foliar-feeding species, mediated through changes in the host plant. *Oecologia* **81**:38–42.

Gange, A. C., and V. K. Brown (eds.) 1997. *Multitrophic Interactions in Terrestrial Systems*. Oxford, UK: Blackwell Science.

Gómez, J. M., and R. Zamora. 2002. Thorns as induced mechanical defense in a long-lived shrub (*Hormathophylla spinosa*, Cruciferae). *Ecology* **83**:885–890.

Hairston, N. G., F. E. Smith, and L. B. Slobodkin. 1960. Community structure, population control, and competition. *American Naturalist* **94**:421–425.

Harrison, S., and R. Karban. 1986. Effects of an early-season folivorous moth on the success of a later-season species, mediated by a change in the quality of the shared host, *Lupinus arboreus* Sims. *Oecologia* **69**:354–359.

Haukioja, E., and S. Neuvonen. 1987. Insect population dynamics and induction of plant resistance: the testing of hypotheses, pp. 411–432 in P. Barbosa and J. C. Schultz (eds.) *Insect Outbreaks*. San Diego, CA: Academic Press.

Hendrix, S. D., and E. J. Trapp. 1989. Floral herbivory in *Pastinaca sativa*: do compensatory responses offset reductions in fitness? *Evolution* **43**:891–895.

Holt, R. D., and J. H. Lawton. 1994. The ecological consequences of shared natural enemies. *Annual Review of Ecology and Systematics* **25**:495–520.

Hunter, M. D. 1987. Opposing effects of spring defoliation on late season oak caterpillars. *Ecological Entomology* **12**:373–382.

Hunter, M. D., and P. W. Price. 1992. Playing chutes and ladders: heterogeneity and the relative roles of bottom-up and top-down forces in natural communities. *Ecology* **73**:724–732.

Ishihara, M., T. Hayashi, and T. Ohgushi. 1999. Life cycle of the willow leaf beetle, *Plagiodera vericolora* (Coleoptera: Chrysomelidae) in Ishikari (Hokkaido, Japan). *Entomological Science* **2**:57–60.

Jones, C. G., J. H. Lawton, and M. Shachak. 1994. Organisms as ecosystem engineers. *Oikos* **69**:373–386.

Jones, C. G., J. H. Lawton, and M. Shachak. 1997. Positive and negative effects of organisms as physical ecosystem engineers. *Ecology* **78**:1946–1957.

Jones, C. G., R. S. Ostfeld, M. P. Richard, E. M. Schauber, and J. O. Wolff. 1998. Chain reactions linking acorns to gypsy moth outbreaks and lyme disease risk. *Science* **279**:1023–1026.

Kagata, H., and T. Ohgushi. 2004. Leaf miner as a physical ecosystem engineer: secondary use of vacant leaf-mines by other arthropods. *Annals of the Entomological Society of America* **97**:923–927.

Karban, R., and I. T. Baldwin. 1997. *Induced Responses to Herbivory*. Chicago, IL: University of Chicago Press.

Karban, R., and J. H. Myers. 1989. Induced plant responses to herbivory. *Annual Review of Ecology and Systematics* **20**:331–348.

Krupnick, G. A., A. E. Weis, and D. R. Campbell. 1999. The consequences of floral herbivore for pollinator service to *Isomeris arborea*. *Ecology* **80**:125–134.

Larsson, S., H. Häggström, and R. F. Denno. 1997. Preference for protected feeding site by larvae of the willow-feeding leaf beetle *Galerucella lineola*. *Ecological Entomology* **22**:445–452.

Lawton, J. H., and D. R. J. Strong. 1981. Community patterns and competition in folivorous insects. *American Naturalist* **118**:317–338.

Leather, S. R. 1993. Early season defoliation of bird cherry influences autumn colonization by the bird cherry aphid, *Rhopalosiphum padi*. *Oikos* **66**:43–47.

Lehtilä, K., and S. Y. Strauss. 1997. Leaf damage by herbivores affects attractiveness to pollinators in wild radish, *Raphanus raphanistrum*. *Oecologia* **111**:396–403.

Leibold, M. A., J. M. Chase, J. B. Shurin, and A. L. Downing. 1997. Species turnover and the regulation of trophic structure. *Annual Review of Ecology and Systematics* **28**:467–494.

Lill, J. T., and R. J. Marquis. 2003. Ecosystem engineering by caterpillars increases insect herbivore diversity on white oak. *Ecology* **84**:682–690.

Martinsen, G. D., K. D. Floate, A. M. Waltz, G. M. Wimp, and T. G. Whitham. 2000. Positive interactions between leafrollers and other arthropods enhance biodiversity on hybrid cottonwoods. *Oecologia* **123**:82–89.

Maschinski, J., and T. G. Whitham. 1989. The continuum of plant responses to herbivory: the influence of plant association, nutrient availability, and timing. *American Naturalist* **134**:1–19.

Masters, G. J., and V. K. Brown. 1997. Host-plant mediated interactions between spatially separated herbivores: effects on community structure, pp. 217–237 in A. C. Gange and V. K. Brown (eds.) *Multitrophic Interactions in Terrestrial Systems*. Oxford, UK: Blackwell Science.

Mauricio, R., M. D. Bowers, and F. A. Bazzaz. 1993. Pattern of leaf damage affects fitness of the annual plant *Raphanus sativus* (Brassicaceae). *Ecology* **74**:2066–2071.

McCann, K., A. Hastings, and G. R. Huxel. 1998. Weak trophic interactions and the balance of nature. *Nature* **395**:794–798.

Menge, B. A. 1995. Indirect effects in marine rocky intertidal interaction webs: patterns and importance. *Ecological Monographs* **65**:21–74.

Milewski, A. V., T. P. Young, and D. Madden. 1991. Thorns as induced defenses: experimental evidence. *Oecologia* **86**:70–75.

Mopper, S., J. Maschinski, N. Cobb, and T. G. Whitham. 1991. A new look at habitat structure: consequences of herbivore-modified plant architecture, pp. 260–280 in S. S. Bell, E. D. McCoy, and H. R. Mushinsky (eds.) *Habitat Structure*. London: Chapman and Hall.

Morris, R. J., O. T. Lewis, and H. C. J. Godfray. 2004. Experimental evidence for apparent competition in a tropical forest food web. *Nature* **428**:310–313.

Mutikainen, P., and L. F. Delph. 1996. Effects of herbivory on male reproductive success in plants. *Oikos* **75**:353–358.

Myers, J. H., and D. Bazely. 1991. Thorns, spines, prickles, and hairs: are they stimulated by herbivory and do they deter herbivores? pp. 325–344 in D. W. Tallamy and M. J. Raupp (eds.) *Phytochemical Induction by Herbivores*. New York: John Wiley.

Myers, J. H., and B. J. Post. 1981. Plant nitrogen and fluctuations of insect populations: a test with the cinnabar moth – tansy ragwort system. *Oecologia* **48**:151–156.

Nakamura, M., and T. Ohgushi. 2003. Positive and negative effects of leaf shelters on herbivorous insects: linking multiple herbivore species on a willow. *Oecologia* **136**:445–449.

Nakamura, M., Y. Miyamoto, and T. Ohgushi. 2003. Gall initiation enhances the availability of food resources for herbivorous insects. *Functional Ecology* **17**:851–857.

Neutel, A. M., J. A. P. Heesterbeek, and P. C. de Ruiter. 2002. Stability in real food webs: weak links in long loops. *Science* **296**:1120–1123.

Nozawa, A., and T. Ohgushi. 2002. How does spittlebug oviposition affect shoot growth and bud production in two willow species? *Ecological Research* **17**:535–543.

Nykänen, H., and J. Koricheva. 2004. Damage-induced changes in woody plants and their effects on insect herbivore performance: a meta-analysis. *Oikos* **104**:247–268.

Ohgushi, T. 1997. Plant-mediated interactions between herbivorous insects, pp. 115–130 in T. Abe, S. A. Levin, and M. Higashi (eds.) *Biodiversity: An Ecological Perspective*. New York: Springer-Verlag.

Ohgushi, T. 2005. Indirect interaction webs: herbivore-induced effects through trait change in plants. *Annual Review of Ecology, Evolution, and Systematics* **36**:81–105.

Oksanen, L., S. D. Fretwell, J. Arruda, and P. Niemelä. 1981. Exploitation ecosystems in gradients of primary productivity. *American Naturalist* **118**:240–261.

Olff, H., V. K. Brown, and R. H. Drent (eds.) 1999. *Herbivores: Between Plants and Predators*. Oxford, UK: Blackwell Science.

Pace, M. L., J. J. Cole, S. R. Carpenter, and J. F. Kitchell. 1999. Trophic cascades revealed in diverse ecosystems. *Trends in Ecology and Evolution* **14**:483–488.

Paine, R. T. 1980. Food webs: linkage, interaction strength and community infrastructure. *Journal of Animal Ecology* **49**:667–685.

Pilson, D. 1992. Aphid distribution and the evolution of goldenrod resistance. *Evolution* **46**:1358–1372.

Pimm, S. L. 1982. *Food Webs*. London: Chapman and Hall.

Polis, G. A. 1999. Why are parts of the world green? Multiple factors control productivity and the distribution of biomass. *Oikos* **86**:3–15.

Polis, G. A., A. L. W. Sears, G. R. Huxel, D. R. Strong, and J. Maron. 2000. When is a trophic cascade a trophic cascade? *Trends in Ecology and Evolution* **15**:473–475.

Polis, G. A., and K. O. Winemiller (eds.) 1996. *Food Webs: Integration of Patterns and Dynamics*. New York: Chapman and Hall.

Power, M. E., D. Tilman, J. A. Estes, *et al.* 1996. Challenges in the quest for keystones. *BioScience* **46**:609–620.

Preszler, R. W., and P. W. Price. 1993. The influence of *Salix* leaf abscission on leaf-miner survival and life history. *Ecological Entomology* **18**:150–154.

Price, P. W. 2002. Species interactions and the evolution of biodiversity, pp. 3–25 in C. M. Herrera and O. Pellmyr (eds.) *Plant–Animal Interactions: An Evolutionary Approach*. Oxford, UK: Blackwell Science.

Pullin, A. S., and J. E. Gilbert. 1989. The stinging nettle, *Urtica dioica*, increases trichome density after herbivore and mechanical damage. *Oikos* **54**:275–280.

Quesada, M., K. Bollman, and A. G. Stephenson. 1995. Leaf damage decreases pollen production and hinders pollen performance in *Cucurbita texana*. *Ecology* **76**:437–443.

Roininen, H., P. W. Price, and J. P. Bryant. 1997. Response of galling insects to natural browsing by mammals in Alaska. *Oikos* **80**:481–486.

Schmitz, O. J., V. Krivan, and O. Ovadia. 2004. Trophic cascades: the primacy of trait-mediated indirect interactions. *Ecology Letters* **7**:153–163.

Schoener, T. W. 1983. Field experiments on interspecific competition. *American Naturalist* **122**:240–285.

Scriber, J. M., and F. J. Slansky. 1981. The nutritional ecology of immature insects. *Annual Review of Entomology* **26**:183–211.

Stanton, M. L. 2003. Interacting guilds: moving beyond the pairwise perspective on mutualisms. *American Naturalist* **162**:S10–S23.

Strauss, S. Y. 1991a. Indirect effects in community ecology: their definition, study and importance. *Trends in Ecology and Evolution* **6**:206–210.

Strauss, S. Y. 1991b. Direct, indirect, and cumulative effects of three native herbivores on a shared host plant. *Ecology* **72**:543–558.

Strauss, S. Y. 1997. Floral characters link herbivores, pollinators, and plant fitness. *Ecology* **78**:1640–1645.

Strauss, S. Y., and A. A. Agrawal. 1999. The ecology and evolution of plant tolerance to herbivory. *Trends in Ecology and Evolution* **14**:179–185.

Strauss, S. Y., J. K. Conner, and S. L. Rush. 1996. Foliar herbivory affects floral characters and plant attractiveness to pollinators: implications for male and female plant fitness. *American Naturalist* **147**:1098–1107.

Strong, D. R. Jr., D. Simberloff, L. G. Abele, and A. B. Thistle (eds.) 1984. *Ecological Communities: Conceptual Issues and the Evidence*. Princeton, NJ: Princeton University Press.

Thompson, J. N. 1996. Evolutionary ecology and the conservation of biodiversity. *Trends in Ecology and Evolution* **11**:300–303.

Tilman, D. 1999. The ecological consequences of changes in biodiversity: a search for general principles. *Ecology* **80**:1455–1474.

Tscharntke, T. 1989. Changes in shoot growth of *Phragmites australis* caused by the gall maker *Giraudiella inclusa* (Diptera: Cecidomyiidae). *Oikos* **54**:370–377.

Tscharntke, T., and B. A. Hawkins (eds.) 2002. *Multitrophic Level Interactions*. Cambridge, UK: Cambridge University Press.

Tuomi, J., P. Niemelä, E. Haukioja, S. Siren, and S. Neuvonen. 1984. Nutrient stress: an explanation for plant anti-herbivore responses to defoliation. *Oecologia* **61**:208–210.

Turlings, T. C. J., S. Gouinguené, T. Degen, and M. E. Fritzsche-Hoballah. 2002. The chemical ecology of plant–caterpillar–parasitoid interactions, pp. 148–173 in T. Tscharntke and B. A. Hawkins (eds.) *Multitrophic Level Interactions*. Cambridge, UK: Cambridge University Press.

Van der Putten, W. H., L. E. M. Vet, J. A. Harvey, and F. L. Wackers. 2001. Linking above- and belowground multitrophic interactions of plants, herbivores, pathogens, and their antagonists. *Trends in Ecology and Evolution* **16**:547–554.

Vázquez, D. P., and D. Simberloff. 2003. Changes in interaction biodiversity induced by an introduced ungulate. *Ecology Letters* **6**:1077–1083.

Vet, L. E. M., and M. Dicke. 1992. Ecology of infochemical use by natural enemies in a tritrophic context. *Annual Review of Entomology* **37**:141–172.

Wardle, D. A. 2002. *Communities and Ecosystems: Linking the Aboveground and Belowground Components*. Princeton, NJ: Princeton University Press.

Werner, E. E., and S. D. Peacor. 2003. A review of trait-mediated indirect interactions in ecological communities. *Ecology* **84**:1083–1100.

Whitham, T. G., and S. Mopper. 1985. Chronic herbivory: impacts on architecture and sex expression of pinyon pine. *Science* **228**:1089–1091.

Whitham, T. G., J. Maschinski, K. C. Larson, and K. N. Paige. 1991. Plant responses to herbivory: the continuum from negative to positive and underlying physiological mechanisms, pp. 227–256 in T. M. Lewinsohn, G. W. Fernandes, W. W. Benson, and P. W. Price (eds.) *Plant–Animal Interactions: Evolutionary Ecology in Tropical and Temperate Regions*. New York: John Wiley.

Williams, A. G., and T. G. Whitham. 1986. Premature leaf abscission: an induced plant defense against gall aphids. *Ecology* **67**:1619–1627.

Williams, K. S., and J. H. Myers. 1984. Previous herbivore attack of red alder may improve food quality for fall webworm larvae. *Oecologia* **63**:166–170.

Wimp, G. M., and T. G. Whitham. 2001. Biodiversity consequences of predation and host-plant hybridization on an aphid–ant mutualism. *Ecology* **82**:440–452.

Wootton, J. T. 1994. The nature and consequences of indirect effects in ecological communities. *Annual Review of Ecology and Systematics* **25**:443–466.

Worm, B., and J. E. Duffy. 2003. Biodiversity, productivity and stability in real food webs. *Trends in Ecology and Evolution* **18**:628–632.

11

Effects of arthropods as physical ecosystem engineers on plant-based trophic interaction webs

ROBERT J. MARQUIS AND JOHN T. LILL

Introduction

Insect herbivores and other arthropods create a variety of constructs on their host plants: silk webs, leaf shelters, galls, and stem cavities. This state change, often modifying resource availability for species other than the construct-builder (constructor, hereafter), is an example of allogenic physical ecosystem engineering. Physical ecosystem engineers are "organisms that directly or indirectly control the availability of resources to other organisms by causing physical state changes in biotic or abiotic materials" (Jones et al. 1994, 1997). In the case of allogenic engineering (in contrast to autogenic engineering), the physical state change is caused by the engineer, but the engineer is not part of the new physical state.

The presence of constructs can impact the species richness, food web structure, and trophic interactions of the community of arthropods associated with engineered plants. The impact of engineering will depend on the responses of individual species to the presence of the constructs and the resulting interactions with all other species. One approach to understanding the nature of these responses is to view the engineered plant as a mosaic of engineered and nonengineered habitats. The response of a particular animal species to a plant that has been colonized by a constructor will depend on the relative value of the engineered versus nonengineered habitat to that animal. This value will be a function of differences between the two habitat types in the intensity of abiotic stress, the relative quality

Ecological Communities: Plant Mediation in Indirect Interaction Webs, ed. Takayuki Ohgushi, Timothy P. Craig, and Peter W. Price. Published by Cambridge University Press. © Cambridge University Press 2007.

of food resources, competitive and mutualistic interactions (with other species in nonengineered habitats and with other secondary inhabitants in engineered habitats), and responses of natural enemies to engineering. Nonengineering species may respond positively (as "specialists" of the engineered habitat), negatively (as "specialists" of the nonengineered habitat), or indiscriminantly (as generalist species that use both habitat types).

Our overall goal with this chapter is to review what is known, and to speculate about what is not, about the effects of plant constructors on the community of arthropods found on plants, and the resulting impacts on plants. We first describe briefly the natural history of constructors and the species that use these constructs. This natural history sets the stage for understanding the impacts of engineers on plant-based food webs. We next review the benefits and costs of building constructs, and secondarily inhabiting them. These benefits and costs represent selective forces for, and constraints on, the evolution of engineering and the secondary use of those constructs. By clarifying the factors involved, we can understand more clearly the forces that define these relationships. We also discuss the plant traits that determine the level of colonization by constructors. In understanding the contribution of plant traits, our goal is to be able to predict how variation in plant traits, both within and among plant species, contributes to variation in the interactions between engineering and nonengineering species. Finally, and most relevant to the subject of this volume, we review what is known of the impacts of construct building on arthropod communities and impacts for the plant. We limit our discussion to herbivores that create structures on plants but that do not kill them in the process.

Overview of plant constructs, their builders, and secondary inhabitants

A variety of constructs are built by arthropods on plants, and can be secondarily occupied by individuals other than the original constructor (Fig. 11.1). These secondary inhabitants include arthropods classified as inquilines, as well as parasitoids and hyperparasitoids of the original construct builder. The term inquiline has been traditionally used to describe secondary inhabitants of galls (e.g., Sanver and Hawkins 2000), but can be extended to include secondary inhabitants of leaf constructs, leaf mines, and stems that have been attacked by stem-borers. In the case of galls, inquilines range in impact from true commensals (those that feed on gall tissue but do not harm the galler), to predators that kill the galler facultatively, and finally to true predators.

Figure 11.1. Various constructs formed by engineering insect herbivores in a deciduous oak–hickory forest of Missouri, USA (all photos taken at Cuivre Rivev State Park, Troy, Missouri, by R. J. Marquis). Upper right-hand corner, clockwise: leaf fold on *Acer saccharum*; leaf web by *Pococera expandens* (Pyralidae) on *Quercus alba* (arrow indicates the caterpillar); leaf tie on *Q. alba*; leaf mine on *Q. velutina*; bored hole in trunk of *Carya* sp.; bud gall of *Q. velutina*; leaf roll of *Carya* sp.; tent made by a colony of the fall webworm, *Hyphantria cunea* (Arctiidae), on *Diospyros virginiana*.

Leaf constructs (and webs)

Herbivorous and predatory arthropods make various constructs using single or multiple leaves in combination with silk. Leaves (or leaf analogs, phyllodes: Crespi 1992, Mound and Morris 2000) are folded, cut to form tents, rolled, tied together, and included in silk nests. In some cases, feeding itself (by aphids, midges, sawflies: Kopelke 2003; psyllids: Young 2003; and mites: Fournier *et al.* 2003) may cause the leaf to deform during development to form partially enclosed cavities. Lepidoptera families that commonly build constructs include the Hesperiidae, Gelechiidae, Oecophoridae, Tortricidae, Stenomidae, Pyralidae, Choreutidae, Crambidae, and Thyrididae, while a few species of Geometridae and Noctuidae also do so. Tettigoniidae (Oecanthinae) form leaf tents from single leaves, and thrips tie together phyllodes, modified leaf petioles of *Acacia*. Some sawflies (Hymeoptera) spin webs into which they incorporate leaves, roll leaves (both Pamphiliidae), or make leaf folds (Tenthredinidae: *Phyllocolpa*). Secondary inhabitants include larvae of other Lepidoptera, herbivorous beetle larvae and adults, Homoptera and herbivorous and predaceous Hemiptera, herbivorous thrips, predaceous Coleoptera, Hemiptera, Neuroptera, and centipedes, spiders, parasitoid larvae and adults, and scavengers, including bark lice, thrips, pseudoscorpions, mites, springtails, and beetles (click and rove). In the case of leaf-tying caterpillars on *Quercus*, many secondary inhabitants of leaf ties are leaf-tying caterpillars that occupy previously constructed leaf ties in lieu of constructing new ones. As many as nine different leaf-tying caterpillars from six different species have been observed to occupy a single leaf tie over the course of its life (Lill 2004). By contrast, leaf tents made by Hesperiidae and Tettigoniidae are constructed (tents are just large enough to allow occupation only by the tent-maker) and destructed (Hesperiidae larvae cut the silk guy wires of their tents upon leaving) in such a way that they are rarely secondarily occupied.

A few species (tent caterpillars in the genus *Malacosoma* and the fall webworm, *Hyphantria cunea*) build communal webs around branches, typically including some leaves. These webs may be secondarily occupied by other species (Morris 1972). Also, many Pyralidae and Tortricidae create loose webs, incorporating additional leaves as the larvae develop.

Galls

Galls are made in stems, twigs, leaves (including the blade, veins, and petioles), meristems, flowers, and fruits. Gall-makers have been termed "microhabitat engineers" (Sanver and Hawkins 2000). They include aphids, beetles, flies (particularly cecidomyiids, or gall midges as well as tephritid flies), gall wasps (Cynipidae), psyllids, coccids, thrips, chalcid wasps, sawflies, mites (Ananthakrishnan 1984a), Lepidoptera (e.g., Miller 2004), and beetles (e.g., Thakur *et al.* 1996, de Souza *et al.*

2001). The resulting galled tissue provides both a resource that is higher in quality than nongalled tissue and protection against natural enemy attack. The gall is also a resource to various secondary inhabitants that use it as food and enemy-free space, including herbivores that feed on the gall tissue, secondary gallers that use the original gall to build a new one, parasitic inquilines that feed both on the gall and parasitize the original gall-maker, fungal pathogens, and consumers of the fungus. Sanver and Hawkins (2000) provide a review of the impacts of galling on these inquilines. Stem galls, in particular because they are formed in more persistent tissue than leaf galls, may be colonized long after the original gall-maker has exited. Eliason and Potter (2000) found ants, spiders, gall midges, and various beetles inside woody stem galls on *Quercus palustris*, and collembolans, psocopterans, and mites living on the gall surface.

Stem-bored cavities

Cavities made by stem-borers can also be secondarily occupied, although study of secondary occupants of this type of construct seems to be limited. In red mangrove (*Rhizophora mangle*), cavities made by stem-borers are secondarily colonized by other stem-borers and inquilines (Feller and Mathis 1997). Certain wood-boring species (primary xylovores) must first attack the tree before other species of wood-borers (secondary xylovores) can colonize (Feller and Mathis 1997). Cavities arising from boring also are colonized by a number of species associated with sap, frass, and boring dust (Larson and Harman 2003) and those that take advantage of the formation of a domicile (the cavity resulting from boring) (Feller and Mathis 1997). Feller and Mathis (1997) report 15 families of insects plus spiders, pseudoscorpions, and scorpions colonizing stem-bored cavities in red mangrove. In the case of trunk-boring beetles, beetles not only excavate a cavity used by secondary occupants, but they also may carry secondary occupants with them (e.g., fungi and mites: Lombardero *et al.* 2003). In a related type of damage, egg nests (slits in twigs into which eggs are deposited) of periodical cicadas (*Magicicada* spp.) are secondarily occupied by Pseudococcidae and Eriococcidae, Cleridae, and Coccinellidae (Russell and Stoetzel 1991). In addition, parasitoid wasps in the families Torymidae and Scelionidae oviposit in the nests as do some species of crickets (Gryllidae), all of which presumably consume *Magicicada* spp. eggs or nymphs after hatching (Russell and Stoetzel 1991). Secondary inhabitants varied in their timing of colonization: mealy bugs (Pseudococcidae) entered nests with unhatched cicada eggs, while all other species occupied nests containing only cicada egg shells.

Leaf mines

There appears to be a single published example of abandoned leaf-mines providing habitat for secondary colonizers: mines made by *Phyllonorycter pastorella* on *Salix eriocarpa* were colonized by six species of arthropods, with springtails selectively colonizing them as feeding and reproductive sites (Kagata and Ohgushi 2004). The springtails feed on fungi growing on the frass of the miner. Freshly abandoned mines of *Cameraria* (which contain frass) on *Quercus alba* in Missouri are commonly occupied by plant lice (R. J. Marquis personal observation).

Predator constructs

Predators also build constructs on plants, and can influence the abundance of herbivores as a result. Predaceous weaver ants make leaf nests (Hölldobler 1983). Spiders build webs on plants (e.g., Louda 1982), make egg nests using leaves, and crab spiders fold leaves and flowers (Morse 1992, Evans 1997, Ott *et al.* 1998, Romero and Vasconcellos-Neto 2004). On *Quercus alba*, spider egg nests, composed of an egg cluster and surrounding web between two leaves, are often secondarily occupied by herbivorous insects (R. J. Marquis and J. T. Lill, personal observations).

Benefits and costs of construct formation

Benefits

Potential benefits of construct formation are five: amelioration of harsh abiotic factors, protection from physical dislodgment, modification of resource quality, escape from natural enemies, and reduced effects of competition. Fukui (2001) reviewed the benefits for leaf shelters (leaf constructs in this chapter). Hunter and Willmer (1989) demonstrated that leaf-tying caterpillars quickly desiccate when excluded from leaf shelters, and Larsson *et al.* (1997) showed that the main benefit of occupying natural leaf rolls for *Galerucella lineola* on *Salix viminalis* is decreased desiccation. Loeffler (1996) demonstrated that the main advantage of leaf fold construction by *Dichomeris* larvae on *Solidago* is reduced dislodgment from physical forces. Leaf constructs can be lower in defensive compounds (Berenbaum 1978, Lewis 1979, Sandberg and Berenbaum 1989, Sagers 1992, Oki 2000, Fukui *et al.* 2002) or leaf toughness (Lewis 1979, Fukui *et al.* 2002). In contrast, Costa and Varanda (2002) found no differences in leaf quality between leaves in shelters and not. When such effects occur, presumably lower light levels within constructs reduce allocation to lignin (thus lowering leaf toughness) and to defensive compounds, including phenolics. Heads and Lawton (1985) and Fowler and MacGarvin (1985) demonstrated that predation on free-feeding herbivore species by ants and spiders is higher than on leaf-tying or leaf-rolling species. Cappuccino (1993) showed that leaf tiers prevented from tying leaves disappear at a higher rate than those in ties while Jones *et al.* (2002) demonstrated that leaf tents provide protection against predatory wasps. In addition, shelter-building Tortricidae escaped predation by birds but free-feeding geometrids did not (Atlegrim 1989). Finally, web construction by *Depressaria pastinacella* (Oecophoridae) in the infructescences of the host *Pastinaca sativa* decreased aggressive encounters with conspecifics, resulting in greater pupal mass (Green et al. 1998).

Galling is the quintessential example of benefits gained from construct formation on plants. Within the gall, the gall-maker is essentially free of exposure to stressful abiotic conditions (Hodkinson 1984), dislodgment is literally impossible, the inner tissue is generally of higher quality than nongalled tissue (e.g., Ananthakrishnan 1984b, Price and Louw 1996), and tough gall tissue provides protection from many natural enemies (e.g., Weis et al. 1985). Stem-boring species would appear to enjoy the same advantages, with the exception that stem-borers probably have less positive impact on plant quality as a food source.

Costs

Costs of construct formation include the energy and resources necessary for building the construct itself, potential competition with secondary inhabitants, and vulnerability to natural enemy attack due to increased "signaling" (constructs appear to stand out against unmodified plant parts, at least to the human observer). Costs of building leaf constructs would include the seemingly high cost of silk, which consists largely of protein (Fitzgerald et al. 1991, Fitzgerald and Clark 1994). Surprisingly, Loeffler (1996) found no growth effect of forcing *Dichomeris* larvae to rebuild leaf folds multiple times, in part because they rebuilt on younger leaves. For the leaf-tying caterpillar *Psilocorsis quercicella* on *Quercus alba*, Lill et al. (J. T. Lill unpublished data) also found no growth effect of forcing larvae to rebuild leaf ties every 3 days in the laboratory, while Weiss (2003) found a negative effect on pupal size only when *Eparygyreus clarus* (Hesperiidae) caterpillars were forced to construct three times as many tents as they would normally.

When leaf constructs are secondarily colonized by other members of the same guild (e.g., herbivores in constructs made by leaf-tying caterpillars, predators in constructs made by spiders), there is the potential for resource competition. The effects of these potential interactions have not been investigated.

Because constructs would appear to be relatively obvious to a searching predator or parasitoid, their construction may lead to higher rates of attack by natural enemies than those experienced by closely related species that do not build constructs. For example, both galls (Dickson and Whitham 1996) and leaf rolls (Murakami 1999) attract birds that apparently develop a search image for them. Rather than offer protection against vertebrate predators, one of three species of leaf-rolling caterpillars (that which made the most conspicuous leaf roll) was in higher abundance when birds were excluded from trees (Murakami 1999). In addition, both bird predation and parasitoid attack is a function of gall size, with intermediate gall sizes enjoying the lowest level of combined attack by both (Weis et al. 1985, Redfern and Cameron 1994, Van Hezewijk and Roland 2003). Parasitism is also related to gall morphology (Plantard and Hochberg 1998) and toughness (Craig et al. 1990). Occupants of different construct types suffer different average levels of parasitism compared with free-feeding species (Hawkins 1994, Le Corff et al. 2000), with parasitism rates in the following order: leaf miners > leaf rollers = leaf webbers = free feeders > stem-borers. However, these comparisons do not control for the confounding effects of phylogeny, timing of occurrence within a season, and size of the host.

Galling species in particular are vulnerable to inquilines that generally feed on gall tissue, but may kill the galler when coming in contact with them. Inquilines can kill 90–100% of inducers (Seibert 1993, Shorthouse 1994). For example, in 61 of the 76 cases reviewed by Sanver and Hawkins (2000), the presence of inquilines resulted in the death of the galler, either by direct predation or through the consumption of gall tissue and subsequent exposure of the galler to desiccation or disease (e.g., Wilson 1995). Timing of oviposition of secondary inhabitants is critical: gall-feeders that colonize after the original gall-maker has completed development will have no influence on that insect's fitness (e.g., Sugiura et al. 2004). Laboratory experiments that manipulate the presence of secondary occupants are needed to test their potential competitive effects on the primary construct builder, especially in the case of leaf constructs.

In most cases, it is difficult to define the costs and benefits of the construction lifestyle because the constructors do not normally occur outside of the construct. In lieu of finding species that are polymorphic in this behavior or change behavior with instar, a more complete definition of the potential costs and benefits of the evolution of the constructor lifestyle awaits comparisons among multiple pairs of closely related species that differ in feeding habit (see Gaston et al. 1991).

Benefits and costs of secondary inhabitation of plant constructs

All benefits and costs associated with formation of the original construct apply to the secondary inhabitants. In fact, benefits and costs of secondary occupation also may apply to the constructor itself if the construct is simultaneously occupied. Timing will be an important factor: at least in leaf ties on Missouri *Quercus*, the tie can be maintained by secondary colonists long past the time the original constructor has completed its life cycle (R. J. Marquis and J. T. Lill personal observations). On the benefit side, per capita investment in construction may be reduced when multiple individuals contribute to tie maintenance. Production of silk may be lower for the leaf construct builder and secondary occupants if the amount produced depends on the number of inhabitants. For galls that require exogenous production of hormones (Mapes and Davies 2001a) or other gall-inducing agents (Mapes and Davies 2001b, Sopow et al. 2003) by the gall-maker in order to induce gall formation, the occurrence of multiple individuals in the gall capable of producing these compounds may reduce the associated energetic costs borne by any one individual. Secondary occupation of constructs may be detrimental to both the constructor and to secondary inhabitants, in the case of leaf constructs, if sharing a construct results in reduced quantity or quality of leaf material. Cohabitation also may increase the likelihood of contracting a disease, or being parasitized or attacked by a predator.

A few benefits of secondary occupation of constructs will accrue only to the secondary occupants. Presumably the cost of searching for a suitable oviposition site may be saved for all four types of construct builders if they selectively oviposit in existing structures. For detritivores, many leaf constructs accumulate frass, cast skins, and dead bodies of the inhabitants (R.J. Marquis and J.T. Lill personal observations), and therefore provide a concentrated and dependable source of these resources. Finally, because initial stem penetration does not need to be made by secondary stem-borers, "excavation costs" may be minimized.

Role of plant traits as determinants of construct formation

Plant traits can influence the level of construct formation at various scales: within plants, among plants within a species but in different habitats, and among plant species within the same habitat. Traits that contribute to plant quality and architecture have been implicated, although there have been no attempts to test the relative importance of plant quality and architecture for constructors on plants (see Kaitaniemi et al. (2004), however, for an example of free-feeding species).

The distribution of constructs within plants is likely to be nonrandom, but the causes of such distribution patterns can be difficult to uncover (Brown et al. 1997). Leaf rolls (higher in sun-exposed branches: Riihimaki et al. 2003), leaf ties (higher in mid canopy: Carroll and Kearby 1978), and some (Whitham 1983, Kampichler and Teschner 2002) but not all gall species (Eliason and Potter 2001, Kampichler and Teschner 2002) show differential distributions within tree canopies. In *Quercus robur*, within-canopy distribution of *Mikiola fagi* galls is related to gradients in microclimate (Kampichler and Teschner 2002). In contrast, in *Q. alba* and *Q. macrocarpa*, greater density in the lower canopy appears to be a function solely of distance from the forest floor, whence adults emerge from pupae (Brown et al. 1997).

Among plants within a plant species, gall attack is positively correlated with plant vigor (length of potentially galled shoots) for many (Price 1991, 2003, Roininen and Danell 1997) but not all (Rehill and Schultz 2001) galling species, for stem-borers (Price et al. 1990, Feller and Mathis 1997), and for at least one leaf roller (Seyffarth *et al.* 1996). In some cases, specific plant traits have been identified. Tannin chemistry, lignins, and nitrogen influence interspecific differences in gall species richness for species of *Quercus* (Abrahamson *et al.* 2003), while levels of salicylic acid determine attack by a gall midge attacking *Salix viminalis* (Ollerstam and Larsson 2003).

Plant architecture is also likely to be important in determining attack levels within and among plant species. For example, attack by *Pemphigus betae* aphids on cottonwood was greatest on genotypes with fewer buds: experimentally reduced bud number resulted in higher establishment success due to lowered source–sink competition for resources (Larson and Whitham 1997). Level of attack by leaf-tying microlepidoptera caterpillars on understory saplings of *Quercus alba* is determined by the number of touching leaves prior to attack (Marquis *et al.* 2002). Plants with more touching leaves are more heavily attacked. Because leaf ties consist of two or more neighboring leaves tied together with silk, apparently newly eclosed larvae, because of their small size, can only attack leaves that are actually touching or nearly so. The community of leaf-tying caterpillars varies by *Quercus* host plant species, presumably in part due to differences in leaf quality and plant architecture (R.J. Marquis and J.T. Lill unpublished data). However, architecture may not always constrain leaf construct formation: in the weaver ant *Oecophylla smaragdina*, ants form chains with their bodies, and in so doing, pull leaves together to build nests (Hölldobler and Wilson 1990).

Differences in leafing phenology are also likely to be important. For instance, leaf rolling can be accomplished only when leaves are young and still pliable. Differences in attack levels among individual trees are tied to leafing phenology in one species of gall-former on *Quercus robur* (Csóka 1994), two species of leaf-tying caterpillars on *Q. robur* (Hunter 1992), and a leaf-mining moth on *Q. geminata* (Mopper and Simberloff 1995). Potentially, these phenological differences are likely to influence total arthropod community structure on individual plants through their impact on individual engineers.

By examining plant–galler relationships across a moisture gradient, Price *et al.* (1998) found that the species richness of gallers was greatest in xeric sites, suggesting that abiotic stress may drive the evolution of this engineering habit. Similarly, Feller and Mathis (1997) found that tall mangrove trees fertilized by a frigate bird/brown boobie rookery were more heavily attacked by stem- and twig-borers than smaller trees in nutrient-poor sites. If abiotic factors also play an adaptive role in the evolution of other concealed feeding strategies (e.g., leaf constructs), then their abundance also should vary with gradients in abiotic factors. There appear to be no systematic surveys for geographic trends in plant construct abundance other than that for gall-makers conducted by Price and his associates (Price *et al.* 1998).

Natural enemies often impose high levels of mortality on insect herbivore species on plants. Given this, we would predict that intra- and interspecific variation in plant traits that mediate such interactions could influence within- and between-species variation in the abundance of engineers (Marquis and Whelan 1996). Damman (1987) demonstrated that leaf constructs provide protection against natural enemies, and that such constructs are only made with mature leaves. These findings suggest that natural enemies may mediate the impact of plant traits on constructor abundance. These plant traits include the timing of leaf maturation, leaf toughness, and the proximity of leaves to one another.

Impacts of constructs on arthropod community structure

The impacts of constructs on arthropod community structure will be a function of the interactions among constructors and other herbivores (Fig. 11.2, pathway 1), among arthropods that colonize constructs and those that secondarily colonize those constructs (pathway 2), and among secondary occupants of engineered habitats and those of nonengineered habitats (pathway 3). Finally, natural enemies foraging in nonengineered habitats may modify the impact of constructs on arthropod community structure (pathway 4). One or more of these pathways may be important in determining the impact of engineering on a plant's arthropod diversity. True generalist species, those that do not specialize on either habitat type, may have impacts on arthropod community structure of a plant, but would not do so preferentially with respect to habitat type (therefore generalist species are not depicted in Fig. 11.2).

Interactions between constructors and other herbivores (pathway 1)

In pathway 1, constructors interact indirectly with other arthropods by changing the availability of the shared resource, the plant. Resource availability can be modified either through consumption of the plant resource, or by altering the quality of nonconsumed plant resources via induced responses (Karban and Baldwin 1997). This pathway views traditional resource competition as an indirect interaction (Goldberg 1990, Morin 1999).

The presence of the construct can influence the quality and abundance of the nonengineered habitat, and species that specialize on it. Previous studies demonstrate that the constructor can have both positive and negative effects on other herbivores. Stem- and twig-borers kill everything distal to the site of attack, thus making that portion of the plant unavailable to species that would otherwise use it (e.g., Letourneau 1998). However, Rathcke (1976) found that competition, presumably through pre-emption of the resource, occurred infrequently in a guild of stem-borers attacking prairie plants. Attack by the oak galler *Callirhytis*

Arthropods as physical ecosystem engineers 257

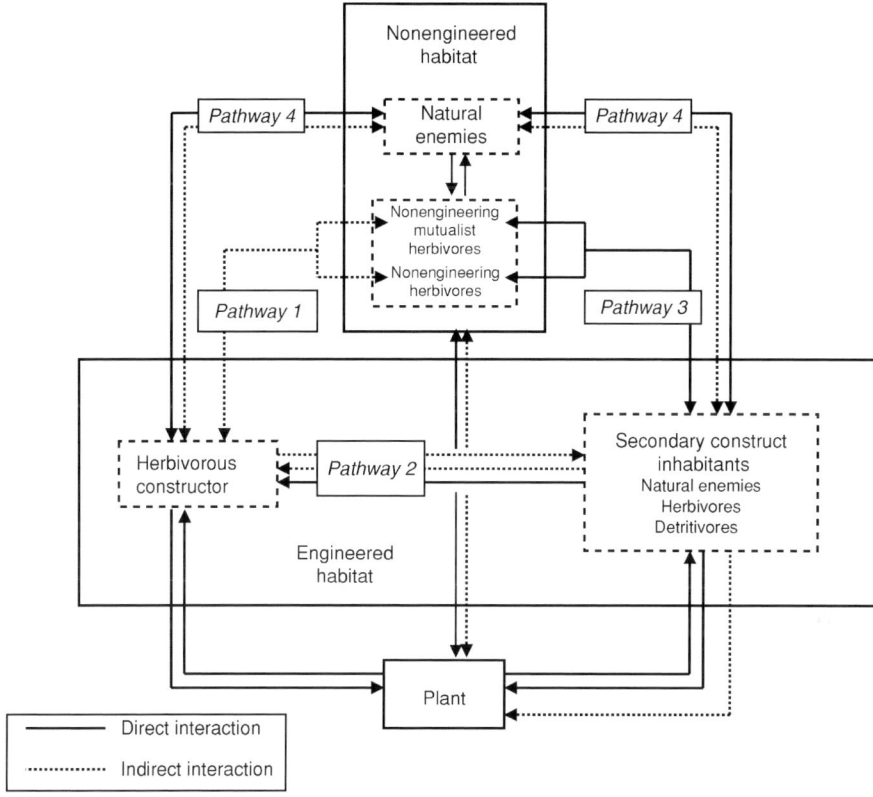

Figure 11.2. A simplified interaction web (see also Fukui 2001: Fig. 11.1) for the direct and indirect effects that ecosystem engineers on plants have on the trophic structure of the plant's food web (see text for further explanation).

cornigera increased foliar tannin levels, perhaps accounting for the preference of nongalled leaves over galled leaves by second instar gypsy moth larvae (*Lymantria dispar*) (Foss and Rieske 2004). Leaf mines of *Eriocrania* spp. on *Betula pendula* when overlapping with the central leaf vein decrease the ability of aphids to attack those leaves (Johnson *et al.* 2002). Positive effects can occur as well. Gall presence on cottonwood increased colonization of the branch by other, nongalling herbivores (Dickson and Whitham 1996). The authors suggest that the presence of the galls increased quality of the branch and its leaves for free-feeding herbivores by acting as a photosynthetic sink. Positive effects also can occur if engineering stimulates axillary meristems, providing additional shoots for attack by nonengineering species (Pilson 1992, Nakamura *et al.* 2003).

When predators are the constructors, herbivore abundance can be negatively affected. Shelter building by crab spiders reduced the abundance of seed predators (Romero and Vasconcellos-Neto 2004), while the weaver ant, *Oecophylla smaragdina*, affects some herbivore species negatively and others positively. These ants build leaf nests from which they make forays to the rest of the plant (and multiple plants) (Hölldobler 1983), reducing herbivore numbers and the damage they cause (Offenberg et al. 2004), but also tend aphids and nectar-producing larvae (Bluthgen and Fiedler 2002). Finally, interactions between constructors and other herbivore species are not mandatory: for *Quercus alba*, there was little impact on species occupying nonengineered habitat (Lill and Marquis 2003). Almost all of the effect of the presence of leaf ties on herbivore abundance at the whole plant level was due to colonization of the ties themselves.

Specialists of nonengineered habitat may also influence the abundance of constructors. For example, the abundance of the leaf rolls made by *Deporaus betulae* beetles (Curculionidae) was reduced as a result of previous defoliation by *Epirrita autumnata* larvae (a free-feeding species) in the previous year due to induction of reduced leaf quality (Riihimaki et al. 2003). Early-season attack by the defoliating species positively influenced the ability of *Diurnella* to construct leaf rolls later in the season despite the reduced quality of those leaves (Hunter 1987). In addition, early-season herbivory by elk reduced the abundance of gall-makers on aspen trees (Bailey and Whitham 2003), and herbivory by the beetle *Trirhabda baccharidis* decreased the density of leaf-mining flies and leaf galls on *Baccharis halimifolia* (Hudson and Stiling 1997), apparently through an induced decrease in leaf quality.

Interactions within constructs (pathway 2)

In a second indirect pathway (Fig. 11.2), constructors influence the abundance and species richness of arthropods that respond positively to the presence of the plant construct itself. Because multiple species often respond to the presence of plant constructs, and because plant constructs typically persist well beyond the time they are occupied by the original constructor, the impacts on the host plant's arthropod community may extend beyond that due solely to the addition of the constructor (i.e., pathway 1). Pathway 2, and its effects on community diversity and structure, are the essence of the impact of ecosystem engineering in plant–constructor systems. Secondary inhabitants can affect the original constructor either directly (by eating them, in the case of gall inquilines and predators), or indirectly, through changes in food quality and abundance or by affecting natural enemy attack.

Building the construct may change the plant's value as a resource for species that specialize on that habitat. Obviously, the presence of a gall will have a positive effect on species that only occur in galls. For example, Dickson and Whitham (1996) demonstrate that the presence of galls on cottonwoods increases diversity on those branches in part due to the colonization of the galls by inquilines. On average, the number of inquiline species was 2.4 for the studies reviewed by Sanver and Hawkins (2000). Experimental addition or removal of leaf constructs has demonstrated that arthropod diversity at the branch (Cappuccino and Martin 1994, Martinsen et al. 2000) and the whole plant level (Lill and Marquis 2003) is increased as result of the presence of leaf constructs. In *Quercus alba*, herbivores, detritivores, and predators are greater in abundance in the presence of ties (J. T. Lill and R. J. Marquis unpublished data). By adding artificial leaf rolls to cottonwoods, Martinsen et al. (2000) concluded that the influence of the leaf roll on arthropod species richness was one of providing shelter rather than a source of prey. In comparing the impact of leaf shelters without a tie-maker to those with a tie-maker, Lill and Marquis (2003) demonstrated that the increase in herbivore species diversity is due to the presence of tie itself, with little additional influence of the tie-maker. For the webber *Pococera expandens* (Pyralidae), however, colonization of previously occupied leaf ties (with accompanying leaf damage and frass) on *Q. alba* was greater than for newly formed ties. This result suggests that some species are attracted to leaf ties by olfactory and/or visual cues of damage (Lill and Marquis 2004).

Perhaps our best understanding of interactions within constructs comes from studies of galls. Within galls, inquilines can reduce the fitness of gall-makers or kill them outright, leading to a direct negative interaction between the colonist and the constructor (Fig. 11.2, pathway 2). When galls made by *Pachypsylla celtidismamma* were colonized by a *Pachypsylla* sp. inquiline, the gall-maker was smaller, presumably because of interspecific competition (Heard and Buchanan 1998). There are also numerous examples of inquilines killing gall-makers (Sanver and Hawkins 2000). Gall size influences the likelihood of interactions occurring within galls in some cases. For example, inquiline abundance was higher in larger galls of the alien gall-maker *Andricus quercuscalicus* on *Q. robur* (Schonrogge et al. 1996) and that of the cynipid *Callirhytis cornigera* on *Q. palustris* (Eliason and Potter 2000). Gall size itself can be a function of interactions among gall-makers in different galls: gall-makers sharing a leaf grew larger (Heard and Buchanan 1998) perhaps because of the increased physiological sink due to multiple galls. Roininen et al. (1996) proposed a scenario by which bottom–up effects drive density of galls, parasitoids, and inquilines (and presumably their interactions within galls) in *Salix pentandra*. Young trees, previously unattacked by *Euura* gall-makers, when colonized have low levels of inquilines and low parasitism rates. Gall abundance increases over time, resulting in increased abundance of inquilines and parasitoid attack. Finally, host plant resistance increases, and the abundance of the gall-maker declines, bringing with it a crash of the associated community.

Studies of interactions within leaf constructs are few. Lill (1999) found that neither the abundance of conspecifics or heterospecifics affected the risk of parasitism of the oak leaf-tying caterpillar *Psilocorsis quercicella*, but the presence of additional tie-makers decreased pupal size in one of two species of tie-makers reared in the laboratory (J. T. Lill and R. J. Marquis unpublished data). The impact of other inhabitants (non-tie-making herbivores, predators, and detritivores) on the tie-maker has yet to be assessed.

Constructs may provide refuges for species during population lows. In *Q. alba*, there was a significant positive effect of leaf tie presence on the abundance of the Asiatic oak weevil (*Cyrtepistomes castaneus*) and the sawfly *Caliroa* sp., both of which are capable of using engineered and nonengineered habitat. This positive effect occurred only in the years when populations of these two species were low (Lill and Marquis 2003). This result suggests two phenomena: (1) leaf ties are preferred sites, and (2) these sites can be filled when populations are high, forcing other individuals to colonize nonengineered portions of the plant.

Interactions between secondary construct inhabitants and other herbivores (pathway 3)

The value of nonengineered habitat on a plant will also depend on a third pathway (Fig. 11.2) in which secondary inhabitants of constructs interact with specialists of nonengineered parts of the plant. In foraging outside of constructs, species that specialize on constructs can positively (through mutualisms) or negatively (through predation) influence the quality of the nonengineered portion of the plant.

Engineered habitat may serve as a resource for predators that have a negative effect on specialists of nonengineered habitat. Ant species tend nectar-secreting galls (e.g., Seibert 1993, Fernandes *et al.* 1999), colonize old galls (Araujo *et al.* 1995, Carver *et al.* 2003) and stem-bored cavities (Feller and Mathis 1997), and tend Homoptera species that colonize leaf constructs (Nakamura and Ohgushi 2003). As a result, these ants can affect species external to the constructs. Nakamura and Ohgushi (2003) demonstrate that ants tending aphids in leaf rolls and leaf ties reduce the survival rate of the leaf beetle *Plagiodera versicolora* on *Salix miyabeana* living outside of the leaf constructs.

Influence of natural enemies (pathway 4)

A fourth pathway involves natural enemies that do not colonize constructs (Fig. 11.2). Despite not utilizing constructs, natural enemies may influence the distribution and abundance of constructs and their occupants within plants, and influence the relative value of engineered versus nonengineered plant parts for other organisms.

Predators that do not use constructs may nevertheless influence the abundance of those constructs. Ants living on *Acacia deprenalobium* reduce the abundance of galls on these plants (Young *et al.* 1997), and ants on *Endospermum labios* reduce the number of stem-borers (Letourneau and Barbosa 1999). Ants also influence the structure of gall communities on willows, decreasing the abundance of leaf- and petiole-gallers while having no influence on stem-gallers (Woodman and Price 1992). Phenology is important: stem-gallers overwinter in galls, emerging in the spring when ants are not active, thus escaping predation, while larvae of the former species are subject to predation by ants when they exit galls in late summer.

Natural enemy attack may increase with increasing abundance of constructs. Birds, for example, are attracted to galls: Dickson and Whitham (1996) demonstrate that branches of cottonwood from which aphids (and therefore the galls they form) are excluded are visited by fewer insectivorous birds. Gall size is one determinant of such post-colonization attack. Larger galls are more susceptible to avian predation (Weis *et al.* 1985) while parasitism rate can be higher (Schonrogge *et al.* 1996), lower (Weis *et al.* 1985, Eliason and Potter 2000), or show no relation with increasing gall size (Schonrogge *et al.* 1996).

Attraction of predators to the gall itself may influence interactions within galls. Exclusion of ants on nectar-secreting galls increases parasitism of the gall-maker (Washburn 1984) and inquiline colonization of the gall (Seibert 1993, Fernandes *et al.* 1999) because ants normally protect galls against these natural enemies. Nectar-producing galls are quite common (Felt 1940); the impact that ants visiting these galls have on species that do not inhabit constructs is apparently untested.

Predators other than ants also may be affected. For example, leaf rolls on *Carica papaya* induced by feeding by herbivorous mites dramatically changed specialist and generalist predator abundance and distribution (Fournier *et al.* 2003). Specialist predators of the mites moved into leaf rolls seeking prey, whereas artificially rolled leaves without mites were not colonized (see also Martinsen *et al.* 2000).

Overall effects on community species richness

As a starting hypothesis, we propose that species richness of arthropod communities, one component of community structure, will be greatest at intermediate levels of engineering (Fig. 11.3). At low levels of colonization by constructors, there will be few constructs present to increase the plant-level species richness, due to colonization by secondary inhabitants that are specialists of constructs. At high levels of engineering, there will be little remaining habitat for species that

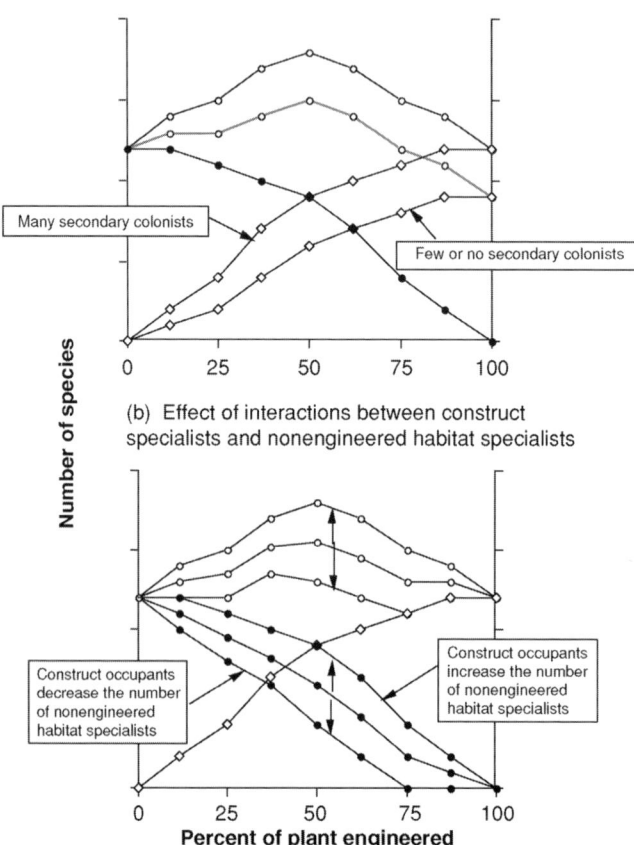

Figure 11.3. Hypothesized relationship between the abundance of the constructors on a plant, proportion of the plant that is engineered, and total species richness of the plant's arthropod fauna. Total species richness is hypothesized to peak at intermediate levels of plant engineering. (a) An increase in the number of secondary colonists of constructs will increase the total number of species on the plant. The secondary colonists include other herbivores, natural enemies, and detritivores. (b) Interactions between constructors and secondary colonizers of constructs, on the one hand, and specialists of nonengineered parts of the plant, on the other, can either increase or decrease the number of the latter group. Predators that make forays outside of constructs will have a negative effect on specialists of nonengineered parts of the plant, while mutualists that do the same will have a positive effect (see text for detailed discussion of each pathway).

are specialists of nonengineered habitat, effectively reducing species richness at the plant level. At intermediate levels, richness will be increased due to the presence of the constructs, and specialists of nonengineered habitat will still have sufficient habitat to be maintained on the plant.

The impact that an individual constructor has on species richness will depend not only on how much of the plant is already engineered (i.e., the number of constructors) but also on the identity of the new colonist. Constructors whose engineered habitats provide little or no resource for secondary inhabitants will have a smaller effect on species richness than those who build constructs that provide habitat to multiple species (Fig. 11.3a). Identity of the constructor also will be important because interactions between constructors and the secondary inhabitants of the constructs they build vary in specificity. For example, initial evidence suggests that the secondary inhabitants of galls are much more specialized to specific species of gall-maker than is true for secondary inhabitants of leaf constructs. Thus, we might predict that the addition of a novel species of gall-maker will increase plant-level arthropod species richness more than the addition of a unique species of leaf constructor.

The impact that addition of constructs has on species richness also will depend on interactions between secondary inhabitants of constructs (and ants attracted to nectar-secreting galls) and specialists of nonengineered habitat (Fig. 11.3b). Mutualistic interactions between predators in constructs and specialists of nonengineered habitats (e.g., Bluthgen and Fiedler 2002) and positive inductive effects on plant quality (e.g., induced susceptibility: Dickson and Whitham 1996) will augment the effects of increasing numbers of constructs on overall species richness. In contrast, predation on nonconstruct inhabitants by predators that make forays out from constructs (e.g., Nakamura and Ohgushi 2003, Offenburg et al. 2004, Romero and Vasconellos-Neto 2004) will reduce overall species richness at an even faster rate as the number of constructs increases.

Consequences for plant fitness

The impact on the plant of colonization by constructors also will depend upon all four pathways of interactions (Fig. 11.2). Total numbers of herbivores associated with the plant may be affected, as well as their identity, distribution within the plant, and quantity of tissues eaten, all of which can modify the impact on plant growth and reproduction. First, the influence of engineers on other herbivores and engineers through resource competition will contribute to the amount and type of tissue consumed. Second, as a consequence of the impact of construct building on species that secondarily inhabit those constructs, we would expect the impacts on the plant to extend well beyond those attributable to the engineers alone. Third, an indirect effect on the plant will result from interactions between secondary inhabitants of constructs and herbivores that feed on nonengineered plant parts (Fig. 11.2). This indirect effect on the plant will be positive if the interactions between herbivore species are negative, and negative if they are mutualistic. Finally, natural enemies that do not colonize constructs may influence damage to the plant by reducing attack by herbivores, including constructors, secondary colonizers of constructs, and free-feeding species. Impacts by natural enemies on the plant via herbivores constitute a top–down trophic cascade (this indirect interaction is not depicted in Fig. 11.2).

Impacts of these interactions on plant growth and reproduction will be a complex function of their effects on the numbers of herbivorous arthropods, both in engineered and nonengineered habitats, and the effects on the behavior and abundance of natural enemies. As the abundance of engineers increases to moderate levels, specialist herbivores of nonengineered habitat could increase (Dickson and Whitham 1996); as nonengineered habitat becomes scarce, however, the associated specialists are likely to decrease. At the highest levels of infestation by constructors, little or no habitat would be available to species specialized on nonengineered habitat, resulting in decreased total species richness (Fig. 11.3). Free-feeding species, in avoiding leaf constructs, may actually increase the amount of damage on leaves not involved in leaf constructs (Marquis et al. 2002). Such redistribution of damage among different parts of a plant can have a significant impact on plant fitness (Marquis 1992, 1996). Constructs on plants, when they lead to secondary colonization by herbivore species that would not otherwise colonize the plant (in the case of leaf ties: Cappuccino and Martin 1994, Martinsen et al. 2000, Lill and Marquis 2003; and stem-borers: Feller and Mathis 1997), will likely result in increased damage. For example, experimentally increased numbers of leaf ties on Q. alba resulted in greater herbivore numbers and species, increased damage, and reduced plant growth (R.J. Marquis and J.T. Lill unpublished data). When the constructor is a predator, overall leaf damage (Offenberg et al. 2004) and seed predation (e.g., Romero and Vasconcellos-Neto 2004) can be reduced, thus having a net positive effect on plant fitness.

Impacts of the presence of constructs on the abundance and kinds of herbivores will be modified by the impact on natural enemies. Ants in particular can reduce overall abundance of engineers (e.g., Woodman and Price 1992). Ants, when attracted to constructs either as a source of food or shelter, may influence the distribution and abundance of herbivores in nonengineered habitats (Nakamura and Ohgushi 2003). As a result, damage by herbivores specialized on nonengineered habitats could be reduced. Because predators of constructors can be attracted to and utilize the same constructs made by those constructors (Fournier et al. 2003), the potential impact of the constructor on plant fitness can be reduced. Studies measuring the impacts of engineers on plant fitness in the presence and absence of predators are needed.

In the case of leaf constructs, modification of plant architecture itself will likely affect plant growth aside from any effect on leaf area loss through herbivory. Superposition of entire leaves and parts of leaves in leaf ties, leaf folds, and leaf tents and modification of leaf orientation in leaf webs both can increase leaf self-shading and change angle of light interception. Both may reduce photosynthetic rate. The negative effects on plant growth by spiders that modify plant architecture (Clark and Clark 1985) serve as initial estimates of the effects of leaf constructs, but potentially underestimate these effects because spiders also will reduce herbivory through their consumption of herbivorous prey.

As frass builds up in constructs, particularly leaf constructs and stem cavities, there is the potential for microbial breakdown and uptake back into the plant. In a sense, such uptake would allow recovery of some of the resources lost through folivory. Carnivorous plants can absorb nutrients from insects and leaves can absorb nutrients from nitrogen-fixing epiphylls (Bentley 1987), both demonstrating that plants are capable of such nutrient uptake. Study of the role that fungi, bacteria, and detritivores play in such interactions is completely open for future study.

Future directions

Understanding the impacts of physical ecosystem engineers on the arthropod trophic structure of plants requires an understanding of the effects on herbivores, their natural enemies, and detritivores. Members of all three trophic levels use constructs made by engineers. The impact of constructors, as described here, is nontrophic and mediated through the host plant. However, in creating constructs, the engineers affect the trophic interactions of their associated community by way of the trophic dependency of the secondary inhabitants.

We predict that the impact of shoot-borers as ecosystem engineers will be relatively limited compared to that for gall-makers and leaf constructors. Leaf constructors, compared to gall-makers, will have a greater impact on species richness of a community per added construct of the same species of constructor. This will be true because leaf constructs are generally larger, and thus would house more individuals than galls. Leaf constructs are also a more open habitat than galls, and therefore could attract a wide variety of generalists unlike galls that may be colonized only by specialists (except for secondary occupants of "opened" galls). In contrast, adding more species of galls to a plant will have a greater effect on community diversity than adding more species of leaf constructors, because of the differences in specificity of secondary inhabitants of the two construct types. We predict greatest diversity at intermediate levels of engineering on plants (Fig. 11.3). These hypotheses await testing. There appear to be no studies of interactions between engineers of different types (e.g., between leaf construct makers and gall-makers). These interactions could be both negative (through resource pre-emption) and positive (leaves that are galled and deformed as a result may be more susceptible to attack by leaf-tying species: R.J. Marquis and J.T. Lill personal observations). Studies of the density effects of engineers are limited to gall-makers (e.g., Heard and Buchanan 1998).

Specific aspects of the host may influence these interactions. For example, inquiline diversity is greater in galls (particularly those made by Hymenoptera) of trees than those of shrubs and herbs (Sanver and Hawkins 2000), suggesting that interactions between inquilines and gall-makers will be greater on trees. This pattern may occur because trees support more galls, thus attracting more inquilines. In contrast, parasitism levels are reduced on trees compared to herbs for gallers and stem-borers (Hawkins 1994: Fig. 5.4). Leaf size also might affect these interactions, but at least for the leaf gall-maker *Pontania* spp. on *Salix lasiolepis*, parasitism is inconsistently related to leaf size (Clancy et al. 1993).

Initial evidence suggests that the response of predators to the presence of engineered habitat is extremely varied. Ants in particular are attracted to some constructs, affecting interactions within constructs (Seibert 1993, Fernandes et al. 1999) and outside of constructs (Nakamura and Ohgushi 2003). In cases such as these, it is not known whether constructs simply shift the distribution of these predators within the plant or increase their overall abundance. In turn, the effect of predators on the distribution of engineers and free-feeding herbivore species has not been investigated.

Impacts of these interactions on plant fitness are mostly unknown. It is clear that impacts will not be a simple function of the total number of herbivores. Carefully designed experiments are needed that manipulate plant quality and the presence of certain key species in order to test the relative importance of these factors for the interactions outlined in Fig. 11.2. An important challenge will be to discriminate between the effects of the construct itself and those of construct-induced changes (architecture, nutrient status, secondary metabolites, and secondary growth) in damaged plants.

The focus of the chapter has been on the nature of the biotic interactions, but it is important to recognize that these interactions occur in the context of the physical environment. Because the physical environment can directly influence the arthropods involved (through desiccation, for example) or indirectly, through changes in plant quality (Agrawal and Van Zandt 2003), future research efforts should consider studying these interactions under contrasting environmental conditions.

Acknowledgments

We thank Drs. Ohgushi, Craig, and Price for the invitation to write this chapter, and Luis Abdala, June Jeffries, Rebecca Forkner, Pete Van Zandt, and two anonymous reviewers for constructive criticism and encouragement. Research leading to this chapter was supported by the US Department of Agriculture Forest Service grant 99-35302-8017.

References

Abrahamson, W.G., M.D. Hunter, G. Melika, and P.W. Price. 2003. Cynipid gall-wasp communities correlate with oak chemistry. *Journal of Chemical Ecology* **29**:208–223.

Agrawal, A.A., and P.A. Van Zandt. 2003. Ecological play in the coevolutionary theater: genetic and environmental determinants of attack by a specialist weevil on milkweed. *Journal of Ecology* **91**:1049–1059.

Ananthakrishnan, T.N. (ed.) 1984a. *Biology of Gall Insects*. London: Edward Arnold.

Ananthakrishnan, T.N. 1984b. Adaptive strategies in cecidogenous insects, pp. 1–9 in T.N. Ananthakrishnan (ed.) *Biology of Gall Insects*. London: Edward Arnold.

Araujo, L.M., A.C.F. Lara, and G.W. Fernandes. 1995. Utilization of *Apion* sp. (Coleoptera, Apionidae) galls by an ant community in southeastern Brazil. *Tropical Zoology* **8**:319–324.

Atlegrim, O. 1989. Exclusion of birds from bilberry stands: impact on insect larval density and damage to the bilberry. *Oecologia* **79**:136–139.

Bailey, J.K., and T.G. Whitham. 2003. Interactions among elk, aspen, galling sawflies and insectivorous birds. *Oikos* **101**:127–134.

Bentley, B.L. 1987. Nitrogen fixation by epiphylls in a tropical rainforest. *Annals of the Missouri Botanical Garden* **74**:234–241.

Berenbaum, M.R. 1978. Toxicity of furanocoumarin to armyworms: a case of biosynthetic escape from insect herbivores. *Science* **201**:532–534.

Bluthgen, N., and K. Fiedler. 2002. Interactions between weaver ants *Oecophylla smaragdina*, homopterans, trees and lianas in an Australian rain forest canopy. *Journal of Animal Ecology* **71**:793–801.

Brown, J.L., S. Vargo, E.F. Connor, and M.S. Nuckols. 1997. Causes of vertical stratification in the density of *Cameraria hamadryadella*. *Ecological Entomology* **22**:16–25.

Cappuccino, N. 1993. Mutual use of leaf-shelters by lepidopteran larvae on paper birch. *Ecological Entomology* **18**:287–292.

Cappuccino, N., and M.-A. Martin. 1994. Eliminating early-season leaf-tiers of paper birch reduces abundance of mid-summer species. *Ecological Entomology* **19**:399–401.

Carroll, M.R., and W.H. Kearby. 1978. Microlepidopterous leaf tiers (Lepidoptera: Gelichioidea) in central Missouri. *Journal of the Kansas Entomological Society* **51**:457–471.

Carver, M., N. Bluethgen, J.F. Grimshaw, and G.A. Bellis. 2003. *Aphis clerodendri* Matsumura (Hemiptera: Aphididae), attendant ants (Hymenoptera: Formicidae) and associates on *Clerodendrum* (Verbenaceae) in Australia. *Australian Journal of Entomology* **42**:109–113.

Clancy, K.M., P.W. Price, and C.F. Sacchi. 1993. Is leaf size important for a leaf-galling sawfly (Hymenoptera: Tenthredinidae)? *Environmental Entomology* **22**:116–126.

Clark, D. B., and D. A. Clark. 1985. Seedling dynamics of a tropical tree: impacts of herbivory and meristem damage. *Ecology* **66**:1884–1892.

Costa, A. A., and E. M. Varanda. 2002. Building of leaf shelters by *Stenoma scitiorella* Walker (Lepidoptera: Elachistidae): manipulation of host plant quality? *Neotropical Entomology* **31**:537–540.

Craig, T. P., J. K. Itami, and P. W. Price. 1990. The window of vulnerability of a shoot-galling sawfly to attack by a parasitoid. *Ecology* **71**:1471–1482.

Crespi, B. J. 1992. Behavioral ecology of Australian gall thrips (Insecta, Thysanoptera). *Journal of Natural History* **26**:769–809.

Csóka, G. 1994. Variation in *Quercus robur* susceptibility to galling wasps (Hymenoptera: Cynipidae) linked to tree phenology, pp. 148–152 in P. W. Price, W. J. Mattson, and Y. N. Baranchov (eds.) *The Ecology and Evolution of Gall-Forming Insects*. St. Paul, MN: US Department of Agriculture Forest Service.

Damman, H. 1987. Leaf quality and enemy avoidance by the larvae of a pyralid moth. *Ecology* **68**:88–97.

de Souza, A. L. T., G. W. Fernandes, and J. E. C. Figueira. 2001. Host plant response and phenotypic plasticity of a galling weevil (*Collabismus clitellae*: Curculionidae). *Austral Ecology* **26**:173–178.

Dickson, L. L., and T. G. Whitham. 1996. Genetically based plant resistance traits affect arthropods, fungi, and birds. *Oecologia* **106**:400–406.

Eliason, E. A., and D. A. Potter. 2000. Biology of *Callirhytis cornigera* (Hymenoptera: Cynipidae) and the arthropod community inhabiting its galls. *Environmental Entomology* **29**:551–559.

Eliason, E. A., and D. A. Potter. 2001. Spatial distribution and parasitism of leaf galls induced by *Callirhytis cornigera* (Hymenoptera: Cynipidae) on pin oak. *Environmental Entomology* **30**:280–287.

Evans, T. A. 1997. Distribution of social crab spiders in eucalypt forests. *Australian Journal of Ecology* **22**:107–111.

Feller, I. C., and W. N. Mathis. 1997. Primary herbivory by wood-boring insects along an architectural gradient of *Rhizophora mangle*. *Biotropica* **29**:440–451.

Felt, E. P. 1940. *Plant Galls and Gall Makers*. London: Hafner.

Fernandes, G. W., M. Fagundes, R. L. Woodman, and P. W. Price. 1999. Ant effects on three-trophic level interactions: plant, galls, and parasitoids. *Ecological Entomology* **24**:411–415.

Fitzgerald, T. D., and K. L. Clark. 1994. Analysis of leaf-rolling behavior of *Caloptilia serotinella* (Lepidoptera: Gracillaridae). *Journal of Insect Behavior* **7**:859–872.

Fitzgerald, T. D., K. L. Clark, R. Vanderpool, and C. Phillips. 1991. Leaf shelter-building caterpillars harness forces generated by axial retraction of stretched and wetted silk. *Journal of Insect Behavior* **4**:21–32.

Foss, L. K., and L. K. Rieske. 2004. Stem galls affect oak foliage with potential consequences for herbivory. *Ecological Entomology* **29**:273–280.

Fournier, V., J. A. Rosenheim, J. Brodeur, L. O. Laney, and M. W. Johnson. 2003. Herbivorous mites as ecological engineers: indirect effects on arthropods inhabiting papaya foliage. *Oecologia* **135**:442–450.

Fowler, S. V., and M. MacGarvin. 1985. The impact of hairy wood ants, *Formica lugubris*, on the guild structure of herbivorous insects on birch, *Betula pubescens*. *Journal of Animal Ecology* **54**:847–856.

Fukui, A. 2001. Indirect interactions mediated by leaf shelters in animal–plant communities. *Population Ecology* **43**:31–40.

Fukui, A., M. Murakami, K. Konno, M. Nakamura, and T. Ohgushi. 2002. A leaf-rolling caterpillar improves leaf quality. *Entomological Science* **5**:263–266.

Gaston, K. J., D. Reavey, and G. R. Valladares. 1991. Changes in feeding habit as caterpillars grow. *Ecological Entomology* **16**:339–344.

Goldberg, D. E. 1990. Components of resource competition in plant communities, pp. 27–49 in J. B. Grace and D. Tilman (eds.) *Perspective on Plant Competition*. San Diego, CA: Academic Press.

Green, E. S., A. R. Zangerl, and M. R. Berenbaum. 1998. Reduced aggressive behavior: a benefit of silk-spinning in the parsnip webworm, *Depressaria pastinacella* (Lepidoptera: Oecophoridae). *Journal of Insect Behavior* **11**:761–772.

Hawkins, B. A. 1994. *Pattern and Process in Host–Parasitoid Interactions*. Cambridge, UK: Cambridge University Press.

Heads, P. A., and J. H. Lawton. 1985. Bracken, ants and extrafloral nectaries. III. How insect herbivores avoid ant predation. *Ecological Entomology* **10**:29–42.

Heard, S. B., and C. K. Buchanan. 1998. Larval performance and association within and between two species of hackberry nipple gall insects, *Pachypsylla* spp. (Homoptera: Psyllidae). *American Midland Naturalist* **140**:351–357.

Hodkinson, I. D. 1984. The biology and ecology of the gall-forming Psylloidea (Homoptera), pp. 59–77 in T. N. Ananthakrishnan (ed.) *Biology of Gall Insects*. London: Edward Arnold.

Hölldobler, B. 1983. Territorial behavior in the green tree ant (*Oecophylla smaragdina*). *Biotropica* **15**:241–250.

Hölldobler, B., and E. O. Wilson. 1990. *The Ants*. Cambridge, MA: Harvard University Press.

Hudson, E. E., and P. Stiling. 1997. Exploitative competition strongly affects the herbivorous insect community on *Baccharis halimifolia*. *Oikos* **79**:521–528.

Hunter, M. D. 1987. Opposing effects of spring defoliation on late season oak caterpillars. *Ecological Entomology* **12**:373–382.

Hunter, M. D. 1992. A variable insect–plant interaction: the relationship between tree budburst phenology and population levels of insect herbivores among trees. *Ecological Entomology* **16**:91–95.

Hunter, M. D., and P. G. Wilmer. 1989. The potential for interspecific competition between two abundant defoliators on oak: leaf damage and habitat quality. *Ecological Entomology* **14**:267–277.

Johnson, S. N., P. J. Mayhew, A. E. Douglas, and S. E. Hartley. 2002. Insects as leaf engineers: can leaf-miners alter leaf structure for birch aphids? *Functional Ecology* **16**:575–584.

Jones, C. G., J. H. Lawton, and M. Shachak. 1994. Organisms as ecosystem engineers. *Oikos* **69**:373–386.

Jones, C. G., J. H. Lawton, and M. Shachak. 1997. Positive and negative effects of organisms as physical ecosystem engineers. *Ecology* **78**:1946–1957.

Jones, M.T., I. Castellanos, and M.R. Weiss. 2002. Do leaf shelters always protect caterpillars from invertebrate predators? *Ecological Entomology* **27**:753–757.

Kagata, H., and T. Ohgushi. 2004. Leaf miner as a physical ecosystem engineer: secondary use of vacant leaf mines by other arthropods. *Annals of the Entomological Society of America* **97**:923–927.

Kaitaniemi, P., H. Vehvilainen, and K. Ruohomaki. 2004. Movement and disappearance of mountain birch defoliators are influenced by the interactive effects of plant architecture and induced resistance. *Ecological Entomology* **29**:437–446.

Kampichler, C., and M. Teschner. 2002. The spatial distribution of leaf galls of *Mikiola fagi* (Diptera: Cecidomyiidae) and *Neuroterus quercusbaccarum* (Hymenoptera: Cynipidae) in the canopy of Central European mixed forest. *European Journal of Entomology* **99**:79–84.

Karban, R., and I.T. Baldwin. 1997. *Induced Responses to Herbivory*. Chicago, IL: University of Chicago Press.

Kopelke, J.P. 2003. Natural enemies of gall-forming sawflies on willows (*Salix* spp.) (Hymenoptera: Tenthredinidae: *Euura, Phyllocolpa, Pontania*). *Entomologia Generalis* **26**:277–312.

Larson, K.C., and T.G. Whitham. 1997. Competition between gall aphids and natural plant sinks: plant architecture affects resistance to galling. *Oecologia* **109**:575–582.

Larson, K.A., and D.M. Harman. 2003. Subcortical cavity dimension and inquilines of the larval locust borer (Coleoptera: Cerambycidae). *Proceedings of the Entomological Society of Washington* **105**:108–119.

Larsson, S., H.E. Haggstrom, and R.F. Denno. 1997. Preference for protected feeding sites by larvae of the willow-feeding leaf beetle *Galerucella lineola*. *Ecological Entomology* **22**:445–452.

Le Corff, J., R.J. Marquis, and J.B. Whitfield. 2000. Temporal and spatial variation in a parasitoid community associated with the herbivores that feed on Missouri *Quercus*. *Environmental Entomology* **29**:181–194.

Letourneau, D.K. 1998. Ants, stem borers, and fungal pathogens: experimental tests of fitness advantage in Piper ant plants. *Ecology* **79**:593–603.

Letourneau, D.K., and P. Barbosa. 1999. Ants, stem borers, and pubescence in *Endospermum* in Papua New Guinea. *Biotropica* **31**:295–302.

Lewis, A.C. 1979. Feeding preferences for diseased and wilted sunflower in the grasshopper, *Melanoplus differentialis*. *Entomologica Experimentalis et Applicata* **26**: 202–207.

Lill, J.T. 1999. Structure and dynamics of a parasitoid community attacking larvae of *Psilocorsis quercicella* (Lepidoptera: Oecophoridae). *Environmental Entomology* **28**:1114–1123.

Lill, J.T. 2004. Seasonal dynamics of leaf-tying caterpillars on white oak. *Journal of the Lepidopterists' Society* **58**:1–6.

Lill, J.T., and R.J. Marquis. 2003. Ecosystem engineering by caterpillars increases insect herbivore density on white oak. *Ecology* **84**:682–690.

Lill, J.T., and R.J. Marquis. 2004. Leaf ties as colonization sites for forest arthropods. *Ecological Entomology* **29**:300–308.

Loeffler, C. C. 1996. Adaptive trade-offs of leaf folding in *Dichomeris* caterpillars on goldenrods. *Ecological Entomology* **21**:34–40.

Lombardero, M. J., M. P. Ayres, R. W. Hofstetter, J. C. Moser, and K. D. Lepzig. 2003. Strong indirect interactions of *Tarsonemus* mites (Acarina: Tarsonemidae) and *Dendroctonus frontalis* (Coleoptera: Scolytidae). *Oikos* **102**:243–252.

Louda, S. M. 1982. Inflorescence spiders: a cost/benefit analysis for the host plant, *Haplopappus venetus* Blake (Asteraceae). *Oecologia* **55**:185–191.

Mapes, C. C., and P. J. Davies. 2001a. Indole-3-acetic acid and ball gall development on *Solidago altissima*. *New Phytologist* **151**:195–202.

Mapes, C. C., and P. J. Davies. 2001b. Cytokinins in the ball gall of *Solidago altissima* and in the gall forming larvae of *Euostoma solidaginis*. *New Phytologist* **151**:203–212.

Marquis, R. J. 1992. Selective impact of herbivores, pp. 301–325 in R. S. Fritz and E. L. Simms (eds.) *Plant Resistance to Herbivores and Pathogens*. Chicago, IL: University of Chicago Press.

Marquis, R. J. 1996. Plant architecture, sectoriality, and tolerance to herbivory. *Vegetatio* **127**:85–97.

Marquis, R. J., and C. J. Whelan. 1996. Plant morphology and recruitment of the third trophic level: subtle and little-recognized defenses? *Oikos* **75**:330–334.

Marquis, R. J., J. T. Lill, and A. Piccini. 2002. Effect of plant architecture on colonization and damage by leaf-tying caterpillars of *Quercus alba*. *Oikos* **99**:531–537.

Martinsen, G. D., K. D. Floate, A. M. Waltz, G. M. Wimp, and T. G. Whitham. 2000. Positive interactions between leafrollers and other arthropods enhance biodiversity on hybrid cottonwoods. *Oecologia* **123**:82–89.

Miller, W. E. 2004. Host breadth and voltinism in gall-inducing Lepidoptera. *Journal of the Lepidopterists' Society* **58**:44–47.

Mopper, S., and D. Simberloff. 1995. Differential herbivory in an oak population: the role of plant phenology and insect performance. *Ecology* **76**:1233–1241.

Morin, P. J. 1999. *Community Ecology*. Oxford, UK: Blackwell Science.

Morris, R. F. 1972. Predation by insects and spiders inhabiting colonial webs of *Hyphantria cunea*. *Canadian Entomologist* **104**:1197–1207.

Morse, D. H. 1992. Predation on dispersing *Misumena vattia* spiderlings and its relationship to maternal foraging decisions. *Ecology* **73**:1814–1819.

Mound, L. A., and D. C. Morris. 2000. Inquilines or kleptoparasites? New phlaeothripine Thysanoptera associated with domicile-building thrips on *Acacia* trees. *Australian Journal of Entomology* **39**:130–137.

Murakami, M. 1999. Effect of avian predation on survival of leaf-rolling lepidopterous larvae. *Researches on Population Ecology* **41**:135–138.

Nakamura, M., and T. Ohgushi. 2003. Positive and negative effects of leaf shelters on herbivorous insects: linking multiple herbivore species on a willow. *Oecologia* **136**:445–449.

Nakamura, M., Y. Miyamoto, and T. Ohgushi. 2003. Gall initiation enhances the availability of food resources for herbivorous insects. *Functional Ecology* **17**:851–857.

Offenberg, J., S. Havanon, S. Aksornkoae, D. J. MacIntosh, and M. G. Nielsen. 2004. Observations on the ecology of weaver ants (*Oecophylla smaragdina* Fabricius) in a

Thai mangrove ecosystem and their effect on herbivory of *Rhizophora mucronata* Lam. *Biotropica* **36**:344–351.

Oki, Y. 2000. Herbivoria por lepidópteros em *Byrsonima intermedia* Juss. (Malpighiaceae) na ARIE Pé-de-Gigante, Santa Rita do Passa Quatro, SP. M.S. thesis, University of São Paulo, Ribeirão Preto, Brazil.

Ollerstam, O., and S. Larsson. 2003. Salicylic acid mediates resistance in the willow *Salix viminalis* against the gall midge *Dasineura marginemtorquens*. *Journal of Chemical Ecology* **29**:163–174.

Ott, J. R., J. A. Nelson, and T. Caillouet. 1998. The effect of spider-mediated flower alteration on seed production in golden-eye phlox. *Southwestern Naturalist* **43**:430–436.

Pilson, D. 1992. Aphid distribution and the evolution of goldenrod resistance. *Evolution* **46**:1358–1372.

Plantard, O., and M. E. Hochberg. 1998. Factors affecting parasitism in the oak-galler *Neuroterus quercusbaccarum* (Hymenoptera: Cynipidae). *Oikos* **81**:289–298.

Price, P. W. 1991. The plant vigor hypothesis and herbivore attack. *Oikos* **62**:244–251.

Price, P. W. 2003. *Macroevolutionary Theory on Macroecological Patterns*. Cambridge, UK: Cambridge University Press.

Price, P. W., and S. Louw. 1996. Resource manipulation through architectural modification of the host plant by a gall-forming weevil *Urodontus scholtzi* Louw (Coleoptera: Anthribidae). *African Entomology* **4**:103–110.

Price, P. W., N. Cobb, T. P. Craig, et al. 1990. Insect herbivore population dynamics on trees and shrubs: new approaches relevant to latent and eruptive species and life table development, pp. 1–38 in E. A. Bernays (ed.) *Insect–Plant Interactions*. Boca Raton, FL: CRC Press.

Price, P. W., G. W. Fernandes, A. C. F. Lara, et al. 1998. Global patterns in local number of insect galling species. *Journal of Biogeography* **25**:581–591.

Rathcke, R. J. 1976. Competition and coexistence within a guild of herbivorous insects. *Ecology* **57**:76–87.

Redfern, M., and R. Cameron. 1994. Risk of parasitism on *Taxomyia taxi* (Diptera: Cecidomyiidae) in relation to the size of its galls on yew, *Taxus baccata*, pp. 213–230 in M. A. J. Williams (ed.) *Plant Galls*. Oxford, UK: Clarendon Press.

Rehill, B. J., and J. C. Schultz. 2001. *Hormaphis hamamelidis* and gall size: a test of the plant vigor hypothesis. *Oikos* **95**:94–104.

Riihimaki, J., P. Kaitaniemi, and K. Ruohomaki. 2003. Spatial responses of two herbivore groups to a geometrid larva on mountain birch. *Oecologia* **134**:203–209.

Roininen, H., and K. Danell. 1997. Mortality factors and resource use of the bud-galling sawfly, *Euura mucronata* (Hartig), on willows (*Salix* spp.) in arctic Eurasia. *Polar Biology* **18**:325–330.

Roininen, H., P. W. Price, and J. Tahvanainen. 1996. Bottom-up and top-down influences in the trophic system of a willow, a galling sawfly, parasitoids and inquilines. *Oikos* **77**:44–50.

Romero, G. Q., and J. Vasconcellos-Neto. 2004. Beneficial effects of flower-dwelling predators on their host plant. *Ecology* **85**:446–457.

Russell, L. M., and M. B. Stoetzel. 1991. Inquilines in egg nests of periodical cicadas (Homoptera: Cicadidae). *Proceedings of the Entomological Society of Washington* **93**:480–488.

Sagers, C. L. 1992. Manipulation of host plant quality: herbivores keep leaves in the dark. *Functional Ecology* **6**:741–743.

Sandberg, S. L., and M. R. Berenbaum. 1989. Leaf-tying by tortricid larvae as an adaptation for feeding on phototoxic *Hypericum perforatum*. *Journal of Chemical Ecology* **15**:875–885.

Sanver, D., and B. A. Hawkins. 2000. Galls as habitats: the inquiline communities on insect galls. *Basic and Applied Ecology* **1**:3–11.

Schonrogge, K., G. N. Stone, and M. J. Crawley. 1996. Abundance patterns and species richness of the parasitoids and inquilines of the alien gall-former *Andricus quercuscalicis* (Hymenoptera: Cynipidae). *Oikos* **77**:507–518.

Seibert, T. F. 1993. A nectar-secreting gall wasp and ant mutualism: selection and counter-selection shaping gall wasp phenology, fecundity and persistence. *Ecological Entomology* **18**:247–253.

Seyffarth, J. A. S., A. M. Colouro, and P. W. Price. 1996. Leaf rollers in *Ouratea hexasperma* (Ocnaceae): fire effect and plant vigor hypothesis. *Revista Brasileira de Biologia* **56**:135–137.

Shorthouse, J. D. 1994. Host shift of the leaf galler *Diplolepis polita* (Hymenoptera: Cynipidae) to the domestic shrub rose *Rosa rugosa*. *Canadian Entomologist* **126**:1499–1503.

Sopow, S. L., J. D. Shorthouse, W. Strong, and D. T. Quiring. 2003. Evidence for long-distance, chemical gall induction by an insect. *Ecology Letters* **6**:102–105.

Sugiura, S., K. Yamazaki, and Y. Fukasawa. 2004. Weevil parasitism of *Ambrosia* galls. *Annals of the Entomological Society of America* **97**:184–193.

Thakur, S. S., N. P. Kashyap, and P. K. Mehta. 1996. New record of a stem boring weevil, *Cypricerus emarginatus* Fst. (Curculionidae: Coleoptera) on rajmash in Himachal Pradesh. *Journal of Insect Science* **9**:183.

van Hezewijk, B. H., and J. Roland. 2003. Gall size determines the structure of the *Rhabdophaga strobiloides* host–parasitoid community. *Ecological Entomology* **28**:593–603.

Washburn, J. O. 1984. Mutualism between a cynipid gall wasp and ants. *Ecology* **65**:654–656.

Weis, A. E., W. G. Abrahamson, and K. D. McCrea. 1985. Host gall size and oviposition success by the parasitoid *Eurytoma gigantea*. *Ecological Entomology* **10**:341–348.

Weiss, M. R. 2003. Good housekeeping: why do shelter-dwelling caterpillars fling their frass? *Ecology Letters* **6**:361–370.

Whitham, T. G. 1983. Host manipulation of parasites: within-plant variation as a defense against rapidly evolving pests, pp. 15–41 in R. F. Denno and M. S. McClure (eds.) *Variable Plants and Herbivores in Natural and Managed Systems*. New York: Academic Press.

Wilson, D. 1995. Fungal endophytes which invade insect galls: insect pathogens, benign saprophytes, or fungal inquilines. *Oecologia* **103**:255–260.

Woodman, R. L., and P. W. Price. 1992. Differential larval predation by ants can influence willow sawfly community structure. *Ecology* **73**:1028–1037.

Young, G. R. 2003. Life history, biology, host plants and natural enemies of the psyllid *Trioza eugeniae* Froggatt (Hemiptera: Triozidae). *Australian Entomologist* **30**:31–38.

Young, T. P., C. H. Stubblefield, and L. A. Isbell. 1997. Ants on swollen-thorn acacias: species coexistence in a simple system. *Oecologia* **109**:98–107.

12

Host plants mediate aphid–ant mutualisms and their effects on community structure and diversity

GINA M. WIMP AND THOMAS G. WHITHAM

Introduction

Much of the emphasis in studying mutualisms has been placed on defining the strength of these associations and the conditions that cause their collapse (Bronstein 1994, 1998). Yet, very few studies of aphid–ant mutualisms have linked the importance of host plant traits with the establishment and persistence of these mutualisms. Aphid performance, as well as the quality and quantity of their honeydew, may be affected by differences in host plant genetics or through environmentally induced effects on host plant quality. Differences among host plants that influence the attractiveness of aphids to tending ants can therefore alter the nature and strength of this association. The importance of host plants in determining the establishment and persistence of aphid–ant mutualisms could have consequences for biodiversity if these aphid–ant mutualisms play an important role in the structure and diversity of ecological communities.

The idea that mutualisms are important components of ecological communities arose 130 years ago (van Beneden 1875, French paper cited in Boucher 1985). However, much of the theoretical and empirical work in ecology for the past 70 years has supported the view that antagonistic interactions among species are more important than positive interactions in determining community organization. Yet, empirical data on an array of different mutualisms has shown that they can be important to community structure and diversity. For example, mycorrhizal mutualists have been shown to affect plant diversity (Grime *et al.* 1987, van

---

Ecological Communities: Plant Mediation in Indirect Interaction Webs, ed. Takayuki Ohgushi, Timothy P. Craig, and Peter W. Price. Published by Cambridge University Press. © Cambridge University Press 2007.

der Heijden *et al.* 1998), plant succession (Gange *et al.* 1990), plant competitive interactions (Moora and Zobel 1996), and interactions with herbivores (Rabin and Pacovsky 1985, Gange and West 1993, Gehring *et al.* 1997). Foliar endophytic fungi can also alter plant competitive relationships (Marks *et al.* 1991, Clay and Holah 1999). Likewise, ants that engage in ant–plant mutualisms can alter community structure by reducing populations of chewing herbivores and vines, while increasing populations of hemipterans and cecidomyiid flies on trees where they are present (Schupp 1986, Fiala 1994). The number of species that are negatively affected by ants attracted to extrafloral nectaries seems to outweigh the number of species that are positively affected, because trees with extrafloral nectaries tended by ants support lower diversity in the surrounding arthropod community (Rudgers and Gardener 2004). Mutualisms can therefore affect many different facets of the biological world and it is important to understand the role of host plants in mediating these interactions.

In this chapter we will focus on the mutualism between aphids (Hemiptera: Aphididae) and ants (Hymenoptera: Formicidae). In this mutualism, aphids feed on plant phloem and excrete sugar-rich honeydew that is a source of carbohydrate for ants (Way 1963). In return, the ants protect aphids from predators (Tilles and Wood 1982), and also create a competitive advantage for aphids by removing other herbivores (Seibert 1992, Wimp and Whitham 2001). Protection of aphids by ants can be essential to aphid survival because their ability to protect themselves from predators is limited, and predators can decimate aphid populations (Way 1963, Bradley and Hinks 1968, Sanders and Knight 1968, Addicott 1978, 1979, Buckley 1987, Müller and Godfray 1999, Wimp and Whitham 2001). Although top–down effects from predators can be strong in these mutualisms, bottom–up factors from host plants could also contribute to aphid establishment and survival by affecting honeydew, and thereby affecting aphids' mutualism with ants.

Although terrestrial plants play a prominent role in mediating food web interactions and represent a major factor structuring biodiversity in ecological communities (Price 2002), their role in mediating aphid–ant interactions has often been under-appreciated. The effects of host plant resources on aphid performance have been well documented in the agricultural literature, but very few studies in the wild have shown how host plants affect aphid performance and subsequent tending by ants. Because host plants can alter both the quality and quantity of aphid honeydew (Bristow 1991, Völkl *et al.* 1999), they can influence when and where aphid–ant mutualisms will form. If aphid–ant mutualisms influence the structure of the surrounding arthropod community, host plant factors that determine the establishment of the mutualism on a local scale (i.e., an individual plant scale) then have landscape-level implications for arthropod biodiversity on a larger scale (i.e., when we consider all of the plants in a given area). Here, we ask two primary questions: "What are the different host plant factors that affect aphid–ant mutualisms?" and "What role do aphid–ant mutualisms play in structuring ecological communities?"

To address these questions, we have organized this chapter into four major sections that emphasize the effects of host plants on aphid–ant mutualisms and community interactions in natural systems. First, we examine how host plants may influence aphid performance and honeydew output, and how this influence leads to differences in tending by ants. Second, we examine how the presence of the aphid–ant mutualism affects stability (i.e. species turnover rates) in the surrounding arthropod community. Third, we examine whether aphid–ant mutualisms affect the structure of the surrounding arthropod community on a local scale (i.e., individual plant scale). Our fourth section extends findings at a local scale to examine the effects of aphid–ant mutualisms on the surrounding arthropod community at a landscape-level scale. We then develop the concept of a mutualism mosaic in which the greatest biodiversity is achieved in a landscape interlaced with mutualism and mutualism-free space.

Host plant traits affect aphid–ant mutualisms

Because of their importance as agricultural pests, numerous studies in the agricultural literature have documented aphid performance on different host plant genotypes (e.g., Painter 1951) in an attempt to discover and breed resistant genotypes. Many of the plant species that aphids feed on in natural systems are not of economic importance, and far less intensity of effort has been placed on determining the effects of host plant traits on aphids in the wild. Such studies are important because host plant traits that affect aphid fecundity are likely to also affect their mutualism with ants. Host plants can affect aphids via differences in host plant quality or through a suite of host plant traits that vary among different host plant species, within hybridizing systems, or within a single plant genotype. Genetic differences among host plants affect phenology (Tsarouhas et al. 2003), the production of physical barriers such as trichomes (Kumar 1992), and the production of secondary plant defenses (Dungey et al. 2000, Orians 2000, Osier and Lindroth 2001), all of which can subsequently affect aphid performance. Tables 12.1 and 12.2 list a number of different host plant traits (differences in: metabolic pathway, quality on different soil types, foliar nitrogen content, vigor, degree of branching, endophyte infection, host plant genotype, host plant species, host plant cross type, and condensed tannins), and their effects on aphid performance in natural systems. These traits have been divided into environmentally induced host plant traits that affect

Table 12.1. *The effects of host plants on aphid performance: environmental effects*

Aphid and host plant	Host plant parameter measured	Effects on aphid performance[a]	Aphid performance parameter measured	Reference
Uroleucon jaceae on *Centaurea jacea*	Quality on two different soil types	+	Number and size class of embryos	Stadler 1995
Four aphid species on *Tanacetum vulgare*	Quality on different soil types	+	Aphid density	Stadler 2004
Schizolachnus pineti on *Pinus sylvestris*	Foliar nitrogen content	+	Relative growth rate	Kainulainen et al. 1996
Capitophorus hippophaes on *Polygonum pennsylvanicum*	Percent leaf nitrogen	–	Aphid density	Mabry et al. 1997
Uroleucon ambrosiae on *Iva frutescens*	Quality via nitrogen addition	+	Aphid density	Levine et al. 1998
Cinara cupressi on two species of *Cupressus*	Quality via nutrient addition	O	Development time, reproduction, and survival	Kairo and Murphy 1999
Four aphid species on *Tanacetum vulgare*	Quality via nutrient addition	+	Maximum daily fecundity	Stadler et al. 2002
Elatobium abietinum on two *Picea* species	Quality via nutrient addition	+	Aphid population growth, aphid density	Straw and Green 2001
Aphis varians on *Epilobium angustifolium*	Quality via plant height	+	Aphid population growth	Breton and Addicott 1992
Euceraphis betulae on *Betula pendula*	Host plant vigor	+/–	Aphid density	Johnson et al. 2003a
Sitobion avenae on 13 grass species	Relative growth rate, presence of hairs	+/O	Aphid density	Fraser and Grime 1999

Species	Trait	Effect	Response	Reference
Two *Uroleucon* species on *Solidago altissima*	Degree of branching	+	Aphid density	Pilson and Rausher 1995
Aphis varians and *Macrosiphum valeriani* on *Epilobium angustifolium*	Shoot height	+	Preferential colonization	Antolin and Addicott 1991
Symydobius oblongus on *Betula pendula* and *B. pubescens*	Quality following moose herbivory	O	Aphid density	Danell and Huss-Danell 1985
Uroleucon tissoti on *Solidago altissima*	Quality following insect herbivory	+	Aphid density and distribution	Pilson 1992
Rhopalosiphum padi on *Prunus padus*	Quality following artificial defoliation	–	Aphid colonization	Leather 1993
Aphis farinosa on *Salix eriocarpa*	Quality following insect gall colonization	+	Aphid colonization	Nakamura et al. 2003
Euceraphis betulae on *Betula pendula*	Quality following insect leaf mine colonization	–	Aphid mortality	Johnson et al. 2002
Chaitophorus saliniger on seven species of *Salix*	Quality following insect leaf mine colonization	O/–	Aphid colonization, aphid colony size	Kagata and Ohgushi 2004
Euceraphis betulae on *Betula pendula*	Quality via fungal endophyte infection	+	Aphid density, colonization, weight, tibial length, embryo development, population growth	Johnson et al. 2003b
Drepanosiphum platanoidis and *Periphyllus acericola* on *Acer pseudoplatanus*	Quality via fungal endophyte infection	+	Aphid density, adult weight, and reproductive potential	Gange 1996

Table 12.1. (cont.)

Aphid and host plant	Host plant parameter measured	Effects on aphid performance[a]	Aphid performance parameter measured	Reference
Myzus ascalonicus and *M. persicae* on *Plantago lanceolata*	Quality via fungal endophyte infection	+/O	Aphid weight, fecundity, nymphal development time, growth rate, longevity, and duration of reproductive life	Gange et al. 1999
Rhopalosiphum padi and *Metopolophium festucae* on *Lolium multiflorum*	Quality via fungal endophyte infection	+/O	Aphid density	Omacini et al. 2001

Note:
[a] Effects were categorized as: positive (+), negative (−), showing mixed positive and negative responses (+/−), or showing no response (O).

Table 12.2. *The effects of host plants on aphid performance: genetic effects*

Aphid and host plant	Host plant parameter measured	Effects on aphid performance[a]	Aphid performance parameter measured	Reference
Seventeen aphid species on 120 plant species	Host plant species	A/O	Adult weight in 88% of alate morphs (+), adult weight in 77% of apterous morphs (+), embryo content in 63% of alate adults (+), and embryo content in 75% of apterous morphs (+)	Llewellyn and Brown 1985
Uroleucon gravicorne on three plant species	Host plant species	A/O	Reproductive performance	Moran 1983
Pterocallis alni on two *Alnus* species and their hybrids	Host plant cross type	A	Aphid density	Gange 1995
Rhopalosiphum padi on 89 species of grasses	Grasses with C_3 versus C_4 metabolic pathways	A	Aphid preference	Weibull 1990
Chaitophorus populicola on two *Populus* species and their hybrids	Host plant cross type	A	Aphid fecundity	Wimp and Whitham 2001
Capitophorus hippophaes on *Polygonum pennsylvanicum*	Host plant genotype	O	Aphid density	Mabry et al. 1997
Elatobium abietinum on two *Picea* species	Host plant genotype	A	Aphid population growth, aphid density	Straw and Green 2001
Aphis craccivora on *Arachis hypogaea*	Condensed tannins on different genotypes	A	Aphid fecundity	Grayer et al. 1992

Table 12.2. (cont.)

Aphid and host plant	Host plant parameter measured	Effects on aphid performance[a]	Aphid performance parameter measured	Reference
Two *Uroleucon* species on *Solidago altissima*	Host plant genotype	A	Aphid density	Maddox and Root 1987
Uroleucon caligatum on three *Solidago* species	Host plant genotype	A	Development time, size of first nymphal instar, adult weight, and total colony weight	Moran 1981
Uroleucon tissoti on *Solidago altissima*	Host plant genotype	A	Development time, colony growth rate, and total colony weight	Pilson 1992
Uroleucon redbeckiae on *Rudbeckia laciniata*	Host plant genotype	A	Aphid fitness	Service 1984

Note:
[a] Aphid performance was classified as either being affected (A) or unaffected (O) by differences in host plant genetics.

aphid performance (Table 12.1), and host plant genetic factors affecting aphid performance (Table 12.2). It should be noted that some of these plant "traits" could encompass both environmental and genetic factors (e.g., plant genotype could affect foliar nitrogen content and architecture). Overall, we found that in 26 out of 32 cases (81%) host plant attributes affected some measure of aphid performance, while aphid performance was unaffected in 6 out of 32 cases (19%).

Environmentally induced differences in host plant nutrients, either via vigor, availability of nitrogen resources, quality of the soil, or indirect effects of fungi and herbivores on host plants all affected aphid fecundity (Table 12.1). When we specifically looked at environmental effects on host plant quality, we found that aphid performance was positively affected in 17 out of 28 cases (61%), negatively affected in 5 out of 28 cases (18%), and unaffected in 6 out of 28 cases (21%). Many of these studies considered differences in host plant nutrients that affected aphid performance, but fewer studies have examined herbivore-induced changes in host plant quality that affected aphids.

Herbivore-induced changes in host plant quality that affect aphid abundance, fecundity, and/or survival have been demonstrated in both agricultural systems and in the wild and may be manifested through changes in host plant growth and defensive chemistry. Removal of leaf tissue by herbivores has been shown to affect aphid colonization via changes in plant architecture that make plants less suitable for oviposition (Leather 1993). When herbivores remove the apical meristems of plants, aphid responses to plant regrowth are generally positive (Danell and Hus-Danell 1985, Pilson 1992), though not always significantly so (Danell and Hus-Danell 1985). The response of aphids to previous herbivore feeding is not solely due to loss of leaf material and subsequent regrowth, but may also be through the release of plant volatiles. In agricultural systems, the application of jasmonic acid to simulate herbivore feeding has been shown to negatively affect aphid abundance, preference, reproduction, and survival (Omer et al. 2001, Thaler et al. 2001, Thaler 2002) or have no effect on aphids (Thaler et al. 2001, Black et al. 2003). Aphids can also affect one another's performance via alterations in their shared phloem resource. Greenhouse experiments have shown that previous feeding by conspecific aphids negatively affects aphid species that feed later in the season, but this only occurs when the aphid species that feeds first on the host plant alters the amino acid content of the phloem (Peterson and Sandström 2001). While the previous examples relate to free-living herbivores, leaf- and stem-modifying herbivores may also affect aphid performance.

Leaf- and stem-modifying herbivores may positively affect aphids through the creation of favorable microhabitats with enhanced nutrient quality (Sandberg and Berenbaum 1989, Sagers 1992). Gall midges have been shown to positively affect aphid colonization by stimulating the production of lateral shoots which are preferred by aphids due to their high nitrogen and water content (Nakamura et al. 2003). Aphids also directly benefit from the structures created by leaf-modifying herbivores. On willow plants, lepidopterans that built leaf shelters had a positive effect on aphid numbers, as well as a positive effect on their ant mutualists (Nakamura and Ohgushi 2003). While aphid response to galling and shelter-building herbivores is generally positive, the presence of leaf miners negatively affects aphid colony size (Kagata and Ohgushi 2004) and survivorship (Johnson et al. 2002). Leaf mines may either be too small to accommodate sufficient numbers of aphids (Kagata and Ohgushi 2004), or mines may sever leaf veins making phloem inaccessible to aphids (Johnson et al. 2002).

Alterations to host plant quality may also be indirect via symbiotic associations between the host plant and fungal species. Fungal endophytes can positively affect aphid density, adult weight, and reproductive potential because leaves infected with fungal endophytes contain greater levels of free amino acids (Johnson et al. 2003a), total nitrogen, and carbon relative to uninfected leaves (Gange 1996). Aphids may also indirectly benefit from plants infected with fungal endophytes because these endophytes have been shown to negatively affect populations of aphid parasitoids (Omacini et al. 2001). However, fungal endophytes can also protect plants from herbivory resulting in decreased aphid population sizes (Omacini et al. 2001). Endophytic fungi may also interact with chewing herbivores to negatively affect aphid densities. Loline alkaloids that deter aphid feeding are only produced in plants with endophytic fungi, and the production of these alkaloids increases following herbivory (Bultman et al. 2004). Similarly, mycorrhizae may indirectly affect aphid fecundity, but these effects may vary depending on the ways in which mycorrhizae affect host plant quality, chemistry, or growth (Gehring and Whitham 2002). For example, cottonwoods form both ectomycorrhizal (EM) as well as arbuscular mycorrhizal (AM) associations, and aphid fecundity is greater on plants with EM fungi relative to uninoculated controls, and lower on plants with AM fungi relative to uninoculated controls (Gehring and Whitham 2002). In addition, abiotic factors (e.g., nutrient availability or water stress) can alter mycorrhizal effects on the host plant, which can subsequently affect the attractiveness of the host plant to aphids. One example is phosphorus availability; when phosphorus is at low to intermediate levels, mycorrhizae have an indirect positive effect on aphid weight and fecundity, while there is no effect of mycorrhizae on aphids at high levels of phosphorus (Gange et al. 1999). Plants may therefore derive different benefits from their associated mycorrhizae under different conditions, which can in turn affect aphid performance. While the above examples demonstrate the indirect effects of herbivores and fungi on aphid performance and survival, very few studies have then demonstrated how these indirect effects on aphids could in turn affect their mutualistic relationship with ants (but see Nakamura and Ohgushi 2003).

Aphid performance was also affected by suites of host plant traits that exist among different host plant species, cross types, or genotypes. Genetic differences among host plants affected aphid performance in 11 out of 14 cases (79%) and had no effect on aphid performance in 3 out of 14 cases (21%) (Table 12.2). In the future, it will be important to quantify the impact of genetic versus environmental factors on aphid performance, as the relative strength of these factors will determine aphid persistence on a particular host plant, which then has implications for both tending ants and the surrounding arthropod community.

These differences in host plant quality could translate into differences in the amount and quality of aphid honeydew, and thereby affect their mutualistic relationship with ants. Honeydew quality is assayed by ants as the relative proportion of their preferred sugars, melezitose and glucosucrose (Kiss 1981), and ants show a preference for trisaccharides over disaccharides and monosaccharides (Völkl et al. 1999). The quality and quantity of aphid honeydew available to ants then alters ant attendance of aphid colonies (Völkl et al. 1999). For example, aphids feeding on the more nutritive floral tips of their host plant were more likely to attract tending ants because differences in the quality of the phloem affected differences in the quality of the honeydew produced by aphids (Bristow 1991). In fact, the quality of the host plant can have such an important effect on aphid honeydew production that tending ants have been shown to move aphids from low- to high-quality host plants (Collins and Leather 2002). In addition to honeydew quality, ants also respond to honeydew quantity, and one of the major determinants of honeydew production is the size of an aphid colony. Simulations have predicted that the number of ant foragers present in an aphid colony should be directly related to the number of aphids in the colony, and thereby the amount of "global honeydew production" (Mailleux et al. 2003). Empirical studies have corroborated the predictions made by simulations; aphids that produce higher-quality honeydew in greater quantities were more likely to be tended by ants (Völkl et al. 1999, Fischer et al. 2001). This leads to greater ant residence time in larger aphid colonies (Katayama and Suzuki 2002) as well as differences in ant attendance among aphid species that differ in honeydew productivity (Völkl et al. 1999). The consequences of decreased honeydew production may be much direr than changes in tending by ants, however. When ants are presented with alternative sugar sources to aphid honeydew, they shift from aphid tending to aphid consumption (Offenberg 2001). Differences in host plant quality that affect aphid honeydew quality and quantity could therefore change friend to foe and result in the elimination of aphids from less suitable host plants.

Such host plant factors that affect aphid performance, and thereby their ability to attract ant mutualists, are important because predation can have a huge impact on aphid populations. Predation leads to the local extinction of aphid species from an area (Müller and Godfray 1999) and makes mutualistic associations with ants a necessity. While host plant mediated effects on the aphid–ant mutualism may indirectly affect predation by altering the tending rate of ants, host plants can also affect predation or parasitism of aphids via a more direct route. For example, plant architecture can affect the parasitoids of aphids by altering their attack patterns and thereby oviposition rates (Weisser 1995).

In a hybridizing cottonwood system, host plant cross type (*Populus fremontii*, *P. angustifolia*, F_1 hybrid, or backcross hybrid) has been shown to affect aphid fecundity, and differences in aphid fecundity among cross types determined aphid distribution throughout the landscape. When aphids (*Chaitophorus populicola*) growing naturally on backcross hybrids were transferred onto different cottonwood cross types in a common garden, aphid numbers were greatest on narrowleaf cottonwoods (*P. angustifolia*), lowest on Fremont cottonwood (*P. fremontii*), intermediate on hybrids in one year of study, and exhibited a dominance effect on hybrids in another year of study (i.e., aphid numbers on hybrids were not significantly different from that found on Fremont or narrowleaf cottonwoods: Wimp and Whitham 2001) (Fig. 12.1). While differences in aphid numbers among cottonwood cross types growing in the common garden were small, these differences among hosts had a large impact on aphid distribution in the wild. Aphids were only found on backcross hybrids and narrowleaf cottonwoods along the ~500 km of the Weber River, and did not occur on Fremont or F_1 hybrid cottonwoods (Wimp and Whitham 2001).

In summary, the quality or type of host plant resources can determine aphid performance, interactions with mutualistic ants, and can affect the distribution of the aphid–ant mutualism on a landscape-level scale. It is therefore important to understand the indirect role that host plants play in structuring community interactions by mediating aphid–ant mutualisms. We will explore the indirect role of host plants on three aspects of arthropod community organization via aphid–ant mutualisms: effects on community stability, structure, and biodiversity, and then discuss how the spatial distribution of aphid–ant mutualisms may affect overall (or beta) diversity in a habitat.

The effect of aphid–ant mutualisms on community stability

Understanding the genetic and environmental host plant factors that affect the persistence of aphid–ant mutualisms is important because, where they

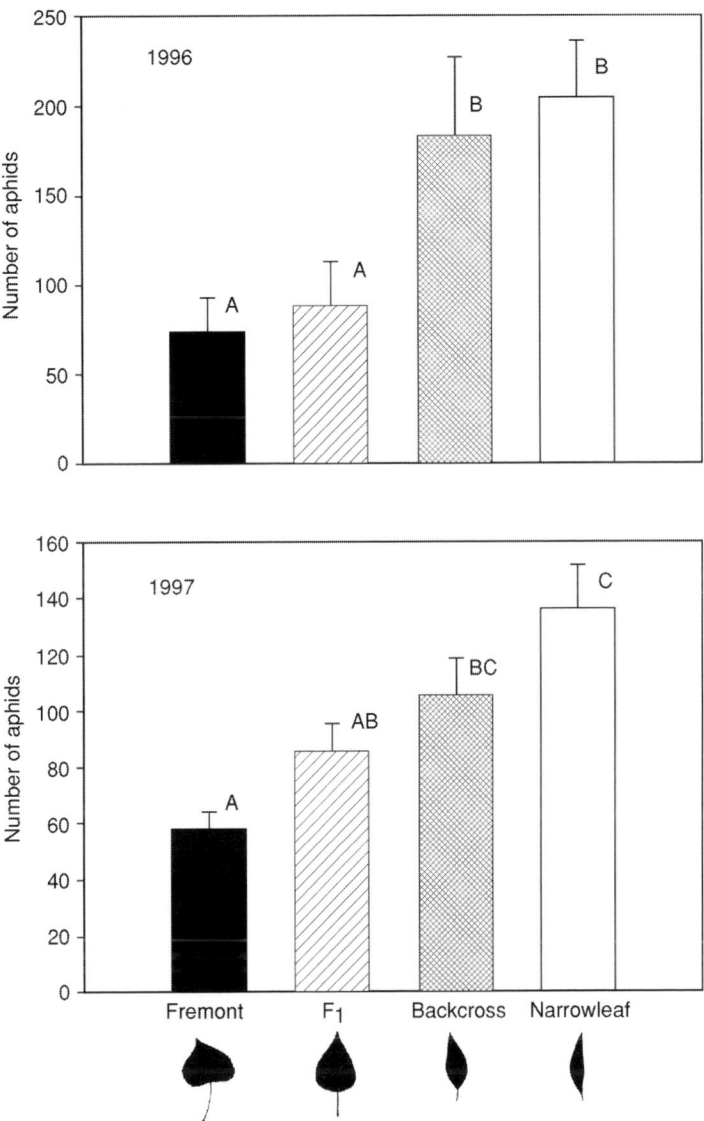

Figure 12.1. Mean number of aphids on the different cottonwood cross types (Fremont, F_1 hybrid, backcross hybrid, and narrowleaf) found growing in a common garden for 1996 and 1997. Shown are means ± 1 standard error; different letters indicate significant differences among means. (Adapted from Wimp and Whitham 2001.)

are present, mutualisms can affect the stability of the surrounding community. Models that explore the effects of mutualisms on multiple species in the community have shown that mutualisms lead to a depauperate but stable community. Models by Kumar and Freedman (1989) showed that when mutualisms were embedded within an interacting food chain, one result was predator extinction. They also showed that mutualisms had a negative effect on other food chain populations as the carrying capacity of the mutualists' population was increased. Regarding the effects of mutualism on community stability, Ringel et al. (1996) found that when mutualisms were embedded within a community matrix using a Jacobian matrix of interactions, there was a stabilizing effect on communities when stability and feasibility were simultaneously considered. Any consideration of stability without feasibility is not biologically realistic because it means that we are considering equilibria with negative population values. Using a model of a mutualism embedded in a matrix of interactions, communities with mutualisms were more likely to show resilience, persistence, and stability compared to communities without mutualisms (Ringel et al. 1996).

Much of the work on the relationship between mutualisms and community stability over the past 30 years has been theoretical and we are not aware of any empirical studies that have examined this relationship. We examined an aphid–ant mutualism embedded within a community of interacting organisms to look at the effects of this mutualism on the stability of the surrounding arthropod community. The following species-richness data were adapted from Wimp and Whitham (2001) and consider a mutualism between a free-feeding aphid, *Chaitophorus populicola*, and its tending ant, *Formica propinqua* (for comprehensive methods involving data collection, see Wimp and Whitham 2001). The establishment of this aphid species is affected by host plant cross type (Fig. 12.1); here we show the indirect effects of host plants on community stability via this aphid–ant mutualism. Species-richness values on aphid–ant trees excluded *Chaitophorus populicola* and *Formica propinqua* because we were interested in the effect of the mutualism on the stability of the surrounding arthropod community and not on the mutualism itself. Aphids and ants were experimentally removed from one set of trees and left intact on a second set of trees, so we were able to measure community responses on paired aphid–ant removal and control trees over the course of a season. At no time during the censuses were aphids allowed to recolonize the removal trees, and ants largely abandoned the trees where aphids were removed, so we were able to establish an effective aphid–ant removal. To examine the effects of the mutualism on community stability, we measured arthropod species turnover across four censuses within a single year. Turnover was calculated as $(E + C)/2$ after the method used by Williamson (1978), where E = species extinction and C = species colonization. If communities with aphid–ant mutualists are more stable through time, then we would expect a low rate of species turnover relative to removal trees; this means that species associated with the mutualism are more likely to be found repeatedly across the different census periods. Data were transformed to meet normality and equality of variance assumptions using a square root transformation and analyzed using a repeated-measures ANOVA.

The presence of an aphid–ant mutualism on cottonwoods increased the seasonal stability of the surrounding arthropod community. We found a significant treatment ($F = 23.761$, $n = 15$, $p < 0.001$) and time effect ($F = 17.609$, $n = 15$, $p < 0.001$) in which species turnover rates on trees with aphid–ant mutualists were significantly less than trees where aphids and ants were removed (Fig. 12.2a). The decreasing rate of species turnover through time on aphid–ant trees also does not appear to reflect a decrease in the density of aphids and ants (Fig. 12.2b) because aphid and ant numbers increased over time. In fact, when we looked at the correlation between aphid/ant density and the rate of species turnover in the surrounding arthropod community, we found a negative relationship between ant density and species turnover rate (Pearson $r = -0.532$, $n = 15$, $p = 0.021$), but not between aphid density and species turnover rate (Pearson $r = -0.293$, $n = 15$, $p = 0.145$). Therefore, we found greater stability (lower species turnover rates) when ant densities were high. While turnover rates were significantly lower on aphid–ant trees relative to removal trees, overall species richness was also lower (Wimp and Whitham 2001). Aphid–ant trees therefore support a more stable, but depauperate community of specialists that are adapted to living among aphids and not being detected by ants (see the following section).

The effects of aphid–ant mutualisms on community structure

When looking at the effects of aphid–ant mutualisms on communities, it is important to examine the effects of mutualisms on the community as a whole, rather than simply the participating guilds, because the effects of mutualists on the surrounding community may extend through multiple trophic levels and different feeding groups may respond differently to the mutualism (Table 12.3). For example, although few studies have examined the effects of aphid–ant mutualisms on the entire community of aboveground arthropods, Wimp and Whitham (2001) found that different feeding groups were differentially affected by an aphid–ant mutualism (Fig. 12.3). Using the same aphid–ant example as the previous section, we found that the aphid–ant mutualism negatively affected other herbivores, generalist predators, and subdominant species of tending ants. These effects were driven largely by the aggressive tending ant

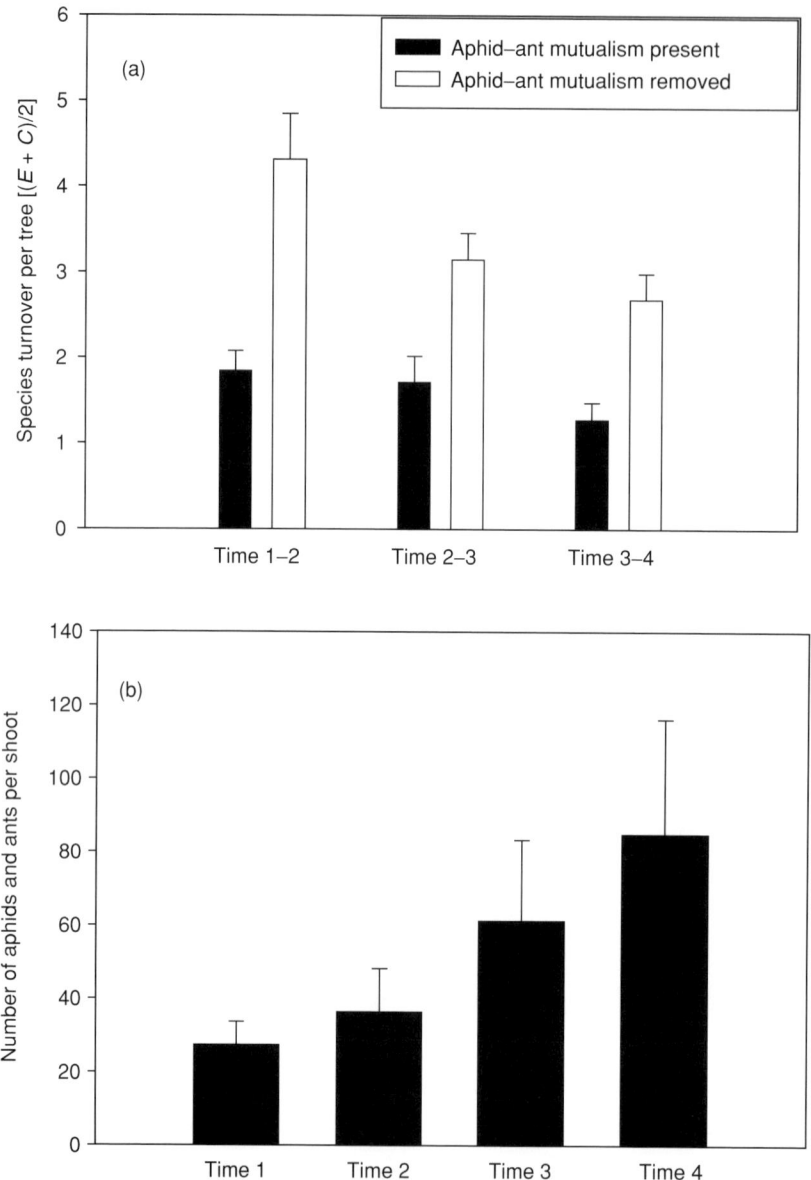

Figure 12.2. The effect of aphid–ant removal on community stability. (a) Species turnover rates for aphid–ant and removal trees. Turnover rates are calculated as $(E+C)/2$ (Williamson 1978), where E = species extinction and C = species colonization. (b) Mean abundance of aphids and ants on trees where the mutualism was left intact. Shown are means ± 1 standard error.

Table 12.3. *The effects of aphid–ant mutualisms on arthropods from different trophic levels*

Mutualists	Effects on herbivores	Effects on other ant mutualists	Effects on generalist predators/parasites[a]	Effects on specialist predators/parasites[b]	Reference
Spissistilus festinus (aphid) and five ant species	(−)				Nickerson et al. 1977
Lachnus allegheniensis (aphid) and *Formica obscuripes* (ant)	(−)				Seibert 1992
Tuberculatus quercicola (aphid) and *Formica yessensis* (ant)	(−)				Ito and Higashi 1991
Symydobius oblongus (aphid) and *Formica lububris* (ant)	(−) Free-living (+) Leaf miners				Fowler and MacGarvin 1985
Schizolachnus pineta and *Cinara* sp. (aphids) and *Formica rufa* (ant)			(−) − 4 spp. **(0)** − 1 sp.	(+)	Sloggett and Majerus 2000
Three aphid species and *Lasius niger* (ant)			(−)	(+)	Fischer et al. 2001
Aphis fabae (aphid) and *Lasius niger* (ant)				(+)	Völkl et al. 1994
Four aphid species and *Lasius niger* (ant)				(+)	Völkl 1995
Aphis fabae (aphid) and *Lasius niger* (ant)			(−)	(+)	Völkl 1992

Table 12.3. (cont.)

Mutualists	Effects on herbivores	Effects on other ant mutualists	Effects on generalist predators/parasites[a]	Effects on specialist predators/parasites[b]	Reference
Aphis fabae (aphid) and three ant species	(−)			(−)	Hübner 2000
Symydobius oblongus (aphid) and *Formica aquilonia*			(−)		Karhu 1998
Cinara banksiana (aphid) and *Formica exsectoides* (ant)	(−)	(−)	(−)		Bishop and Bristow 2001
Chaitophorus populicola (aphid) and *Formica propinqua* (ant)	(−)	(−)	(−)	(+)	Wimp and Whitham 2001

Notes:
[a] The term "generalist" refers to predators and parasites that do not have specific adaptations for feeding and/or ovipositing in aphid colonies that are tended by ants.
[b] The term "specialist" refers to predator/parasite species that do have such adaptations.

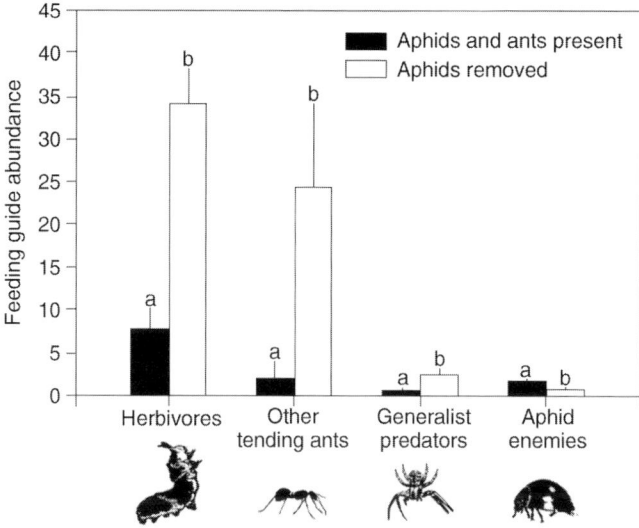

Figure 12.3. The effect of an aphid–ant mutualism on the abundance of different feeding groups. Icons represent different feeding groups affected by the mutualism: herbivores, other species of tending ants, generalist predators, and aphid-specific enemies. (Reproduced from Wimp and Whitham (2001) with permission.)

Formica propinqua, which not only removed potential predators of the aphids, but also potential competitors (i.e., other herbivore species). The negative effect of the aphid–ant mutualism on other tending ants was driven by the competitive superiority of *Formica propinqua* with respect to other ant species (G. M. Wimp personal observation). While the aphid–ant mutualism negatively affected herbivores, generalist predators, and other species of tending ants, they had a positive effect on species that were specialist predators and parasites of aphids. These predators and parasites have adaptations for living cryptically within an aphid colony without being detected by ants and were able to circumvent ant defenses to exploit their aphid prey. This and other studies from the literature show that aphid–ant mutualisms commonly have negative effects on other herbivore species (in seven out of eight cases or 88% of the time), on subordinate species of tending ants (in two out of two cases or 100% of the time), and on generalist predators and parasites (in six out of seven cases or 86% of the time), but positively affect predators and parasites that are specialists on ant-tended aphids (in six out of seven cases or 86% of the time) (Table 12.3).

Aphids living in colonies tended by ants are often more stable through time and are larger than untended colonies (Addicott 1979); therefore, adaptations that allow predators and parasites to feed in these colonies could have fitness rewards for such aphid-specialist species of predators and parasites. For example, different ladybird beetle species show different levels of association with aphid–ant mutualisms depending on their relative degree of myrmecophily, or adaptations for living amongst ants without being detected (Sloggett and Majerus 2000). Ladybird beetles avoid being killed or removed by tending ants through a number of different behavioral and mechanical means. Behaviorally, ladybird beetles remain undetected in aphid colonies by moving in a slow and inconspicuous manner (Völkl 1995). Ladybird beetles may also defend themselves via mechanical structures such as dense, stiff hairs that make it difficult for ants to grasp ladybird beetles or attack them (Völkl 1995). In contrast to ladybird beetles, green lacewing larvae use physical structures to provide concealment. They pluck wax from the wooly alder aphids that they feed upon and place it on their own bodies in what is truly a "wolf-in-sheep's-clothing" method of evading ant defenses (Eisner et al. 1978).

Adaptations for living within aphid–ant colonies also benefit parasites and even the hyperparasites of aphids. Three major parasitoid groups (Ephedrini, Praini, and Aphidiini) have evolved various mechanisms that facilitate oviposition in aphids that are tended by ants (Völkl and Mackauer 2000). Wasps in these groups evade ant defenses through stealthy behavior, mimicry, "sneak" oviposition, chemical camouflage, and mimicking the cuticular hydrocarbon signature of aphids (Liepert and Dettner 1996, Völkl and Mackauer 2000). Some parasitoids will even tap the aphids to elicit honeydew in a manner similar to ants; in this way, aphids are not alarmed by the presence of the parasitoids and do not attempt to escape (Völkl and Mackauer 2000). Not only do ant-tended aphids provide a stable resource, but aggressive ants defending their aphids from attackers can also reduce the number of hyperparasitoids attacking the primary parasitoids of aphids. Hyperparasitism in primary aphid parasitoids was 70% when they were found in aphid colonies that were not tended by ants, but was only 17% when they were found in ant-tended aphid colonies (Völkl 1992). In order to receive indirect benefits from ant tending during the course of their development, aphid parasitoids have mechanisms for maintaining honeydew production in the aphids they attack. Honeydew output in parasitized aphids decreases for the first day after the aphid has been parasitized, but then increases in the days following (Völkl 1992). This is because the parasitized aphids show a reduction in efficient food assimilation, so they intake more phloem and output more honeydew than aphids that have not been parasitized (Cloutier and Mackauer 1979). In addition to aggressive removal of hyperparasitoids, tending ants also remove aphid honeydew, which is used as a search kairomone by hyperparasitoids (Budenberg 1990). However, some species of hyperparasitoids also have mechanisms for evading ant defenses and attacking aphid primary parasitoids. For example, the wasp hyperparasitoid *Alloxysta brevis* secretes a defensive compound from its mandibles that repels ants (Völkl et al. 1994). This allows the hyperparasitoid to oviposit into parasitized aphids where it attacks the aphids' primary parasitoid. Additionally, tending by ants provides protection for *Alloxysta brevis*, an endohyperparasitoid, from attacks made by an ectohyperparasitoid (Völkl et al. 1994). Therefore, adaptations for cryptic living among aphid–ant mutualists not only provide a stable food resource for parasites, but these effects can even extend to secondary consumers that are hyperparasites of aphid parasitoids.

The effects of spatial variation in aphid–ant mutualisms on biodiversity

As we have described previously, the presence of suitable host plant resources affects the distribution of the aphid–ant mutualism. Host plant resources will therefore be key to determining the degree of patchiness in the distribution of aphid–ant mutualists, and their distribution on a landscape-level scale. For example, the degree of aphid clumping on goldenrods is affected by both host plant genotype and local host plant density (Pilson and Rausher 1995). Host plant quality related to pathogenic stress can also affect spatial patterns in aphid distribution (Johnson et al. 2003b). For aphid species that are obligate myrmecophiles, the distribution of high- and low-quality host plants directly affects aphid abundance; they are more abundant in areas with a large proportion of high-quality host plants, whereas unattended species are more abundant in areas with a large proportion of low-quality patches (Stadler 2004). The quality of host plant patches, and subsequent effects on aphid honeydew production, can be so important to tending ants that they will actually move aphids from low- to high-quality patches, thereby influencing their spatial distribution (Collins and Leather 2002). Additionally, the spatial distribution of ant species on a landscape can not only affect the aphid–ant mutualism, but species associated with the mutualism as well. Ants may differ in their guarding behavior, with some species being more aggressive in recognizing and attacking aphid parasitoids (Hübner 2000). These differences in ant aggression may then determine the types of species associated with aphids. For example, aphids that are facultative myrmecophiles tend to have the highest rates of parasitism and support the highest number of specialist parasitoids relative to aphids that are obligate myrmecophiles (Stadler 2002). The spatial distribution of different ant species could therefore indirectly affect community structure if different ant species occupy different areas and have different effects on aphid-specialist predators and parasites.

Although few studies have taken into account both host plant quality and the distribution of preferred ant species to describe the distribution of aphids on a landscape, Wimp and Whitham (2001) found that both host plant cross type and the presence of suitable ant mutualists contributed to spatial patterns in aphid abundance. Along the $\sim$500 km of the Weber River, aphids were positively associated with both ants and suitable host plant types. In combining these two factors, the realized distribution of aphids was only 21% of their potential habitat space, thereby creating a mosaic of occupied and unoccupied habitat.

Because the distribution of aphid–ant mutualists can be so patchy, and because different feeding groups respond differently to the aphid–ant mutualism, we hypothesized that this mutualism mosaic may then lead to a community mosaic with different species assemblages found in areas with and without aphid–ant mutualists. To examine this hypothesis, we analyzed experimental community data from Wimp and Whitham (2001) using nonmetric multidimensional scaling (NMDS) and a subsequent analysis of similarity (ANOSIM). NMDS is a robust ordination technique for analyzing community data (Kruskal 1964, Fasham 1977, Kenkel and Orloci 1986, Minchin 1987), and ANOSIM analysis allows for statistical comparisons in community composition between treatment groups (in this case, trees with aphid–ant mutualists and aphid–ant removal trees). We found that arthropod communities were significantly different on trees with aphid–ant mutualists compared to trees where aphid–ant mutualists were removed ($R = 0.3969$, $p < 0.0001$) (Fig. 12.4). When we examined the number of species uniquely associated with aphid–ant trees and removal trees, we found that of the 90 species we examined, 56% were found only on trees without aphid–ant mutualists, 12% were found only on trees with aphid–ant mutualists, and 32% were common to both (Wimp and Whitham 2001). The presence of the aphid–ant mutualists in a mosaic community of trees with and without the mutualists can therefore lead to an overall increase in arthropod diversity relative to cottonwood stands where the mutualism is absent. Therefore, although aphid–ant mutualists decrease biodiversity on a per tree basis (or alpha diversity), it is possible that they could enhance overall biodiversity (or beta diversity) on a stand level by attracting species not found elsewhere in the community.

We suggest that when mutualists alter community composition, biodiversity will be influenced by the mosaic distribution of mutualists. The presence of mutualists in an environment may have such an effect on communities that their presence represents an essential habitat type for organisms that are uniquely associated with the mutualists (e.g., parasites and predators with adaptations for cryptic living among aphid–ant mutualists). Conversely, areas they do

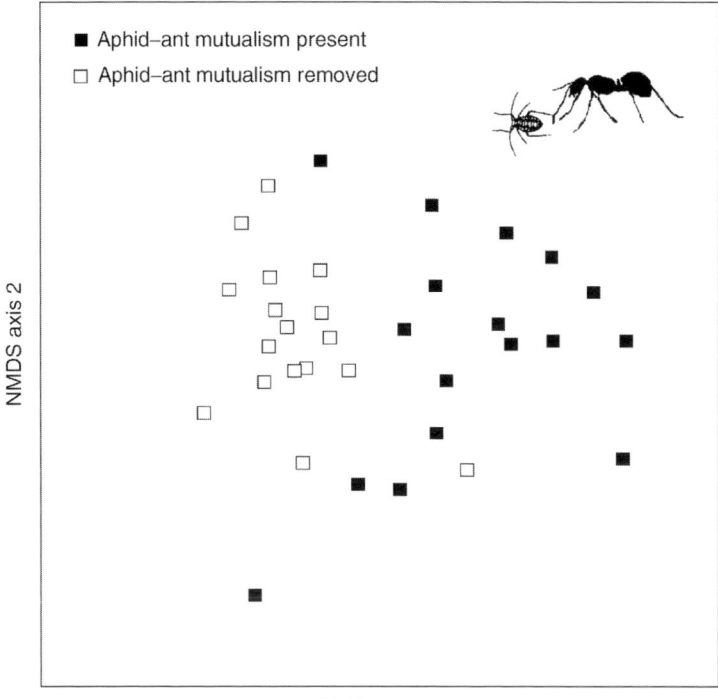

Figure 12.4. NMDS ordination shows that community composition for aphid–ant trees is different from aphid–ant removal trees (ANOSIM $r = 0.3969$, $p < 0.0001$). Closed squares represent trees where aphids were present, and open squares represent trees where aphids were removed. The axes on the NMDS figure represent the communities in two-dimensional space.

not occupy may represent essential habitat for organisms that cannot survive in their presence (e.g., other herbivores, subordinate species of tending ants, and generalist predators). If clusters of aphids and ants on different trees attract a different set of species, then we would expect that a heterogeneous mixture of occupied and unoccupied area would lead to the greatest overall (or beta) diversity, similar to the way in which structural complexity leads to greater species diversity (MacArthur and MacArthur 1961, Recher 1969). Another keystone species, the beaver, has been shown to increase beta diversity in mosaics of beaver-modified and unmodified habitat due to their effects on plant species composition (Wright et al. 2002), and could create dynamic mosaics as their populations change through time (Wright et al. 2004). It is therefore possible that aphid–ant mutualists could have similar effects in mosaic habitats as they are also capable of altering arthropod species composition in their presence.

Conclusions and future directions

We have found that host plant traits can affect the establishment of aphid–ant mutualisms, and thereby the effects of this mutualism on the surrounding arthropod community. Host plant quality (i.e., genetic differences among host plants and environmental variation in nutrient availability) influences aphid performance and honeydew production, which in turn affect the mutualistic relationship between aphids and tending ants. In the future, it will be important to design studies that distinguish between the genetic factors that influence aphid–ant mutualisms relative to environmental effects. When examining the genetic and environmental factors that affect aphid–ant mutualisms, it will be particularly important to look at the effects of host plant constitutive versus induced secondary chemistry on aphids, as such studies will be important in determining the persistence of aphid–ant mutualisms on particular host plants. Finally, a knowledge of the host plant genetic and environmental factors that determine the distribution of aphid–ant mutualisms could have important implications in conservation (e.g., Wimp et al. 2004, Bangert et al. 2005), as well as the emerging field of community genetics (e.g., Whitham et al. 2003).

Once the aphid–ant mutualism is in place, it can enhance community stability and alter community structure. Because aphid–ant mutualisms attract a distinct assemblage of arthropods where they are present, we hypothesize they may increase diversity on a landscape-level scale where we find mutualism mosaics of occupied and unoccupied plant patches. To test the generality of our findings, it is important to not only include studies from other systems, but also extend these studies to include ant–membracid and ant–lycaenid mutualisms. We are aware of no studies of ant–membracid and ant–lycaenid mutualisms that have specifically examined their effects on arthropod species composition. Such studies are important because host plant quality has been shown to affect ant–membracid (e.g., de Assis Dansa and Duarte Rocha 1992) and ant–lycaenid (e.g., Fiedler 1995) mutualisms. Therefore, differences in host plant quality could affect the establishment of these mutualisms, with subsequent effects on arthropod diversity. Host plant resources that influence the distribution of ant–membracid and ant–lycaenid mutualists may affect landscape-level patterns of arthropod diversity in a similar manner to aphid–ant mutualisms. Future studies that consider the effects of the mutualism mosaic on landscape-level arthropod diversity should also consider the environmental and species-specific factors that lead to clumped distributions in tending ants. For example, both aphid occurrence (Hopkins and Thacker 1999) and population density (Bishop and Bristow 2001) are positively influenced by ant density. If differences in aphid density affect their ability to recruit specialist species of predators and parasites, then the clumped distributions of ants may promote further patchiness in the mutualism mosaic and thereby alter their effects on landscape-level patterns of arthropod diversity.

In this chapter, we have seen how genetic and environmentally induced differences in host plant quality not only shape the nature of the interaction between aphids and their ant mutualists, but can also result in differences in their distribution on a landscape. Because these aphid–ant mutualists alter community structure where they are present, it is now important to consider how spatial patterns in their distribution affect arthropod beta diversity. Host plant traits (both genetic and environmentally induced) therefore have consequences not only for the herbivores that feed upon them, but also have the potential to shape landscape patterns of arthropod diversity via aphid–ant mutualisms.

Acknowledgments

We thank T. Ohgushi, T. Craig, and P. Price for their invitation to contribute to this book. Additionally, we thank J. Bailey, R. Bangert, K. Gehring, K. Haskins, N. Johnson, G. Martinsen, P. McIntyre, P. Morrow, T. Ohgushi, L. Ries, K. Rowell, J. Ruel, J. Schweitzer, T. Trotter, and two anonymous reviewers for their thoughtful comments on this chapter. This research was funded by National Science Foundation grant DEB-9726648 and US Department of Agriculture grant 95-37302-1801.

References

Addicott, J. F. 1978. The population dynamics of aphids on fireweed: a comparison of local populations and metapopulations. *Canadian Journal of Zoology* **56**:2554–2564.

Addicott, J. F. 1979. A multispecies aphid–ant association: density dependence and species-specific effects. *Canadian Journal of Zoology* **57**:558–569.

Antolin, M. F., and J. F. Addicott. 1991. Colonization, among shoot movement, and local population neighborhoods of two aphid species. *Oikos* **61**:45–53.

Assis Dansa, C. V. de, and C. F. Duarte Rocha. 1992. An ant–membracid–plant interaction in a cerrado area of Brazil. *Journal of Tropical Ecology* **8**:339–348.

Bangert, R. K., R. J. Turek, G. D. Martinsen, *et al.* 2005. Benefits of conservation of plant genetic diversity on arthropod diversity. *Conservation Biology* **19**:379–390.

Bernard, E. C., K. D. Gwinn, C. D. Pless, and C. D. Williver. 1997. Soil invertebrate species diversity in endophyte-infected tall fescue pastures, pp. 125–135 in C. W. Bacon and N. S. Hill (eds.) *Neotyphodium/Grass Interactions*. New York: Plenum Press.

Bishop, D. B., and C. M. Bristow. 2001. Effect of Allegheny mound ant (Hymenoptera: Formicidae) presence on homopteran and predator populations in Michigan jack pine forests. *Annals of the Entomological Society of America* **94**:33–40.

Black, C. A., R. Karban, L. D. Godfrey, J. Granett, and W. E. Chaney. 2003. Jasmonic acid: a vaccine against leafminers (Diptera: Agromyzidae) in celery. *Environmental Entomology* **32**:1196–1202.

Boucher, D. H. 1985. The idea of mutualism, past and future, pp. 1–28 in D. H. Boucher (ed.) *The Biology of Mutualism*. London: Croom Helm.

Bradley, G. A., and J. D. Hinks. 1968. Ants, aphids, and jack pine in Manitoba. *Canadian Entomologist* **100**:40–50.

Breton, L. M., and J. F. Addicott. 1992. Does host plant quality mediate aphid–ant mutualism? *Oikos* **63**:253–259.

Bristow, C. M. 1991. Are ant–aphid associations a tritrophic interaction? Oleander aphids and Argentine ants. *Oecologia* **87**:514–521.

Bronstein, J. L. 1994. Our current understanding of mutualism. *Quarterly Review of Biology* **69**:31–51.

Bronstein, J. L. 1998. The contribution of ant–plant protection to our understanding of mutualism. *Biotropica* **30**:150–161.

Buckley, R. C. 1987. Interactions involving plants, homoptera, and ants. *Annual Review of Ecology and Systematics* **18**:119–135.

Budenberg, W. 1990. Honeydew as a contact kairomone for aphid parasitoids. *Entomologia Experimentalis et Applicata* **55**:139–148.

Bultman, T. L., G. Bell, and W. D. Martin. 2004. A fungal endophyte mediates reversal of wound-induced resistance and constrains tolerance in a grass. *Ecology* **85**:679–685.

Clay, K., and J. Holah. 1999. Fungal endophyte symbiosis and plant diversity in successional fields. *Science* **285**:1742–1744.

Cloutier, C., and M. Mackauer. 1979. The effect of parasitism by *Aphidius smithi* (Hymenoptera: Aphidiidae) on the food budget of the pea aphid, *Acyrthosiphon pisum* (Homoptera: Aphididae). *Canadian Journal of Botany* **57**:1605–1611.

Collins, C. M., and S. R. Leather. 2002. Ant-mediated dispersal of the black willow aphid *Pterocomma salicis* L.: does the ant *Lasius niger* L. judge aphid–host quality? *Ecological Entomology* **27**:238–241.

Danell, K., and K. Hus-Danell. 1985. Feeding by insects and hares on birches earlier affected by moose browsing. *Oikos* **44**:75–81.

Dungey, H. S., B. M. Potts, T. G. Whitham, and H.-F. Li. 2000. Plant genetics affects arthropod community richness and composition: evidence from a synthetic eucalypt hybrid population. *Evolution* **54**:1938–1946.

Eisner, T., K. Hicks, M. Eisner, and D. S. Robson. 1978. "Wolf-in-Sheep's-Clothing" strategy of a predaceous insect larva. *Science* **199**:790–794.

Fasham, M. J. R. 1977. A comparison of nonmetric multidimensional scaling, principal components analysis and reciprocal averaging for the ordination of simulated coenoclines and coenoplanes. *Ecology* **58**:551–561.

Fiala, B., H. Grunsky, U. Maschwitz, and K. E. Linsenmair. 1994. Diversity of ant–plant interactions: protective efficacy in *Macaranga* species with different degrees of ant association. *Oecologia* **97**:186–192.

Fiedler, K. 1995. Lycaenid butterflies and plants: is myrmecophily associated with particular host plant preferences? *Ethology Ecology and Evolution* **7**:107–132.

Fischer, M. K., K. H. Hoffmann, and W. Völkl. 2001. Competition for mutualists in an ant–homopteran interaction mediated by hierarchies of ant attendance. *Oikos* **92**:531–541.

Fowler, S. V., and M. MacGarvin. 1985. The impact of hairy wood ants, *Formica lugubris*, on the guild structure of herbivorous insects on birch, *Betula pubescens*. *Journal of Animal Ecology* **54**:847–855.

Fraser, L. H., and J. P. Grime. 1999. Aphid fitness on 13 grass species: a test of plant defence theory. *Canadian Journal of Botany* **77**:1783–1789.

Gange, A. C. 1995. Aphid performance in an alder (*Alnus*) hybrid zone. *Ecology* **76**:2074–2083.

Gange. A. C. 1996. Positive effects of endophyte infection on sycamore aphids. *Oikos* **75**:500–510.

Gange, A., and H. West. 1993. Interactions between foliar-feeding insects and VA mycorrhizae. *Bulletin of the British Ecological Society* **24**:72–76.

Gange, A. C., V. K. Brown, and L. M. Farmer. 1990. A test of mycorrhizal benefit in an early successional plant community. *New Phytologist* **115**:85–91.

Gange, A. C., E. Bower, and V. K. Brown. 1999. Positive effects of an arbuscular mycorrhizal fungus on aphid life history traits. *Oecologia* **120**:123–131.

Gehring, C. A., and T. G. Whitham. 2002. Mycorrhizae–herbivore interactions: population and community consequences, pp. 295–320 in M. G. A. van der Heijden and I. Sanders (eds.) *Mycorrhizal Ecology*. Berlin, Germany: Springer-Verlag.

Gehring, C. A., N. S. Cobb, and T. G. Whitham. 1997. Three-way interactions among ectomycorrhizal mutualists, scale insects, and resistant and susceptible pinyons. *American Naturalist* **149**:824–841.

Grayer, R. J., F. M. Kimmins, D. E. Padgham, J. B. Harborne, and D. V. R. Rao. 1992. Condensed tannin levels and resistance of groundnuts (*Arachis hypogaea*) against *Aphis craccivora*. *Phytochemistry* **31**:3795–3800.

Grime, J. P., J. M. L. Mackey, S. H. Hillier, and D. J. Read. 1987. Floristic diversity in a model system using experimental microcosms. *Nature* **328**:420–422.

Heijden, M. G. A. van der, T. Boller, A. Wiemken, and I. R. Sanders. 1998. Different arbuscular mycorrhizal fungal species are potential determinants of plant community structure. *Ecology* **79**:2082–2091.

Hopkins, G. W., and J. I. Thacker. 1999. Ants and habitat specificity in aphids. *Journal of Insect Conservation* **3**:25–31.

Hübner, G. 2000. Differential interactions between an aphid endohyperparasitoid and three honeydew-collecting ant species: a field study of *Alloxysta brevis* (Thomson) (Hymenoptera: Alloxystidae). *Journal of Insect Behavior* **13**:771–784.

Ito, F., and S. Higashi. 1991. An indirect mutualism between oaks and wood ants via aphids. *Journal of Animal Ecology* **60**:463–470.

Johnson, S. N., P. J. Mayhew, A. E. Douglas, and S. E. Hartley. 2002. Insects as leaf engineers: can leaf-miners alter leaf structure for birch aphids? *Functional Ecology* **16**:575–584.

Johnson, S. N., A. E. Douglas, S. Woodward, and S. E. Hartley. 2003a. Microbial impacts on plant–herbivore interactions: the indirect effects of a birch pathogen on a birch aphid. *Oecologia* **134**:388–396.

Johnson, S. N., D. A. Elston, and S. E. Hartley. 2003b. Influence of host plant heterogeneity on the distribution of a birch aphid. *Ecological Entomology* **28**:533–541.

Kagata, H., and T. Ohgushi. 2004. Leaf miner as a physical ecosystem engineer: secondary use of vacant leaf mines by other arthropods. *Annals of the Entomological Society of America* **97**:923–927.

Kainulainen, P., J. Holopainen, V. Palomäki, and T. Holopainen. 1996. Effects of nitrogen fertilization on secondary chemistry and ectomycorrhizal state of Scots pine seedlings and on growth of grey pine aphid. *Journal of Chemical Ecology* **22**:617–636.

Kairo, M.T. K., and S.T. Murphy. 1999. Temperature and plant nutrient effects on the development, survival, and reproduction of *Cinara* sp. nov., an invasive pest of cypress trees in Africa. *Entomologia Experimentalis et Applicata* **92**:147–156.

Karhu, K.J. 1998. Effects of ant exclusion during outbreaks of a defoliator and a sap-sucker on birch. *Ecological Entomology* **23**:185–194.

Katayama, N., and N. Suzuki. 2002. Cost and benefit of ant attendance for *Aphis craccivora* (Hemiptera: Aphididae) with reference to aphid colony size. *Canadian Entomologist* **134**:241–249.

Kenkel, N.C., and L. Orloci. 1986. Applying metric and nonmetric multidimensional scaling to ecological studies: some new results. *Ecology* **67**:919–928.

Kiss, A. 1981. Melezitose, aphids and ants. *Oikos* **37**:382.

Kruskal, J.B. 1964. Multidimensional scaling by optimizing goodness of fit to a nonmetric hypothesis. *Psychometrika* **29**:1–27.

Kumar, H. 1992. Inhibition of ovipositional responses of *Chilo partellus* (Lepidoptera: Pyralidae) by the trichomes on the lower leaf surface of a maize cultivar. *Journal of Economic Entomology* **85**:1736–1739.

Kumar, R., and H.I. Freedman. 1989. A mathematical model of facultative mutualism with populations interacting in a food chain. *Mathematical Biosciences* **97**:235–261.

Leather, S.R. 1993. Early season defoliation of bird cherry influences autumn colonization by the bird cherry aphid, *Rhopalosiphum padi*. *Oikos* **66**:43–47.

Levine, J.M., S.D. Hacker, C.D.G. Harley, and M.D. Bertness. 1998. Nitrogen effects on an interaction chain in a salt marsh community. *Oecologia* **117**:266–272.

Liepert, C., and K. Dettner. 1996. Role of cuticular hydrocarbons of aphid parasitoids in their relationship to aphid-attending ants. *Journal of Chemical Ecology* **22**:695–707.

Llewellyn, M., and V.K. Brown. 1985. The effect of host-plant species on adult weight and the reproductive potential of aphids. *Journal of Animal Ecology* **54**:639–650.

Mabry, C.M., M. Jasieński, J.S. Coleman, and F.A. Bazzaz. 1997. Genotypic variation in *Polygonum pennsylvanicum*: nutrient effects on plant growth and aphid infestation. *Canadian Journal of Botany* **75**:546–551.

MacArthur, R.H., and J.W. MacArthur. 1961. On bird species diversity. *Ecology* **42**:594–598.

Maddox, G.D., and R.B. Root. 1987. Resistance of 16 diverse species of herbivorous insects within a population of goldenrod, *Solidago altissima*: genetic variation and heritability. *Oecologia* **72**:8–14.

Mailleux, A.-C., J.-L. Deneubourg, and C. Detrain. 2003. Regulation of ants' foraging to resource productivity. *Proceedings of the Royal Society of London Series B* **270**:1609–1616.

Marks, S., K. Clay, and G. P. Cheplick. 1991. Effects of fungal endophytes on interspecific and intraspecific competition in the grasses *Festuca arundinacea* and *Lolium perenne. Journal of Applied Ecology* **28**:194–204.

Minchin, P. R. 1987. An evaluation of the relative robustness of techniques for ecological ordination. *Vegetatio* **69**:89–107.

Moora, M., and M. Zobel. 1996. Effect of arbuscular mycorrhiza on inter- and intraspecific competition of two grassland species. *Oecologia* **108**:79–84.

Moran, N. A. 1981. Intraspecific variability in herbivore performance and host quality: a field study of *Uroleucon caligatum* (Homoptera: Aphididae) and its *Solidago* hosts (Asteraceae). *Ecological Entomology* **6**:301–306.

Moran, N. A. 1983. Seasonal shifts in host usage in *Uroleucon gravicorne* (Homoptera: Aphididae) and implications for the evolution of host alternation in aphids. *Ecological Entomology* **8**:371–382.

Müller, C. B., and H. C. J. Godfray. 1999. Predators and mutualists influence the exclusion of aphid species from natural communities. *Oecologia* **119**:120–125.

Nakamura, M., and T. Ohgushi. 2003. Positive and negative effects of leaf shelters on herbivorous insects: linking multiple herbivore species on a willow. *Oecologia* **136**:445–449.

Nakamura, M., Y. Miyamoto, and T. Ohgushi. 2003. Gall initiation enhances the availability of food resources for herbivorous insects. *Functional Ecology* **17**:851–857.

Nickerson, J. C., C. A. Rolph KayKay, L. L. Buschman, and W. H. Whitcomb. 1977. The presence of *Spissistilus festinus* as a factor affecting egg predation by ants in soybeans. *Florida Entomologist* **60**:193–199.

Offenberg, J. 2001. Balancing between mutualism and exploitation: the symbiotic interaction between *Lasius* ants and aphids. *Behavioral Ecology and Sociobiology* **49**:304–310.

Omacini, M., E. J. Chaneton, C. M. Ghersa, and C. B. Müller. 2001. Symbiotic fungal endophytes control insect host–parasite interaction webs. *Nature* **409**:78–81.

Omer, A. D., J. Granett, R. Karban, and E. M. Villa. 2001. Chemically induced resistance against multiple pests in cotton. *International Journal of Pest Management* **47**:49–54.

Orians, C. M. 2000. The effects of hybridization in plants on secondary chemistry: implications for the ecology and evolution of plant–herbivore interactions. *American Journal of Botany* **87**:1749–1756.

Osier, T. L., and R. L. Lindroth. 2001. Effects of genotype, nutrient availability, and defoliation on aspen phytochemistry and insect performance. *Journal of Chemical Ecology* **27**:1289–1313.

Painter, R. H. 1951. *Insect Resistance in Crop Plants*. New York: Macmillan.

Petersen, M. K., and J. P. Sandström. 2001. Outcome of indirect competition between two aphid species mediated by responses in their common host plant. *Functional Ecology* **15**:525–534.

Pilson, D. 1992. Aphid distribution and the evolution of goldenrod resistance. *Evolution* **46**:1358–1372.

Pilson, D., and M. D. Rausher. 1995. Clumped distribution patterns in goldenrod aphids: genetic and ecological mechanisms. *Ecological Entomology* **20**:75–83.

Price, P. 2002. Resource-driven terrestrial interaction webs. *Ecological Research* **17**:241–247.

Rabin, L. B., and R. S. Pacovsky. 1985. Reduced larva growth of two Lepidoptera (Noctuidae) on excised leaves of soybean infected with a mycorrhizal fungus. *Journal of Economic Entomology* **78**:1358–1363.

Recher, H. F. 1969. Bird species diversity and habitat diversity in Australia and North America. *American Naturalist* **103**:75–80.

Ringel, M. S. 1996. The stability and persistence of mutualisms embedded in community interactions. *Theoretical Population Biology* **50**:281–297.

Rudgers, J. A., and M. C. Gardener. 2004. Extrafloral nectar as a resource mediating multispecies interactions. *Ecology* **85**:1495–1502.

Sagers, C. L. 1992. Manipulation of host plant quality: herbivores keep leaves in the dark. *Functional Ecology* **6**:741–743.

Sandberg, S. L., and M. R. Berenbaum. 1989. Leaf-tying by tortricid larvae as an adaptation for feeding on phototoxic *Hypericum perforatum*. *Journal of Chemical Ecology* **15**:875–885.

Sanders, C. J., and F. B. Knight. 1968. Natural regulation of the aphid *Pterocomma poulifoliae* on bigtooth aspen in northern lower Michigan. *Ecology* **49**:234–244.

Schupp, E. W. 1986. *Azteca* protection of *Cercropia*: ant occupation benefits juvenile trees. *Oecologia* **70**:379–385.

Seibert, T. F. 1992. Mutualistic interactions of the aphid *Lachnus allegheniensis* (Homoptera: Aphididae) and its tending ant *Formica obscuripes* (Hymenoptera: Formicidae). *Annals of the Entomological Society of America* **85**:173–178.

Service, P. 1984. Genotypic interactions in an aphid–host plant relationship: *Uroleucon rudbeckiae* and *Rudbeckia laciniata*. *Oecologia* **61**:271–276.

Sloggett, J. J., and M. E. N. Majerus. 2000. Aphid-mediated coexistence of ladybirds (Coleoptera: Coccinellidae) and the wood ant *Formica rufa*: seasonal effects, interspecific variability and the evolution of a coccinellid myrmecophile. *Oikos* **89**:345–359.

Stadler, B. 1995. Adaptive allocation of resources and life-history trade-offs in aphids relative to plant quality. *Oecologia* **102**:246–254.

Stadler, B. 2002. Determinants of the size of aphid–parasitoid assemblages. *Journal of Applied Entomology* **126**:258–264.

Stadler, B. 2004. Wedged between bottom–up and top–down processes: aphids on tansy. *Ecological Entomology* **29**:106–116.

Stadler, B., A. F. G. Dixon, and P. Kindlmann. 2002. Relative fitness of aphids: effects of plant quality and ants. *Ecology Letters* **5**:216–222.

Straw, N. A., and G. Green. 2001. Interactions between green spruce aphid (*Elatobium abietinum* (Walker)) and Norway and Sitka spruce under high and low nutrient conditions. *Agricultural and Forest Entomology* **3**:263–274.

Thaler, J. S. 2002. Effect of jasmonate-induced plant responses on the natural enemies of herbivores. *Journal of Animal Ecology* **71**:141–150.

Thaler, J. S., M. J. Stout, R. Karban, and S. S. Duffey. 2001. Jasmonate-mediated induced plant resistance affects a community of herbivores. *Ecological Entomology* **26**:312–324.

Tilles, D. A., and D. L. Wood. 1982. The influence of carpenter ant (*Camponotus modoc*) (Hymenoptera: Formicidae) attendance on the development and survival of aphids (*Cinara* sp.) (Homoptera: Aphididae) in a giant sequoia forest. *Canadian Entomologist* **114**:1133–1142.

Tsarouhas, V., U. Gullberg, and U. Lagercrantz. 2003. Mapping of quantitative trait loci controlling timing of bud flush in *Salix*. *Hereditas* **138**:172–178.

Völkl, W. 1992. Aphids and their parasitoids: who actually benefits from ant-attendance? *Journal of Animal Ecology* **61**:273–281.

Völkl, W. 1995. Behavioral and morphological adaptations of the coccinellid, *Platynaspis luteorubra* for exploiting ant-attended resources (Coleoptera: Coccinellidae). *Journal of Insect Behavior* **8**:653–670.

Völkl, W., and M. Mackauer. 2000. Oviposition behavior of aphidiine wasps (Hymenoptera: Braconidae, Aphidiinae): morphological adaptations and evolutionary trends. *Canadian Entomologist* **132**:197–212.

Völkl, W., G. Hübner, and K. Dettner. 1994. Interactions between *Alloxysta brevis* (Hymenoptera, Cynipoidea, Alloxystidae) and honeydew-collecting ants: how an aphid hyperparasitoid overcomes ant aggression by chemical defense. *Journal of Chemical Ecology* **20**:2901–2915.

Völkl, W., J. Woodring, M. Fischer, M. W. Lorenz, and K. H. Hoffman. 1999. Ant–aphid mutualisms: the impact of honeydew production and honeydew sugar composition on ant preferences. *Oecologia* **118**:483–491.

Way, M. J. 1963. Mutualism between ants and honeydew-producing Homoptera. *Annual Review of Entomology* **8**:307–344.

Weibull, J. 1990. Host plant discrimination in the polyphagous aphid *Rhopalosiphum padi*: the role of leaf anatomy and storage carbohydrate. *Oikos* **57**:167–174.

Weisser, W. W. 1995. Within-patch foraging behaviour of the aphid parasitoid *Aphidius funebris*: plant architecture, host behaviour, and individual variation. *Entomologia Experimentalis et Applicata* **76**:133–141.

Whitham, T. G., W. Young, G. D. Martinsen, et al. 2003. Community and ecosystem genetics: a consequence of the extended phenotype. *Ecology* **84**:559–573.

Williamson, G. B. 1978. A comment on equilibrium turnover rates for islands. *American Naturalist* **112**:241–243.

Wimp, G. M., and T. G. Whitham. 2001. Biodiversity consequences of predation and host plant hybridization on an aphid–ant mutualism. *Ecology* **82**:440–452.

Wimp, G. M., W. P. Young, S. A. Woolbright, et al. 2004. Conserving plant genetic diversity for dependent animal communities. *Ecology Letters* **7**:776–778.

Wright, J. P., C. Jones, and A. S. Flecker. 2002. An ecosystem engineer, the beaver, increases species richness at the landscape scale. *Oecologia* **132**:96–101.

Wright, J. P., W. S. C. Gurney, and C. Jones. 2004. Patch dynamics in a landscape modified by ecosystem engineers. *Oikos* **105**:336–348.

13

Biodiversity is related to indirect interactions among species of large effect

JOSEPH K. BAILEY AND THOMAS G. WHITHAM

Introduction

Because communities are structured by the interactions among species, indirect interactions (i.e., effects of one species on another mediated by a third) are likely to play a major role in determining community composition. Through indirect interactions with plants, herbivores can have large effects on community composition by creating habitats and conditions to which other species respond. For example, beaver herbivory of cottonwoods increases phytochemical defensive compounds in resprout cottonwoods that positively affect the abundance of a leaf-chewing chrysomelid beetle (Martinsen *et al.* 1998). Herbivores can create these habitats or conditions by modifying plant architecture (Nakamura and Ohgushi 2003), secondary chemistry (Karban and Baldwin 1997), plant species composition (Johnston and Naiman 1990, Chadde and Kay 1991), building of structures (Cappuccino 1993, Jones *et al.* 1994, Dickson and Whitham 1996, Martinsen *et al.* 2000, Bailey and Whitham 2003), changes to the spatial distribution of habitat (Chadde and Kay 1991), or some combination of these effects, any of which can influence community composition. When herbivores are dominant species, keystone species (Hunter 1992) and/or ecosystem engineers, they can have strong positive or negative effects on associated species (Jones *et al.* 1997, Wimp and Whitham 2001, Bailey and Whitham 2002). Hereafter, we refer to such organisms as *species of large effect*, i.e., species which create ecological conditions to which other species respond resulting in a change in community composition.

Ecological Communities: Plant Mediation in Indirect Interaction Webs, ed. Takayuki Ohgushi, Timothy P. Craig, and Peter W. Price. Published by Cambridge University Press. © Cambridge University Press 2007.

Regardless of the sign of the interaction (i.e., positive or negative), novel species are often recruited to the habitats and conditions created by species of large effect such that overall biodiversity increases. For example, Wimp and Whitham (2001) showed that an aphid–ant mutualism negatively affected arthropod richness on individual trees, but recruited novel specialist arthropod species that could survive in the presence of aggressive ants. However, because not all trees support this mutualism, at the stand level, which includes trees *with and without* this aphid–ant interaction, the increased heterogeneity increased overall richness.

This chapter addresses how the individual and combined effects of species interactions (e.g., plant–herbivore, mutualism, and parasite–host) create the habitats or environments that are crucial for other species and positively affect biodiversity. As the number of interactions among species of large effect increase so do the number of potential niches for other species, i.e., increased heterogeneity results in increased diversity. We use plant–herbivore–community interactions from two *Populus* systems – quaking aspen (*P. tremuloides*) and narrowleaf cottonwood (*P. angustifolia*) – to explore this relationship for two reasons. First, both trees characterize major habitat types in western North America (mountain forests and riparian habitat, respectively) that contribute disproportionately to biodiversity (Noss *et al.* 1995). Second, both are greatly affected by the individual and combined effects of plant–herbivore interactions. Browsing by elk (*Cervus canadensis*) on aspen, and beavers (*Castor canadensis*) on narrowleaf cottonwood greatly affect the survival and population dynamics of these dominant trees, which in turn greatly affect the rest of the community that depends on them for their survival (e.g., Bailey and Whitham 2002, Bailey *et al.* 2004).

Conditional responses of species to ecological factors across scales of space and time

Although we often study the short-term interactions among species and report the interaction as positive or negative, most community interactions are conditional (Lawton 1999). In other words, community composition depends upon the ecological factors that structure the system and their relative importance. In natural systems organisms rarely exist in isolation (McArthur and Wilson 1967, Hunter and Price 1992, Brown *et al.* 2001). Over their lifetimes, species interact, directly and indirectly, through many mechanisms that affect habitat, resource quality, quantity, and their spatial distribution (Root 1973, Brown and Davidson 1977, Brown and Heske 1990, Price 1991, Thompson *et al.* 1991). Interactions among species can also change in relation to distribution, abundance, and effects through space and time (Faeth 1986, Hunter 1987). For example, if a species is very abundant in space and time, we generally expect its effects to be large relative to a species that is rare. Due to changing environmental conditions, disturbance regimes, herbivory, predation, competition, nutrient pulses, and other factors, direct and indirect interactions among species and their environment should be common (Strauss 1991, Wootton 1994). Thus, it is important to understand how common indirect interactions are and how multiple factors interact to create unique ecological conditions that affect the responses of associated species. These responses vary along a continuum from negative to neutral to positive. Importantly, the sign of these interactions can reverse such that given one set of conditions the interactions are negative, whereas, under another set of conditions they can shift to become positive. Five examples from diverse systems and taxa serve to illustrate these shifts or reversals in major interactions.

First, mistletoe and their host plants are considered classic examples of a parasite–host interaction where mistletoe negatively affects its host plant. Van Ommeren and Whitham (2002) found that juniper infected with mistletoe attracted significantly more avian seed dispersal agents than trees without mistletoe. Because these birds disperse the seeds of both mistletoe and juniper, mistletoe could have a positive effect on juniper by attracting more birds that would more effectively disperse juniper seeds. Consistent with this hypothesis, stands with mistletoe attracted three times more seed-dispersing birds and recruited almost two-and-a-half times more juniper seedlings than stands without mistletoe. Thus, under some conditions, mistletoe appears to serve as a mutualist, which positively affects juniper by attracting common seed-dispersal agents. Therefore, by adding just one more species (i.e., an avian seed-dispersal agent) to the mistletoe–juniper interaction, the relationship between mistletoe and juniper can switch from parasitism to mutualism.

Second, Orians and Fritz (1996) found that some willow genotypes were twice as resistant to insect herbivores as other susceptible genotypes. However, when fertilizer was added to simulate good environmental conditions, the positive effects of the resistant genotypes switched to become nearly three times more susceptible. These results demonstrated a genetic × environment interaction where herbivores respond to willow genotypes differently due to environmental effects on genetically based host plant quality.

Third, Johnson *et al.* (1997) showed beneficial effects of mycorrhizae on the reproduction of grasses when the soil they grew in was unmanaged and they received normal light regimes. When soils were fertilized to simulate agricultural practices the mutualistic benefits disappeared and mycorrhizae became parasitic and negatively affected plant fitness. This example shows how parasitism can shift to mutualism, or vice versa, as mycorrhizae interact with their host plants and the environment. These two fertilizer examples also illustrate how the widespread use of fertilizer could have unintended consequences in agricultural systems (i.e., increased susceptibility to insects in the willow example and a switch to parasitic fungi in the grass example).

Fourth, pattern reversals can also occur along spatial and temporal scales. Brown and Davidson (1977) and Brown and Heske (1990) demonstrated that ants and rodents indirectly compete for similar-sized seeds in the short term and at localized scales, thus having a negative effect on each other. In the long term, and at larger scales of investigation, these keystone rodents improved grassland habitat and increased seed abundance to have a positive effect on ants. With this rodent–ant interaction, scaling up to long-term landscape-level studies based on short-term local studies would have resulted in conclusions that were opposite of those that were actually observed.

Fifth, plant ontogeny or the change in gene expression as a function of plant age can result in a reversal of resistant and susceptible traits demonstrating how plant genetics and time interact. Kearsley and Whitham (1989, 1998) showed that the aphid *Pemphigus betae* is concentrated on mature branches of hybrid cottonwood trees, whereas the beetle *Chrysomela confluens* is concentrated on juvenile branches and juvenile ramets of the same clone. Aphid and beetle transfer experiments onto different aged ramets of the same clone showed that the preferences of these insects were adaptive and related to differences in survival and reproduction. Furthermore, when cuttings from juvenile and mature zones of the same tree were rooted and planted out in a common garden for 3 years, trees derived from juvenile tissues remained resistant to aphids, while trees derived from mature tissues remained susceptible (Kearsley 1991). This stability of resistance and susceptibility traits after asexual propagation argues for an ontogenetic basis to resistance (Kearsley and Whitham 1989, 1998). Thus, as individual trees or ramets in a clone age, the expression of genes associated with resistance and susceptibility can change such that a young tree resistant to aphids becomes susceptible later in life, whereas the reverse pattern occurs for beetles. It is important to emphasize that the height and age at which this shift from juvenile to mature traits occurs within an individual tree can be under strong genetic control and can be highly heritable (Jordan *et al.* 1999). Most importantly, scaling up from what is observed on juvenile trees could be completely erroneous for mature trees and could have important economic consequences in commercial plantations (see also Lawrence *et al.* 2003).

In conclusion, from microbes to birds, these examples illustrate that as ecological conditions change, species responses can change both quantitatively and qualitatively. Because the individual species responses can vary depending upon multiple ecological factors, these examples suggest that whole communities might also differ in response to the same ecological factors. For example, Dirzo and Miranda (1991) showed that the local extinction of large mammalian herbivores in tropical forests and the subsequent loss of their herbivory as a community-structuring factor resulted in a 66% decline of plant and insect biodiversity. This study suggests that when species interactions are lost from systems, the effect can be even greater losses in associated diversity. Similar logic would suggest that as the number of species interactions increase: (1) novel community compositions can emerge, and (2) diversity should increase. These predictions are based on the premise that the interactions of species play a major role in structuring communities. If these interactions are lost, then we should expect major community-level consequences. These interactions are likely to be most important when they involve dominant and/or keystone species that are thought to define and structure communities and ecosystems (e.g., Power et al. 1996).

Trait-mediated effects of plants on community composition

Biotic and abiotic ecological factors can result in induced plant traits and indirect interaction webs (Bailey and Whitham 2002). We focus on the indirect community effects of plant–herbivore interactions. Studies have shown that on average only 20% of the annual net primary productivity (ANPP) in terrestrial systems is consumed by herbivores (Cyr and Pace 1993), and it is likely that much of the remaining 80% of ANPP is affected by herbivory. Because herbivores are a component of all ecosystems (Cyr and Pace 1993, Cebrian 1999, Polis 1999), and their effects are common and diverse, we argue that the effects of herbivory on plant traits such as productivity, reproduction, and defense can have large effects on community composition and diversity.

There are at least six major ways plants and herbivores can indirectly interact to affect arthropod community composition: (1) selective herbivory can alter plant community composition (Johnston and Naiman 1990, Bailey et al. 2004); (2) herbivory can affect the chemical composition of the host plant through induction of secondary compounds (Bryant 1981, Martinsen et al. 1998); (3) herbivory can result in architectural changes in the host plant that alter microclimate and resource quality (Nakamura and Ohgushi 2003); (4) herbivory can result in changes to the spatial distribution of habitat (Tilman 1988, 1994); (5) herbivores can build structures such as galls and lodges that create refuges from predation (Cappuccino 1993, Bailey and Whitham 2003); and (6) herbivores can interact with other herbivores resulting in novel combinations of the previous effects (Bailey and Whitham 2003). For example, galling herbivores create shelters that attract unique species and induce phytochemical defenses that can have novel combined effects.

The study of direct effects has dominated ecological research and only recently has the importance of indirect effects become more appreciated in both the ecological and genetics literature (e.g., Strauss 1991, Wootton 1994, Wolf et al. 1998). In the following case studies we demonstrate that direct and indirect effects of plant–herbivore interactions result in distinct communities which support novel species and positively affect biodiversity.

Case studies

We explore two studies from the aspen (*Populus tremuloides*) forests of northern Arizona and one study in a riparian forest dominated by narrowleaf cottonwood (*P. angustifolia*). These studies examine how plant–herbivore interactions indirectly structure arthropod and avian communities, which in turn affect biodiversity.

Aspen–fire–elk

Aspen has the largest distribution of any North American deciduous tree species and is a dominant tree of montane and boreal forests. Aspen plays a complex role in supporting a diverse community (Bartos and Cambell 1998), and is a major food and habitat source for a wide range of animals (Clausen et al. 1989, Basey et al. 1990, Kay 1990, Romme et al. 1995, Baker et al. 1997). Since post-European settlement, aspen regeneration has declined across the western USA principally due to fire suppression and mammalian herbivory (Jones and Debyle 1985, Cantor and Whitham 1989, Boyce 1989, Bartos et al. 1994, Romme et al. 1995, Baker 1997, Bailey and Whitham 2002). Because aspen forests are structured by multiple factors including fire and herbivory by elk (*Cervus canadensis*) they represent an ideal system for examining how multiple factors impact aspen, which in turn affects other community members.

Fire and mammalian herbivory can have diverse consequences for associated communities by modifying plant resources (Danell and Huss-Danell 1985, Neuvonen and Danell 1987, Stein et al. 1992, Bailey and Whitham 2002). Abiotic and biotic factors can modify plant resources in a variety of ways including changes to chemistry and architecture. Because species differentially respond to these changes in aspen, a community-wide response is predicted. To examine these community responses we quantified the arthropod communities on individual trees that had experienced one of four different treatments (i.e., high burn and no elk browse, high burn and browsed, low burn and no browse, low burn and browsed). To analyze this data we used nonmetric multidimensional scaling (NMDS), which is a robust ordination technique for community analysis (Minchin 1987). NMDS has been used to analyze differences in community composition (Bailey and Whitham 2002, 2003, Wimp et al. 2005), and was chosen for its ability to handle some of the issues commonly faced with community data. It does not make any assumptions about the nature of the data; species need not have normal distributions across an environmental gradient, and it can accommodate narrow and skewed distributions (Minchin 1987). Significant differences in community composition among treatment groups were obtained using analysis of similarity (ANOSIM), which uses 1000 random reassignments of species to groups and determines if the generated dissimilarity matrix is significantly different than chance (Warwick et al. 1990). It is important to emphasize that points that are close together have similar arthropod communities, while points that are far apart have dissimilar communities. With NMDS, the axes are not used for interpretation as in other ordination techniques.

Using these methods of community analysis, Bailey and Whitham (2002) showed that arthropod community composition was significantly different on unbrowsed aspen ramets growing in different burn treatments (Fig. 13.1a). In the absence of elk, fire severity alone resulted in aspen ramets that differed in productivity (Bailey and Whitham 2002), chemistry (J.K. Bailey unpublished data) and architecture (Bailey and Whitham 2003). In turn, these impacts on the plant significantly altered arthropod community composition.

The same factors that affect the arthropod community directly can also affect the foraging of vertebrate herbivores whose browsing can indirectly affect the arthropod community. For example, elk heavily browse aspen in high-burn sites and moderately browse aspen in moderate-burn sites. These interactions created a mosaic of arthropod communities related to the intensity of burning and level of elk browsing on aspen saplings (Bailey and Whitham 2002). Elk herbivory also induced changes in phytochemistry and altered architecture beyond the effects of fire alone. Depending upon the quality of aspen being browsed, elk herbivory can either positively or negatively affect arthropod species richness (Bailey and Whitham 2002).

When the effects of elk herbivory and the interaction of herbivory and fire severity on arthropod community composition were considered, NMDS ordination followed by ANOSIM analysis showed that arthropod community composition changed with each of the four possible stand conditions (i.e., high-severity burn/unbrowsed, high-severity burn/browsed, intermediate-severity/unbrowsed, intermediate-severity burn/browsed; ANOSIM: $r = 0.212$, $p < 0.05$) (Fig. 13.1b). These analyses showed that the interaction of fire severity and elk browsing supports four significantly different arthropod communities.

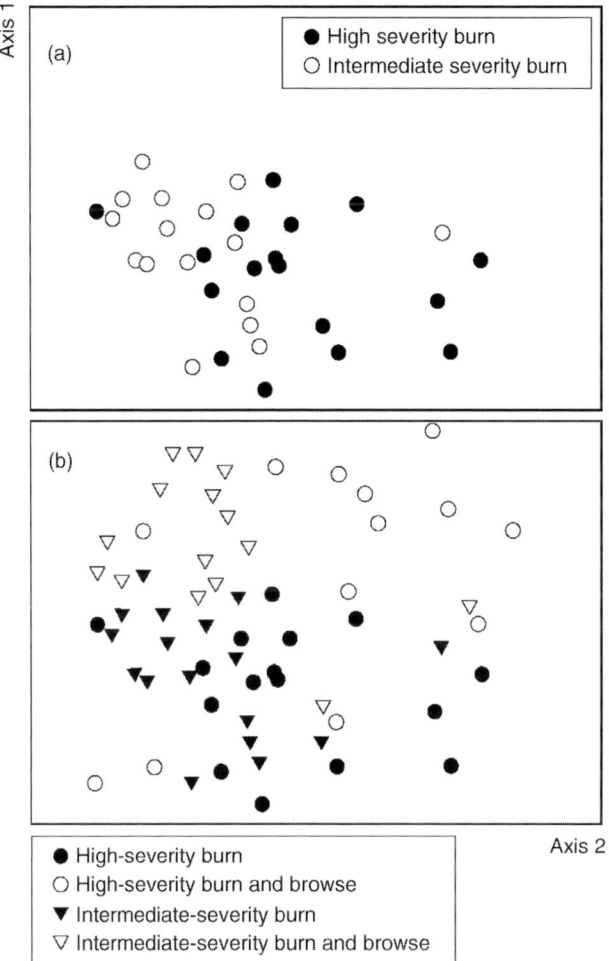

Figure 13.1. Fire and herbivory interact to structure arthropod communities in aspen forests. Nonmetric multidimensional scaling (NMDS), ordination using 33 different species of arthropods distributed across four different treatments of aspens (i.e., high burn and no elk browse, high burn and browsed, low burn and no browse, low burn and browsed) shows that the arthropod communities associated with each treatment type are significantly different (see text). Each symbol represents the community of arthropods on a single aspen ramet. (a) Arthropod community composition differs on aspen that vary in plant quality due to burn severity. (b) Arthropod community composition can further differentiate when elk browse aspen, altering phytochemistry, architecture, and the spatial distribution of habitat creating greater habitat heterogeneity. (Modified from Bailey and Whitham 2002.)

Because species responses to ecological conditions are conditional and changing ecological factors such as browse and burn intensity often recruit novel species, the net effect is the creation of greater habitat heterogeneity or a mosaic of habitat types. When the species present in these different combinations of fire and browsing were added together, the net effect was an overall increase in diversity (for species list see Bailey and Whitham 2002). For example, 12% of the arthropod species responded to fire severity alone while nearly 30% of arthropod species responded to the combination of fire and browsing. So the addition of another ecological factor increased the number of species present by nearly three times. Furthermore, these data showed that the loss of ecological conditions related to indirect interactions of plants and herbivores can result in the loss of distinct community compositions. While the above example shows how fire, elk, and aspen interact to affect the arthropod community, the second example shows how a keystone arthropod affects the arthropod community and avian predators.

One herbivore commonly found on many species of *Populus* is the galling sawfly, *Phyllocolpa* spp. *Phyllocolpa bozemanii* is commonly found on aspen in mountain forests of northern Arizona. *Phyllocolpa* is a keystone species that creates a fold on the margin of a leaf. This gall provides habitat for many other species such as aphids and spiders that utilize the space within the fold for shelter and food. In aspen forests, the presence of this sawfly can cause a two-fold increase in arthropod richness and a four-fold increase in abundance relative to unbrowsed aspen alone (Bailey and Whitham 2003). Using NMDS ordination of individual species standardized to species maxima, Bailey and Whitham (2003) found that arthropod community composition differed significantly when galls were present on aspen ramets ($n = 16$, $r = 0.8413$, $p < 0.05$) than when galls were absent (Fig. 13.2). These findings argue that localized shelters created by sawflies have community-wide effects. For example, at least 90% of the abundances for each of five herbivore species and six predator species were found within galls (Bailey and Whitham 2003). These findings indicate that sawflies represent a "species of large effect" because they create novel ecological conditions to which other species respond resulting in a change in community composition.

Phyllocolpa bozemanii is also an important food item for insectivorous birds (Bailey and Whitham 2003). However, elk herbivory negatively affects the abundance of *P. bozemanii* through changes to aspen (i.e., architecture, chemistry) and thus could indirectly and negatively affect insectivorous birds. Two hypotheses could account for this pattern. Elk negatively affect birds via their browsing, which reduces the abundance of the galls that birds prey upon. Alternatively, birds selectively forage on unbrowsed ramets relative to browsed ramets independent of the effects of elk browsing on gall abundance. When the number of

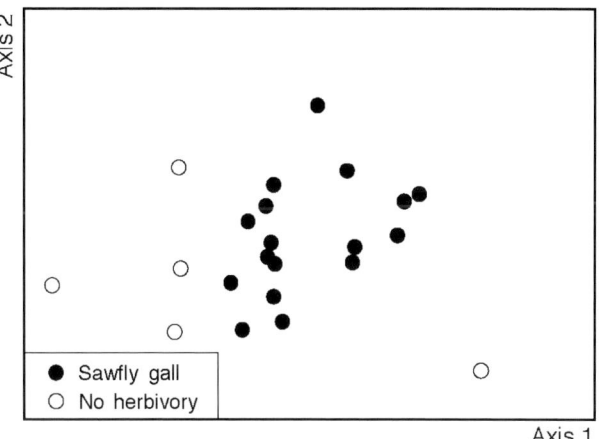

Figure 13.2. Galling sawflies alter arthropod community composition. Using NMDS ordination analyses, we found that the composition of the arthropod community was significantly different on ramets with galls than on ramets without galls. Community composition was based on combined arthropod richness and abundance on aspen ramets with and without sawfly galls. ANOSIM compared within and between group differences. (Modified from Bailey and Whitham 2003.)

galls was experimentally held constant on browsed and unbrowsed aspen we found there was no difference in bird predation (Bailey and Whitham 2003). This demonstrated that elk browsing negatively affected insectivorous bird foraging on aspen ramets by negatively affecting the abundance of galls.

Even at the stand level, fire and elk browsing interacted to affect avian communities. In the spring of 2000, 22 migratory bird species were surveyed inside and outside five similar-aged aspen stands that were protected from elk herbivory by an exclosure. NMDS followed by ANOSIM analyses showed that the avian community was significantly different inside and outside of the exclosure ($r = 0.23$, $p < 0.05$) (Fig. 13.3). Not only do elk alter avian community composition, but they indirectly recruit novel species that do not occur inside the exclosure. For example, seven species were unique to aspen stands without herbivory and five were only found outside of the exclosure. This effect is similar to the aphid–ant mutualism example of Wimp and Whitham (2001) in which the aggressive behavior of aphid-tending ants negatively affected arthropod community richness and abundance to favor specialized species that could circumvent the aggressive behavior of ants. Because not all trees were colonized by this aphid–ant mutualism, the overall heterogeneous habitat of trees with and without this aphid–ant interaction supported the highest arthropod diversity.

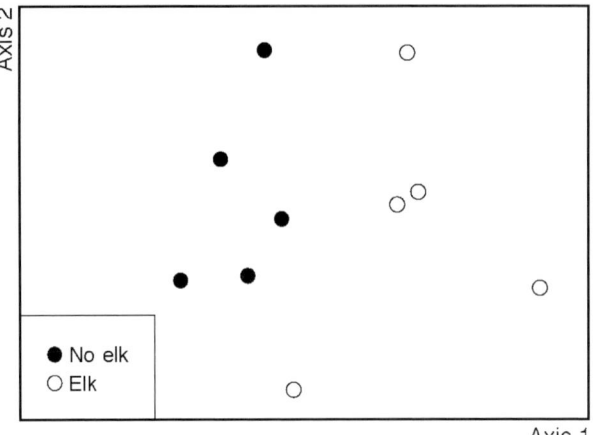

Figure 13.3. Elk alter migrating avian community composition. Using NMDS ordination analyses, we found that the composition of the avian community was significantly different on aspen inside elk exclosures than outside elk exclosures. Community composition was based on combined avian richness and abundance inside and outside of five elk exclosures. ANOSIM analyses was use to compare within- and between-group differences.

The above examples suggest that elk–aspen interactions contribute to the observed differences in both the avian and arthropod communities. These studies in an aspen forest system also show that the individual and combined effects of indirect interactions among "species of large effect" positively affect diversity. Regardless of the sign of an interaction (i.e., positive, negative) as the number of interactions among species of large effect increases, biodiversity tends to increase. Because there has been no explicit test of this hypothesis, future research should focus on this important issue, which we believe has important biodiversity and conservation implications.

Cottonwoods–beavers–galling sawflies

Using a regional species pool of arthropods that occur on *P. angustifolia*, we tested two major hypotheses: (1) arthropod community composition was related to the individual and combined effects of species interactions among species of large effect (i.e., beaver *Castor canadensis*, sawfly *Phyllocolpa* spp.), and (2) arthropod diversity increases as the number of community structuring species interactions increase.

Cottonwood trees are recognized as a dominant species of riparian forests along most river systems of western North America (Rood and Mahoney 1990). Beavers are recognized as keystone species and ecosystem engineers along riparian forests of the USA (Naiman et al. 1986, 1988, 1994, Jones et al. 1994, 1997, Power et al. 1996, Wright et al. 2002, 2003, Bailey et al. 2004) due to their dam-building, tree-felling, and other impacts on the environment that affect many other species (Johnston and Naiman 1990, Chadde and Kay 1991, Martinsen et al. 1998, Wright et al. 2003). In cottonwood riparian forests along the Weber River, we found that beaver felling of trees stimulated cottonwood regeneration through stump and root sprouting. These asexually derived cottonwoods express juvenile traits rather than the mature traits of the larger trees from which they were derived, and they exhibit very different phytochemical and architectural traits (Martinsen et al. 1998). These beaver-induced changes in cottonwoods have major indirect effects on a diverse community of organisms.

The leaf-galling sawfly, *Phyllocolpa* spp., is also an ecosystem engineer that creates a fold on the margin of a leaf. This gall provides habitat for many other species such as aphids and spiders that utilize the space within the fold for shelter and food (examples of other leaf modifiers are given in Cappuccino 1993, Dickson and Whitham 1996, Martinsen et al. 2000, Bailey and Whitham 2003, Nakamura and Ohgushi 2003). Along the Weber River, beaver herbivory positively affected the abundance of *Phyllocolpa* through changes to host plant architecture (i.e., increased shoot length). We then examined how the interactions among cottonwoods, beavers, and sawflies affected the arthropod community.

The observational data support the hypothesis that arthropod community composition was related to both the individual and combined effects of species interactions among a dominant plant and two keystone herbivores (Bailey and Whitham 2006). Furthermore, we found that arthropod diversity increased as the number of community interactions increased. In total we found 23 families and 33 species of arthropods associated with these treatments ($r = 0.51$, $p < 0.05$) (Fig. 13.4). Using NMDS and ANOSIM analyses, our surveys showed that significantly different arthropod communities occupied control cottonwoods (one component – i.e., trees only with no beaver activity or sawfly galls), cottonwoods with sawfly galls (two components – i.e., trees and galls), trees that had resprouted after being felled by beavers (two components – i.e., trees and beavers), and cottonwoods that had resprouted after being felled by beavers that also supported galls (three components – i.e., trees, beavers, and sawfly galls).

To examine how the number of community structuring species interactions affects community composition we used the vector-fitting procedure in the program DECODA. Vectors allowed us to identify sample variables that were correlated with the community similarity matrix and were responsible for separating communities in multidimensional space. In this study, vectors were used as a test to determine if increasing numbers of species interactions resulted

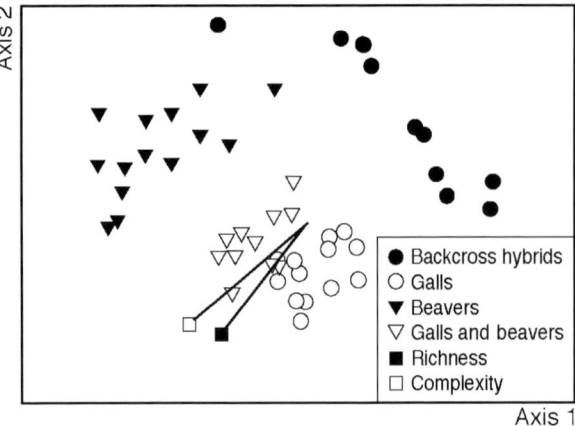

Figure 13.4. Using NMDS ordination analyses, we found that the composition of the arthropod community was significantly different on control cottonwoods, cottonwoods with sawfly galls, beaver-cut cottonwoods without sawfly galls, and when all three (i.e., cottonwoods, beavers, and sawflies) occurred together. Community composition was based on combined arthropod richness and abundance among the treatments. Vector analysis showed that as the number of interactions among cottonwoods, beavers, and/or sawflies increased, the richness of the arthropod community also increased (i.e., vector with a solid square). Importantly, we also detected a significant vector of increasing interactions (i.e., cottonwoods, beavers, and/or sawflies) along the same axis (i.e., vector with an open square). The fact that the two vectors are on the same axis argues that species richness increases with increasing complexity of interactions.

in greater diversity. Vector analysis showed that the arthropod community composition significantly shifted in response to the number of interactions among species of large effect (i.e., cottonwoods, beavers, and/or sawflies; vector illustrated in Fig. 13.4 with an open square; $r = 0.62$, $p < 0.05$) and that total species richness increased along this same axis (vector illustrated in Fig. 13.4 with a solid square; $r = 0.49$, $p < 0.05$). In other words, as the number of interactions increases among species of large effect, the diversity of the dependent arthropod community also increases.

These results suggest that: (1) arthropod communities on control cottonwoods respond to some underlying factors of host plant quality, (2) arthropod communities respond differently to the unique effects of beaver-cut cottonwoods and cottonwoods with sawfly galls, and (3) arthropod communities respond differently to habitat mosaics created by the unique interactions of cottonwoods, beavers, and sawflies. In summary, these case studies demonstrate how the direct and indirect effects of plant–herbivore interactions can influence patterns of diversity. In each case, the interactions of species of large effect have resulted in distinct communities. They have supported novel species and have positively affected arthropod diversity. Furthermore, while the effects of these interactions might result in a local decline in diversity (e.g., on an individual tree), at the stand or landscape level, the effects on diversity are positive as long as a mosaic of interactions is maintained.

Review of interaction effects and pattern reversals

Our studies suggest that indirect effects of plant–herbivore interactions on associated communities appear to be conditional, can commonly reverse, and vary in space and time. Because biological systems are complex and structured by multiple interacting ecological factors whose effects vary across scales of space and time, we made two major predictions. Indirect effects should be common and the response of most dependent variables (e.g., species) to multiple ecological factors should be conditional. In other words, reversals in patterns (e.g., parasitism to mutualism) should be common.

Furthermore, it is important to know if the case studies presented above represent rare occurrences, or if they might be so common that researchers should be made more aware of them and/or if experimental designs might need to incorporate more factors to minimize the probability of a reversal occurring. This could be especially important when major issues such as costly management decisions are at stake. To examine this hypothesis, we conducted a preliminary survey of the journal *Ecology* from April 1999 to November 2000 for studies that had significant results from two-, three-, and four-factor full model ANOVAs (all studies that met these criteria were used in our analyses). Just as in our examples above, these factors included time, space, number of species, and different treatments such as fertilizer.

Overall, there were 85 studies using full model, multifactor ANOVAs, and 417 tests from those studies. For each test, we determined whether the result was a single factor or an interaction effect using ANOVA tables or reported results. We categorized effects as either $0 =$ single factor or $1 =$ interaction effect. Due to the continuous nature of the independent variable (e.g., number of factors) and the categorical nature of the response variable (e.g., single factor or interaction) we used logistic regression for analysis (Sokal and Rohlf 1995).

For the relationship between the number of ecological factors used in the analysis and the probability of interaction effects, we characterized two general classes of interaction effects (*sensu* Orians and Fritz 1996):

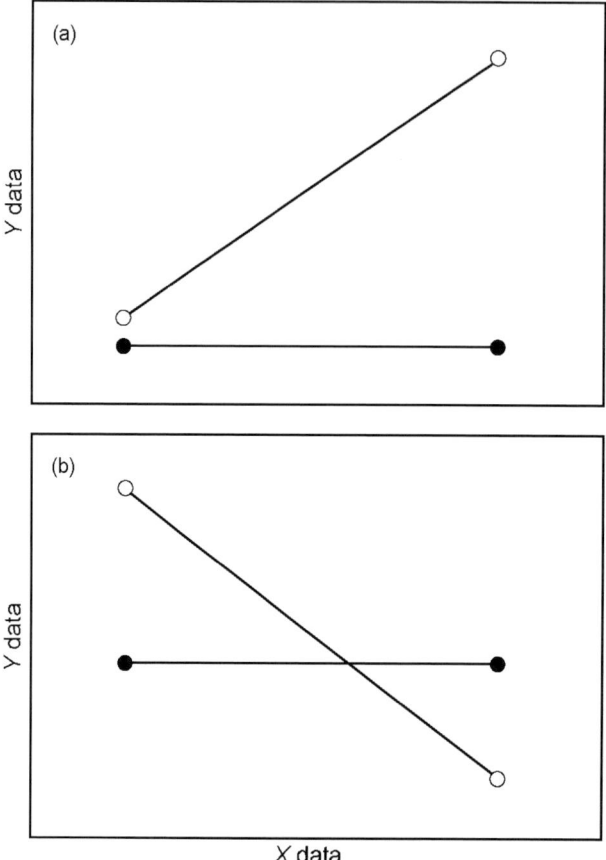

Figure 13.5. Hypothetical model of two-factor interaction plots: (a) represents a significant interaction term along some gradient (e.g., density or environmental stress) whose effects could potentially reverse if the lines were to cross; (b) represents a significant interaction term whose effects have reversed to fundamentally change the sign of an interaction from negative to positive or vice versa.

1. Interaction effects that do not result in pattern reversals (Fig. 13.5a). In this case the lines differ in slope demonstrating a quantitative interaction in which the response was greater to one factor than another, but the sign of the interaction remained the same over the range of studied values.
2. When the lines of an interaction plot cross, this demonstrates that factors not only interacted, but there was a qualitative change in the outcome; i. e., a switch from positive to negative or vice versa (Fig. 13.5b). These reversal data were more difficult to gather, but were based on means when they were reported and figures when means were unavailable.

Using logistic regression which compares a continuous variable to a categorical variable, our analyses of these published studies shows that there was a positive relationship between the number of factors studied and the probability of interaction effects ($n = 417$ tests, $p < 0.05$) (Table 13.1; Fig. 13.6). If a study examined only one factor, the outcome is limited by the design and interaction effects cannot be detected. However, as we increased the number of factors examined to two, in our survey of 85 studies with 336 two-factor tests, significant interaction effects (i.e., quantitative changes) were observed in 53% of the tests (i.e., 178 tests). Similarly, out of 64 three-factor tests surveyed, significant interaction effects occurred in 71% of the tests (i.e., 45 tests). Out of 17 four-factor tests surveyed, significant interaction terms were observed in 94% of the tests (i.e., 16 tests). Thus, there is a clear pattern of increasing significant interactions with the number of factors in the studies.

Most importantly, there was also a significant positive logistic relationship between the number of factors studied and the percentage of reversals (i.e., a qualitative change in the sign of the interaction from positive to negative or vice versa; $n = 417$, $p < 0.05$) (Table 13.1; Fig. 13.6). Similar to interaction terms, pattern reversals cannot be detected in single-factor design studies. However, when a second factor is added to a study, 11% (38 out of 336) of the tests resulted in a qualitative change in the sign of an effect. With three factors, 31% (20 out of 64) of tests resulted in pattern reversals and with four factors, 75% of the tests (13 out of 17) exhibited pattern reversals.

Together these findings show that interaction effects are common and that dependent variables of investigation were very responsive to changes in the independent variables. Furthermore, as the number of factors in the study increased, the probability of reversals occurring also increased. It is also important to note that these reversals were not one way. Shifts could be from positive to negative or negative to positive. These data suggested that when spatial and

Table 13.1. *Results of logistic regression compare the relationship of number of factors studied, spatial scale of study, and duration of study to the percent of interaction effects and pattern reversals*

Scaler	Wald χ^2	β	p
Number of factors (interactions)	116.55	1.74	<0.05
Number of factors (reversals)	90.54	1.67	<0.05
Spatial	20.99	0.69	<0.05
Time	6.82	0.17	<0.05

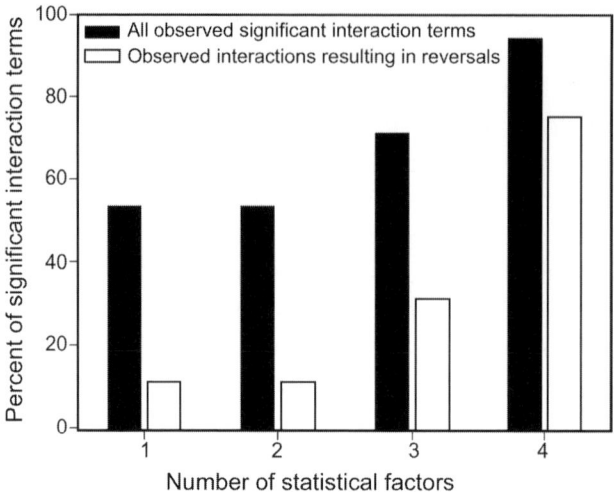

Figure 13.6. Survey of the literature shows interaction effects and pattern reversals are common. Solid bars show that as the number of factors in a study increases, the observed percentage of significant interaction effects also increases (i.e., both quantitative and qualitative interactions such as shown in panels (a) and (b) of Fig. 13.5). Open bars show that as the number of factors in a study increases, the observed percentage of reversals also increases (i.e., a change in the sign of an interaction such as shown in panel (b) of Fig. 13.5).

temporal scales are randomized, biological systems can be highly linked through direct and indirect effects.

There was also a positive logistic relationship between the scale of a study (e.g., the progression from individual leaves to shoots, to whole plants, to a stand, to a forest) and the percentage of two-factor interactions that occurred (i.e., only two-factor studies were used in the scale analyses). Single factor effects are more common at small spatial scales, but two factors are more likely to interact at larger spatial scales ($n = 180$ tests, $p < 0.05$) (Table 13.1). Studies examining leaf- or flower-level processes detected interaction effects in only 39% of tests, which suggests that at small scales, single factors are likely to be most important. However, studies examining stand- or forest-level processes detected interaction effects nearly two times more often (72% of tests). There was also a positive logistic relationship between study duration (1–5 years) and the percentage of two-factor interactions that occurred: 1-year studies detected interaction effects in 57% of the tests, but 5-year studies detected interaction effects in 83% of the tests ($n = 485$, $p < 0.05$) (Table 13.1).

These analyses suggest that indirect effects can be more important than direct effects in structuring biological systems. Overall, these data show that interaction effects among multiple factors are common, occurring up to 94% of the time and suggest that it is rare for any single factor to operate alone. Furthermore, up to 75% of the time the effects of multiple ecological factors on an associated response variable are conditional in which there is a change in the sign of an effect. Regardless of whether response variables change in value quantitatively or qualitatively, these data suggest that the independent and combined effects of ecological factors create novel conditions to which associated species can respond. We are aware of no other study that has attempted to quantify the frequency of such reversals in biological systems. Although we were not surprised by the fact that reversals occur, we were surprised that the frequency of their occurring could be so high (i.e., 75% in a four-factor study).

The basic results of this preliminary interaction survey reflect what many ecologists have recognized for a long time: much of ecology is conditional. What is potentially important is that if the high frequency of switches or reversals that emerged in our preliminary findings is generally correct, then ecologists need to be more aware of how common reversals can be and the need for more multivariate experimental approaches in the study of species interactions. Other studies have made similar arguments about reversals in effects (e.g., Hobbs and Huenneke 1992, Johnson *et al.* 1997, van Ommeren and Whitham 2001, Bailey and Whitham 2002). Whether or not an interaction is mutualism or parasitism is fundamentally important to our understanding of a system and has important management implications (e.g., Johnson *et al.* 1997, van Ommeren and Whitham 2001). We have emphasized that indirect effects are likely to be very common in biological systems. These findings are important because they show how multiple ecological factors can interact to create different ecological conditions that can influence the response of a species or community. These data provide generality to the hypothesis that as the number of interactions among species of large effect increase, habitat heterogeneity is also likely to increase, which in turn will have an overall positive effect on biodiversity.

Conclusions and synthesis

Our studies and examples demonstrate three major patterns that emerge from the interactions of species. First, species interactions create the habitats and conditions to which whole communities of organisms respond. Second, regardless of the sign of an interaction, species interactions often recruit novel species such that at the community level, biodiversity tends to increase as the number of species interactions increase. Third, as we scale up from one to multiple species, local to regional levels, or across time, reversals in the sign of interactions appear to be common. These patterns are important for two reasons.

First, they show that interactions have a positive effect on biodiversity, regardless of the sign of the interaction, as long as a heterogeneous habitat is maintained. Second, because interactions are often conditional, these data suggest that conclusions drawn from single-species interactions or single-factor studies can change as greater complexity is incorporated. This realization is important for conservation biologists and land managers so they can avoid making decisions that might have consequences opposite of what was initially intended.

Acknowledgments

We appreciate reviews by Jen Schweitzer, Peter Price, Tim Craig, and Takayuki Ohgushi and several anonymous reviewers which made this a better chapter. Special thanks to the Whitham laboratory group for discussions of interaction effects and pattern reversals. This research was funded by National Science Foundation grants FIBR-0425908 and IRCEB-0078280.

References

Bailey, J.K., and T.G. Whitham. 2002. Interactions among fire, aspen and elk affect insect biodiversity: reversal of a community response. *Ecology* **83**:1701–1712.

Bailey, J.K., and T.G. Whitham. 2003. Interactions among elk, aspen, galling sawflies, and insectivorous birds. *Oikos* **101**:127–134.

Bailey, J.K., and T.G. Whitham. 2006. Interactions between cotton wood and beavers positively affect sawfly abundance. *Ecological Entomology* **31**:294–297.

Bailey, J.K., J.A. Schweitzer, B. Rehill, R. Lindroth, and T.G. Whitham. 2004. Beavers as molecular geneticists: a genetic basis to the foraging of an ecosystem engineer. *Ecology* **85**:603–608.

Baker, W.L., J.A. Munroe, and A.E. Hessl. 1997. The effects of elk on aspen in the winter range in Rocky Mountain National Park. *Ecography* **20**:155–165.

Bartos, D.L., and R.B. Cambell, Jr. 1998. Water depletion and other ecosystem values forfeited when conifer forests displace aspen communities. In D.F. Potts (ed.) *Rangeland Management and Water Resources*. Herndon, VA: America Water Resources Association.

Bartos, D.L., J.K. Brown, and G.D. Booth. 1994. Twelve years biomass response in aspen communities following fire. *Journal of Range Management* **47**:79–83.

Basey, J.M., S.H. Jenkins, and G.C. Miller. 1990. Food selection by beaver in relation to inducible defenses of *Populus tremuloides*. *Oikos* **59**:57–62.

Boyce, M.S. 1989. *The Jackson Elk Herd: Intensive Wildlife Management in North America*. New York: Cambridge University Press.

Brown, J.H., and D.W. Davidson. 1977. Competition between seed-eating rodents and ants in desert ecosystems. *Science* **196**:880–882.

Brown, J.H., and E.J. Heske. 1990. Control of a desert-grassland transition by a keystone rodent guild. *Science* **250**:1705–1707.

Brown, J.H., T.G. Whitham, S.K.M. Ernest, and C.A. Gehring. 2001. Complex species interactions and the dynamics of ecological systems: long-term experiments. *Science* **293**:643–650.

Bryant, J. P. 1981. Phytochemical deterrence of snowshoe hare browsing by adventitious shoots of four Alaskan trees. *Science* **213**:889–890.

Cantor, L. F., and T. G. Whitham. 1989. Importance of belowground herbivory: pocket gophers may limit aspen to rock outcrop refugia. *Ecology* **70**:962–970.

Cappuccino, N. 1993. Mutual use of leaf-shelters by lepidopteran larvae on paper birch. *Ecological Entomology* **19**:399–401.

Cebrian, J. 1999. Patterns in the fate of production in plant communities. *American Naturalist* **154**:449–468.

Chadde, S. W., and C. E. Kay. 1991. Tall willow communities on Yellowstone's northern range: a test of the "natural regulation" paradigm, pp. 231–262 in R. B. Keiter and M. S. Boyce (eds.) *The Greater Yellowstone Ecosystem*. New Haven, CT: Yale University Press.

Clausen, T. P., P. B. Reichardt, J. P. Bryant, *et al.* 1989. Chemical model for short-term induction in quaking aspen (*Populus tremuloides*) foliage against herbivores. *Journal of Chemical Ecology* **15**:2335–2346.

Cyr, H., and M. L. Pace. 1993. Magnitude and patterns of herbivory in aquatic and terrestrial ecosystems. *Nature* **361**:148–150.

Dannell, K., and K. Huss-Danell. 1985. Feeding by insects and hares on birches earlier affected by moose browsing. *Oikos* **44**:75–81.

Dickson, L. L., and T. G. Whitham. 1996. Genetically-based plant resistance traits affect arthropods, fungi, and birds. *Oecologia* **106**:400–406.

Dirzo, R., and A. Miranda. 1991. Altered patterns of herbivory and diversity in the forest understory: a case study of the possible consequences of contemporary defaunation, pp. 273–287 in P. W. Price, T. M. Lewinsohn, G. W. Fernandez, and W. W. Benson (eds.) *Plant–Animal Interactions: Evolutionary Ecology in Tropical and Temperate Regions*. New York: Wiley Interscience.

Faeth, S. H. 1986. Indirect interactions between temporally separated herbivores mediated by the host plant. *Ecology* **67**:479–494.

Hobbs, R. J., and L. F. Huenneke. 1992. Disturbance, diversity, and invasion: implications for conservation. *Conservation Biology* **6**:324–337.

Hunter, M. D. 1987. Opposing effects of spring defoliation on late-season oak caterpillars. *Ecological Entomology* **12**:373–382.

Hunter, M. D. 1992. Interactions within herbivore communities mediated by the host plant: the keystone herbivore concept, pp. 287–325 in M. D. Hunter, T. Ohgushi, and P. W. Price (eds.) *Effects of Resource Distribution on Animal–Plant Interactions*. San Diego, CA: Academic Press.

Hunter, M. D., and P. W. Price. 1992. Playing chutes and ladders: heterogeneity and the relative role of bottom-up and top-down forces in natural communities. *Ecology* **73**:724–732.

Johnson, N. C., J. H. Graham, and F. A. Smith. 1997. Functioning of mycorrhizal associations along the mutualism–parasitism continuum. *New Phytologist* **135**:575–585.

Johnston, C. A., and R. J. Naiman. 1990. Browse selection by beaver: effects on riparian forest composition. *Canadian Journal of Forest Research* **20**:1036–1043.

Jones, C.G., J.H. Lawton, and M. Shachak. 1994. Organisms as ecosystem engineers. *Oikos* **69**:373–386.

Jones, C.G., J.H. Lawton, and M. Shachak. 1997. Positive and negative effects of organisms as physical ecosystem engineers. *Ecology* **78**:1946–1957.

Jones, J.R., and N.V. Debyle. 1985. Aspen: ecology and management in the Western United States, pp. 77–81 in N.V. Debyle and R.P. Winokur (eds.) US Department of Agriculture Forestry Service Gen. Technical Report RM-119. Fort Collins, CO: Rocky Mountain Forest Range Experimental Station.

Jordan, G.J., B.M. Potts, and R.J. Wiltshire. 1999. Strong, independent, quantitative genetic control of the timing of vegetative phase change and first flowering in *Eucalyptus globulus* ssp. *globulus* (Tasmanian Blue Gum). *Heredity* **83**:179–187.

Karban, R., and I.T. Baldwin. 1997. *Induced Responses to Herbivory*. Chicago, IL: University of Chicago Press.

Kay, C.E. 1990. Yellowstone northern elk herd: a critical evaluation of the natural regulation paradigm. Ph.D. Dissertation, Logan, UT, USA: Utah State University.

Kearsley, M.J.C. 1991. The effects of host development on herbivores of narrowleaf cottonwoods (*Populus angustifolia* L.). Ph.D. dissertation, Flagstaff, AZ, USA: Northern Arizona University.

Kearsley, M.J.C., and T.G. Whitham. 1989. Developmental changes in resistance to herbivory: implications for individuals and populations. *Ecology* **70**:422–434.

Kearsley, M.J.C., and T.G. Whitham. 1998. The developmental stream of cottonwoods affects ramet growth and resistance to herbivory by galling aphids. *Ecology* **79**:178–191.

Lawrence, R., B.M. Potts, and T.G. Whitham. 2003. Relative importance of plant ontogeny, host genetic variation, and leaf age for a common herbivore. *Ecology* **84**:1171–1178.

Lawton, J.H. 1999. Are there general laws in ecology? *Oikos* **84**:177–192.

Martinsen G.D., E.M. Driebe, and T.G. Whitham. 1998. Indirect interactions mediated by changing plant chemistry: beaver browsing benefits beetles. *Ecology* **79**:192–200.

Martinsen, G.D., K.D. Floate, A.M. Waltz, G.M. Wimp, and T.G. Whitham. 2000. Positive interactions between leaf-gallers and other arthropods enhance biodiversity on hybrid cottonwoods. *Oecologia* **123**:82–89.

McArthur, R., and E.O. Wilson. 1967. *The Theory of Island Biogeography*. Princeton, NJ: Princeton University Press.

Minchin, P.R. 1987. An evaluation of the relative robustness of techniques for ecological ordination. *Vegetatio* **69**:89–107.

Naiman, R.J., J.M. Melillo, and J.E. Hobbie. 1986. Ecosystem alteration of boreal forest streams by beaver (*Castor canadensis*). *Ecology* **67**:1254–1269.

Naiman, R.J., C.A. Johnston, and J.C. Kelley. 1988. Alteration of North American streams by beaver. *BioScience* **38**:753–762.

Naiman, R.J., G. Pinay, C.A. Johnston, and J. Pastor. 1994. Beaver influences on the long-term biogeochemical characteristics of boreal forest drainage networks. *Ecology* **75**:905–921.

Nakamura, M., and T. Ohgushi. 2003. Positive and negative effects of leaf shelters on herbivorous insects: linking multiple herbivore species on a willow. *Oecologia* **136**:445–449.

Neuvonen, S., and K. Danell. 1987. Does browsing modify the quality of birch forage for *Epirrita autumnata* larvae? *Oikos* **49**:156–160.

Noss, R. F., E. T. LaRoe III, and J. M. Scott. 1995. *Endangered Ecosystems of the United States: A Preliminary Assessment of Loss and Degradation*. Washington, DC: US Department of the Interior, National Biological Service.

Orians, C. M., and R. S. Fritz. 1996. Genetic and soil-nutrient effects on the abundance of herbivores on willow. *Oecologia* **105**:388–396.

Polis, G. A. 1999. Why are parts of the world green? Multiple factors control productivity and the distribution of biomass. *Oikos* **86**:3–15.

Power, M. E., D. Tilman, J. A. Estes, et al. 1996. Challenges in the quest for keystones. *BioScience* **46**:609–620.

Price, P. W. 1991. The plant vigor hypothesis and herbivore attack. *Oikos* **62**:244–251.

Romme, W. H., M. G. Turner, L. L. Wallace, and J. S. Walker. 1995. Aspen, elk, and fire in northern Yellowstone National Park. *Ecology* **76**:2097–2106.

Rood, S. B., and J. H. Mahoney. 1990. Collapse of riparian poplar forests downstream from dams in western prairies: probable causes and prospects for mitigation. *Environmental Management* **14**:451–464.

Root, R. B. 1973. Organization of a plant–arthropod association in simple and diverse habitats: the fauna of collards, *Brassica oleracea*. *Ecological Monographs* **43**:95–124.

Sokal, R. R., and F. J. Rohlf. 1995. *Biometry: The Principles and Practice of Statistics in Biological Research*. New York: W. H. Freeman.

Stein, J. S., P. W. Price, W. G. Abrahmson, and C. F. Sacchi. 1992. The effect of fire on stimulating willow regrowth and subsequent attack by grasshoppers and elk. *Oikos* **65**:190–196.

Strauss, S. Y. 1991. Indirect effects in community ecology: their definition, study, and importance. *Trends in Ecology and Evolution* **6**:206–210.

Thompson, D. B., J. H. Brown, and W. D. Spencer. 1991. Indirect facilitation of granivorous birds by desert rodents: experimental evidence from foraging patterns. *Ecology* **72**:852–863.

Tilman, D. 1988. *Plant Strategies and the Dynamics of Structure of Plant Communities*. Princeton, NJ: Princeton University Press.

Tilman, D. 1994. Competition and biodiversity in spatially structured habitats. *Ecology* **75**:2–16.

Van Ommeren, R. J., and T. G. Whitham. 2002. Changes in interactions between juniper and mistletoe mediated by shared avian frugivores: parasitism to potential mutualism. *Oecologia* **130**:281–288.

Warwick, R. M., K. R. Clarke, and Suharsono. 1990. A statistical analysis of coral community responses to the 1982–1983 El Niño in the Thousand Islands, Indonesia. *Coral Reefs* **8**:171–179.

Wimp, G. M., and T. G. Whitham. 2001. Biodiversity consequences of predation and host plant hybridization on an aphid–ant mutualism. *Ecology* **82**:440–452.

Wimp, G. M., G. D. Martinsen, K. D. Floate, R. K. Bangert, and T. G. Whitham. 2005. Plant genetic determinants of arthropod community structure and diversity. *Evolution* **59**:61–69.

Wolf, J. B., E. D. Brodie III, J. M. Cheverud, A. J. Moore, and M. J. Wade. 1998. Evolutionary consequences of indirect genetic effects. *Trends in Ecology and Evolution* **13**:64–69.

Wootton, J. T. 1994. The nature and consequences of indirect effects in ecological communities. *Annual Review of Ecology and Systematics* **25**:443–466.

Wright, J. P., C. G. Jones, and A. S. Flecker. 2002. An ecosystem engineer, the beaver, increases species richness at the landscape scale. *Oecologia* **132**:96–101.

Wright, J. P., A. S. Flecker, and C. G. Jones. 2003. Local vs. landscape controls on plant species richness in beaver meadows. *Ecology* **84**:3162–3173.

Part V EVOLUTIONARY CONSEQUENCES OF
        PLANT-MEDIATED INDIRECT EFFECTS

14

Evolution of plant-mediated interactions among natural enemies

TIMOTHY P. CRAIG

Introduction

Indirect evolutionary effects occur when selection by one species on a mediating species influences selection in future generations of a third species. Evolutionary interactions can be pictured in an evolutionary interaction web (Fig. 14.1) just as indirect ecological interactions can be represented in an ecological interaction web. In common with ecological interaction webs these do not involve direct trophic interaction webs. A difference is that ecological interaction webs involve alterations in mediating species that influence other species in the current generation, while in evolutionary interaction webs selection to alter the mediating species affects other species in subsequent generations. The study of the evolution of indirect interactions is not new. For example, character displacement is an indirect interaction where by competition between species for limiting resources mediates the evolution of divergent traits among competitors (Brown and Wilson 1956, Schluter et al. 1985).

The intensive study of tritrophic and multitrophic interactions has shown that interactions between species separated by at least one trophic level are widespread and important (Price et al. 1980, Barbosa and Letourneau 1988).

An example of this kind of interaction evolutionary effect occurs when plant chemical cues attract parasitoids that attack herbivores feeding on the plant. Many other kinds of indirect evolutionary effects occur in other interactions, such as the mediation of the evolution of two species on the same trophic level by a species on another trophic level, remain relatively unexplored.

Ecological Communities: Plant Mediation in Indirect Interaction Webs, ed. Takayuki Ohgushi, Timothy P. Craig, and Peter W. Price. Published by Cambridge University Press. © Cambridge University Press 2007.

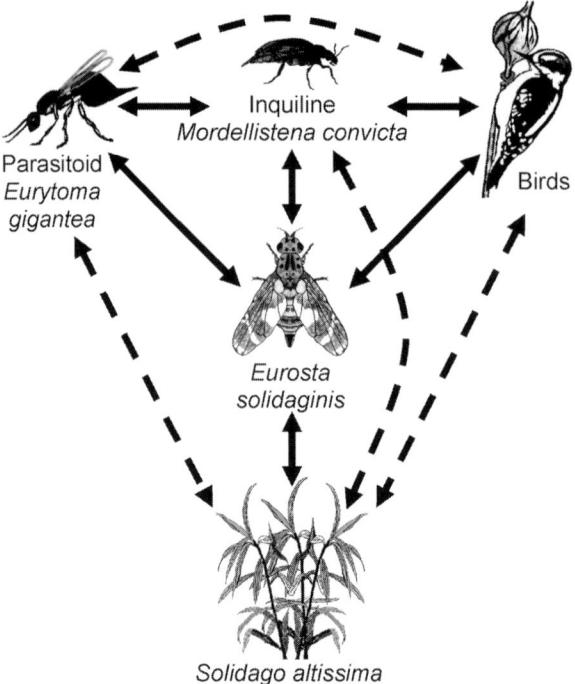

Figure 14.1. The simplified *E. solidaginis* interaction web with the host plant *Solidago altissima*, the galler *Eurosta solidaginis*, the parasitoid *Eurytoma gigantea*, the inquiline *Mordellistena convicta*, and birds represented by the downy woodpecker (*Picoides pubescens*). Direct interactions are represented by the solid lines, and indirect interactions by the dashed lines. Other interactions are possible; only those discussed in the text are included.

The coevolution of conifers and their seed predators provides a well-documented example of how the evolution of characters in two indirectly interacting species on the same trophic level can be mediated by a third species (reviewed by Benkman 2003). Lodgepole pines have cones evolved to protect their seeds from two different species: squirrels and crossbills. These species attack the pine cones in different ways. In most locations squirrels are the dominant seed predators and pine cones have evolved to protect their seeds from their feeding strategy. In a few isolated mountain areas squirrels are lacking for historical reasons and in these areas cone morphology and crossbills have coevolved. This pattern indicates that both squirrels and crossbills are exerting selection on cone morphology. Indirectly squirrels are exerting selection on crossbills, since in their absence cones differ and crossbill bill shape differs. Thus it can be said that squirrel jaw muscles exert strong selection on bill morphology although bird bills and squirrel jaws never directly interact.

In this chapter I will focus on how natural enemies, using a common host, influence each other's evolution: two species on one trophic level have their interactions mediated by a species on a lower trophic level. A failure to consider indirect evolutionary interactions may lead to a misleading understanding of the forces driving the evolution of an interaction web. The first part of the chapter will focus on the results of research on the evolution of a plant-mediated interaction web of a gall-inducing fly *Eurosta solidaginis*. Next, I will discuss other endophytic, plant-mediated interaction webs. I will conclude by discussing the problems of studying the evolution of indirect interaction webs, and some of the methods that can be used to study them.

The *E. solidaginis* interaction web

The interaction web centered on the gall-inducing fly *E. solidaginis* will be used to illustrate indirect evolutionary interactions (Fig. 14.1). *Eurosta solidaginis* is a fly that has host races which induce galls on two species of goldenrod, *Solidago altissima* and *S. gigantea* (Craig et al. 1993, 1997, 2001), but I will only discuss the host race on *S. altissima* in this chapter. *Eurosta solidaginis* oviposits into the bundle of unfolding leaves on the terminal bud of goldenrod in late May and early June in Minnesota. The larvae burrow into the stem, and their feeding action induces the formation of a gall that reaches its maximum diameter by early August. Gall size is determined by the interaction of plant genotype, fly genotype, and the environment (Weis and Abrahamson 1986).

In Minnesota *E. solidaginis* is attacked by three natural enemies. In early summer *Mordellistena convicta* (Coleoptera: Mordellidae) oviposits eggs on the surface of the gall (Eubanks et al. 2003). The beetle larvae is an inquiline that burrows into gall parenchymal tissue (Ping 1915) and kills the *E. solidaginis* larva about 70% of the time, although it can mature without consuming the fly larva or pupa (Uhler 1951, Abrahamson et al. 1989). Starting in early July a parasitoid, *Eurytoma gigantea* (Hymenoptera: Eurytomidae), oviposits through the gall wall into the *E. solidaginis* larva in the central chamber. There it consumes the host larva and the gall tissue of the inner chamber (Weis et al. 1989) *Eurytoma obtusiventris* is a parasitoid that attacks *E. solidaginis* in the north eastern USA (Uhler 1951), but it does not occur in Minnesota. Starting in the fall larvae are preyed upon by chickadees (*Poecile atricapillus*) and downy woodpeckers (*Picoides pubescens*). Other species may also be part of the web, but they are not discussed in this chapter.

The *E. solidaginis* interaction web has been extensively studied in the deciduous forest biome of north eastern USA (for a review see Abrahamson and Weis 1997). I will compare the interaction web in the forest biome with new data from the prairie biome in the north central USA, where selection differs from that in the forest, to illustrate the importance of indirect evolutionary effects.

I will focus on indirect selection by birds and the inquiline beetle on the evolution of the parasitoid *Eurytoma gigantea*. Natural enemies exert selection on the evolution of *Eurosta solidaginis* gall size (Weis and Abrahamson 1985, 1986, Weis et al. 1992, Weis and Kapelinksi 1994, Weis 1996). *Eurytoma gigantea* exerts selection for large gall size because it causes higher mortality rates on smaller galls (Uhler 1951, Cane and Kurczewski 1976, Weis et al. 1992). The parasitoid's ovipositor is too short to reach larvae in large galls (Weis et al. 1985). Downy woodpeckers and chickadees have higher attack rates on larger galls, exerting strong selection for smaller gall size (Uhler 1951, Cane and Kurczewski 1976, Confer and Paicos 1985, Weis et al. 1992). Neither *Eurytoma obtusiventris* nor *Mordellistena convicta* exerted selection on gall size in the north eastern USA forest biome (Weis and Abrahamson 1986).

Selection by natural enemies for gall size varied among sites and years in the forest due to variation in parasitoid density, bird attack, and climate (Weis et al. 1992). The combination of parasitoid mortality on small galls and bird-induced mortality on large galls creates some stabilizing selection on *E. solidaginis* in some years and sites. Overall, however, there is directional selection for larger gall size because parasitoid induced mortality is higher and more consistent than bird predation (Weis et al. 1992). Bird predation is highly variable and it is concentrated near trees (Confer and Paicos 1985, Abrahamson and Weis 1997).

The indirect evolutionary effects that natural enemies exert on the evolution of gall size also impacts *S. altissima*. Selection on *E. solidaginis* for increased gall size by natural enemies in turn exerts selection on the plant to resist increased allocation to gall tissue that will divert resources from plant growth and reproduction (Abrahamson and Weis 1997). *Mordellistena convicta* and *E. obtusiventris* also exert indirect selection on gall size by mediating the attack rates of birds on galls because birds have lower attack rates on galls containing parasitoids or inquilines. These galls may be avoided either because birds use *E. solidaginis* exit tunnels to locate larvae (Moeller and Thogerson 1978, Abrahamson and Weis 1997) or because they find the *M. convicta* and *E. obtusiventris* larvae distasteful (Cane and Kurczewski 1976, Schlichter 1978). Thus *M. convicta* and *E. obtusiventris* remove galls from bird predation indirectly exerting selection on the plant for gall size by curtailing bird attack.

Geographical variation in the interaction

The indirect interactions in the *E. solidaginis* interaction web vary geographically, and geographic variation can be used to determine how interactions evolve and coevolve (Thompson 1994). My colleagues and I have studied variation in the *E. solidaginis* interaction web in Minnesota in the north central USA on both sides of the border between the prairie and forest biomes. The biomes are differentiated in their flora and fauna, but many species are found on both sides of the prairie–forest border including the *E. solidaginis* interaction web on *S. altissima altissima*. There are many biotic and abiotic differences between the biomes (Tester 1995) that could alter indirect evolutionary interactions. I will discuss the hypothesis that the differences in the ovipositor length of *E. gigantea* populations in the two biomes are due to its indirect interactions with birds and *M. convicta* that are mediated in turn by gall size.

Eurytoma gigantea ovipositors were significantly longer in the prairie than in the forest (T.P. Craig and J.K. Itami unpublished data) (Fig. 14.2). These data showing large differences in ovipositor length are from sites collected near the border between the forest and prairie, over distances of less than 50 km. Weis *et al.* (1985) reported slightly shorter ovipositors from forest sites in Pennsylvania, 1800 km east of our farthest east site in Minnesota than we found in our forest samples although the standard deviations of these forest populations are widely overlapping.

Eurosta solidaginis gall sizes were larger in the prairie than in the forest (Fig. 14.3) with significantly more galls in the larger gall size categories in the prairie than in the forest. Prairie gall characteristics were maintained when prairie flies induced galls on prairie plants in a garden in the forest biome indicating that gall characteristics have a genetic component (T.P. Craig and J.K. Itami unpublished data). Weis *et al.* (1992) report gall sizes from the forest biome in the Pennsylvania area that are intermediate between those we found in the forest and prairie in Minnesota. Thus the general relationship between gall size and ovipositor length seems to be found over a wide geographic area.

The distribution of gall sizes parasitized by *E. gigantea* was significantly different from what would be expected if the parasitoid had attacked the same size categories in the prairie and the forest. *Eurytoma gigantea* attacked larger galls in the prairie than *E. gigantea* in the forest; and this pattern would not be expected if some factor other than ovipositor length limited the size of galls attacked. A simulation model by Weis *et al.* (1985) predicts a sigmoidal decline in *E. gigantea* oviposition success with increasing gall size before reaching the absolute limit set by ovipositor length on the depth of larvae that can be attacked. I lacked all of the information necessary to make estimates of parasitoid success using this model, and so instead I determined the percentage of larvae within reach of the mean ovipositor length in each population. This probably overestimates the vulnerability of larvae. We estimated that forest *E. gigantea* could reach *E. solidaginis* larvae in 84% of the forest galls; if forest *E. gigantea* had

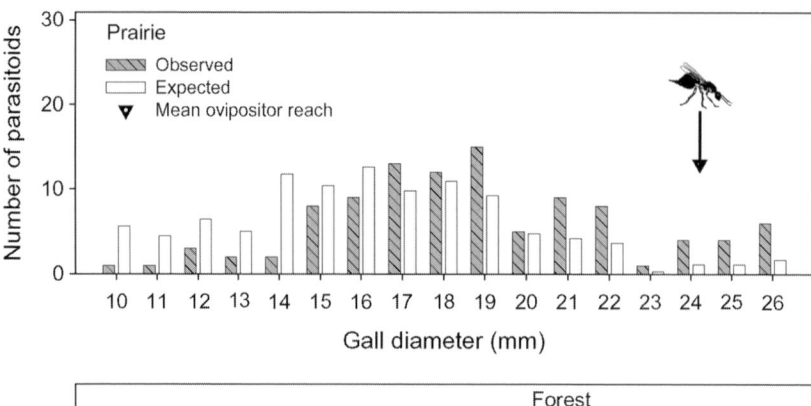

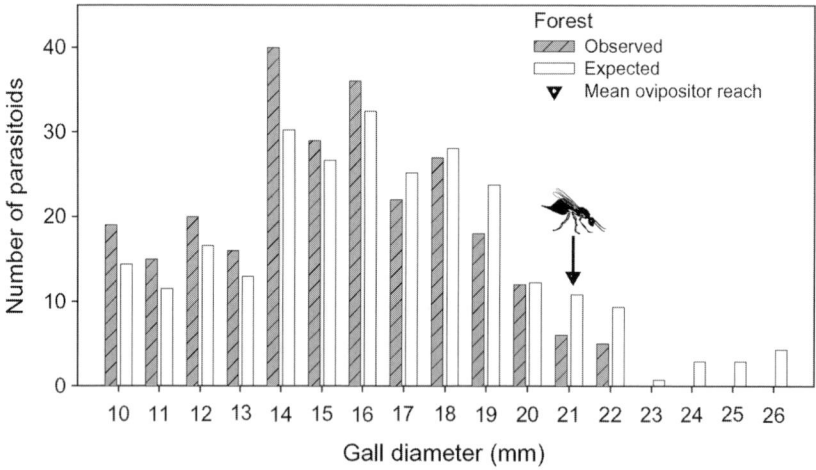

Figure 14.2. The range of gall sizes from which *Eurytoma gigantea* emerged in the forest and prairie, and the expected distribution if *E. gigantea* attacked the same range of gall sizes in both biomes. A goodness of fit test to the expected equal distribution was rejected ($\chi^2 = 87.588$, $p < 0.001$, d.f. = 17). The data are from the pooled results of 5 years of dissection data from several sites in both biomes. The arrows indicate the mean ovipositor length in both biomes from a sample of 225 ovipositors in both biomes.

been exposed to prairie galls they would have been able to reach 33.4% of the larvae. If forest *E. gigantea* were moved to the prairie they would be under strong selection to evolve longer ovipositors. Even with these longer ovipositors we estimate that prairie *E. gigantea* could reach only 57% of larvae in prairie galls. If prairie *E. gigantea* were exposed to forest galls they would be able to reach 97% of the larvae.

Bottom–up cascades and ovipositor lengths

The abiotic differences in the two biomes have effects that cascade upwards to influence *E. gigantea* ovipositor length. Differences in temperature,

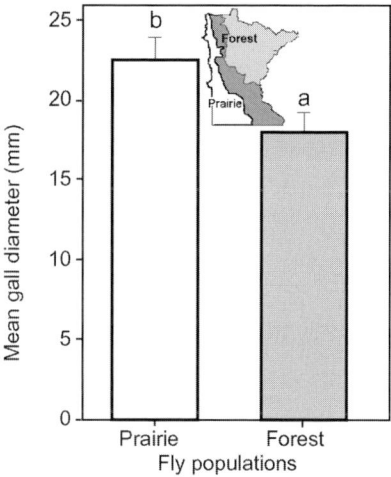

Figure 14.3. The mean gall diameter (± SE) for three populations in the prairie and forest biomes separated by approximately 45 km in north western Minnesota, USA.

rainfall, and soil type as you move from east to west in Minnesota produce a cline in growing-season–soil-water availability that creates a shift from forest to prairie in Minnesota (Tester 1995). The absence of trees in the prairie reduces the number of the woodland dwelling species such as chickadees (Smith 1993) and downy woodpeckers (Jackson and Ouellet 2002) lowering the bird predation on *E. solidaginis* galls. These woodland birds extend their ranges into the prairie wherever trees are found along streams and rivers, or where trees have been planted after European settlement of the prairies. However, birds do not disperse far from woodland edges (Cane and Kurczewski 1976, Abrahamson and Weis 1997) so galls in undisturbed prairies would not be likely to be preyed upon.

In a survey of bird predation on 12 prairie and 12 forest sites that had been protected by the state of Minnesota to preserve their natural vegetation we found no bird predation in the prairie, and high but variable bird predation in the forest (Fig. 14.4a). This produces selection against large galls in the forest (Fig. 14.4b) which was absent in the prairie (Fig. 14.4c). In contrast *E. gigantea* and *M. convicta* mortality in the prairie was higher (Fig. 14.4a) than in the forest. *Eurytoma gigantea* and *M. convicta* caused higher mortality on *E. solidaginis* in small galls in both biomes (Fig. 14.4b,c), and bird predation increased with gall size in the forest (Fig. 14.4b). This combination of size selective attack created strong stabilizing selection in the forest, and strong directional selection for larger gall size in the prairie (Craig and Itami submitted). Our results differ from Weis *et al.* (1992) who did not find that *E. solidaginis* mortality due to *M. convicta* was related to gall size in the north eastern USA.

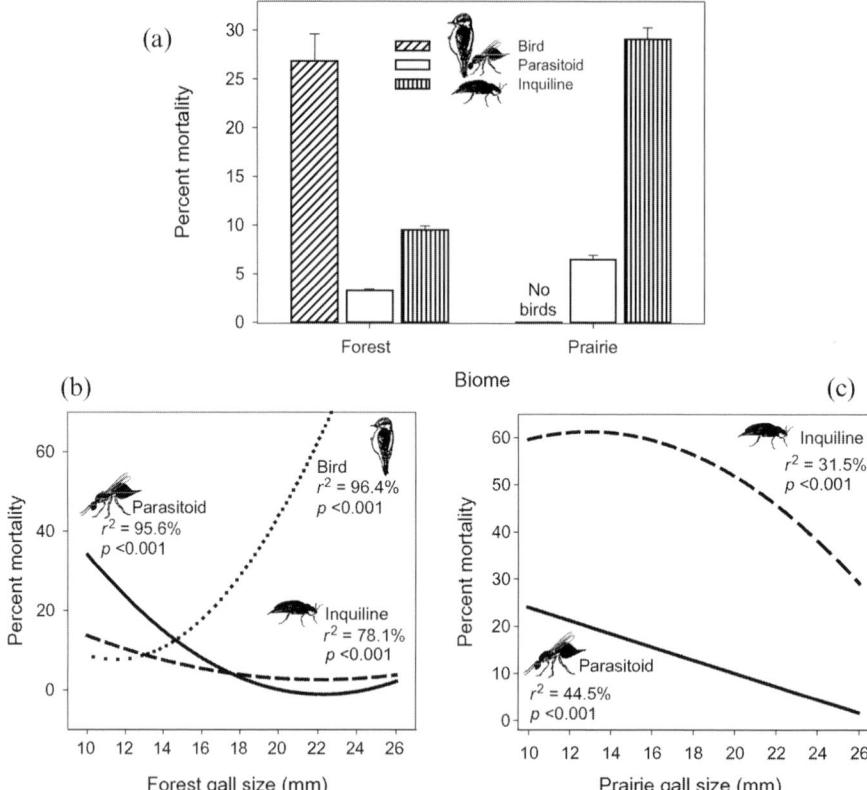

Figure 14.4. (a) The mean percent mortality (± SE) of *Eurosta solidaginis* due to natural enemies in the forest and prairie biomes from a wide range of sites from cohorts initiated between 1998 and 2003. (b) The relationship between gall diameter and *Eurosta solidaginis* mortality caused by: *Eurytoma gigantea*, *Mordellistena convicta*, and bird predation in the forest biome in Minnesota. The data are pooled from 5760 galls collected from 42 sites over a 5-year period. Galls in which larvae had died before attack by any of the natural enemies were excluded from the analysis. (c) The relationship between gall diameter and *Eurosta solidaginis* mortality caused by: *Eurytoma gigantea*, *Mordellistena convicta*, and bird predation in the prairie biome in Minnesota. The data are pooled from 1879 galls from 20 sites over a 5-year period. Galls in which larvae had died before attack by any of the natural enemies were excluded from the analysis.

Mordellistena convicta also selects for larger gall size, and longer *E. gigantea* ovipositors. *Eurosta solidaginis* mortality due to *M. convicta* selection decreases with gall size while *M. convicta* survival increases (T.P. Craig and J.K. Itami unpublished data). *Eurosta solidaginis* in large galls with *M. convicta* may have lower mortality because the inquiline has more gall tissue to feed on without encountering the *E. solidaginis* larva, whereas in a very small gall it is almost inevitable that it will encounter and kill the *E. solidaginis*. This would produce selection for an increase in gall size, while also increasing the density of the *M. convicta*.

Bird predation in the forest thus appears to be the key constraint maintaining selection for shorter ovipositors. It indirectly selects for plants to allocate less tissue to galls through selection on *E. solidaginis*. Small galls reduce the resources available to *M. convicta* keeping their populations low, and therefore diminish their selection for larger gall size. Together the selection against large gall size results in smaller galls, and the evolution of shorter *E. gigantea* ovipositors. When the constraint of bird predation is removed in the prairie it leads to the evolution of larger galls influencing both *E. gigantea* and *M. convicta*.

Decreased bird predation could result in a coevolutionary arms race between *E. gigantea* and *E. solidaginis*: increasing gall size selects for longer ovipositor lengths and vice versa producing an escalating cycle. Weis *et al.* (1989) found, however, that there could be escalation in this arms race between these species without coevolution if ovipositor length is not a heritable trait. They found that *E. gigantea* ovipositor length is correlated with the size of gall from which it emerged because the parasitoid feeds on gall tissue once the *E. solidaginis* has been consumed. Simulation modeling showed that ovipositors could increase due to a nongenetic increase in parasitoid size while selecting for increased gall size (Weis *et al.* 1989). The cycle could not continue indefinitely because at larger gall sizes the gall wall thickness increased faster than ovipositor length increased, and the parasitism rate would decrease. It is important to note that Weis *et al.* did not rule out the possibility that heritable genetic variation for ovipositor length could also produce an increase in gall size. If there is no heritable variation for ovipositor length, then birds still indirectly contribute to determining the phenotype of ovipositor length in a site.

If ovipositor length is a heritable trait then how is the polymorphism in lengths maintained? One possibility is that ovipositor differences are maintained by divergent selection in spite of gene flow between the populations in the two biomes. Alternatively the *E. gigantea* populations may be two partially or completely reproductively isolated populations. These populations may have diverged in sympatry or during a past period of allopatry. The populations show no apparent current geographical isolation: the prairie and forest plant–galler–natural enemy communities exist sympatrically in a complex mosaic distribution along the border of the prairie and forest biomes. *Eurytoma gigantea* evidently have some ability to disperse across at least short distances of unfavorable habitat: I have found *E. gigantea* in galls that are apparently isolated from other populations by several kilometers of plowed fields. Differentiation may have been initiated during a past period of allopatry as populations may have been isolated during the extensive past glaciations in this region. We cannot discriminate among these hypotheses without more information about current reproductive isolation, and past distributions.

Reciprocal evolutionary forces are absent from the indirect interactions among natural enemies of *E. solidaginis*. Birds exert selection on *E. gigantea* ovipositors, but there seems little possibility that *E. gigantea* influences bird characteristics. In contrast to the other members of the interaction web there is no obligate, intimate association of birds with any of the other organisms in the web reducing selection on birds. The birds evidently opportunistically utilize *E. solidaginis* larvae as a small part of a diverse diet, and they show no evidence of being specifically adapted to this interaction (Abrahamson and Weis 1997). *Mordellistena convicta* also indirectly selects for increased ovipositor length in *E. gigantea* by selecting for increased gall size. *Eurytoma gigantea* selection for increased gall size would exert no reciprocal selection on *M. convicta* that oviposit on the surface of the gall so there is no reciprocal selection for morphological changes in the beetle with a change in gall size.

Other factors may also exert selection on gall size and in turn *E. gigantea* ovipositor length. Increased gall size could increase *E. solidaginis* fitness through improved larval nutrition, but this hypothesis has not been supported. Larger galls do not have increased larval survival independent of the effects of natural enemies. The nutritive layer of galls is continuously renewed and it is the rate of production of this layer rather than the size of the pith that determines survival (Abrahamson and Weis 1997). Gall size has also been hypothesized to buffer environmental impacts, but experiments have not shown any impact on temperature regulation (Layne 1993). There is the possibility that some other unmeasured abiotic factor could influence the evolution of gall size.

Despite selection for larger gall size in the north east USA and in the prairie in Minnesota there is no evidence for the evolution of increased gall size during the period when the interaction has been studied (Weis et al. 1992, T.P. Craig and J.K. Itami unpublished data). As Weis et al. (1989) asked: "Why haven't galls evolved to the size of baseballs?" Either there is an arms race where the parasitoid temporarily has the advantage, or there is unidentified stabilizing selection that provides a constraint that prevents larger galls from evolving (Abrahamson and Weis 1997). The most logical candidate is that galls impose a strong cost on the plant that leads to the evolution of resistance to increased gall size by the plant. Although, there is little support for this hypothesis from studies on the impact of increased gall size on *S. altissima* growth and reproduction

(Abrahamson and Weis 1997), other mechanisms by which plant resistance could influence gall size remain to be explored.

In both biomes a large proportion of the larvae cannot be reached by the *E. gigantea* ovipositor, so unless there are other constraints on *E. gigantea* we would expect longer ovipositors to evolve. We do not have long-term data on ovipositor lengths, but if longer ovipositors can evolve in the prairie why don't they evolve in the forest? Or why hasn't the population from the prairie invaded the forest biome where their longer ovipositors would give them advantage in competition with the local population with shorter ovipositors? There may be costs to longer ovipositors that outweigh their benefits which we have not identified.

Abiotic differences between biomes may also cause changes in the host plant characters that have the potential to have effects that cascade upwards. *Solidago altissima* plants in the prairie and forest are differentiated in their phenology and morphology (T. P. Craig and J. K. Itami unpublished data). Alteration in the host plant may also exert selection on *E. solidaginis* for changes in phenology, or gall morphology that may influence natural enemies although this hypothesis has not been tested.

Indirect evolutionary interactions in other endophytic systems

Several characteristics of herbivore interaction webs indicate that the potential for the evolution of indirect interactions among natural enemies is widespread, but the hypothesis has rarely been examined in other interactions. First, the indirect interactions between natural enemies can occur at any trophic level: plants can mediate the evolution of interactions among herbivores, herbivores can mediate the evolution of interactions among natural enemies, herbivores can mediate interactions among parasitoids, and parasitoids can mediate the evolution of hyperparasitoids. Second, these interactions can involve any number of natural enemies ranging from a strongly reciprocally interacting pair to a single species being influenced by the joint selective pressure on a mediating species by many other species. Third, natural enemies may interact by exerting selection on the mediating species in the same direction or opposing directions. For example, in the *E. solidaginis* interaction web *Modellistena* indirectly selected for increased, and birds for decreased, *E. gigantea* ovipositor length. Two traits are necessary for evolution of indirect effects between natural enemies: multiple species that are influenced by herbivore traits, and at least one natural enemy that exerts selection on herbivore traits. The evolution of indirect evolutionary effects could be prevented in a species if it interacts with other species through many mediating species. For example, many parasitoids have a wide host range (Godfray 1994) and opposing selection on such traits as ovipositor length could limit the response to any one species.

Many interaction webs centered on galls other than the *E. solidaginis* web have multiple enemies that have the potential to influence herbivore traits that would influence the evolution of other natural enemies. The divergence of gall morphology has been hypothesized to be due to selection for protection against natural enemies (Cornell 1983, Price et al. 1997, Stone et al. 2002, Stone and Schonrogge 2003), and it seems likely that morphology was influenced by the interaction of selection by multiple enemies. Askew (1961) proposed that the divergence of the morphology of cynipid galls in the community was due to parasitoid-mediated competition resulting in galls that have unique parasitoid communities. Many gall traits could potentially mediate natural enemy interactions including gall size, gall toughness, the number of gall chambers, surface spines, and internal air spaces.

Multiple natural enemies may act to influence the evolution of gall size which has been shown to mediate the interactions of natural enemies in a number of gall-centered interaction webs. Ovipositor length, and attack phenology could evolve in response to changes in gall size due to these indirect interactions. Askew (1961, 1975) detailed how complex interactions within the interaction web of cynipid galls could be modified by the action of gall-makers. Idiobont ectoparasitoids stop gall growth, making the occupants vulnerable to later attacking inquilines and parasitoids, and potentially influence the evolution of these later natural enemies. Van Hezewijk and Roland (2003) found, in an interaction that parallels the *E. solidaginis* interaction web, that the gall midge *Rhabdophaga strobiloides* suffers mortality from birds, and two parasitoids. Each natural enemy had a different mean gall size attacked, and that this could exert selection on gall size with indirect effects on each natural enemy. The composition of a parasitoid community of the gall midge *Asphondylia borrichiae* on three host plants is determined by plant-mediated competition (Stiling and Rossi 1994) that is determined by gall size. The potential exists for indirect evolution among natural enemies because ovipositor length determines the gall size attacked and the dominant competitor *Torymus umbilicatus* is limited by ovipositor length.

The evolution of gall size in the gall midge *Asphondylia atriplicis* on four-wing saltbush *Atriplex canescens* may also have been influenced by the interaction of multiple natural enemies, and this in turn may have influenced the evolution of the indirect interactions among natural enemies (Marchosky and Craig 2004). *Tetrastichus cecidobroter* is the organizer of the natural enemy community in this web (Hawkins and Goeden 1984). This parasitoid has the longest ovipositor among the natural enemies and its parasitism is negatively correlated with parasitism by the other parasitoid species. Marchosky and Craig (2004) found that *A. atriplicis* gall size varied over three orders of magnitude, and that gall size determined parasitism rates by a group of parasitoids from the family Eurytomidae, suggesting that *T. cecidobroter* and this eurytomid group could exert selection on each other through selection on gall size.

The number of chambers in a gall influences gall size and this too may have evolved under selection by multiple parasitoids creating the potential for indirect evolution. Plakidas and Weis (1994) found that the number of chambers in *Asphondylia rudbeckiaeconspicua* galls determined the maximum depth of the larvae, and that larval survival increased with depth. The mean depth of attack by each of the three parasitoids was different indicating the potential for indirect interactions based on selection for host chamber number. Tabuchi and Amano (2004) hypothesized that the number of gall chambers in two closely related species of Cecidomyiidae was determined by the attack rates of two different groups of parasitoids. The ectoparasitoids attacked galls with small numbers of chambers, and endoparasitoids attacked those with large numbers of chambers. These opposing selection pressures would indicate that each species could exert selection on the galler for a change in the number of gall chambers that would in turn influence the other parasitoids. The inquiline *Periclistus* alters the morphology and number of larval chambers in galls formed by the cynipid galler *Diplolepis rosaefolii* (Leblanc and Lacroix 2001). Thus the interaction of the galler and the inquiline could alter the indirect evolutionary interactions of other inquilines and parasitoids.

Gall toughness is another trait that may mediate the evolution of natural enemies, and gall toughness can be influenced by organisms other than the plant and the galler adding further complexity to the evolution of indirect effects. Askew (1961) found that attacks by parasitoids of the genus *Olynx* makes galls harder, influencing attack by later parasitoids. Weis (1982) found that the gall-maker *Astermoyia carbonifera* utilizes a fungus that makes the gall tougher as the gall grows and limits the attack period of the parasitoid *Torymus capite*. Three other parasitoids attack the gall-maker so the potential exists that gall toughness influenced by both the galler and the fungus could mediate the evolutionary interaction between parasitoids.

Gallers can have a window of vulnerability to parasitoid attack that is determined by host development and/or gall development (Washburn and Cornell 1981, Craig et al. 1990, Stiling and Rossi, 1994, Biggs and Latto 1996). Selection for escape on the host by one natural enemy could influence the selection on the attack strategy of other natural enemies. Craig et al. (1990) found that vulnerability of *Euura lasiolepis* to the ichneumonid *Lathrostizus euurae* was limited to the larval period before galls became too tough to be penetrated. Selection on the host to limit the window of vulnerability could act to slow development rate or to increase the rate of gall toughening and growth which could affect other parasitoids.

Alteration of the phenology of the host by natural enemies could also lead to the evolution of indirect effects in an interaction web. Askew and Shaw (1986) describe how two endoparasitoids of *Leucoma salicis* alter the phenology of their host: *Cotesia melanoscelus* advances host emergence times and *Aleiodes pallidator* delays its emergence. This affects the interaction of its host and its host plants and with hyperparasitoids, and this creates the potential of indirect interactions evolving between the primary endoparasitoids and host plants or the host and endoparasites.

The community of agaonid wasps that utilize figs presents interesting possibilities for the evolution of indirect evolutionary interactions. Each fig species generally has one host-specific wasp mutualist that induces galls in the fig ovary, and pollinates the fig (Weiblen 2002). Most figs have many other wasps that also form galls, but are not pollinators, and other wasps that are parasitoids or inquilines of the gallers (Kerdelhue et al. 2000, Weiblen 2002). The community of wasps attack in sequence as the fruit, the galls and the rest of the community develops. Ovipositors are progressively longer in the later-attacking wasps (Kerdelhué and Rasplus 1996). The ability to attack the hosts by all the parasitoids and inquilines is influenced by the size of the developing fruit, and or the size of the developing galls which are derived from the developing ovules. The phenology of the fruit and the pollinator is a coevolved trait (Galil and Eisikowhich 1968), thus the interaction of the rest of the community including the phenology of attack and ovipositor length has mediated this coevolved interaction.

Natural enemies of other endophytic species could also indirectly exert selection on each. Paine et al. (2000) found that the success of Braconidae parasitoids of the longhorned borer *Proacantha semipunctata* were influenced by ovipositor length: two species with shorter ovipositors parasitized larvae under thinner bark, and the two with longer ovipositors parasitized those under thicker bark. Freese (1995) found that the parasitoid community on two species of stem-boring weevils on *Rumex crispus* was determined by their position in the stem. In these interactions the selection pressure by each parasitoid on the host's site selection could influence the evolution of the other parasitoid's ovipositor length or period of attack.

Problems with detecting and evaluating indirect effects

The description and measurement of indirect interactions has lagged behind those of direct interactions as there are inherent difficulties in evaluating the evolution of indirect effects. If the generation time of the mediating species is different from the indirectly interacting species there will be a time lag in the expression of selection, making it difficult to detect the effects of selection. In the *E. solidaginis* interaction web the generation of the *E. gigantea* and *E. solidaginis* are the same so there would be no time lag in the reciprocal selection for gall diameter and ovipositor length. Gall size is also influenced by the plant (Weis and Abrahamson 1986) so that indirect selection by *E. gigantea* for a change in gall size could lag. *Solidago altissima* clones are long-lived, often living for more than 50 years (Abrahamson and Weis 1997), so there would a lag in the expression of any change in gall morphology. This may be a common situation where interaction webs include univoltine or multivoltine insects and long-lived trees such as clonally reproducing aspen and cottonwoods (Whitham 1983). An extreme example of time lags is the interaction of present-day seed predators and extinct megafauna. Janzen and Martin (1982) hypothesized that many large fruits in the Americas have characters that adapted them for dispersal by megafauna that went extinct more than 10 000 years ago. These fruits have retained traits such as large size and tough seed coats even though they are not well adapted for obtaining dispersal services from present-day organisms. These traits must also have evolved in response to seed predators present 10 000 years ago. Because these fruit traits are retained, gomphotheres and mastodons continue to exert selection on present-day seed predators and herbivores. To my knowledge no one has documented specific adaptations that have evolved in the present-day organisms that use these megafauna fruits, but this would offer excellent opportunities for studies of indirect interactions among extinct mutualists and present-day natural enemies.

Indirectly interacting natural enemies may also be separated spatially. Bird predation is often absent in the eastern deciduous forest due to the local distribution of trees and the unpredictable movements of birds (Weis *et al.* 1992). Selection for gall size by birds in other areas and years and gene flow determined gall size in these sites. In the absence of more extensive surveys one would have reached erroneous conclusions about the shape of the selection curve. If gene flow in the mediating species is different from that of indirectly interacting species this would result in maladaptation by the local natural enemies, and confusing results for the researcher. If for example, long distant dispersal of *S. altissima* seeds resulted in prairie population plants being found in the forest region it would result in a mismatch between parasitoid ovipositors and gall size.

Another reason that indirect evolutionary interactions are difficult to detect is that they are not necessarily produced by direct or indirect ecological interactions producing measurable changes within a generation. For example, parasitoid mortality does not alter the resources available to birds in the current generation since gall size is unchanged and birds will consume parasitized *E. solidaginis*. Only in future generations would the larger gall size in subsequent generations increase the number of galls attacked by birds.

Evaluating indirect selection

Tools exist by which the difficulties of evaluating indirect selection can be overcome. There has been an explosion in interest in measuring selection on specific phenotypic traits such as gall size (Brodie et al. 1995), and these techniques can be extended to the study of indirect selection. The simplest approach is to calculate the selection differential, which is the difference between the mean phenotype before and after selection (Falconer 1981). Further insight can be gained by analyzing natural selection as the covariance between the value of a trait, such as gall size, and the expected relative fitness values (Brodie et al. 1995). Lande and Arnold (1983) showed how multiple regression techniques can be used to analyze selection. This approach uses the components of selection functions to determine the strengths of different modes of selection including directional, stabilizing, and disruptive. For example, gall size is under directional selection by *E. gigantea* and this is indicated by the coefficient β in a linear regression. Gall size is also under stabilizing selection in some sites due to the interaction of selection by birds and *E. gigantea* in some sites, and this is indicated by the coefficient γ from a quadratic regression. Because it can be difficult to discriminate the internal peaks and valleys of selection, graphical methods of analyzing the shape of fitness functions have been developed such as the cubic spline technique (Schluter 1988). Weis et al. (1992) used this technique to evaluate the mode of selection on *E. solidaginis* gall size in the *E. solidaginis* interaction web in the eastern forest biome. These techniques could be extended to analysis of indirect effects. For example, if two parasitoids utilized the same gall species the selection of one parasitoid on the ovipositor length of the other could be analyzed.

Path analysis is a refinement of the regression approach to measuring phenotypic selection, and it allows the development of causal hypotheses about relationship of variables in the model predicting fitness (Crespi 1990). The difference in approaches can be illustrated in the situation where correlated gall characters such as gall diameter, gall length, and length of trichomes on a gall surface were all used to predict galler fitness. In the multiple regression approach no assumptions are made about the cause and effect relationships in a model and the covariances of all characters in the model are adjusted to every other character in predicting fitness. In the path analysis approach the researcher could develop a model that hypothesized an underlying factor such as gall size that was correlated with all of these characters. A model where gall size predicted variation in all three of those variables and subsequently galler fitness would be specified with a path diagram. The model would then be evaluated by examining the amount of covariance of the characters and fitness. The model would be subsequently revised to reflect biological reality in each cycle of modeling and analysis. The technique could be used to evaluate the impact of direct and indirect selection on natural enemy traits such as ovipositor length.

Adding geographic variation and experimental manipulations would make the technique more powerful. One problem in measuring selection is that many traits are at equilibrium and there is little remaining heritable variation left for selection to act on. Utilizing geographic variation in an interaction can overcome these problems. First, a comparison of path diagrams in areas with and without the presence of a selective force can indicate its importance. Second, reciprocal transplant experiments between populations where different equilibriums have been reached will allow evaluation of selective forces. For example, reciprocally transplanting large, round prairie galls to a forest site with high bird predation and *E. gigantea* with short ovipositors would test the prediction that there would be strong selection for decreased gall size on *E. solidaginis* and for increased ovipositor length on *E. gigantea*. The reciprocal transplant would be predicted to show that there was strong selection in the prairie for larger galls.

How plant traits mediate indirect evolutionary effects

The trajectory of the evolution between the indirectly interacting organisms discussed in this chapter cannot be understood unless the characteristics of the mediating plant are understood. I have focused on reviewing studies that indicate that in specific cases plant characters can mediate the evolution of indirect evolutionary effects, but as studies continue I predict that general patterns will emerge as to how specific plant characters affect other trophic levels. The "plant mediation" hypothesis is that evolutionary forces shaping the characteristics of the mediating plants will ultimately determine the evolutionary interactions of higher trophic levels. The phylogenetic constraints of the mediating species may determine how the interaction between the indirectly interacting species evolves. Plant growth form can impose many constraints on the way that they respond to selection by either herbivores or natural enemies.

Forces from any trophic level can potentially influence a plant's evolution, but the bottom–up forces are potentially the most important because they are always present and set the template on which other evolutionary forces act (Price 2002). A parasitoid may exert strong selection on a herbivore to induce large, tough galls, but whether such galls evolve depends on the evolutionary constraints on the plant. For example, if an early successional plant evolves a strategy of rapid, herbaceous growth with thin stems to colonize disturbed sites large, tough galls could not evolve. Thin stems could not support large galls, and a lack of lignin in the stem tissues would limit gall toughness. In a woody, later successional plant, large tough galls could be produced. If the same parasitoids attacked related gallers on the woody and herbaceous plants the outcomes of the indirect actions among these natural enemies would be different. On woody plants selection for large tough galls by one parasitoid could place selective pressures on other parasitoids, while on the herbaceous plant large tough galls could not evolve and other characters would mediate the interaction of the parasitoids. A within-plant shift of a galler from a stem to leaf would also change the interaction: large tough galls are also less likely to evolve on leaves than on stems.

The comparative method could be used to test the importance of the plant characters in mediating the interaction of natural enemies. The plant mediation hypothesis predicts that in interaction webs of closely related galler species using similar plant resources similar types of interactions would evolve among natural enemies, while those utilizing different plant resources would have very different indirect interactions. Gallers have undergone adaptive radiation both within parts of the same plant and among plants of related species making tests of such predictions possible. For example, the adaptive radiation of sawflies (Hymenoptera: Tenthredinidae) on the Salicaceae has resulted in closely related species being found on different plant parts (leaves, buds, petioles, stems) on the same host species, and in closely related sawflies forming galls on the same plant parts on different host species (Nyman *et al.* 2000, Price 2003). This provides an ideal opportunity to compare the evolution of indirect interaction webs among natural enemies using closely related galling species on different resources, with different plant-mediated characteristics. Closely related sawfly gallers are found on leaves, petioles, and stems of *Salix lasiolepis*, and the interactions of the natural enemy community on the different resources are very different (Craig 1994); more detailed studies are needed before conclusions about indirect interactions can be drawn. Other rich adaptive radiations including those on oaks by cynipids (Askew 1961, 1975), and figs and fig wasps (Weiblen 2002) would also provide opportunities for such comparative studies.

Thus while the genomes of several other organisms may be involved in the evolution of these indirect interactions it is the plant that ultimately constrains the evolution of these interactions. The plant sets the template on which the indirect interaction initially evolves, and this sets the boundaries on how the interaction can evolve. The outcome of the interaction cannot be predicted based on knowledge of the indirectly interacting organisms alone: it requires a detailed knowledge of the plant.

Conclusion

The *E. solidaginis* interaction web shows that indirect interaction among natural enemies of a mediating species has the potential to influence the evolution of these species. We have focused on physical traits, but many other characteristics such as chemical interactions have the potential to evolve indirectly. These indirect interactions are probably widespread and have different characteristics than direct interactions. The utilization of new approaches is needed to uncover these interesting and complex interactions.

Acknowledgments

I thank K. Carzares, J. V. Craig, J. A. Craig, J. D. Horner, J. K. Itami, K. Johnson, and J. Marincel for helping in data collection. J. K. Itami drew the figures. I also thank J. K. Itami, T. Ohgushi, P. W. Price, and two anonymous reviewers for comments on the manuscript.

References

Abrahamson, W. G., and A. E. Weis. 1997. *Evolutionary Ecology across Three Trophic Levels: Goldenrods, Gallmakers, and Natural Enemies.* Princeton, NJ: Princeton University Press.

Abrahamson W. G., J. F. Sattler, K. D. McCrea, and A. E. Weis. 1989. Variation in selection pressures on the goldenrod gall fly and the competitive interactions of its natural enemies. *Oecologia* **79**:15–22.

Askew, R. R. 1961. On the biology of the inhabitants of oak galls of cynipidae (Hymenoptera) in Britain. *Transactions of the Society for British Entomology* **14**:237–268.

Askew, R. R. 1975. The organization of chalcid-dominated parasitoid communities centred upon endophytic hosts, pp. 130–153 in P. W. Price (ed.) *Evolutionary Strategies of Parasitic Insects and Mites.* New York: Plenum Press.

Askew, R. R., and M. R. Shaw. 1986. Parasitoid communities: their size, structure and development, pp. 225–264 in J. Waage and D. Greathead (eds.) *Insect Parasitoids.* London: Academic Press.

Barbosa, P., and D. K. Letourneau (eds.) 1988. *Novel Aspects of Insect–Plant Interactions.* New York: John Wiley.

Benkman, C. W. 2003. Reciprocal selection causes a coevolutionary arms race between crossbills and lodgepole pine. *American Naturalist* **162**:182–194.

Biggs, C. J., and J. Latto. 1996. The window of vulnerability and its effect on relative parasitoid abundance. *Ecological Entomology* **21**:128–140.

Brodie, E. D. III, A. J. Moore, and F. J. Janzen. 1995. Visualizing and quantifying natural selection. *Trends in Ecology and Evolution* **10**:313–318.

Brown, W. L., Jr., and E. O. Wilson. 1956. Character displacement. *Systematic Zoology* **5**:49–64.

Cane, J. T., and F. E. Kurczewski. 1976. Mortality factors affecting *Eurosta solidaginis* (Diptera; Tephritidae). *Journal of the New York Entomological Society* **84**:275–292.

Confer, J. L., and P. Paicos. 1985. Downy woodpecker predation on golden-rod galls. *Field Ornithology* **56**:56–64.

Cornell, H. V. 1983. The secondary chemistry and complex morphology of galls formed by the Cynipinae (Hymenoptera): why and how? *American Midland Naturalist* **110**:225–234.

Craig, T. P. 1994. Effects of intraspecific plant variation on parasitoid communities, pp. 205–227 in B. D. Hawkins and W. Sheehan (eds.) *Parasitoid Communities*. Oxford, UK: Oxford University Press.

Craig, T. P., J. K. Itami, and P. W. Price 1990. The window of vulnerability of a shoot-galling sawfly to attack by a parasitoid. *Ecology* **71**:1471–1482.

Craig, T. P., J. K. Itami, W. G. Abrahamson, and J. D. Horner. 1993. Behavioral evidence for host-race formation in *Eurosta solidaginis*. *Evolution* **47**:1696–1710.

Craig, T. P., J. D. Horner, and J. K. Itami. 1997. Hybridization studies on the host races of *Eurosta solidaginis*: implications for sympatric speciation. *Evolution* **51**:1552–1560.

Craig, T. P., J. D. Horner, and J. K. Itami. 2001. Genetics, experience and host-plant preference in *Eurosta solidaginis*: implications for host shifts and speciation. *Evolution* **55**:773–782.

Crespi, B. J. 1990. Measuring the effect of natural selection on phenotypic interaction systems. *American Naturalist* **135**:32–47.

Eubanks, M. D., C. P. Blair, and W. G. Abrahamson. 2003. One host shift leads to another? Evidence of host-race formation in a predaceous gall-boring beetle. *Evolution* **57**:168–172.

Falconer, D. S. 1981. *Introduction to Quantitative Genetics*, 2nd edn. New York: John Wiley.

Freese, G. 1995. Structural refuges in two stem-boring weevils on *Rumex crispus*. *Ecological Entomology* **20**:351–358.

Galil, J., and D. Eisikowich, 1968. On the pollination ecology of *Ficus sycomorus* in East Africa. *Ecology* **49**:259–269.

Godfray, H. C. J. 1994. *Parasitoids: Behavioral and Evolutionary Ecology*. Princeton, NJ: Princeton University Press.

Hawkins, B. A., and R. D. Goeden. 1984. Organization of a parasitoid community associated with a complex of galls on *Atriplex* spp. in southern California. *Ecological Entomology* **9**:271–292.

Jackson, J. A., and H. R. Ouellet. 2002. Downy woodpecker (*Picoides pubescens*). In A. Poole and F. Gill (eds.) *The Birds of North America*, no. 613, Philadelphia. Washington, DC: American Ornithologists' Union.

Janzen, D. H., and P. S. Martin. 1982. Neotropical anachronisms: the fruits the gomphotheres ate. *Science* **215**:19–27.

Kerdelhué, C., and J.-Y. Rasplus. 1996. Non-pollinating afrotropical fig wasps affect the fig-pollinator mutualism in *Ficus* within the subgenus *Sycomorus*. *Oikos* **75**:3–14.

Kerdelhué, C., J.-P. Rossi, and J.-Y. Rasplus. 2000. Comparative community ecology studies on Old World figs and fig wasps. *Ecology* **81**:2832–2849.

Lande, R., and S.J. Arnold. 1983. The measurement of selection on correlated characters. *Evolution* **37**:1210–1226.

Layne, J.R., Jr. 1993. Winter microclimate of the goldenrod spherical gall and its effects on the gall inhabitant *Eurosta solidaginis* (Diptera: Tephritidae). *Journal of Thermal Biology* **18**:125–130.

LeBlanc, D.A., and C.R. Lacroix. 2001. Developmental potential of galls induced by *Diplolepis rosaefolii* (Hymenoptera: Cynipidae) on the leaves of *Rosa virginiana* and the influence of *Periclistus* species on the *Diplolepis rosaefolii* galls. *International Journal of Plant Science* **162**:29–46.

Marchosky, R.J., and T.P. Craig. 2004. Gall size-dependent survival for *Asphondylia atriplicis* (Diptera: Cecidomyiidae) on *Atriplex canescens*. *Environmental Entomology* **33**:709–719.

Moeller, R., and M.T. Thogerson. 1978. Predation by the downy woodpecker on the goldenrod gall fly larva. *Iowa Bird Life* **48**:131–136.

Nyman, T., A. Widmer, and H. Roininen. 2000. Evolution of gall morphology and host-plant relationships in willow-feeding sawflies (Hymenoptera: Tenthredinidae). *Evolution* **54**:526–533.

Paine, T.D., E.O. Paine, L.M. Hanks, and J.G. Millar. 2000. Resource partitioning among parasitoids (Hymenoptera: Brachonidae) of *Phoracantha semipunctata* in their native range. *Biological Control* **19**:223–231.

Ping, C. 1915. Some inhabitants of the round gall of goldenrod. *Pomona Journal of Entomology and Zoology* **8**:161–179.

Plakidas, J.D., and A.E. Weis. 1994. Depth associations and utilization patterns in the parasitoid guild of *Asphondylia rudbeckiaeconspicua* (Diptera: Cecidomyiidae). *Environmental Entomology* **23**:115–121.

Price, P.W. 2002. Resource-driven terrestrial interaction webs. *Ecological Research* **17**:241–247.

Price, P.W. 2003. *Macroevolutionary Theory on Macroecological Patterns*. Cambridge, UK: Cambridge University Press.

Price, P.W., C.E. Bouton, P. Gross, et al. 1980. Interactions among three trophic levels: influence of plants on interactions between insect herbivores and natural enemies. *Annual Review of Ecology and Systematics* **11**:41–65.

Price, P.W., G.W. Fernandes, and G.L. Waring. 1997. The adaptive nature of insect galls. *Environmental Entomology* **16**:15–24.

Schluter, D. 1988. Estimating the form of natural selection on a quantitative trait. *Evolution* **42**:849–861.

Schluter, D., T.D. Price, and P.R. Grant. 1985. Ecological character displacement in Darwin's finches. *Science* **227**:1056–1059.

Schlichter, L. 1978. Winter predation by black-capped chickadees and downy woodpeckers on inhabitants of the goldenrod ball gall. *Canadian Field Naturalist* **92**:71–74.

Smith, S. M. 1993. Black-capped chickadee (*Parus atricapillus*). In A. Poole, P. Stettenheim, and F. Gill (eds.) *The Birds of North America*, no. 39, Philadelphia. Washington, DC: *American Ornithologists' Union.*

Stiling, P., and A. M. Rossi. 1994. The window of parasitoid vulnerability to hyperparasitism: template for parasitoid complex structure, pp. 228–244 in B. A. Hawkins and W. Sheehan (eds.) *Parasitoid Community Ecology*. Oxford, UK: Oxford University Press.

Stone, G. N., and J. M. Cook. 1998. The structure of cynipid oak galls: patterns in the evolution of an extended phenotype. *Proceedings of the Royal Society of London Series B* **265**:979–988.

Stone, G. N., and K. Schonrogge. 2003. The adaptive significance of gall morphology. *Trends in Ecology and Evolution* **18**:512–522.

Stone, G. N., K. Schonrogge, R. J. Atkinson, D. Bellido, and J. Pujade-Villar. 2002. The population biology of oak gall wasps (Hymenoptera: Cynipidae). *Annual Review of Entomology* **47**:633–668.

Tabuchi, K., and H. Amano. 2004. Impact of differential parasitoid attack on the number of chambers in multilocular galls of two closely related gall midges (Diptera: Cecidomyiidae). *Evolutionary Ecology Research* **6**:695–707.

Tester, J. R. 1995. *Minnesota's Natural Heritage*. Minneapolis, MN: University of Minnesota Press.

Thompson, J. N. 1994. *The Coevolutionary Process*. Chicago, IL: University of Chicago Press.

Uhler, L. D. 1951. Biology and ecology of the goldenrod gall fly, *Eurosta solidaginis* (Fitch). *Cornell University Agricultural Station Memoir* **300**:1–51.

Van Hezewijk, B. H., and J. Roland. 2003. Gall size determines the structure of the *Rhabdophaga strobiloides* host–parasitoid community. *Ecological Entomology* **28**:593–603.

Washburn, J. O., and H. V. Cornell. 1981. Parasitoids, patches, and phenology: their possible role in the local extinction of a cynipid gall wasp population. *Ecology* **62**:1597–1607.

Weiblen, G. D. 2002. How to be a fig wasp. *Annual Review of Entomology* **47**:299–330.

Weis, A. E. 1982. Use of a symbiotic fungus by the gall maker *Astermoyia carbonifera* to inhibit attack by the parasitoid *Torymus capite*. *Ecology* **63**:1602–1605.

Weis, A. E. 1996. Variable selection on *Eurosta*'s gall size. III. Can an evolutionary response to selection be detected? *Journal of Evolutionary Biology* **9**:623–640.

Weis, A. E., and W. G. Abrahamson. 1985. Potential selective pressures by parasitoids on the evolution of a plant–herbivore interaction. *Ecology* **66**:1261–1269.

Weis, A. E., and W. G. Abrahamson, 1986. Evolution of host plant manipulation by gall makers: Ecological and genetic factors in the *Solidago–Eurosta* interaction. *American Naturalist* **127**:681–695.

Weis, A. E., and A. D. Kapelinksi. 1994. Variable selection on *Eurosta*'s gall size. II. A path analysis of the ecological factors behind selection. *Evolution* **48**:734–745.

Weis, A. E., W. G. Abrahamson, and K. D. Mcrea. 1985. Host gall size and oviposition success by the parasitoid coma *Eurytoma gigantea*. *Ecological Entomology* **10**:341–348.

Weis, A. E., K. D. Mcrea, and W. G. Abrahamson. 1989. Can there be an escalating arms race without coevolution? Implications from a host–parasitoid simulation. *Evolutionary Ecology* **3**:361–370.

Weis, A. E., W. G. Abrahamson, and M. C. Andersen. 1992. Variable selection on *Eurosta*'s gall size. I. The extent and nature of variation in phenotypic selection. *Evolution* **46**:1674–1697.

Whitham, T. G. 1983. Host manipulation of parasites: within-plant variation as a defense against rapidly evolving pests, pp. 15–41 in R. F. Denno and M.S. McClure (eds.) *Variable Plants and Herbivores in Natural and Managed Systems*. New York: Springer-Verlag.

15

Linking ecological and evolutionary change in multitrophic interactions: assessing the evolutionary consequences of herbivore-induced changes in plant traits

DAVID M. ALTHOFF

Introduction

Much of earth's biodiversity is composed of species that feed upon plants, and in turn these herbivores are the prey base for carnivorous species. Food webs have been characteristically used to describe and examine the direct links among these trophic levels. As a consequence of feeding, however, herbivores cause many morphological and chemical changes in their host plants (Karban and Baldwin 1997, Agrawal et al. 1999). The induced changes that herbivores create indirectly provide new links among trophic levels. These indirect links add an additional level of complexity in understanding not only trait evolution within trophic levels, but also the evolutionary connections among levels within plant–herbivore–carnivore communities. In ecological time, we know that herbivore-induced changes in plants can facilitate indirect interactions such as those between plants and carnivores (Botrell et al. 1998). Parasitoid and predator species modify their searching behavior in response to these changes, and can learn to associate plant changes with herbivore location (Turlings et al. 1993, Vet et al. 1995). The use of herbivore-induced changes has been demonstrated for a number of parasitoid and predator species that attack insect pests of agricultural systems. For example, the aphid parasitoid *Aphidius ervi* is more attracted to

Ecological Communities: Plant Mediation in Indirect Interaction Webs, ed. Takayuki Ohgushi, Timothy P. Craig, and Peter W. Price. Published by Cambridge University Press. © Cambridge University Press 2007.

broad bean plants that are infested with the aphid *Acyrthosiphon pisum*. Naïve female parasitoids are attracted to plant volatiles that are released as a consequence of aphid feeding (Du et al. 1998). In comparison, female *A. ervi* that had previous experience foraging on aphid-infested bean plants showed an even greater attraction to the volatiles.

The question is whether these ecological effects of herbivore-induced changes translate into evolutionary change. The ubiquity of herbivore-induced changes in plant–herbivore interactions suggests there is ample opportunity for evolutionary change to occur (Faeth 1994, Dicke and Vet 1999). Many studies have documented the way other trophic levels respond to herbivore-induced changes in plants, but there is a paucity of studies that have shown a link between these ecological effects and evolutionary change in the trophic levels involved. The indirect effects that comprise the ecological complexity of multitrophic communities may also be important in directing the evolution of members of these communities as well. Plants may evolve signals to attract carnivores, herbivores may be under selection pressure to recognize previously damaged plants, and carnivores may be selected to cue into herbivore-induced changes. The evolutionary consequences could be relatively simple, such as changes in specific traits within species, or may even have far-reaching effects that influence speciation and phylogenetic patterns of diversification.

The paucity of studies investigating the evolutionary consequences of herbivore-induced changes in plants requires that this chapter will necessarily be one of a prospectus rather than a synthesis of a wealth of studies. In this chapter I will address the following questions:

1. What are the prerequisites for herbivore-induced changes to facilitate evolution of other trophic levels, and do such prerequisites exist?
2. What evidence is there for evolutionary change in trophic levels as a result of herbivore-induced changes in plants?
3. What future research directions do we need to pursue to examine the evolutionary impact of herbivore-induced changes?

Whenever possible, I will provide concrete examples of the evolutionary consequences of herbivore-induced changes.

The interactions among plants, their insect herbivores, and the carnivores of these herbivores are perhaps one of the best systems to examine the evolutionary impact of herbivore-induced changes in a multitrophic context. There are numerous studies on the ecological consequences of herbivore-induced changes, as well as an understanding of the physiological and genetic mechanisms by which these consequences are mediated. I will use these communities as the basis for this chapter, and in particular, focus on the effect of herbivore-induced plant volatiles on carnivores with a heavy bias towards parasitoid taxa. This stems primarily from the fact there has been more research on the chemical ecology of herbivore-induced changes in these tritrophic interactions than for any other interactions. Further research on the evolutionary consequences of indirect effects can add an evolutionary component to a solid ecological foundation. As is evident throughout this book, however, herbivore-induced changes can affect many other trophic levels. When possible, I have also provided examples from herbivores and predators to illustrate how herbivore-induced changes can also influence the evolution of other taxa. I stress that the ideas and suggested methodologies presented for understanding the evolutionary consequences of herbivore-induced changes in carnivores also apply to other multitrophic systems.

Question 1: What are the prerequisites for herbivore-induced changes to influence the evolution of other trophic levels?

Herbivores induce a myriad of changes in their host plants that range from changes in phenology to morphological defenses to secondary chemistry (e.g., Karban and Baldwin 1997, Botrell *et al.* 1998). No matter the change induced, several criteria must be present for that change to have the potential to influence the evolution of any of the trophic levels. These criteria are in addition to two prerequisites for natural selection, trait variation among individuals within a population and heritability of that trait (Endler 1986). To facilitate evolution within multitrophic communities as a result of herbivore-induced changes, the induced plant changes also have to be:

- consistent with the presence of herbivory
- identifiable by organisms in other trophic levels
- influence the fitness of members of other trophic levels.

A lack of any of these criteria will limit the evolutionary potential of herbivore-induced changes (Fig. 15.1). For example, if plant changes induced by herbivores are not distinctly different from those caused by other nonherbivore and abiotic factors, then there is not a clear signal associated with herbivory. Carnivores would have a difficult time identifying plants that were experiencing herbivory, and this would decrease the effectiveness of natural selection in shaping carnivore populations (Vet and Dicke 1992). Besides being consistent with herbivory, herbivore-induced changes would also have to be identifiable by other trophic levels. In essence, there needs to be a receiver of the signals provided by herbivore-induced changes for those changes to have the potential to generate phenotypic variation among individuals from other trophic levels. Finally, if the herbivore-induced changes do not have fitness consequences for

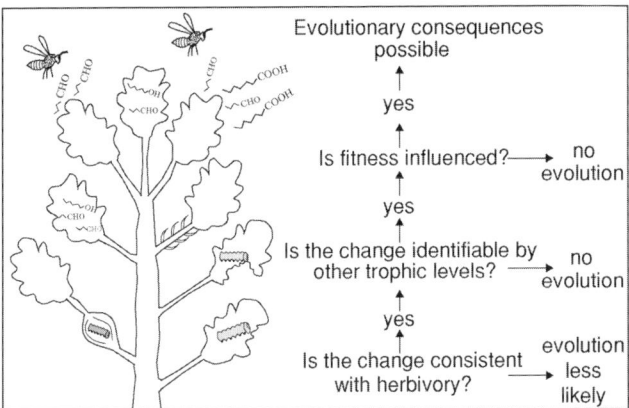

Figure 15.1. Herbivores induce many changes in plants such as galls, thorns, and secondary chemistry. For these changes to have evolutionary consequences through natural selection, three prerequistes must be met.

the phenotypic variation observed among members of other trophic levels, then natural selection will not occur.

These prerequisites restrict the instances in which the ecological effects of herbivore-induced changes have direct evolutionary consequences for multitrophic systems. In many cases, one of these criteria may be lacking and the ecological changes experienced by the current generation will not influence subsequent generations. There are also further complications in terms of identifying the target of evolutionary change – is it directly on trait expression or the shape of the norm of reaction? This could be particularly important for understanding the expression and evolution of induced changes in plants and also their use by other trophic levels. For example, are generalist predators/parasitoids that feed on prey across many different plant species more likely to be plastic in their responses to herbivore-induced changes in comparison to specialists that feed upon only one or two prey species? Plasticity in plant responses is well known (Callaway et al. 2003), but we know less about the plasticity of traits in other trophic levels. Increased complexity is also added by the fact that multiple species within or from different trophic levels can exert selection pressures either additively or nonadditively to determine the overall selection intensity (Strauss and Irwin 2004). The plasticity of herbivore-induced changes, community context, and the need to study multiple species over several generations represent a significant barrier to studying the evolutionary impacts of herbivore-induced changes in plants. This is likely one reason that there are few studies on this topic.

One of the first steps in examining the evolutionary impacts of herbivore-induced changes in plant characteristics is to determine whether each of the prerequisites listed above are present in multitrophic communities. This approach is relatively simplistic because it does not examine the mechanisms of evolutionary change, however, it does address whether the possibility exists for evolutionary change to occur throughout indirect interaction webs. I will assume that trait variation is present within populations and that this variation has a heritable basis. The degree to which this assumption is true may depend in part on the species and the trait examined (e.g., Via 1990, Cipollini 2002, Cipollini et al. 2003, Glawe et al. 2003).

Are induced changes consistent with the presence of herbivory?

Many plant changes are the direct result of herbivory and are only expressed as a consequence of herbivory. For morphologically induced changes such as galls, there is a direct relationship between herbivory and induced changes in plant characteristics. Carnivores that can identify gall shape will therefore have an honest signal of potential prey. This circumstance happens in *Eurosta* flies that induce galls in goldenrod (*Solidago*). Black-capped chickadees and downy woodpeckers use the galls as a visual cue to find *Eurosta* larvae, and in some populations, these birds can inflict up to 60% mortality on the larvae (Abrahamson and Weis 1997).

Plant chemical volatiles can also serve as reliable signals indicating the presence of potential prey. Numerous studies have determined that many plant volatiles are released in a direct response to feeding by herbivore taxa. That is, the plant volatile profile changes after feeding or contact with herbivores. In some cases, this change is the result of mechanical damage to plant cells by feeding, but in others, there are specific chemical cues that come from the herbivores. Turlings et al. (1990) demonstrated that mechanically damaged maize plants release significantly less induced volatiles than those that are mechanically damaged and treated with the regurgitant from armyworm larvae. In particular, the compound volicitin, which is present in some caterpillar regurgitant, is responsible for turning on the induced volatile pathway in maize (Turlings et al. 2000). Although the specific chemical or stimulant that causes induced responses has not been identified for many plant species, Dicke and Vet (1999) demonstrated that herbivore specific changes in plant volatile profiles are present across many phylogenetically distant plant families.

In some cases, plants can have different volatile profiles depending on the herbivore species. De Moraes and Lewis (1999) demonstrated that the volatile profile of both maize and tobacco plants were quantitatively different depending on whether the herbivore feeding upon them was the lepidopteran *Heliothis virescens* or *H. zea*. In this case, carnivores are provided with extremely accurate information about the potential prey species. The parasitoid *Cardiochiles nigriceps* takes advantage of this difference, and can identify maize or tobacco plants that are being attacked by its host *H. virescens* and plants attacked by the non-host *H. zea* (De Moraes and Lewis 1999). A similar result was obtained for the aphid parasitoid *Aphidius ervi*. Female parasitoids are more attracted to plants being fed upon by its host the pea aphid *Acyrthosiphon pisum* than plants being fed upon by a non-host aphid, *Aphis fabae* (Du et al. 1998).

The above examples are a small subset of the induced changes that herbivores can cause in plants. At least initially, any herbivore-induced change in plant characteristics has to be caused exclusively by herbivores in order for other trophic levels to be selected for utilizing those changes. Such honest signaling is important because it provides reinforcement in terms of fitness gains for those individuals that cue into the induced changes. For example, those carnivore individuals that can use induced changes to find more prey are likely to leave more offspring. If induced changes are not linked with the presence of herbivores, then carnivorous taxa will be selected to disregard any plant signals they may detect.

Are induced changes identifiable by other trophic levels?

The fact that there are a number of instances in which other trophic levels respond to herbivore-induced changes in plants demonstrates that many induced changes are detectable. At the herbivore level, there are many studies that have shown how herbivores can recognize and respond to the induced changes in plants – either caused by members of the same species or by other herbivore species (Sabelis *et al.* 1999, Kessler and Baldwin 2001, Hilker and Meiners 2002, Wise and Weinberg 2002, van Zandt and Agrawal 2004). Dicke and Vet (1999) review studies on tritrophic interactions in which the third trophic level utilizes novel compounds emitted by plants in response to herbivory. Based on this list, a phylogenetically diverse set of carnivores from seven families (representing four orders in two arthropod classes) is able to identify herbivore-induced changes in plant volatiles. Although much of the research on the detection of herbivore-induced changes has focused on arthropod taxa, it is likely that other taxonomic groups are able to identify herbivore-induced changes. For example, van Tol *et al.* (2001) demonstrated that the nematode *Heterorhabditis medigis* is attracted to the roots of *Thuja occidentalis* that are being fed upon by larvae of the root-feeding weevil *Otiorhynchus sulcatus*. Mäntylä *et al.* (2004) showed that willow warblers preferentially chose branches from mountain birch trees that had been fed upon by sawfly larvae over those that had not, even when the branches presented to the birds had not come in contact with sawfly larvae. The fact that there are few studies demonstrating the use of herbivore-induced changes by nonarthropod taxa is most likely due to a lack of research effort, rather than a lack of an effect.

Do induced changes impact fitness of the trophic levels?

Addressing this question is central to understanding the importance of herbivore-induced changes to the evolution of multitrophic communities. The ecological changes need to generate fitness differences among individuals in order for natural selection to occur. Surprisingly, there has been comparably little research examining fitness differences associated with the expression of herbivore-induced changes. There have been several studies that have focused on plants and their herbivores, but almost none on the fitness consequences of higher trophic levels.

Four studies in particular have demonstrated how herbivore-induced changes have increased plant fitness; two studies examined the direct effect of induced plant defenses and two examined the fitness effects from the involvement of other trophic levels. Agrawal (1998) showed that young individuals of the wild radish *Raphanus sativa* that were fed upon by a larva of *Pieris rapae* increased trichome density and the concentration of glucosinolates in leaf tissue. These induced defenses resulted in decreased attack from a variety of other species such as aphids, flea beetles, and earwigs. As a consequence, induced plants had 60% higher seed mass than control plants. Baldwin (1998) treated naturally occurring *Nicotiana attenuata* with jasmonic acid to simulate herbivory and induce plant defenses against herbivores. He demonstrated that induced defenses have a fitness cost at low levels of herbivory, a fitness benefit at intermediate levels, and no fitness benefit at high levels of herbivory.

Plants have been assumed to also gain a fitness benefit through the attraction of carnivores that reduce the impact of herbivory. *Arabidopsis thaliana* changes its volatile emissions in response to herbivory; van Loon et al. (2000) showed that *A. thaliana* does indeed benefit from the attraction of carnivores. Plants fed upon by larvae of *Pieris rapae* that had been parasitized by the parasitoid *Cotesia rubecula* matured significantly more seeds and at a faster rate than plants fed upon by unparasitized larvae. Similarly, Fritzsche-Hoballah and Turlings (2001) demonstrated that parasitoids attracted to corn provide a positive effect on corn fitness. They exposed young plants to feeding by a single larva of *Spodoptera littoralis*. The *S. littoralis* larvae parasitized by *Cotesia marginiventris* fed less and consequently gained less weight than unparasitized larvae. The decrease in herbivore damage translated into a 30% increase in seed production. The four studies mentioned above demonstrate that herbivore-induced changes impact plant fitness both through the effects of plant defenses and the attraction of the third trophic level. Several studies have also demonstrated that induced plant changes impact the fitness of the second trophic level.

For herbivores, induced defenses have a negative impact on fitness. Thaler (1999) has shown that induced defenses in tomato decrease the abundances of a suite of herbivores (thrips, aphids, flea beetles, and noctuid larvae). The decreased abundance of larvae of the noctuid *Spodoptera exigua* was the result of increased parasitism by the parasitoid *Hyposoter exiguae*. Tomato plants treated with jasmonic acid (which induces defenses) had more parasitoid pupae, and larvae of *S. exigua* placed under the canopy of treated plants were more likely to be parasitized than those under control plants. Parasitism rates of another noctuid, *Heliothis virescens*, that feeds on tobacco, ground cherry, and cotton are also the result of parasitoid attractiveness to induced plant volatiles (Oppenheim and Gould 2002). Damaged tobacco plants were more attractive to *Cardiochiles nigriceps* parasitoid females than damaged ground cherry plants (*Physalis angulata*), and both were more attractive than damaged cotton plants. Parasitism rates of *H. virescens* larvae in the field followed this pattern of differential parasitoid attractiveness, with the highest rate on tobacco and the lowest on cotton.

The above examples demonstrate that herbivore-induced changes have fitness consequences for both plants and herbivores. Whether there are fitness consequences for other trophic levels, however, has rarely been demonstrated. There are many studies that have demonstrated increased attraction, parasitism, and predation rates on plants with herbivore-induced changes, but it is not known if this results in increased fitness for the individuals responding to the plant changes. Do these individuals find more prey and leave more offspring in the next generation in comparison to individuals that do not use herbivore-induced changes? Research directed at answering this question is much needed, and our understanding of the impact of herbivore-induced changes on the evolution of multitrophic communities will be severely limited until more studies are conducted.

Question 2: What evidence is there for evolutionary change in trophic levels as a result of herbivore-induced changes in plants?

Documenting how herbivore-induced changes in plants influence the ecology of other trophic levels is the first step towards examining their evolutionary consequences. Much of the research on herbivore-induced changes has been concerned with identifying the ecological changes in multitrophic communities. If we are to understand whether these changes are important in evolutionary time, however, we have to conduct additional studies above and beyond documenting the ecological changes. Evidence of an evolutionary effect would entail identifying adaptation for any species in the interaction web. Studies attempting to document such adaptation are rare, and as a consequence we are mostly assuming that herbivore-induced changes have an evolutionary impact on multitrophic communities. In fact, van der Meijden and Klinkhamer (2000) raised concerns about the widely held belief that evolution will favor plants that can utilize herbivore-induced changes to attract natural enemies of herbivores. They highlighted the very limited number of studies identifying fitness benefits for plants or natural enemies of herbivores. Janssen *et al.* (2002) responded by suggesting that there are multiple pathways in which induced changes influence evolution of plant–herbivore–carnivore interactions, and that the criticisms raised by Meijden and Klinkhamer are less applicable to true predators. These two studies highlight the need for research specifically aimed at examining the evolutionary influence of herbivore-induced changes for all trophic levels.

There are a number of approaches that can be taken to identify the evolutionary consequences of herbivore-induced changes. Below I describe three different approaches to examine adaptation: selection experiments, comparisons among closely related species, and phylogenetic surveys of taxa in which herbivore-induced changes in plants might be important evolutionarily. Although these approaches focus on different hierarchical levels, each attempts to determine the presence of adaptations to herbivore-induced changes. The latter two approaches are comparative ones rather than experimental, and the best approach to take will depend on the particular study system. Not all systems are amenable to experimentation, nor is the necessary information available for all members within a group of related taxa. In the following sections I provide examples for each of the three approaches. I ask the reader to keep in mind that advances in quantitative genetics and molecular techniques such as gene expression can be integrated into these approaches to examine the genetic basis of evolutionary changes.

Selection experiments

The use of selection experiments would be the most direct means to investigate the evolutionary implications of herbivore-induced changes via indirect interactions. These experiments could be conducted on single members of a community or, alternatively, whole communities could be subjected to different treatments and each of the trophic levels surveyed for evolutionary changes. This approach would be extremely powerful because it would pinpoint the evolutionary patterns generated across a community and provide an understanding of community genetics (*sensu* Whitham *et al.* 2003). Community manipulations, however, would likely only be feasible for plants, herbivores, and carnivores that are small enough to cultivate pragmatically and may be of limited value for most systems.

More realistically, another possibility is to perform selection experiments on individuals within a species from one trophic level for their response to herbivore-induced changes. This approach has been taken in a series of studies on the braconid parasitoid *Cotesia glomerata*. Female parasitoids of this species respond to the volatiles from cabbage that are induced by feeding by *Pieris brassicae* larvae. Gu and Dorn (2000) documented genetic variation in the response of female wasps to plants infested by *P. brassicae*. All females from isofemale lines responded to herbivore-infested plants, but there were significant differences in their flight orientation behavior and landing success. Based on the differences among isofemale lines, heritability for the flight orientation response was estimated at 0.447 ± 0.011 (SE) which suggests this trait can respond to natural selection. Wang and Dorn (2003) applied bidirectional selection on *C. glomerata* based on the responsiveness of females to hexane-extracted, herbivore-induced plant volatiles. After four generations of selection, there were significant differences between the high and low olfactory response groups in their flight orientation and landing success. Females from the high olfactory response group were significantly better at orienting and landing on the surface from which the odor plume originated. Realized heritability for orientation and landing success was estimated at 0.248 and 0.216, respectively, again demonstrating that additive genetic variance is present for behavioral traits responsive to herbivore-induced changes.

Results from these two studies on *C. glomerata* demonstrate that there is significant genetic variation in responses to herbivore-induced changes in at least some carnivore taxa, and that this variation allows populations to respond to artificial selection. The question remains, however, whether under field conditions there would be fitness differences between individuals that differed in their response to herbivore-induced changes. Wang et al. (2004) provide evidence for such fitness consequences. They examined four measures of fitness, the development time of immature stages, the body size of female wasps, and the number of females produced in a brood as well as the parasitism rate for the high and low olfactory response groups of *C. glomerata*. The two groups did not significantly differ in the three life-history characteristics; however, the high olfactory response group had a significantly greater parasitism rate. Females of this group were much more likely to parasitize larvae distributed across large areas. This result implies that high olfactory response females are much better at orienting to induced volatiles, locating larvae on attacked plants, and hence leaving more progeny than low olfactory response females.

Selection for olfactory response has also been identified in true predators. Margolies *et al.* (1997) demonstrated that the predatory mite *Phytoseiulus persimilis* also responded to artificial selection for sensitivity to herbivore-induced plant volatiles. In this study, *P. persimilis* preyed upon spider mites feeding on bean plants. Increased attraction to induced volatiles was evident after one generation of selection and continued over subsequent generations. Whether this selection could result in increased fitness for *P. persimilis* was unclear because individuals from the high sensitivity line left infested leaves more rapidly, thus consuming fewer spider mites per leaf than individuals from the control population. These individuals, however, may be maximizing energy intake by moving to more prey-rich patches.

Results from these selection studies demonstrate that carnivores can respond to selection for increased sensitivity to herbivore-induced plant volatiles. In these studies, selection produced a change in innate behavior. Some authors have argued that such results are probably rare because carnivores come in contact with many different plant volatiles, and all of these volatiles do not necessarily provide accurate information about herbivory. Perhaps, more importantly, plant species differ in their volatile profile emissions and there is even volatile variation within species depending on genetic and environmental factors (Dicke 1999). The variability in profile emissions has been suggested by Vet and Dicke (1992) to select for a combination of innate responses to volatiles and learning to associate volatiles with the presence of prey individuals. Much of the evolution for using herbivore-induced changes by carnivores, then, may have been selection for learning rather than selection for innate preferences of certain volatile profiles. Parasitoids in particular have the ability to associate colors, shapes, and in particular odors with the presence of host species (Turlings *et al.* 1993, Godfray 1994, Vet *et al.* 1995). Because the emittance and information content of herbivore-induced changes may be dependent on many different factors, associative learning may be one way that parasitoids can make changes in their foraging strategy that maximize fitness. Fukushima *et al.* (2002) demonstrated that *Cotesia kariyai* possesses an innate preference for odors from plants fed upon by herbivores, but can also learn to associate nonspecific volatile cues with hosts as well. This type of learning has also been demonstrated in the aphid parasitoid *Aphidius ervi* (Powell *et al.* 1998). Vet and Dicke (1992) have suggested that associative learning would be particularly beneficial and widely used in generalist carnivores rather than specialists. Steidle and van Loon (2003) surveyed the literature on volatile use by carnivores and demonstrated that learning occurs more frequently in generalist carnivores than specialists, although it is present in both groups and the importance of learning may be species-specific. For our purposes here, the interesting question is whether there has been selection on parasitoids for the ability to learn to use induced volatiles. To the best of my knowledge this has not been tested.

Comparisons among closely related species

A second approach to determine if herbivore-induced changes in plants can influence the evolution of other trophic levels is to make comparisons among closely related species in which the species differ in their use of herbivore-induced changes. In the ideal comparison, two sibling species would be compared and one would use herbivore-induced changes and one would not. This would allow the identification of those traits that have evolved specifically for dealing with herbivore-induced changes. This approach would require correlating behavioral, physiological, or morphological differences with the utilization of herbivore-induced changes.

De Moraes and Mescher (2004) provide a striking example of how herbivores have evolved to circumvent herbivore-induced plant defenses that involve a third trophic level. The noctuid moth *Heliothis subflexa* feeds exclusively on fruits of the solanaceous plant genus *Physalis*. One of its congeners, *H. virescens*, can also feed on *Physalis* species, but feeds on the leaves instead of fruits. De Moraes and Mescher (2004) demonstrated that the fruit of *Physalis angulata* does not contain linolenic acid. This acid is crucial for insect development and is the precursor to forming volicitin (that occurs in the saliva of some herbivorous insects) which is an elicitor of volatile signaling in plants. *Heliothis subflexa* has evolved the ability to develop on fruit, thus avoiding competition with other herbivores and also reducing the plant's ability to attract parasitoids. The specialist parasitoid *Cardiochiles nigriceps* is significantly more attracted to the volatile profile of caterpillar-damaged leaves than damaged fruit. In addition, the lack of linolenic acid makes *H. subflexa* larvae very unsuitable as hosts for developing parasitoid larvae. The number of parasitoids surviving to adults is very low when feeding on larvae from fruits and significantly greater when linolenic acid is artificially added to the diet of *H. subflexa*.

Comparisons among closely related species of the third trophic level provide insights into how other trophic levels have evolved in response to their use of herbivore-induced changes. Takabayashi *et al.* (1998) examined the use of plant volatiles by two species of *Cotesia*, *C. kariyai* and *C. glomerata*. In preference trials, *C. kariyai* is able to differentiate between host and non-host plants, non-host plants and air, and infested and artificially damaged plants. *Cotesia glomerata* also prefers host plants over non-host plants, infested plants over uninfested, but prefers artificially damaged plants over those infested with host larvae. Takabayashi *et al.* (1998) suggest that *C. kariyai* has the innate ability to utilize herbivore-specific induced cues, whereas *C. glomerata* responds to general cues that signify any kind of plant damage. Female *C. glomerata* may then learn to associate these general plant cues with larval presence of many different host species. Geervliet et al. (1998) demonstrated that preference of female *C. glomerata* can be changed through experience with different host–plant complexes. In contrast, females of the specialist *C. rubecula* do not change their preferences in response to experience with different host–plant complexes. For these *Cotesia* species, selection has genetically fixed the response to herbivore-induced cues in one species, but in the other the response has a genetic as well as a large environmental component.

In another comparison of *Cotesia* species, Rutledge and Weidenmann (1999) further demonstrate that the preference for plant volatiles in the laboratory differs among three closely related species and is congruent with host use patterns in the field. Five monocot and four dicot plant species were used to examine the preference hierarchies of naïve female *C. flavipes*, *C. sesamiae*, and *C. chilonus*. The generalists *C. flavipes* and *C. sesamiae* had the following rankings in their preference hierarchies from high to low: normal host plant species, closely related host plant, non-host plant species. These two parasitoid species also responded to the presence of herbivore damage on plants. Females of *C. chilonus*, which has the narrowest host range of the three species, only responded to one of the plant species tested, and did not discriminate between non-damaged and host-damaged plants for any of the nine plant species used in the preference trials. These results again demonstrate that the genetic architecture underlying the use of herbivore-induced traits may be different among closely related species.

The above studies for both herbivores and carnivores suggest that these taxa can respond evolutionarily to herbivore-induced changes in plants. In some cases this response can lead to changes in microhabitat use among closely related species, and in others it changes the degree to which herbivore-induced changes are used in modifying behavior and host use patterns. The very limited number of studies that have made comparisons among closely related species, however, makes it difficult to assess the possible types of evolutionary responses to herbivore-induced plant changes. One main avenue of research to elucidate the evolutionary influence of induced changes could be directed at comparing the effect on generalist versus specialist species of predators/parasitoids. The selective pressures for species along this continuum vary. Generalists will use a variety of prey species across different plant species and frequently modify their behavior to utilize the most abundant prey at a given time. As such, generalists might be selected to respond to a wide variety of herbivore-induced changes that come from many different herbivore/plant complexes. Specialists in contrast would be under selection to locate only a few prey species, and in some cases the prey may not be abundant. For specialists, herbivore-induced changes that are herbivore species-specific will be more valuable in terms of prey location than induced changes that occur from attack by any herbivore species. One interesting approach would be to examine the genetic basis for the use of herbivore-induced changes in generalists and specialists, and assess whether differences in the genetic basis correspond with differences in the specific mechanisms plants may use to attract one group or the other.

Phylogenetic surveys across taxa

The above two approaches focus on how herbivore-induced changes can influence evolution on a microevolutionary scale. Broad surveys of many taxonomic groups at different trophic levels allow an assessment of the possible macroevolutionary consequences of herbivore-induced changes. This approach would be best undertaken in a comparative and phylogenetic framework (e.g., Harvey and Pagel 1986) to identify the origin of behavioral, physiological, or morphological adaptations associated with utilization of herbivore-induced changes. For example, convergent evolution by distantly related taxa to a particular herbivore-induced change would be indicative of adaptation. Correlating the occurrence of specific traits with the use of herbivore-induced changes in a monophyletic group would also provide insight into adaptation.

To my knowledge, there have been no studies that have taken a phylogenetic approach to examine the evolutionary consequences of herbivore-induced changes. This approach requires robust phylogenies for the taxa under investigation as well as the relevant data for each taxon. These data are not yet available for many groups likely to be affected by herbivore-induced changes. The most promising group of carnivores to which this approach might be applied are the parasitoids, mainly because they are used extensively in the biological control of insect pests. For parasitoids, the searching behavior and use of herbivore-induced changes (mainly plant volatiles) has been investigated for many species (Dicke and Vet 1999, Turlings *et al.* 2002, Steidle and van Loon 2003).

If we examine the phylogenetic distribution of a subset of parasitoids that have been shown to respond to herbivore-induced volatiles, we see the distribution includes many phylogenetically distinct lineages and represents egg, larval, and pupal parasitoids (Table 15.1). Under the assumption that use of induced volatiles is not basal for all parasitoids, this distribution suggests that the utilization of herbivore-induced changes via plant volatiles has independently evolved multiple times within parasitoids. Phylogenies of groups at a variety of different taxonomic levels are needed to more rigorously test the phylogenetic distributions of adaptations to herbivore-induced changes.

As a first pass at attempting to use a phylogenetic approach to examine the evolutionary use of herbivore-induced volatiles, I have assembled the phylogenetic and behavioral data currently available for the braconid genus *Cotesia* and

Table 15.1. *Phylogenetic distribution of a subset of parasitoid taxa known to use herbivore-induced plant volatiles in locating hosts (taxa were chosen to highlight the broad taxonomic distribution of this trait)*

Parasitoid family	Parasitoid species	Host stage utilized	Reference
Aphelinidae	Encarsia formosa	Nymphs	Birkett et al. (2003)
Aphidiidae	Aphidius ervi	Nymphs	Guerrieri et al. (1999)
Braconidae	Cotesia rubecula	Larvae	van Poecke et al. (2003)
Encrytidae	Apoanagyrus lopezi	Nymphs	Souissi et al. (1998)
Eulophidae	Chrysonotomyia ruforum	Eggs	Mumm et al. (2003)
Figitidae	Trybliographa rapae	Larvae	Neveu et al. (2002)
Ichneumonidae	Dentichasmias busseolae	Pupae	Gohole et al. (2003)
Scelionidae	Trissolcus basalis	Eggs	Colazza et al. (2004)

for the subfamily Aphidiinae. Michel-Salzat and Whitfield (2004) have provided a preliminary phylogeny of the genus *Cotesia* based on DNA sequences of four genes from the mitochondrial and nuclear genomes. The species used in the phylogeny were chosen because of their widespread use in biological control and laboratory experiments. The phylogenetic analyses identified four monophyletic groups of *Cotesia* species (Fig. 15.2). I have plotted the data currently available on the use of herbivore-induced plant volatiles by *Cotesia* onto the phylogeny. Given this crude method, the use of plant volatiles appears to be widespread throughout the genus and, in fact, may be ancestral. In one group, the "melanoscela," sibling taxa differ in their use of herbivore-induced plant volatiles. *Cotesia chilonus* does not differentiate between damaged and undamaged plants (Rutledge and Wiedenmann 1999). This species is relatively specialized compared to its sibling species, *C. sesamiae* and *C. flavipes*, and the use of herbivore-induced volatiles may not be as important because females only search for host larvae on a few plant species.

Belshaw and Quicke (1997) examined the phylogenetic relationships of genera within the Aphidiinae. A molecular phylogeny based on sequence data from three genes provided a well-resolved picture of generic relationships (Fig. 15.3). For the monophyletic clade that contains the three tribes, Aphidiini, Trioxini, and Praini, the use of plant volatiles in host location appears to be ancestral. For the fourth tribe, Ephedrini, Powell and Wright (1991) showed that, the inclusion

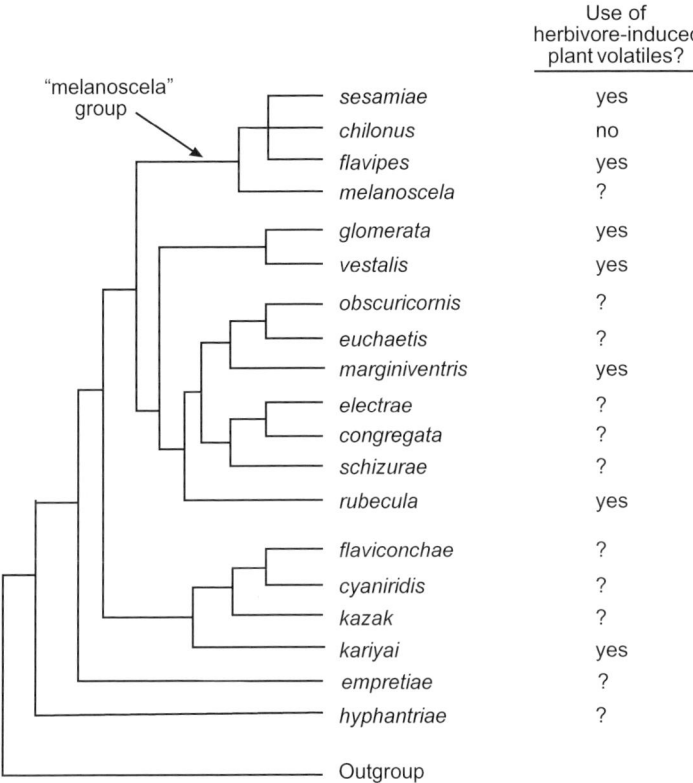

Figure 15.2. Use of herbivore-induced plant volatiles in the braconid parasitoid genus *Cotesia* (phylogeny redrawn from Michel-Salzat and Whitfield 2004). Female parasitoids in a number of species use induced volatiles to locate potential hosts. Within the "melanoscela" clade there has been the loss of using induced volatiles. Although studies documenting the use of induced volatiles have not been conducted for all species, the use of induced volatiles may be ancestral for the genus.

of leaf tissue during host use tests did not increase the parasitism rate by *Ephedrus plagiator* in contrast to other aphidiinid species. Whether the use of plant volatiles is important in this tribe, however, is unclear. For the species in the phylogeny, there was a transition to the use of herbivore-induced plant volatiles in host location that occurred in the Aphidiini. *Aphidius ervi* and other *Aphidius* species use induced plant volatiles to locate hosts, whereas it appears that *Diaeretiella* and *Lysiphlebus* only respond to noninduced plant volatiles (Vaughn et al. 1996, Lo Pinto et al. 2004). Phylogenetic comparisons such as the ones above allow further exploration of behavioral, physiological, and morphological traits involved in herbivore-induced changes.

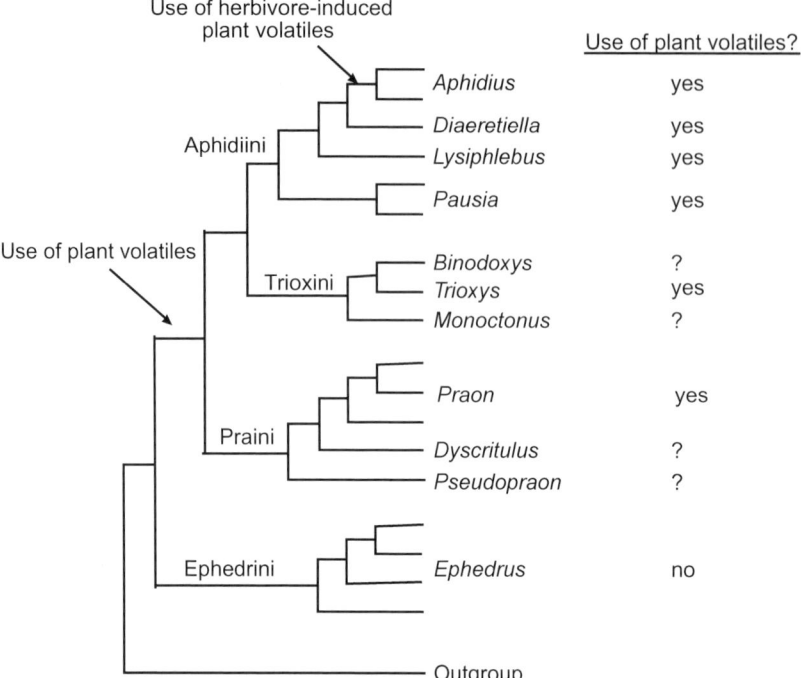

Figure 15.3. Phylogenetic patterns of plant volatile use for aphid parasitoids in the Aphidiinae (phylogeny redrawn from Belshaw and Quicke 1997). Based on the available data, the use of plant volatiles in host location evolved in the lineage leading to the Praini, Trioxini, and Aphidiini. The evolution of using herbivore-induced plant volatiles has only evolved in the genus *Aphidius*.

Another possibility is to examine the phylogenetic distribution of genes and the gene expression patterns responsible for producing those changes. For example, van Poecke et al. (2001) examined gene expression patterns of eight genes in *Arabidopsis thaliana* known to be important in induced defenses. They found increased expression of two genes, *AtSP10* and *AtPAL1*, that are involved in salicylate production and two other genes, *AtLOX2* and *AtHPL*, that are involved in green leaf volatile production in plants that had been fed upon by *Pieris rapae*. Sequences of these genes and their promoters (see Zangerl 2003) can be used to screen many plant taxa to help identify the phylogenetic distribution of genes important in induced defenses rather than conducting separate experiments on many plant taxa to determine if plants have induced defenses. For plants, especially, many inducible genes and proteins have already been identified (Constabel 1999, Roda and Baldwin 2003). A similar approach can be taken in other trophic levels once candidate genes are identified and expression studies establish their role in utilizing herbivore-induced changes. Indeed, corresponding phylogenetic analyses of plant genes used in induced defenses and genes in other trophic levels responsible for responding to induced defenses may provide powerful tests of coevolution in multitrophic communities.

Question 3: What future research directions do we need to pursue to examine the evolutionary impact of herbivore-induced changes?

Much of the research on herbivore-induced changes has focused on two areas: identifying the changes in plants and documenting their ecological effects on multitrophic communities. In particular, research has been heavily biased towards agriculturally important species. Results from these species have demonstrated both the widespread occurrence and the importance of herbivore-induced changes in understanding the ecology of plant-based communities. There is now a wealth of studies that can be used as a stepping stone to place ecological effects into an evolutionary context, both at the microevolutionary and macroevolutionary scales.

At the microevolutionary scale, research should be directed at demonstrating fitness differences for individuals that vary in their response to herbivore-induced changes. As pointed out by van der Meijden and Klinkhamer (2000), fitness effects have been assumed to exist rather than shown to exist for many groups. There are very few studies that have documented such differences even though this is crucial for understanding the potential for natural selection. Further research on agriculturally important species may be the best avenue for pursuing studies on fitness effects. Many of the ecological effects of herbivore-induced changes have already been documented, and experimental protocols and techniques are already developed for studying and rearing these organisms. In some species, there is a wealth of information regarding the genetics of host use as well. Results from selection studies can be readily integrated with previous or ongoing ecological studies to determine the evolutionary impact of herbivore-induced changes in plants.

At the macroevolutionary scale, the incorporation of phylogenetics into a comparative approach will be extremely powerful in identifying the origins of traits that have evolved in response to herbivore-induced changes. Again, studies on agricultural species are a good starting point, but this approach will require examining the use of herbivore-induced changes in species that occur in more natural settings (e.g., Baldwin 1998). The goal of this approach is to combine phylogenetic analyses with surveys of traits across groups of taxa to determine the number of instances in which herbivore-induced changes have been important in influencing evolutionary lineages. This approach, conducted on several trophic levels, will also provide insights into the evolutionary history of multitrophic communities.

The tools and conceptual framework are already in place to begin to study how the ecological effects caused by herbivore-induced changes can translate into evolutionary effects. In essence, studying this phenomenon is studying the process of adaptation. What is required is a shift in perspective from thinking in ecological timescales to thinking in evolutionary timescales.

Acknowledgments

I thank K. Segraves and S. Nuismer for constructive discussion and criticism on the chapter. I also thank T. Craig, M. Sabelis, and two anonymous reviewers for many helpful suggestions for improving the chapter. This work was support by the National Science Foundation DEB-0321293.

References

Abrahamson, W.G., and A.E. Weis. 1997. *Evolutionary Ecology across Three Trophic Levels.* Princeton, NJ: Princeton University Press.

Agrawal, A.A. 1998. Induced responses to herbivory and increased plant performance. *Science* **279**:1201–1202.

Agrawal, A.A., S. Tuzun, and E. Bent (eds.) 1999. *Induced Plant Defenses against Pathogens and Herbivores.* St. Paul, MN: American Phytopathological Society Press.

Baldwin, I.T. 1998. Jasmonate-induced responses are costly but benefit plants under attack in natural populations. *Proceedings of the National Academy of Sciences of the USA* **95**:8113–8118.

Belshaw, R., and D.L.J. Quicke. 1997. A molecular phylogeny of the Aphidiinae (Hymenoptera: Braconiae). *Molecular Phylogenetics and Evolution* **7**:281–293.

Birkett, M.A., K. Chamberlain, E. Guerrieri, et al. 2003. Volatiles from whitefly-infested plants elicit a host-locating response in the parasitoid, *Encarsia formosa. Journal of Chemical Ecology* **29**:1589–1600.

Botrell, D.G., P. Barbosa, and F. Gould. 1998. Manipulating natural enemies by plant variety selection and modification: a realistic strategy? *Annual Review of Entomology* **43**:346–367.

Callaway, R.M., S.C. Pennings, and C.L. Richards. 2003. Phenotypic plasticity and interactions among plants. *Ecology* **84**:115–118.

Cipollini, D.F. 2002. Variation in the expression of chemical defenses in *Alliaria petiolata* (Brassicaceae) in the field and common garden. *American Journal of Botany* **89**:1422–1430.

Cipollini, D.F., J.W. Busch, K.A. Stowe, E.L. Simms, and J. Bergelson. 2003. Genetic variation and relationships of constitutive and herbivore-induced glucosinolates, trypsin inhibitors, and herbivore resistance in *Brassica rapa. Journal of Chemical Ecology* **29**:285–302.

Colazza, S., A. Fucarino, E. Peri, et al. 2004. Insect oviposition induces volatile emission in herbaceous plants that attracts egg parasitoids. *Journal of Experimental Biology* **207**:47–53.

Constabel, C. P. 1999. A survey of herbivore-induced defensive proteins and phytochemicals, pp. 137–166 in A. A. Agrawal, S. Tuzun, and E. Bent (eds.) *Induced Plant Defenses against Pathogens and Herbivores*. St. Paul, MN: American Phytopathological Society Press.

De Moraes, C. M., and W. J. Lewis. 1999. Analyses of two parasitoids with convergent foraging strategies. *Journal of Insect Behavior* **12**:571–583.

De Moraes, C. M., and M. C. Mescher. 2004. Biochemical crypsis in the avoidance of natural enemies by an insect herbivore. *Proceedings of the National Academy of Sciences of the USA* **101**:8993–8997.

Dicke, M. 1999. Evolution of induced indirect defense in plants, pp. 62–88 in R. Tollrain and C. D. Harvell (eds.) *The Ecology and Evolution of Inducible Defenses*. Princeton, NJ: Princeton University Press.

Dicke, M., and L. E. M. Vet. 1999. Plant–carnivore interactions: evolutionary and ecological consequences for plant, herbivore and carnivore, pp. 483–520 in H. Olff, V. K. Brown, and R. H. Drent (eds.) *Herbivores: Between Plants and Predators*. Oxford, UK: Blackwell Scientific Publications.

Du, Y., G. M. Poppy, W. Powell, *et al.* 1998. Identification of semiochemicals released during aphid feeding that attract parasitoid *Aphidius ervi*. *Journal of Chemical Ecology* **24**:1355–1368.

Endler, J. A. 1986. *Natural Selection in the Wild*. Princeton, NJ: Princeton University Press.

Faeth, S. H. 1994. Induced plant responses: effects on parasitoids and other natural enemies of phytophagous insects, pp. 245–260 in B. A. Hawkins and W. Sheehan (eds.) *Parasitoid Community Ecology*. Oxford, UK: Oxford University Press.

Fritzsche-Hoballah, M. E., and T. C. J. Turlings. 2001. Experimental evidence that plants under caterpillar attack may benefit from attracting parasitoids. *Evolutionary Ecology Research* **3**:533–565.

Fukushima, J., Y. Kainoh, H. Honda, and J. Takabayashi. 2002. Learning of herbivore-induced and nonspecific plant volatiles by a parasitoid, *Cotesia kariyai*. *Journal of Chemical Ecology* **28**:579–586.

Geervliet, J. B. F., A. I. Vreugdenhill, M. Dicke, and L. E. M. Vet. 1998. Learning to discriminate between infochemicals from different plant-host complexes by the parasitoids *Cotesia glomerata* and *C. rubecula*. *Entomologia Experimentalis et Applicata* **86**:241–252.

Glawe, G. A., J. A. Zavala, A. Kessler, N. M. van Dam, and I. T. Baldwin. 2003. Ecological costs and benefits correlated with trypsin protease inhibitor production in *Nicotiana attenuata*. *Ecology* **84**:79–90.

Godfray, H. C. J. 1994. *Parasitoids, Behavioral and Evolutionary Ecology*. Princeton, NJ: Princeton University Press.

Gohole, L. S., W. A. Overholt, Z. R. Khan, and L. E. M. Vet. 2003. Role of volatiles emitted by host and non-host plants in the foraging behaviour of *Dentichasmias busseolae*, a pupal parasitoid of the spotted stemborer *Chilo partellus*. *Entomologia Experimentalis et Applicata* **107**:1–9.

Gu, H., and S. Dorn. 2000. Genetic variation in behavioral response to herbivore-infested plants in the parasitic wasp *Cotesia glomerata* (L.) (Hymenoptera: Braconidae). *Journal of Insect Behavior* **13**:141–156.

Guerrieri, E., G. M. Poppy, W. Powell, E. Tremblay, and F. Pennacchio. 1999. Induction and systematic release of herbivore-induced plant volatiles mediating in-flight orientation of *Aphidius ervi*. *Journal of Chemical Ecology* **25**:1247–1261.

Harvey, P. H., and M. D. Pagel. 1986. *The Comparative Method in Evolutionary Biology*. Oxford, UK: Oxford University Press.

Hilker, M., and T. Meiners. 2002. Induction of plant responses to oviposition and feeding by herbivorous insects: a comparison. *Entomologia Experimentalis et Applicata* **104**:181–192.

Janssen, A. R. M., M. W. Sabelis, and J. Bruin. 2002. Evolution of herbivore-induced plant volatiles. *Oikos* **97**:134–138.

Karban, R., and I. T. Baldwin. 1997. *Induced Responses to Herbivory*. Chicago, IL: University of Chicago Press.

Kessler, A., and I. T. Baldwin. 2001 Defensive function of herbivore-induced plant volatile emissions in nature. *Science* **291**:2141–2144.

Lo Pinto, M., E. Wajnberg, S. Colazza, C. Curty, and X. Fauvergue. 2004. Olfactory response of two aphid parasitoids, *Lysiphlebus testaceipes* and *Aphidius colemani*, to aphid-infested plants from a distance. *Entomologia Experimentalis et Applicata* **110**:159–164.

Mäntylä, E., T. Klemola, and E. Haukioja. 2004. Attraction of willow warblers to sawfly-damaged mountain birches: novel function of inducible plant defences? *Ecology Letters* **7**:915–918.

Margolies, D. C., M. W. Sabelis, and J. E. Boyer. 1997. Response of a phytoseiid predator to herbivore-induced plant volatiles: selection on attraction and effect of prey exploitation. *Journal of Insect Behavior* **10**:695–709.

Michel-Salzat, A., and J. B. Whitfield. 2004. Preliminary evolutionary relationships within the parasitoid wasp genus *Cotesia* (Hymenoptera: Braconidae: Microgastrinae): combined analysis of four genes. *Systematic Entomology* **29**:371–382.

Mumm, R., K. Schrank, R. Wegener, S. Schulz, and M. Hilker. 2003. Chemical analysis of volatiles emitted by *Pinus sylvestris* after induction by insect oviposition. *Journal of Chemical Ecology* **29**:1235–1252.

Neveu, N., J. Grandgirard, J. P. Nenon, and A. M. Cortesero. 2002. Systemic release of herbivore-induced plant volatiles by turnips infested by concealed root-feeding larvae *Delia radicum* L. *Journal of Chemical Ecology* **28**:1717–1732.

Oppenheim, S. A., and F. Gould. 2002. Is attraction fatal? The effects of herbivore-induced plant volatiles on herbivore parasitism. *Ecology* **83**:3416–3425.

Powell, W., and A. F. Wright. 1991. The influence of host food plants on host recognition by four aphidiine parasitoids (Hymenoptera, Braconidae). *Bulletin of Entomological Research* **81**:449–453.

Powell, W., F. Pennacchio, G. M. Poppy, and E. Tremblay. 1998. Strategies involved in the location of hosts by the parasitoid *Aphidius ervi* Haliday (Hymenoptera: Braconidae: Aphidiinae). *Biological Control* **11**:104–112.

Roda, A. L., and I. T. Baldwin. 2003. Molecular technology reveals how the induced direct defenses of plants works. *Basic and Applied Ecology* **4**:15–26.

Rutledge, C. E., and R. N. Weidenmann. 1999. Habitat preferences of three congeneric braconid parasitoids: implications for host-range testing in biological control. *Biological Control* **16**:144–154.

Sabelis, M. W., M. A. Janssen, A. Pallini, et al. 1999. Behavioral responses of predatory and herbivorous arthropods to induced plant volatiles: from evolutionary ecology to agricultural applications, pp. 269–296 in A. A. Agrawal, S. Tuzun, and E. Bent (eds.) *Induced Plant Defenses against Pathogens and Herbivores*. St. Paul, MN: American Phytopathological Society Press.

Souissi, R., J. P. Nenon, and B. Le Ru. 1998. Olfactory responses of parasitoid *Apoanagyrus lopezi* to odor of plants, mealybugs, and plant–mealybug complexes. *Journal of Chemical Ecology* **24**:37–48.

Steidle, J. L. M., and J. J. A. van Loon. 2003. Dietary specialization and infochemical use in carnivorous arthropods: testing a concept. *Entomologia Experimentalis et Applicata* **108**:133–148.

Strauss, S. Y., and R. E. Irwin. 2004. Ecological and evolutionary consequences of multispecies plant–animal interactions. *Annual Review of Ecology, Evolution, and Systematics* **35**:435–466.

Takabayashi, J., Y. Sato, M. Horikoshi, et al. 1998. Plant effects on parasitoid foraging: differences between two tritrophic systems. *Biological Control* **11**:97–103.

Thaler, J. S. 1999. Jasmonic acid mediated interactions between plants, herbivores, parasitoids, and pathogens: a review of field experiments in tomato, pp. 319–334 in A. A. Agrawal, S. Tuzun, and E. Bent (eds.) *Induced Plant Defenses against Pathogens and Herbivores*. St. Paul, MN: American Phytopathological Society Press.

Turlings, T. C. J., J. H. Tumlinson, and W. J. Lewis. 1990. Exploitation of herbivore-induced plant odors by host-seeking parasitic wasps. *Science* **250**:1251–1253.

Turlings, T. C. J., F. L. Wäckers, L. E. M. Vet, W. J. Lewis, and J. H. Tumlinson. 1993. Learning of host-finding cues by hymenopterous parasitoids, pp. 51–78 in D. R. Papaj and A. C. Lewis (eds.) *Insect Learning: Ecological and Evolutionary Perspectives*. New York: Chapman and Hall.

Turlings, T. C. J., H. T. Alborn, J. H. Loughrin, and J. H. Tumlinson. 2000. Volicitin, an elicitor of maize volatiles in the oral secretions of *Spodoptera exigua*: isolation and bio-activity. *Journal of Chemical Ecology* **26**:189–202.

Turlings, T. C. J., S. Gouinguene, T. Degen, and M. E. Fritzsche-Hoballah. 2002. The chemical ecology of plant–caterpillar–parasitoid interactions, pp. 148–173 in T. Tscharntke and B. A. Hawkins (eds.) *Multitrophic Level Interactions*. Cambridge, UK. Cambridge University Press.

van der Meijden, E., and P. G. L. Klinkhamer. 2000. Conflicting interests of plants and the natural enemies of herbivores. *Oikos* **89**:202–208.

van Loon, J. J. A., de Boer, J. G., and Dicke, M. 2000. Parasitoid–plant mutualism: parasitoid attack of herbivore increases plant reproduction. *Entomologia Experimentalis et Applicata* **97**:219–227.

van Poecke, R. M. P., M. A. Posthumus, and M. Dicke. 2001. Herbivore-induced volatile production by *Arabidopsis thaliana* leads to attraction of the parasitoid *Cotesia rubecula*: chemical, behavioral, and gene-expression analysis. *Journal of Chemical Ecology* **27**:1911–1928.

van Poecke, R. M. P., M. Roosjen, L. Pumarino, and M. Dicke. 2003. Attraction of the specialist parasitoid *Cotesia rubecula* to *Arabidopsis thaliana* infested by host or non-host herbivore species. *Entomologia Experimentalis et Applicata* **107**:229–236.

van Tol, R. W. H. M., A. T. C. van der Sommers, M. I. C. Boff, et al. 2001. Plants protect their roots by alerting the enemies of grubs. *Ecology Letters* **4**:292–294.

van Zandt, P. A., and A. A. Agrawal. 2004. Community-wide impacts of herbivore-induced responses in milkweed (*Asclepias syriaca*). *Ecology* **85**:2616–2629.

Vaughn, T. T., M. F. Antolin, and L. B. Bjostad. 1996. Behavioral and physiological responses of *Diaeretiella rapae* to semiochemicals. *Entomologia Experimentalis et Applicata* **78**:187–196.

Vet, L. E. M., and M. Dicke. 1992. Ecology of infochemical use by natural enemies in a tritrophic context. *Annual Review of Ecology and Systematics* **37**:141–172.

Vet, L. E. M., W. J. Lewis, and R. T. Cardé. 1995. Parasitoid foraging and learning, pp. 65–101 in R. T. Cardé and W. J. Bell (eds.) *Chemical Ecology of Insects*, vol. 2. New York: Chapman and Hall.

Via, S. 1990. Ecological genetics in herbivorous insects: the experimental study of evolution in natural and agricultural systems. *Annual Review of Entomology* **35**:421–446.

Wang, Q., and S. Dorn. 2003. Selection on olfactory response to semiochemicals from a host-plant complex in a parasitic wasp. *Heredity* **91**:430–435.

Wang, Q., H. Gu, and S. Dorn. 2004. Genetic relationship between olfactory response and fitness in the *Cotesia glomerata* (L.). *Heredity* **92**:579–584.

Whitham, T. G., W. P. Young, G. D. Martinsen, et al. 2003. Community and ecosystem genetics: a consequence of the extended phenotype. *Ecology* **84**:559–573.

Wise, M. J., and A. M. Weinberg. 2002. Prior flea beetle herbivory affects oviposition preference and larval performance of a potato beetle on their shared host plant. *Ecological Entomology* **27**:115–122.

Zangerl, A. R. 2003. Evolution of induced plant responses to herbivores. *Basic and Applied Ecology* **4**:91–103.

Part VI SYNTHESIS

16

Indirect interaction webs propagated by herbivore-induced changes in plant traits

TAKAYUKI OHGUSHI, TIMOTHY P. CRAIG, AND PETER W. PRICE

Introduction

This volume is the first to survey trait-mediated indirect effects in a wide variety of plant-based indirect interaction webs in terrestrial systems. In the introductory chapter we proposed four questions about important patterns in indirect interaction webs: (1) What novel interaction linkages are produced by plant-mediated indirect effects? (2) What complex interactions are generated by plant-mediated indirect effects in multitrophic systems? (3) What are the effects of plant-mediated indirect effects on community structure and biodiversity? and (4) What are the evolutionary consequences of plant-mediated indirect effects? In this chapter we will review the progress that has been made in answering these questions. We will examine the conceptual framework that the study of plant-mediated indirect effects provides for future research on the ecology and evolution of multispecies and multitrophic interactions. We will also explore the impact of these effects that influence the structure of ecological communities and biodiversity.

Ohgushi (2005) surveyed indirect interactions between herbivores through herbivore-induced changes in plants. The key features identified in these interactions are found in the interactions considered in this volume. First, plant-mediated indirect effects are common and widespread because trait changes in plants induced by a range of organisms are ubiquitous in terrestrial plants. Second, plant-mediated interactions occur between temporally separated, spatially separated, and taxonomically distant species. Third, there are often mutualistic

Ecological Communities: Plant Mediation in Indirect Interaction Webs, ed. Takayuki Ohgushi, Timothy P. Craig, and Peter W. Price. Published by Cambridge University Press. © Cambridge University Press 2007.

interactions between species sharing the same plant because of enhanced resource availability due to compensatory regrowth, and changes in sink–source relationships. In addition, ecosystem engineers create new physical structures on plants that mostly benefit later, secondary users. In this way plant-mediated indirect effects can (1) drive bottom–up trophic cascades, (2) alter community organization, (3) increase species richness and interaction diversity, and (4) affect directly and indirectly the evolution and coevolution of species in a community. Understanding these effects requires the linkage of distinct research disciplines in ecology. These issues will be discussed in detail in the following sections.

All organisms interact both directly and indirectly with other species in nature, but only in the past 25 years have indirect effects been recognized as an important force in forming community structure (Price et al. 1980, 1986, Strauss 1991, Holt and Lawton 1994, Wootton 1994, Menge 1995, Abrams et al. 1996, Ohgushi 2005). Almost all interactions between two species involve at least a third species, and therefore have the potential for indirect effects in every ecosystem. Recognizing the importance of indirect effects in natural communities, community ecologists have been conducting empirical studies, and developing a conceptual framework that incorporates indirect effects. Over the past decade, there has been rapid progress in research on indirect effects (Wootton 2002), and indirect effects are now recognized as being central to community dynamics (Polis et al. 2000). Indirect effects can act either by modifying the density of intermediate species (density-mediated indirect effects), or by altering the traits of intermediate species (trait-mediated indirect effects).

Density-mediated indirect effects and their community consequences have recently been extensively explored (Strauss 1991, Wootton 1994, Abrams et al. 1996, Menge and Branch 2001). The impact of density-mediated indirect effects on community structure and ecosystems has been demonstrated by the study of trophic cascades and apparent competition. Trophic cascades result in inverse patterns in abundance or biomass across more than one trophic link in a food web (Carpenter and Kitchell 1993). Recent studies have indicated that trophic cascades are found in a wide variety of systems (Pace et al. 1999, Shurin et al. 2002). Studies of trophic cascades have greatly stimulated the research on a variety of community consequences of species interactions, such as multi-trophic interactions, the relative importance of top–down and bottom–up effects, and the effects of keystone predation (Paine 1980, Carpenter et al. 1985, Hunter and Price 1992, Strong 1992, Polis and Strong 1996, Pace et al. 1999, Polis et al. 2000, Borer et al. 2005). Similarly, research on apparent competition has shown the negative interactions between prey species mediated by a shared natural enemy, which have demonstrated the importance of multiple prey–predator or host–parasite interactions in structuring and stabilizing ecological communities (Price *et al.* 1986, Holt and Lawton 1994, Bonsall and Hassell 1997, Morris *et al.* 2004).

Trait-mediated indirect effects act by mediating behavior, morphology, physiology, phenology, and life histories of intervening species, and they have received less attention than density-mediated indirect effects. Recently, some ecologists have recognized the importance of trait-mediated indirect effects in structuring ecological communities (Werner and Peacor 2003, Schmitz *et al.* 2004). Werner and Peacor (2003) argued that ecological communities are replete with trait-mediated indirect effects that arise from phenotypic plasticity, and that these effects are important in community dynamics. Trait-mediated indirect effects involved in plant–herbivore systems have remained largely unstudied (Ohgushi 2005). Studies of trait-mediated indirect effects in terrestrial systems have mainly been restricted to indirect prey–predator interactions, such as nonlethal effects of predators on prey behavior and physiology (Sih *et al.* 1998, Schmitz *et al.* 1997, 2004). For example, in a recent special feature section on "Linking individual-scale trait plasticity to community dynamics" in the journal *Ecology* (**84**:1081–1157, in 2003), the community consequences of trait-mediated indirect effects were highlighted, but little attention was given to plant-mediated indirect interactions (but see Callaway *et al.* 2003).

The conceptual basis for a network of interacting species in ecological communities is food web theory, which is one of the fundamental concepts in ecology (Lawton 1989, Polis and Winemiller 1996). Since food webs are diagrams depicting the structure of ecological communities by trophic links mainly through prey–predator interactions, nonfeeding relationships between species at the same trophic level or between species across more than one trophic level are not included. There is increasing evidence that such nontrophic, indirect interactions are ubiquitous and can play an important role in structuring arthropod communities by trait changes in plants (see Ohgushi 2005 for references therein). Thus, plant-based food webs that ignore nontrophic, indirect links are an inadequate tool for understanding the structural organization of communities. In this context, Berlow *et al.* (2004) pointed out that the future challenge to develop the theory of food web dynamics is to incorporate nontrophic links into food web structure. Thus, indirect interaction webs that explicitly incorporate nontrophic and indirect links into components of the traditional food webs can provide a useful tool to understand the reality of structure and biodiversity in natural communities. This volume clearly demonstrates how nontrophic and indirect interactions mediated by plant traits can form plant-based interaction webs which enrich species diversity and increase interaction diversity. This indicates the importance of including indirect interactions in interaction webs. In particular, positive interactions, nonfeeding interactions, ecosystem engineering, and above- and belowground interactions have the potential to have inordinately strong impacts on community structure and biodiversity. These keystone interactions have been rarely involved in the traditional food web theory, simply because they are not feeding relationships.

Nonlethal effects of herbivores on plants and phenotypic plasticity of plants

Plants provide a bottom–up template that determines the structure of ecological communities. The importance of bottom–up forces in structuring food webs has long been appreciated (Lindeman 1942, Pimm 2002), and the research summarized in this volume shows that nontrophic forces also flow from the bottom up from primary producers. Plants are the basal component of all food webs, and their phenotypic plasticity creates nontrophic variation that moves upwards to control the dynamics of multitrophic level interactions. The conclusion that ecological communities are structured from the bottom up is a radical departure from "the world is green" scenario of Hairston et al. (1960). They argued that herbivore populations are not limited by resources, because green plants are rarely depleted, and they concluded that predators must control herbivore populations. Their argument is based on the assumption that all plants are edible and available for herbivores. On the other hand, there is abundant evidence that all plants are not equally suitable and available as food because of physical and chemical defenses, and temporal and spatial heterogeneity (Strong et al. 1984, Hunter et al. 1992, Ohgushi 1992), and this book presents abundant data that plants are a difficult resource to exploit for herbivores. The primacy of bottom–up forces in structuring communities is evident because the removal of higher trophic levels leaves lower levels present, whereas the removal of primary producers leaves no system at all (Pimm 1991, Hunter and Price 1992). Bottom–up forces induce changes on multiple trophic levels that set the stage on which top–down forces act: the top–down effects of predation on lower trophic levels can be greatly altered by spatial and temporal variation in plant growth and quality (Hunter and Price 1992, Hartvigsen et al. 1995, Forkner and Hunter 2000, Denno et al. 2002).

Nutrition and a place to live are the essential resources that terrestrial plants provide for herbivores and all higher trophic levels, and so any alteration of these traits can have far-reaching effects. In general, the nutrition that plants provide for herbivores is poor in quality and highly heterogeneous in space and time (Mattson 1980, Strong et al. 1984, White 1993, Hartley and Jones 1997). Herbivores have evolved specialized feeding preferences and feeding strategies in order to efficiently find suitable food among this large mass of nutritionally poor plant material. Herbivores frequently specialize on specific plant parts such as leaves, stems, flowers, buds, seeds, fruits, or roots, and to exploit these diverse resources they use a wide range of feeding modes including chewing, sucking, rolling, mining, and galling. This specialization creates a diversity of niches on each terrestrial plant, and therefore many insect herbivores can coexist on individual terrestrial plants (Strong et al. 1984, Thompson 1994, Price 2002). Therefore, alteration of a single plant trait has the potential to affect the feeding niches of many different herbivores associated with the plant.

A place to live is the second major resource that terrestrial plants provide for members of an interaction web. The architectural variation of terrestrial plants provides a wide range of structures for herbivores to occupy, and herbivores manipulate the plant to produce many more kinds of structures (Ohgushi Chapter 10, Marquis and Lill Chapter 11, this volume). Variation in plant architecture also alters the quality of these domiciles, and this can have a large influence on the preference and performance of herbivores and/or predators (Lawton 1983, Langellotto and Denno 2004) in their indirect interaction webs (Ohgushi Chapter 10, Marquis and Lill Chapter 11, Craig Chapter 14, this volume).

Insect population dynamics and community structure are strongly influenced by variation in host plant quality and architecture. Herbivore preference for host plants and their performance on these resources influence their population dynamics. The variation in responses of different species of herbivores to plant resource variation can determine the composition of the herbivore community, which in turn influences all other community members. Plant resources can also influence insect herbivore quality, development, and behavior with effects that cascade upwards to influence predators and parasitoids on the third trophic level (Price et al. 1980, Masters et al. 2001, Teder and Tammaru 2002, Gratton and Denno 2003, Kagata et al. 2005, Nakamura et al. 2005).

Mature terrestrial plants are usually not killed when attacked by herbivores; instead the quantity and quality of food and habitat available to the community that utilizes the plants are altered. This herbivore-induced phenotypic plasticity of plants provides the mechanistic basis for trait-mediated indirect effects. The prevalence of trait-mediated indirect effects is indicated by the ubiquity of phenotypically plastic plant responses to herbivory. Herbivory can induce changes in a wide variety of traits, such as secondary chemicals, cell structure and growth, physiology, morphology, and phenology (Karban and Baldwin 1997). Induced resistance against insect herbivores and pathogens is common and widespread in both agricultural and natural systems (Haukioja and Neuvonen 1987, Karban and Myers 1989, Karban and Baldwin 1997, Constabel 1999). Compensatory growth following herbivory is another well-established form of phenotypic plasticity (Gómez and Gonzáles-Megías Chapter 5, Ohgushi Chapter 10, Bailey and Whitham Chapter 13, this volume). Recently, it has been recognized that the alteration of architecture and quality of previously damaged plants has important consequences for many members of interaction webs (Mopper *et al.* 1991, Strauss and Agrawal 1999, Nakamura *et al.* 2003). Plant phenotypic plasticity as a response to herbivory means that one insect herbivore can indirectly affect the performance and/or abundance of other insects by changing plant traits (Ohgushi 2005), and that these nonlethal effects of herbivory provide the mechanistic basis for plant-mediated indirect interactions.

What have we learned about induced plant traits and indirect interaction webs?

We have learned of their richness

Indirect interaction webs are ubiquitous and diverse: they involve many plants, many insects, and they produce a wide spectrum of induced effects on plants. There is a three-dimensional matrix of interactions that will inevitably generate many and diverse effects in a community. Plants are mediators of these interactions and we have gained an appreciation that the plasticity of plant responses to a range of interactions is the key to understanding the richness and complexity of these interactions. The plant is chameleon-like in its ability to change; it is protean, contrasting dramatically from the prevalent image of plants being simple, static, and passive. In the study of plant-mediated indirect effects, we enter into a world largely unseen and certainly underappreciated: a world of fascinating complexity compared to direct effects such as prey–predator interactions.

Indirect interaction webs are inherently more complex than direct interaction (food) webs because each interaction involves a minimum of three and frequently many more organisms. The key to understanding the evolution of the rich variation in indirect interaction webs is the realization that an interaction between organisms can be transformed as it is mediated by the plant, and it can affect the wide variety of organisms that use a plant independently. We have seen in this volume that there are complex, and in many cases unpredictable, ways in which an interaction between two organisms can be altered when the plant intervenes in the interaction. Furthermore, since plants are a central interaction point through which interactions are propagated upwards and then back down through the trophic levels, alteration of the plant phenotype can result in the involvement of many more organisms into the indirect interaction web.

The chapters in this book provide a plethora of examples on the diverse ways that the plant responses result in the initiation of indirect interaction webs. Herbivory can alter plants through the production of plant shelters, architectural changes, phytochemical alterations, the induction of galls, mines, and tunnels, and through change in the plant's vascular system. All of these changes have the strong potential to have impacts beyond the plant–herbivore interaction including influences on competing herbivores (Denno and Kaplan Chapter 2), natural enemies (Sabelis *et al.* Chapter 9, Craig Chapter 14), pollinators (Bronstein *et al.* Chapter 4), and fungi (Chaneton and Omacini Chapter 8). Besides herbivores, other plant interactions with mychorrhizal (Gange Chapter 6) and endophytic (Chaneton and Omacini Chapter 8) fungi, and hemi-parasitic plants (Hartley *et al.* Chapter 3) can be initiators of indirect interaction webs. The propagation of these effects may be easily detectable because they occur rapidly and over small spatial scales, or they may be difficult to study because they occur slowly over a large spatial scale as they influence species on the same and different trophic levels.

> Indirect plant mediation facilitates interactions among distantly related species

Many strong interactions among taxonomically distant species are reported in this volume. They include interactions between fungi and insects (Gange Chapter 6, Chaneton and Omacini Chapter 8), phytopathogens and herbivorous arthropods (Sabelis *et al.* Chapter 9), mammals and insects (Gómez and Gonzáles-Megías Chapter 5, Bailey and Whitham Chapter 13), birds and insects (Bailey and Whitham Chapter 13, Craig Chapter 14), and hemi-parasitic plants and insects (Hartley *et al.* Chapter 3), and they are summarized below. Plant mediation greatly increases the probability of interactions among distantly related species because it provides pathways allowing interactions between species that would otherwise be isolated spatially and temporally, by size differences, and by modes of interaction. This counters the generalization from direct interactions that closely related species taxonomically interact most strongly and these interactions are most common in nature (Denno and Kaplan Chapter 2). Hochberg and Lawton (1990) argued, however, that organisms in different phyla or even kingdoms may compete for the same resources and that such interactions may be one of the most pervasive forms of interspecific competition in nature, and yet such interactions are still among the most poorly understood.

> Indirect plant mediation creates interactions among species isolated by space

Plants can mediate interactions between species on several different spatial scales: within a plant part, such as a leaf, between different parts of a plant, such as between shoot and root feeders, between the plant and belowground biota, and even between organisms on different plants. The variation in outcomes of indirect interactions on a plant can create a spatial mosaic of resources that influence interactions in communities at the landscape scale. Indirect interaction webs can be induced among organisms using different parts within a plant by many different mechanisms. Herbivory can induce resistance that affects other herbivores utilizing the same or different parts of the plant (Denno and Kaplan Chapter 2). Herbivory on any part of the plant can alter plant architecture influencing other herbivores (Gómez and Gonzáles-Megías Chapter 5, Ohgushi Chapter 10), and this can affect mutualists by altering the production of flowers for pollinators (Bronstein et al. Chapter 4) or the interaction with mycorrhizae (Gange Chapter 6). Herbivores, such as sap-sucking insects, can also interact at a distance by altering the flow of xylem and phloem (Denno and Kaplan Chapter 2). Finally, herbivores and hemi-parasitic plants can alter source–sink relationships influencing resources available to other organisms. Gallers (Inbar et al. 1995, Larson and Whitham 1997) and hemi-parasitic plants (Hartley et al. Chapter 3) can divert resources from other parts of the plant for their use. This redistribution of resources can have effects that cascade up the trophic levels. These interactions between organisms using different plant parts are frequently asymmetric, with one of the interacting species strongly affecting a second with no reciprocal effect on the other (Denno and Kaplan Chapter 2, Hartley et al. Chapter 3).

Plant-mediated interactions can link above- and belowground communities. Growing evidence is that interactions between leaf- and sap-feeding insects and mycorrhizae mediated by the plant can occur frequently (Masters et al. 1993, Van der Putten et al. 2001, Bardgett and Wardle 2003, Gange Chapter 6, Poveda et al. Chapter 7). Herbivore-induced changes in plant traits can indirectly affect soil organisms, including soil invertebrates, fungi, and bacteria. Endophytic fungi may also link above- and belowground food webs (Chaneton and Omacini Chapter 8).

Genetic and phenotypic variation among plants generates variation in direct and indirect interactions that can create spatial mosaics of resources for other species which influences community structure. Spatial mosaics resulting from plant-mediated indirect interactions include the interaction of plants, ants, and aphids (Wimp and Whitham Chapter 12), and the interaction of elk and aspen (Bailey and Whitham Chapter 13), the interaction of plants and mycorrhiza (Gange Chapter 6), and the interaction of plants, insects, and hemi-parasitic plants (Hartley et al. Chapter 3). These studies show that indirect interactions on individual plants can have influences at the landscape scale (Wimp and Whitham Chapter 12, Bailey and Whitham Chapter 13).

These interactions among organisms in spatially isolated niches challenge the traditional assumption that organisms using different niches do not interact strongly (Denno and Kaplan Chapter 2), and that the tradition of focusing on interactions among organisms utilizing the same plant part may provide an inadequate and misleading understanding of interactions. To gain a better understanding of community ecology and evolution, we need to examine the mechanisms by which species interact and not just their location: a root feeder and a leaf feeder may interact more strongly than two leaf feeders. Several authors point out that we need to develop a better understanding of how plant physiology mediates the interaction of such seemingly disparate groups using different plant parts or having different functional interactions with the plant such as pollination and herbivory (Bronstein *et al.* Chapter 4), above- and belowground organisms (Poveda *et al.* Chapter 7), and herbivores and natural enemies (Sabelis *et al.* Chapter 9).

> Indirect plant mediation creates interactions among species isolated by time

Indirect interactions can link temporarily isolated species, and greatly increase the diversity of interactions in a web. Indirect effects caused by herbivory can create interactions between temporally isolated species via at least four pathways. First, an earlier attacking herbivore can induce chemical defenses that influence later herbivores (Denno and Kaplan Chapter 2), and these induced chemical changes may also affect natural enemies (Sabelis *et al.* Chapter 9, Althoff Chapter 15). Second, nutritional quality of plants can be altered by herbivores and these effects can influence later-feeding herbivores (Denno and Kaplan Chapter 2). For example, Ohgushi (Chapter 10) showed that early-season herbivory altered later-season sap and nitrogen quality which influenced other herbivores. Gange (Chapter 6) showed that herbivory decreased mycorrhiza infections and this effect persisted over years. Third, herbivore damage can have a time-lagged effect by altering plant growth or architecture. Herbivores can induce compensatory growth (Gómez and Gonzáles-Megías Chapter 5, Ohgushi Chapter 10, Bailey and Whitham Chapter 13) or physical defenses (Denno and Kaplan Chapter 2) that affect later-feeding herbivores. Hemi-parasitic plants can also affect source–sink relationships, and in some cases alter root : shoot ratios (Hartley *et al.* Chapter 3) influencing subsequent herbivory. Mutualists are also affected as Bronstein *et al.* (Chapter 4) have shown with the example that early herbivory can limit the number and size of flowers impacting later pollinators. Fourth, ecosystem engineers that produce new structures, such as leaf shelters, can result in the assemblage of entirely new communities at a later time (Ohgushi Chapter 10, Marquis and Lill Chapter 11, Bailey and Whitham Chapter 13).

The temporal separation of indirectly interacting species means that both the ecological and evolutionary effects are highly asymmetric (Denno et al. 1995, Denno and Kaplan Chapter 2, Hartley et al. Chapter 3). This would be particularly true for herbivore species sharing the same plant at different times during the growing season if the effect of later species on plant traits does not carry over to the next year. In this case, the earlier species has a strong impact on the later species, but the later species will have little impact on the earlier species. This frequently results in the competitive superiority of earlier species due to the pre-emption of resources, induced resistance, and a superior ability to tolerate physiological changes caused by feeding (Denno and Kaplan Chapter 2). This asymmetrical relationship would seem to preclude coevolutionary interactions, and reciprocal effects are clearly lacking when there is a time lag between when one species impacts a plant and when the altered plant character affects another species. Since long-term studies on the asymmetric interactions between temporally separated species are lacking, we need more evidence of how long the herbivore-induced changes in plant traits are maintained.

Ecologically engineered structures can linger and be maintained by other species long after the initiator is gone (Marquis and Lill Chapter 11), as is the case when a caterpillar constructs and abandons a leaf shelter that is later used by a community of new species (Ohgushi Chapter 10). Frequently, in interactions where plant architecture is altered the impact of the initiator will influence later generations of organisms because the effects on the plant may persist over many years. For example, both insect herbivores (Ohgushi Chapter 10) and mammalian grazers (Gómez and Gonzáles-Megías Chapter 5, Bailey and Whitham Chapter 13) can induce compensatory regrowth in plants that has long-term impacts on the community.

Time lags create the potential for two strongly interacting species to experience very different environments exerting very different selective effects on the two species, and this may explain why indirectly interacting organisms are often distantly related. In contrast, directly interacting species experience the same environments. This contrast is illustrated by comparing herbivores involved in exploitation or interference competition, and those indirectly interacting by using the same plant at a different time. The early-season herbivores encounter soft, rapidly growing leaves with high water content, and low concentrations of defensive compounds, while later herbivores are adapted to deal with tough, slow-growing leaves that are strongly defended, and with many plant characters that have been modified by the earlier herbivores. Convergent selection would create similar characters in the directly interacting species, while divergent selection would produce different characters in the indirectly interacting species. Resources can be modified by whole series of intervening organisms so

We have learned of complexity of interactions and its effects on community structure and biodiversity

Compared to direct interaction (food) webs, indirect interaction webs are more complex and biologically diverse. These community characteristics are all the result of an expanded number of potential pathways for interactions that are provided by plant mediation. The increased number of pathways increases the complexity of interactions and the number of species involved.

Indirect interactions can increase complexity of both their component community and the overall community in several ways, and the complexity is higher than in traditional food webs. We define the community as all the organisms that interact within an area (Price 1997) and define the plant-based component community as all of the directly and indirectly interacting species associated with an individual plant. This is analogous to a source food web based on a plant (Pimm 1991, Pimm *et al.* 1991). The complexity of ecological communities is usually measured in four ways: species richness, link density, connectance, and the number of trophic levels (Morin and Lawler 1995). Link density is the number of links per species, and connectance is defined as the number of realized links out of the total number of possible links (Pimm *et al.* 1991). Indirect interactions can increase complexity by all of these measures both by increasing the links among species already involved in a community via direct interactions, and by involving new species in a community. Plant-mediated indirect interactions can increase species richness by adding new species into the plant-based community (Denno and Kaplan Chapter 2, Sabelis *et al.* Chapter 9). Ecosystem engineers that alter plant resources by such mechanisms as the construction of shelters, and the initiation of galls, by boring stems, or producing leaf mines create niches for additional species that would not otherwise be found in the community (Ohgushi Chapter 10, Marquis and Lill Chapter 11, Bailey and Whitham Chapter 13). Mutualism can also strongly influence species richness. Wimp and Whitham (Chapter 12) found that the presence of aphid–ant mutualism had complex impacts on species richness on individual plants but greatly increased biodiversity at the community level.

Indirect interactions between species generate new links among species in a community, and this can increase link density. Indirect interactions by plant mediation have the great potential to increase link density because of the rich array of interaction pathways that intersect at the plant. There is a complex array of interaction initiators, including a range of herbivores (free-feeders, gallers, miners, borers, and shelter-makers) and mutualists (endophytic fungi and mycorrhiza) that elicit a corresponding complex array of plant responses (Sabelis *et al.* Chapter 9), and these effects can move up the trophic system. Authors in this volume provide many examples of how indirect interactions initiate new links among species not included in food webs, including interactions between: herbivores (Denno and Kaplan Chapter 2, Gómez and Gonzáles-Megías Chapter 5), plants and predators/parasitoids (Denno and Kaplan Chapter 2, Sabelis *et al.* Chapter 9), predators/parasitoids (Craig Chapter 14), herbivores and pollinators (Bronstein *et al.* Chapter 4), mutualistic fungi of plants and herbivores (Gange Chapter 6, Chaneton and Omacini Chapter 8), and plants and insect mutualists (Wimp and Whitham Chapter 12). Looking just at the interactions among herbivores we see mediation of competition via induced chemical resistance, nutritional effects, induced changes in physical morphology such as trichome density, and altered risk of enemy attack (Denno and Kaplan Chapter 2). Also, it is important to recognize that plant-mediated indirect effects can generate new links of facilitation between herbivores (Denno and Kaplan Chapter 2, Gómez and Gonzáles-Megías Chapter 5, Ohgushi Chapter 10, Marquis and Lill Chapter 11, Bailey and Whitham Chapter 13). Thus, link density is much higher in indirect interaction webs than in direct interaction webs, because each species has more possible nontrophic and indirect linkages. For food webs, link density has been reported as roughly two for all webs (Pimm *et al.* 1991), and a mean of 2.2 for insect-dominated food webs (Schoenly *et al.* 1991). The increase in link density in general or hypothetical terms is shown in the indirect interaction webs in several chapters (Hartley *et al.* Chapter 3, Gómez and Gonzáles-Megías Chapter 5, Gange Chapter 6, Poveda *et al.* Chapter 7, Chaneton and Omacini Chapter 8, Ohgushi Chapter 10, Marquis and Lill Chapter 11, Wimp and Whitham Chapter 12, Bailey and Whitham Chapter 13, Craig Chapter 14).

We predict that the link density in indirect interaction webs increases as the number of species in the web increases. Food webs have been hypothesized to have scale-invariant link density, with the number of links per species staying constant as the number of species increases except in very large food webs (Cohen and Briand 1984, Lawton and Warren 1988, Pimm *et al.* 1991; but see Martinez 1992). The relationships between species richness and link density will differ in food webs and indirect interaction webs because the species in these webs interact in different ways. In a direct food web a change in the number of links requires that a change in the predator's feeding choices occurs: if a new prey species is utilized then a predator must specifically adapt to finding and utilizing this prey. For example, in a food web, if a predator that has been feeding on chrysomelid beetles adds membracids to its diet it must change: its search image, its search area, and its interaction with the defensive mutualists of the membracids. This would increase its link density by one. In contrast, in a plant-mediated interaction web a herbivore can induce a change in a plant that alters the interaction with other herbivore species without any specific change in the initiating species in response to the other species. For example, if a leaf-webbing moth colonizes a new plant it may undergo no change in behavior and continue to web the leaf as it did on other plant species, and yet this webbing may have a strong effect on many other species on the plant such as aphids, membracids, and the mutualists and natural enemies of these herbivores. Link density would be increased dramatically without any alteration in the webbing behavior of the herbivore. In indirect interactions webs, the effect can multiply as the impact moves through the interaction web as each change potentially affects the rest of the web. Thus, as the number of species initiating responses of a plant can increase the link density rapidly as every species is potentially affected by the initiating species. Indirect interaction webs will also have increased connectance compared to food webs. In contrast to food webs where only one kind of interaction is possible, a higher trophic level feeding on a lower one, the number of possible links in interaction webs is unknown as we are constantly recognizing novel links in interaction webs. As we noted earlier, we have entered into a world largely unseen and certainly underappreciated.

Indirect interactions can also increase the third measure of complexity: the number of trophic levels involved in interactions. A basic characteristic of food chains is that they are short, being limited to four or five trophic levels by either energetic or stability limitations (Elton 1958, Pimm and Lawton 1977, Oksanen et al. 1981, Morin and Lawler 1995, Pimm 2002). Schoenly et al. (1991) found that in insect dominated food webs the mean chain length was 2.4 in the tropics, 2.8 in deserts, and 3.2 in the temperate regions. There has been no attempt yet to determine the mean length of indirect interaction chains. However, a review of the interaction webs in this book indicates that interaction chains can be very long and reticulate moving up and down through the trophic levels. For example, the indirect interaction web on willow (Ohgushi Chapter 10) contains interaction links up to ten steps long, not counting possible feedback loops. These chain lengths in indirect interaction webs are not limited by the energetic constraints of food webs, since energy transfer is not an obligate characteristic of nontrophic, indirect interactions. For the same reason, stability also may not be a constraint of indirect interaction chain length.

The outcomes of the plant-mediated interactions are conditional, creating spatial and temporal environmental heterogeneity, and vary depending on the genotypic variation of individuals in the web or the number and strength of interactions (Wimp and Whitham Chapter 12, Bailey and Whitham Chapter 13). Bailey and Whitham (Chapter 13) have shown that the sign of an interaction between organisms in a web may change as the number of species in the interaction changes. The conditional nature of indirect interactions adds complexity not found in food webs.

Indirect interaction diversity may increase community stability, just as species diversity has long been hypothesized to increase ecosystem stability (Morin and Lawler 1995, Polis and Strong 1996, Tilman 1999, 2000, Hooper et al. 2005). Ecological stability has been defined and measured in several different ways (McNaughton 1977, Pimm 1991, McCann 2000). Dynamic stability occurs when there is low variation in population densities or community processes such as biomass production, and it is usually measured as the coefficient of variation. Resilience and resistance stability increases when the return time to an equilibrium/nonequilibrium solution decreases after a perturbation (McCann 2000). Persistence is how long a variable lasts before it is changed to a new value (Pimm 1991). All or some of these measures of stability may be influenced by species or interaction diversity.

The past mathematical models that assumed equilibrium population dynamics have suggested that diversity tends to destabilize community dynamics (May 1973, Pimm and Lawton 1977), but these models have not been supported empirically (McNaughton 1977, 1985, Tilman 1999, 2000). Recently, several authors (McCann et al. 1988, Tilman 1999, 2000, McCann 2000, Kondoh 2003) have argued that both new theoretical work, including the development of nonequilibrium models, and empirical research have strongly supported the hypothesis of earlier ecologists (Elton 1958, Odum 1953, MacArthur 1955) that species and interaction diversity increases community stability. Field studies (McNaughton 1977, 1985) and experimental work (Tilman 1996, 1999, 2000) have shown a positive relationship between species diversity and stability of ecosystem processes such as biomass production in plant communities. Stability in these studies was determined as a decrease in variability of biomass production as measured by the coefficient of variation. Other authors, however, maintain that more experimental evidence is needed to substantiate the positive relationship between diversity and stability (Morin and Lawler 1995, Polis and Strong 1996, Hooper et al. 2005).

Two processes by which diversity may increase stability are the averaging or "portfolio" effect and the negative covariance effect (Tilman 1999, 2000, McCann 2000), and both could be enhanced by plant-mediated indirect effects. The portfolio effect is based on the well-known economic principle that diversified stock portfolios are less volatile than a single stock, because individual stocks react individually to perturbations in the economic environment. As a result, the mean variation of many stocks through time is lower than that of a single stock. By analogy the stability in diverse communities increases because species respond individually to environmental variation dampening the mean response to environmental disturbances. Indirect interaction webs such as those involving ecosystem engineers that add species to a community may therefore have a stabilizing effect on their communities. In addition, the increased link density could increase stability if each interaction responds independently to environmental variability.

The potential impact of a diversity of indirect interactions on community stability can be illustrated by comparing two species: one involved in only a single indirect interaction that influenced its population density, and one where multiple interactions influenced its population density. The indirect interactions could be positive or negative. If the single indirect interaction controlling the first species density were removed due to an environmental perturbation, it could lead to a population density change that could have a strong impact on its own abundance and the abundance of its prey, predators, or competitors. These changes could impact community processes such as biomass production. In contrast, if multiple indirect interactions controlled population density then the removal of any one indirect interaction would be less likely to alter the population density of that species and therefore community stability. For example, if the population density of a species were controlled by one negative interaction and this interaction were removed, other negative interactions might compensate for the missing interaction and there would be little change in the density of that species. The direction of change and the degree of impact on community stability would depend on the number of indirect interactions and the balance of positive and negative interactions. A failure to consider the effect of indirect interactions on community stability may explain why adding or removing a link to a food web sometimes radically alters a food web and sometimes does not. For example, sometimes adding a species to an upper trophic level may or may not induce a trophic cascade (Hunter and Price 1992).

Negative covariance of species densities can also increase food web stability (McCann 2000). Negative covariance occurs when the abundance of two species on the same trophic level compete with each other, so that when one declines the other increases, lowering the overall variation in abundance. This negative covariance reduces the variability of a community as a whole. A key finding is that a weak interaction effect generates negative covariances and promotes community-level stability (McCann et al. 1988, Berlow 1999, McCann 2000). This effect occurs in the case where a predator influences the densities of two competing prey species differently: it interacts strongly with one prey species and weakly with the other. If predation decreases the density of the strongly interacting competitor, it will release the weakly interacting competitor. This indirect effect of the predator through competing prey species would tend to dampen the destabilizing potential of strong prey–predator interactions. Since many plant-mediated competitive interactions are asymmetric (Denno and Kaplan Chapter 2, Hartley et al. Chapter 3) the impact of negative covariance would stabilize the community. For example, negative covariance would occur when an early herbivore induced a plant defense that had a strong asymmetrical effect on a later herbivore. Decreases or increases in the early herbivore would be compensated by the opposite response in the later herbivore. It should be also noted that the increased link density, i.e., the number of interactions per species, in indirect interaction webs has the great potential to enhance community stability by increasing the number of weak interactions involved in a web. Therefore, the recognition of the plethora of weak interactions created by plant mediation could explain how interaction diversity increases stability.

Wimp and Whitham (Chapter 12) directly addressed this question showing that on individual plants the presence of aphid–ant mutualism increased community stability in terms of species turnover based on extinction and colonization of species involved. To cope with the recent development of theoretical and empirical work on the diversity–stability relationship, we should concentrate on how increased species richness and interaction diversity in the plant-based indirect interaction webs influence stability of communities of higher trophic levels.

We have learned of evolutionary implications of indirect effects

Studies of the evolutionary impacts of indirect effects are in their infancy. Most of the work in the field as summarized in this book has been to establish the existence of indirect effects in ecological interactions, and we have only begun to assess whether these ecological effects have evolutionary implications (Althoff Chapter 15). Several methods for measuring the evolutionary impact of indirect effects are suggested in this volume. Althoff (Chapter 15) demonstrates how selection experiments, comparisons of traits of closely related species, and phylogenetic analysis can be used, and Craig (Chapter 14) suggests using phenotypic selection analysis and path analysis to assess the strength of indirect selection.

There are at least three important evolutionary effects of indirect interactions. First, indirect interactions create new links between species that do not exert selection on each other directly, creating the potential for new selective forces in an interaction web. Second, at least three and frequently more species are involved in selection on any trait, increasing the complexity in the effects of selection and complicating the analysis of selection. Third, indirect selection through changes in plant traits means that selective forces are transformed by the mediating plant before selection acts on a third species. As a result, selection on both species in an indirect interaction acts through responses to selection by plant traits and not traits of the other species.

The transformation of selection pressures by the mediation of the plant means that selection pressures are more complex than in direct pair-wise interactions. Selection for a change in a trait involves at least three genomes: the plant and the two indirectly interacting species. Thus, the outcome of the interaction depends on the response to selection by the plant, and the constraints and opposing selective forces on the plant will shape the evolutionary trajectory of the two indirectly interacting species. For example, the evolution of gall size that mediates the interaction between a parasitoid and a bird in the *Eurosta* interaction web may be constrained by the plant (Craig Chapter 14). Because plants are at the center of complex interaction webs, their traits are potentially under selection by many other species, and any response to selection that alters plant traits may alter selective pressures on many other species. Selection on the plant due to indirect interactions may lead to a sequential radiation of new species. Divergent selection on *Eurosta solidaginis* has led to the evolution of host races on two species of goldenrod (Craig et al. 1993, 1997, 2001). Evolution of the plant and the herbivore has led to selection for genetic divergence in other species in the interaction including the inquiline *Mordellistena convicta* (Abrahamson et al. 2003, Eubanks et al. 2003). Indirect interactions mediated by plant traits can produce complex evolutionary forces influenced by many species, and in turn influencing many species. Gómez and Gonzáles-Megías (Chapter 5) discuss how these forces may act in concert as when two herbivores both utilize the same plant cues and result in selection for a reduction in this cue, or they can create opposing selection forces that constrain the evolution of a trait.

Several authors have hypothesized how specific traits may evolve in response to indirect selection. Sabelis et al. (Chapter 9) consider a number of unexplored pathways by which indirect interactions involving plant chemistry could evolve to influence the insect community. For example, the evolution of herbivore-induced plant volatiles is influenced by a complex interplay of forces since the plant cannot control who uses these signals: they could attract beneficial natural enemies or other herbivores. Plants not suffering herbivory could also evolve to mimic these signals. Althoff (Chapter 15) discusses a number of other intriguing specific pathways by which chemical interactions could evolve indirectly. One example is the evolution of biochemical crypsis where a plant lacks crucial chemicals used in volatile signaling involving herbivores and parasitoids. He also discusses how closely related parasitoids using different strategies have evolved different responses to plant cues. Craig (Chapter 14) has shown that natural enemies can interact indirectly by influencing the evolution of the physical characteristics of galls. Identifying and measuring the strength of specific indirect evolutionary forces is a potentially rich area for future research.

Although previous studies of coevolution have chiefly investigated pair-wise interactions between two species, recent appreciation of diffuse coevolution has highlighted the important indirect effects of other species on evolution of pair-wise interactions in multispecies systems; directions and pressures of selective forces by one species can change in the presence or absence of other species in a community (Thompson 1994, Inouye and Stinchcombe 2001, Craig Chapter 14). It should be noted that the evolutionary changes in any particular pair-wise interaction would influence community structure through networks of species interactions. Thus, the indirect interaction webs are not a static but dynamic system in an evolutionary context. Evolutionary changes in traits mediated by reciprocal interactions may alter composition and structure of natural communities. For example, the strength and direction of natural selection on plant resistance to a particular herbivore species can be modified by the presence and/or absence of other herbivorous species in a community (Pilson 1996). Since plant resistance traits have a large impact on insect community structure (Fritz 1992, Thaler *et al.* 2001, Hochwender and Fritz 2004), such indirect evolutionary effects would modify the structure of plant-based indirect interaction webs.

> *We have learned that plant-mediated indirect effects act as an interface of separated disciplines*

This book provides many examples of how plants provide the center of direct and indirect interactions, and how species in different guilds, taxa, and habitats are connected indirectly in plant-based indirect interaction webs. Since ecology of species in these different categories has been explored in different disciplines, we need to merge these research fields to establish a conceptual framework that describes the consequences of indirect interactions among all possible interactions. In particular, we need to integrate below- and aboveground ecology, herbivory and pollination ecology, and insect and microbial ecology.

> Interface of below- and aboveground ecology

Several chapters show that plant-mediated interactions between leaf-feeding and root-feeding insects and those between leaf- and sap-feeding insects and mycorrhizae commonly occur, and can link above- and belowground communities (Hartley *et al.* Chapter 3, Gange Chapter 6, Poveda *et al.* Chapter 7, Chaneton and Omacini Chapter 8). This strengthens an emerging view that we must link above- and belowground approaches to community and ecosystem ecology to enhance our understanding of the regulation and functional significance of diversity and ecosystem functioning (Van der Putten *et al.* 2001, Wardle

2002, Bardgett and Wardle 2003, Wardle *et al.* 2004). Herbivore-induced changes in plant traits indirectly affect soil organisms, including soil invertebrates, fungi, and bacteria, all of which play an important role in key soil processes of decomposition and nutrient mineralization. There is increasing evidence that aboveground herbivory can change root carbon allocation, root exudation, root biomass, and morphology (Bardgett *et al.* 1998). Thus, the quantity and quality of organic matter input from plants damaged by herbivores have the potential to greatly influence abundance, species composition, and activity of the soil organisms in the rhizosphere by altering interactions in soil food webs. The role of variation in individual plants in mediating these interactions needs to be evaluated as differences among plant species can have important effects on soil biota and the processes that they regulate because plant species differ in quality and quantity of resources that they return to soil (Rehill *et al.* 2005). Even variation within plant species impact ecosystem processes (Schweitzer *et al.* 2004, 2005a, 2005b).

Interface of herbivory and pollination ecology

Herbivory and pollination ecology long have developed as separate disciplines, and until recently few studies have tried to merge the two fields. A growing literature has shown that leaf herbivory by insects often decreases considerably the quantity and/or quality of floral traits, such as flower number, flower size, pollen production, pollen performance, and nectar production, suggesting that because these floral traits provide various pollinator services, early-season herbivory can strongly alter plant relationships with pollinators (Strauss 1997, Strauss and Irwin 2004, Bronstein *et al.* Chapter 4). Recent accumulated evidence has shown that leaf herbivory reduces pollination success through reduced investment of resources into floral traits or increased plant defense against herbivory results in decreased investment into floral traits. In contrast, positive effects may occur when attacked plants can overcompensate for damaged tissues, thereby increasing floral traits that attract pollinators. The plant-mediated indirect interactions between herbivores and pollinators also include root feeders and decomposers underground (Poveda *et al.* Chapter 7). While indirect effects of early herbivory on the pollination process are ubiquitous in a wide variety of systems, underlying mechanisms of how and why these interactions occur remain poorly understood. To link the herbivory and pollination ecology, Bronstein *et al.* (Chapter 4) argue that future research should be directed to answer the following questions: (1) By what routes does the reduction of pollination success through previous herbivory take place? (2) Does reduced pollination success necessarily lead to reduced fitness, or can plants compensate to some extent for their losses? (3) Are defenses against herbivory costly in terms of pollination? (4) Are herbivores ever beneficial to pollination? (5) How does pollination alter herbivory? In addition, we need to extend indirect effects of herbivory to the seed and fruit dispersal processes. This is because early herbivory also can indirectly influence interactions between seed and seed dispersers by altering resource allocation to fruit and/or seed maturation that has a large impact on seed dispersal success (Strauss and Irwin 2004).

Interface of insect and microbial ecology

There are many kinds of microorganisms utilizing a wide variety of plant tissues. Recent studies have started to uncover plant-mediated indirect interactions between microorganisms, including pathogens, endophytes, and mycorrhizae, and insect herbivores sharing a host plant (Gehring and Whitham 1994, Clay 1997, Hatcher and Ayres 1997). The association between mycorrhizal fungi and host plants is generally conditional (Johnson et al. 1997, Swaty et al. 2004), but whether the interaction is positive or negative, the relationship provides a mechanistic basis for altering plant quality by mycorrhizal colonization, thereby initiating plant-mediated indirect effects on herbivorous insects. Effects of arbuscular mycorrhizal fungi on herbivorous insects vary from negative to positive (Gange Chapter 6), depending on diet breadth of insects or soil nutrients. In general, mycorrhizae have negative effects on root-feeding insects and negative or positive effects on foliar-feeding insects. Foliar herbivores, in turn, reduce mycorrhizal abundance, while root herbivores are more likely to enhance mycorrhizal colonization (Gange Chapter 6). Since systemic endophytes are well known for increasing host plant defenses against insect herbivores and pathogens by producing mycotoxins, endophytes can reduce insect feeding and deter oviposition, and alter individual growth, survival and developmental rates, and reproductive rates of insects. Likewise, interactions between pathogens and herbivores sharing the same host plant have recently received increasing attention (Hatcher and Ayres 1997, Sabelis et al. Chapter 9). Plant-mediated indirect interactions between pathogens and herbivorous insects vary from negative (Hatcher et al. 1994, Simon and Hilker 2003) to positive (Johnson et al. 2003, Belliure et al. 2005). For example, infection of a rust fungus was significantly reduced on yellow dock leaves when previously damaged by a leaf beetle (Hatcher et al. 1994). In contrast, a fungal pathogen of silver birch increased preference, performance, and population growth of an aphid (Johnson et al. 2003).

Since the nature and strength of plant–insect interactions mediated by microorganisms depend on the specific plant–microorganism association and strain, insect species identity, and environmental conditions, we need to incorporate accumulated knowledge about how such microbes can change plant traits.

Furthermore the microorganisms can affect not only herbivorous insects but also the third trophic level of predators and parasitoids by bottom–up trophic cascades, as clearly shown by Chaneton and Omacini (Chapter 8).

When and where are trait-mediated indirect interactions via plants important?

Plant-mediated indirect effects can be generated by the two key features involved in plant–herbivore interactions: (1) nonlethal effects of herbivory on plants, and (2) phenotypic plasticity of plant traits. Based on these features concerning terrestrial plants, we can predict when and where these trait-mediated indirect interactions via plants occur more frequently and their importance in influencing community structure and biodiversity.

In terrestrial systems trait-mediated indirect effects should predominate in plant–herbivore interactions, while density-mediated indirect effects should occur frequently in prey–predator interactions, such as top–down trophic cascades and apparent competition between two prey species sharing a common predator (Sih et al. 1998, Polis et al. 2000, Shurin et al. 2002). This is because predators kill individuals of their prey and change prey density while herbivores only alter plant traits after feeding. Thus, indirect effects through numerical changes due to mortality by consumers, i.e., density-mediated indirect effects, occur infrequently in plant–herbivore systems, compared to prey–predator systems. There is increasing evidence of trait-mediated indirect effects resulting from changes in prey behavior to avoid predation risk in terrestrial systems, i.e., the nonlethal effects of predators (Schmitz et al. 1997, 2004, Losey and Denno 1998, Schmitz 1998, Gastreich 1999). However, it is important to recognize a large difference in indirect effects through trait mediation between prey–predator and plant–herbivore systems in terms of when herbivores or plants respond to their enemies. The indirect effects mediated by changes in behavior of a herbivore prey result from the presence of a predator *before* feeding, whereas the trait-mediated indirect effects of plants occur *after* feeding by herbivores. In other words, trait-mediated indirect effects in plant–herbivore interactions emerge in the post-feeding process, while those in prey–predator interactions appear in the pre-feeding process.

In contrast to direct competition among herbivores, which frequently requires strong resource limitation due to a high level of herbivory, plant-mediated indirect interactions can occur at a low level of herbivory. One reason for this is that plant responses following herbivore attack that mediate indirect interactions are often rapidly induced at a low level of herbivory before herbivores cause plant mortality. Conversely, heavy defoliation can actually decrease indirect interactions. For example, plants that are heavily exploited by herbivores cannot compensate for lost tissue, and maintain habitats that were previously created by ecosystem engineers, less well for other herbivores. A lack of visible depletion of green plants, therefore, does not mean that interspecific interactions between herbivores rarely occur. Instead, limited herbivory greatly increases the likelihood of indirect interactions mediated by changes in plant traits. Thus, it is inferred that plant-mediated indirect effects predominate at a low level of herbivory.

Indirect effects mediated by changes in plant traits should be more common in terrestrial than in aquatic, particularly pelagic systems. This is because the level of herbivory is much lower in terrestrial than aquatic systems. In terrestrial systems, the average consumption rate by herbivores varies from 4% to 18% of aboveground plant biomass, while in aquatic systems herbivore consumption often exceeds 50% of primary production (Cyr and Pace 1993, Hairston and Hairston 1997, Polis 1999). The low level of herbivory in terrestrial plants enhances a predominance of plant-mediated indirect effects. In contrast in aquatic systems phytoplankton forms the primary producers and individuals are killed by zooplankton, and this high level of lethal predation produces more direct effects than indirect effects. On the other hand, recent studies have suggested that some green algal species can change their morphology or colony formation in the presence of grazers (Van Donk et al. 1999, Lürling 2003). Thus, trait-mediated indirect effects may occur in algal–zooplankton interactions in aquatic systems.

Some ideas on future studies

We cannot ignore plant-mediated indirect effects

This volume has illustrated that plant-mediated indirect effects, through trait changes caused by herbivory, are common and widespread in many plant-based interaction webs, and that they provide keystone interactions that strongly influence species richness and interaction diversity. Therefore, they could affect stability of community structure. This strengthens an emerging view of community ecology that ubiquitous indirect effects are an important force shaping ecological communities. The prevalence of indirect effects by trait mediation of terrestrial plants also indicates the importance of trait-mediated indirect effects on community organization (Werner and Peacor 2003, Schmitz et al. 2004), which contrasts with the importance of density-mediated indirect effects such as trophic cascades and apparent competition in food web dynamics (Polis et al. 2000). Herbivores are a principal component of all ecosystems, and their effects are widespread and diverse. They inevitably propagate plant-mediated indirect effects, and increase the heterogeneity of basal resources in terrestrial systems, thereby generating feedback loops from plants to higher trophic levels through bottom–up effects. These bottom–up cascades can have repercussions through entire communities, and have the potential to increase abundance of each species, species richness, and interaction diversity. Thus, herbivory is not just a process consuming primary production, it is also a pervasive force organizing the complexity and biodiversity of plant-based interaction webs in multitrophic systems.

From pair-wise to multispecies/multitrophic perspectives

Organisms have to cope simultaneously with multiple species in ecological communities. However, studies on species interactions have concentrated on pair-wise interactions, which have resulted in the importance of indirect effects in nature being overlooked. More recent appreciation has developed on how ecological and evolutionary consequences of these single interactions are altered by the presence of other species through indirect effects. These include the effects of multitrophic interactions (Gange and Brown 1997, Tscharntke and Hawkins 2002), multiple predators on prey (Sih *et al.* 1998), multispecies mutualisms (Stanton 2003), multispecies plant–animal interactions (Strauss and Irwin 2004), geographical variation of coevolution (Thompson 1994), trophic cascades (Strong 1992, Pace *et al.* 1999, Polis *et al.* 2000, Borer *et al.* 2005), and apparent competition (Holt and Lawton 1994, Bonsall and Hassell 1997). Furthermore, recent arguments on biodiversity have emphasized the incorporation of multitrophic dynamics into the research of the relationship between biodiversity and ecosystem processes (Raffaelli *et al.* 2002, Hooper *et al.* 2005). For example, the relationship between plant species diversity and ecosystem functioning is largely weakened by the impact of insect herbivores (Mulder *et al.* 1999). Most chapters in this volume have clearly illustrated how plant-mediated indirect effects can greatly influence the strength and direction of outcomes of other single interactions, which strongly supports the multispecies/multitrophic perspective. Future research should extend beyond a focus on pair-wise interactions, and examine plant-mediated effects on multiple species at multiple spatial and temporal scales to understand the ecological and evolutionary consequences of species interactions in structuring communities.

New hypotheses to be tested in future research

This volume has provided several hypotheses that are important to understanding the consequences of ubiquitous plant-mediated indirect effects on complex community structure and biodiversity, and they are open to experimental tests in future research. These hypotheses are that plant-mediated indirect effects can *increase* (1) species richness by creating new niches for herbivores and predators, (2) interaction diversity by generating new nontrophic and indirect linkages, (3) the opportunity for interactions among organisms that are temporally separated, spatially separated, and distantly related organisms occupying plants in differing times and spaces, and different niches, (4) facilitation between herbivores when enhancing resource availability, and between plants and predators when plants provide cues to predators to find their prey, (5) biodiversity in species richness and relative abundance of a species by creating habitat mosaics at large scale regardless of the sign (positive or negative) of the interaction involved, (6) feedback cascading effects upwards through trophic levels by enhancing resource heterogeneity in each trophic level, and (7) community complexity by increasing species richness, link density, connectance, and the number of trophic levels through an increased number of pathways for each species to interact with other members in a community. In addition, exploring whether increased species richness and interaction diversity in indirect interaction webs can enhance community stability would greatly contribute to the mechanistic understanding of the diversity–stability relationship, which is one of the most important issues for the conservation of biodiversity.

To test these hypotheses, we should develop approaches to evaluate important plant-mediated indirect effects. First, carefully designed field experiments are needed to investigate herbivore-induced changes in traits of plants. In particular, removal and/or addition of key herbivore species will illustrate how such species alter community structure and biodiversity by propagating plant-mediated indirect effects. This experimental approach can uncover underlying mechanisms of how species indirectly interact with other species through changes in plant traits. Second, the comparative approach should be used to examine what kind of plant-mediated indirect effects are important in forming networks of interactions in different ecological communities. Phylogenetic analysis is another useful tool to investigate how plant-mediated indirect effects have shaped evolution of insect traits in multispecies systems (Althoff Chapter 15). Third, the behavioral and physiological approach, such as using the resource allocation model (Bronstein *et al.* Chapter 4) and the hormonal control of plant defense specifically aimed at pathogens or herbivores (Sabelis *et al.* Chapter 9), will provide an important insight to understanding how plants respond to herbivory in a wide variety of ways and how herbivores respond to altered plant traits. Fourth, the same needs and advances in experiments on species interactions apply to statistical methods. More advanced multivariate analysis will reveal the ecological and evolutionary consequences of plant-mediated indirect interactions in a community context. In particular, path analysis provides a powerful tool to quantify causal relationships between direct and indirect effects and their relative strengths within a network of interactions (Wootton 1994, Bronstein *et al.* Chapter 4, Craig Chapter 14).

This volume has documented the situation that nontrophic, indirect effects, and positive interactions are frequently created by herbivore-induced plant changes in terrestrial systems. These interactions that are not included in traditional food webs are common and widespread in natural communities. Therefore, food web theory that occupies a central position in community and ecosystem ecology today will be fundamentally altered by the realization that many species interactions are nontrophic, indirect, and positive. Understanding of ecological and evolutionary consequences of plant-mediated indirect effects will revise ecological theory, lead to the discovery of fundamental principles of ecology, and therefore provide an important guide for the management and conservation of natural communities and biodiversity.

Acknowledgments

We thank Maurice Sabelis, Robert Denno, and Joanne Itami for their critical comments which improved the manuscript.

References

Abrahamson, W.G., C.P. Blair, M.D. Eubanks, and S.A. Morehead. 2003. Sequential radiation of unrelated organisms: the gall fly *Eurosta solidaginis* and the tumbling flower beetle *Mordellistena convicta*. *Journal of Evolutionary Biology* **16**:781–789.

Abrams, P.A., B.A. Menge, G.G. Mittelbach, D.A. Spiller, and P. Yodzis. 1996. The role of indirect effects in food webs, pp. 371–395 in G.A. Polis and K.O. Winemiller (eds.) *Food Webs: Integration of Patterns and Dynamics*. New York: Chapman and Hall.

Bardgett, R.D., and D.A. Wardle. 2003. Herbivore-mediated linkages between aboveground and belowground communities. *Ecology* **84**:2258–2268.

Bardgett, R.D., D.A. Wardle, and G.W. Yeates. 1998. Linking above-ground and below ground interactions: how plant responses to foliar herbivory influence soil organisms. *Soil Biology and Biochemistry* **30**:1867–1878.

Belliure, B., A. Janssen, P.C. Maris, D. Peters, and M.W. Sabelis. 2005. Herbivore arthropods benefit from vectoring plant viruses. *Ecology Letters* **8**:70–79.

Berlow, E.L. 1999. Strong effects of weak interactions in ecological communities. *Nature* **398**:330–334.

Berlow, E.L., A.M. Neutel, J.E. Cohen, *et al.* 2004. Interaction strengths in food webs: issues and opportunities. *Journal of Animal Ecology* **73**:585–598.

Bonsall, M.B., and M.P. Hassell. 1997. Apparent competition structures ecological assemblages. *Nature* **388**:371–373.

Borer, E.T., E.W. Seabloom, J.B. Shurin, *et al.* 2005. What determines the strength of a trophic cascade? *Ecology* **86**:528–537.

Callaway, R.M., S.C. Pennings, and C.L. Richards. 2003. Phenotypic plasticity and interactions among plants. *Ecology* **84**:1115–1128.

Carpenter, S.R., and J.F. Kitchell. 1993. *The Trophic Cascade in Lakes.* Cambridge, UK: Cambridge University Press.

Carpenter, S.R., J.F. Kitchell, and J.R. Hodgson. 1985. Cascading trophic interactions and lake productivity. *BioScience* **35**:634–639.

Clay, K. 1997. Fungal endophytes, herbivores and the structure of grassland communities, pp. 151–169 in A.C. Gange and V.K. Brown (eds.) *Multitrophic Interactions in Terrestrial Systems.* Oxford, UK: Blackwell Science.

Cohen, J.E., and F. Briand. 1984. Trophic links of community food webs. *Proceedings of the National Academy of Sciences of the USA* **81**:4105–4109.

Constabel, C.P. 1999. A survey of herbivore-inducible defensive proteins and phytochemicals, pp. 137–166 in A.A. Agrawal, S. Tuzun, and E. Bent (eds.) *Induced Plant Defenses against Pathogens and Herbivores:.* St. Paul, MN: American Phytopathological Society Press.

Craig, T.P., J.K. Itami, W.G. Abrahamson, and J.D. Horner. 1993. Behavioral evidence for host-race formation in *Eurosta solidaginis*. *Evolution* **47**:1696–1710.

Craig, T.P., J.D. Horner, and J.K. Itami. 1997. Hybridization studies on the host races of *Eurosta solidaginis*: implications for sympatric speciation. *Evolution* **51**:1552–1560.

Craig, T.P., J.D. Horner, and J.K. Itami. 2001. Genetics, experience and host-plant preference in *Eurosta solidaginis*: implications for host shifts and speciation. *Evolution* **55**:773–782.

Cyr, H., and M.L. Pace. 1993. Magnitude and patterns of herbivory in aquatic and terrestrial ecosystems. *Nature* **361**:148–150.

Denno, R.F., M.S. McClure, and J.R. Ott. 1995. Interspecific interactions in phytophagous insects: competition reexamined and resurrected. *Annual Review of Entomology* **40**:297–331.

Denno, R.F., C. Gratton, M.A. Peterson, et al. 2002. Bottom-up forces mediate natural-enemy impact in a phytophagous insect community. *Ecology* **83**:1443–1458.

Elton, C.S. 1958. *The Ecology of Invasions by Animals and Plants.* London: Methuen.

Eubanks, M.D., C.P. Blair, and W.G. Abrahamson. 2003. One host shift leads to another? Evidence of host-race formation in a predaceous gall-boring beetle. *Evolution* **57**:168–172.

Forkner, R.E., and M.D. Hunter. 2000. What goes up must come down? Nutrient addition and predation pressure on oak herbivores. *Ecology* **81**:1588–1600.

Fritz, R.S. 1992. Community structure and species interactions of phytophagous insects on resistant and susceptible host plants: ecology, evolution, and genetics, pp. 240–277 in R.S. Fritz and E.L. Simms (eds.) *Plant Resistance to Herbivores and Pathogens.* Chicago, IL: University of Chicago Press.

Gange, A.C., and V.K. Brown. (eds.) 1997. *Multitrophic Interactions in Terrestrial Systems.* Oxford, UK: Blackwell Science.

Gastreich, K.R. 1999. Trait-mediated indirect effects of a theridiid spider on an ant–plant mutualism. *Ecology* **80**:1066–1070.

Gehring, C.A., and T.G. Whitham. 1994. Interactions between aboveground herbivores and the mycorrhizal mutualists of plants. *Trends in Ecology and Evolution* **9**:251–255.

Gratton, C., and R. F. Denno. 2003. Inter-year carryover effects of a nutrient pulse on *Spartina* plants, herbivores, and natural enemies. *Ecology* **84**:2692–2707.

Hairston, N. G., Jr., and N. G. Hairston, Sr. 1997. Does food web complexity eliminate trophic-level dynamics? *American Naturalist* **149**:1001–1007.

Hairston, N. G., F. E. Smith, and L. B. Slobodkin. 1960. Community structure, population control, and competition. *American Naturalist* **94**:421–425.

Hartley, S. E., and C. G. Jones. 1997. Plant chemistry and herbivory, or why the world is green, pp. 284–324 in M. Crawley (ed.) *Plant Ecology*. Oxford, UK: Blackwell Science.

Hartvigsen, G., D. A. Wait, and J. S. Coleman. 1995. Tri-trophic interactions influenced by resource availability: predator effects on plant performance depend on plant resources. *Oikos* **74**:463–468.

Hatcher, P. E., and P. G. Ayres. 1997. Indirect interactions between insect herbivores and pathogenic fungi on leaves, pp. 133–149 in A. C. Gange and V. K. Brown (eds.) *Multitrophic Interactions in Terrestrial Systems*. Oxford, UK: Blackwell Science.

Hatcher, P. E., N. D. Paul, P. G. Ayres, and J. B. Whittaker. 1994. Interactions between *Rumex* spp., herbivores and a rust fungus: *Gastrophysa viridula* grazing reduces subsequent infection by *Uromyces rumicis*. *Functional Ecology* **8**:265–272.

Haukioja, E., and S. Neuvonen. 1987. Insect population dynamics and induction of plant resistance: the testing of hypotheses, pp. 411–432 in P. Barbosa and J. C. Schultz (eds.) *Insect Outbreaks*. San Diego, CA: Academic Press.

Hochberg, M. E., and J. H. Lawton. 1990. Competition between kingdoms. *Trends in Ecology and Evolution* **5**:367–371.

Hochwender, C. G., and R. S. Fritz. 2004. Plant genetic differences influence herbivore community structure: evidence from a hybrid willow system. *Oecologia* **138**:547–557.

Holt, R. D., and J. H. Lawton. 1994. The ecological consequences of shared natural enemies. *Annual Review of Ecology and Systematics* **25**:495–520.

Hooper, D. U., F. S. Chapin III, J. J. Ewel, *et al.* 2005. Effects of biodiversity on ecosystem functioning: a consensus of current knowledge. *Ecological Monographs* **75**:3–35.

Hunter, M. D., and P. W. Price. 1992. Playing chutes and ladders: heterogeneity and the relative roles of bottom-up and top-down forces in natural communities. *Ecology* **73**:724–732.

Hunter, M. D., T. Ohgushi, and P. W. Price (eds.) 1992. *Effects of Resource Distribution on Animal–Plant Interactions*. San Diego, CA: Academic Press.

Inbar, M., A. Eshel, and D. Wool. 1995. Interspecific competition among phloem-feeding insects mediated by induced host-plant sinks. *Ecology* **76**:1506–1515.

Inouye, B., and J. R. Stinchcombe. 2001. Relationships between ecological interaction modifications and diffuse coevolution: similarities, differences, and casual links. *Oikos* **95**:353–360.

Johnson, N. C., J. H. Graham, and F. A. Smith. 1997. Functioning of mycorrhizal associations along the mutualism–parasitism continuum. *New Phytologist* **135**:575–585.

Johnson, S. N., A. E. Douglas, S. Woodward, and S. E. Hartley. 2003. Microbial impacts on plant–herbivore interactions: the indirect effects of a birch pathogen on a birch aphid. *Oecologia* **134**:388–396.

Kagata, H., M. Nakamura, and T. Ohgushi. 2005. Bottom-up cascade in a tri-trophic system: different impacts of host-plant regeneration on performance of a willow leaf beetle and its natural enemy. *Ecological Entomology* **30**:58–62.

Karban, R., and I. T. Baldwin. 1997. *Induced Responses to Herbivory*. Chicago, IL: University of Chicago Press.

Karban, R., and J. H. Myers. 1989. Induced plant responses to herbivory. *Annual Review of Ecology and Systematics* **20**:331–348.

Kondoh, M. 2003. Foraging adaptation and the relationship between food-web complexity and stability. *Science* **299**:1388–1391.

Langellotto, G. A., and R. F. Denno. 2004. Responses of invertebrate natural enemies to complex-structured habitats: a meta-analytical synthesis. *Oecologia* **139**:1–10.

Larson, K. C., and T. G. Whitham. 1997. Competition between gall aphids and natural plant sinks: plant architecture affects resistance to galling. *Oecologia* **109**:575–582.

Lawton, J. H. 1983. Plant architecture and the diversity of phytophagous insects. *Annual Review of Entomology* **28**:23–39.

Lawton, J. H. 1989. Food webs, pp. 43–78 in J. M. Cherrett (ed.) *Ecological Concepts: The Contribution of Ecology to an Understanding of the Natural World*. Oxford, UK: Blackwell Scientific Publications.

Lawton, H. L., and P. H. Warren. 1988. Static and dynamic explanations for patterns in food webs. *Trends in Ecology and Evolution* **9**:242–245.

Lindeman, R. L. 1942. The trophic–dynamic aspect of ecology. *Ecology* **23**:399–418.

Losey, J. E., and R. F. Denno. 1998. Interspecific variation in the escape responses of aphids: effect on risk of predation from foliar-foraging and ground-foraging predators. *Oecologia* **115**:245–252.

Lürling, M. 2003. Phenotypic plasticity in the green algae *Desmodesmus* and *Scenedesmus* with special reference to the induction of defensive morphology. *International Journal of Limnology* **39**:85–101.

MacArthur, R. 1955. Fluctuations of animal populations and a measure of community stability. *Ecology* **36**:533–536.

Martinez, N. D. 1992. Constant connectance in community food webs. *American Naturalist* **139**:1208–1218.

Masters, G. J., V. K. Brown, and A. C. Gange. 1993. Plant mediated interactions between above- and below-ground insect herbivores. *Oikos* **66**:148–151.

Masters, G. J., T. H. Jones, and M. Rogers. 2001. Host-plant mediated effects of root herbivory on insect seed predators and their parasitoids. *Oecologia* **127**:246–250.

Mattson, W. J. J. 1980. Herbivory in relation to plant nitrogen content. *Annual Review of Ecology and Systematics* **11**:119–161.

May, R. M. 1973. *Stability and Complexity in Model Ecosystems*. Princeton, NJ: Princeton University Press.

McCann, K. S. 2000. The diversity–stability debate. *Nature* **405**:228–233.

McCann, K., A. Hastings, and G. R. Huxel. 1988. Weak trophic interactions and the balance of nature. *Nature* **395**:794–798.

McNaughton, S. J. 1977. Diversity and stability of ecological communities: a comment on the role of empiricism in ecology. *American Naturalist* **111**:515–525.

McNaughton. S. J. 1985. Ecology of a grazing ecosystem: the Serengeti. *Ecological Monographs* **55**:259–294.

Menge, B. A. 1995. Indirect effects in marine rocky intertidal interaction webs: patterns and importance. *Ecological Monographs* **65**:21–74.

Menge, B. A., and G. M. Branch. 2001. Rocky intertidal communities, pp. 221–251 in M. D. Bertness, S. D. Gaines, and M. E. Hay (eds.) *Marine Community Ecology*. Sunderland, MA: Sinauer Associates.

Mopper, S., J. Maschinski, N. Cobb, and T. G. Whitham. 1991. A new look at habitat structure: consequences of herbivore-modified plant architecture, pp. 260–280 in S. S. Bell, E. D. McCoy, and H. R. Mushinsky (eds.) *Habitat Structure*. London: Chapman and Hall.

Morin, P. J., and S. P. Lawler. 1995. Food web architecture and population dynamics: theory and empirical evidence. *Annual Review of Ecology and Systematics* **26**:505–529.

Morris, R. J., O. T. Lewis, and H. C. J. Godfray. 2004. Experimental evidence for apparent competition in a tropical forest food web. *Nature* **428**:310–313.

Mulder, C. P. H., J. Koricheva, K. Huss-Danell, P. Högberg, and J. Joshi. 1999. Insects affect relationships between plant species richness and ecosystem processes. *Ecology Letters* **2**:237–246.

Nakamura, M., Y. Miyamoto, and T. Ohgushi. 2003. Gall initiation enhances the availability of food resources for herbivorous insects. *Functional Ecology* **17**:851–857.

Nakamura, M., S. Utumi, T. Miki, and T. Ohgushi. 2005. Flood initiates bottom-up cascades in a tri-trophic system: host plant regrowth increases densities of a leaf beetle and its predators. *Journal of Animal Ecology* **74**:683–691.

Odum, E. P. 1953. *Fundamentals of Ecology*. Philadelphia, PA: W. A. Saunders.

Ohgushi, T. 1992. Resource limitation on insect herbivore populations, pp. 199–241 in M. D. Hunter, T. Ohgushi, and P. W. Price (eds.) *Effects of Resource Distribution on Animal-Plant Interactions*. San Diego, CA: Academic Press.

Ohgushi, T. 2005. Indirect interaction webs: herbivore-induced effects through trait change in plants. *Annual Review of Ecology, Evolution, and Systematics* **36**:81–105.

Oksanen, L., S. D. Fretwell, J. Arruda, and P. Niemelä. 1981. Exploitation ecosystems in gradients of primary productivity. *American Naturalist* **118**:240–261.

Pace, M. L., J. J. Cole, S. R. Carpenter, and J. F. Kitchell. 1999. Trophic cascades revealed in diverse ecosystems. *Trends in Ecology and Evolution* **14**:483–488.

Paine, R. T. 1980. Food webs: linkage, interaction strength and community infrastructure. *Journal of Animal Ecology* **49**:667–685.

Pilson, D. 1996. Two herbivores and constraints on selection for resistance in *Brassica rapa*. *Evolution* **50**:1492–1500.

Pimm, S. L. 1991. *Balance of Nature: Ecological Issues in the Conservation of Species and Communities*. Chicago, IL: University of Chicago Press.

Pimm, S. L. 2002. *Food Webs*. Chicago, IL: University of Chicago Press.

Pimm, S. L., and J. H. Lawton. 1977. Number of trophic levels in ecological communities. *Nature* **268**:329–331.

Pimm, S. L., J. H. Lawton, and J. E. Cohen. 1991. Food web patterns and their consequences. *Nature* **350**:669–674.

Polis, G. A. 1999. Why are parts of the world green? Multiple factors control productivity and the distribution of biomass. *Oikos* **86**:3–15.

Polis, G. A., and D. R. Strong. 1996. Food web complexity and community dynamics. *American Naturalist* **147**:813–846.

Polis, G. A., and K. O. Winemiller (eds.) 1996. *Food Webs: Integration of Patterns and Dynamics*. New York: Chapman and Hall.

Polis, G. A., A. L. W. Sears, G. R. Huxel, D. R. Strong, and J. Maron. 2000. When is a trophic cascade a trophic cascade? *Trends in Ecology and Evolution* **15**:473–475.

Price, P. W. 1997. *Insect Ecology,* 3rd edn. New York: John Wiley.

Price, P. W. 2002. Species interactions and the evolution of biodiversity, pp. 3–25 in C. M. Herrera and O. Pellmyr (eds.) *Plant–Animal Interactions: An Evolutionary Approach*. Oxford, UK: Blackwell Science.

Price, P. W., C. E. Bouton, P. Gross, et al. 1980. Interactions among three trophic levels: influence of plants on interactions between insect herbivores and natural enemies. *Annual Review of Ecology and Systematics* **11**:41–65.

Price, P. W., M. Westoby, B. Rice, et al. 1986. Parasite mediation in ecological interactions. *Annual Review of Ecology and Systematics* **17**:487–505.

Raffaelli, D., W. H. van der Putten, L. Persson, et al. 2002. Multi-trophic dynamics and ecosystem processes, pp. 147–154 in M. Loreau, S. Naeem, and P. Inchausti (eds.) *Biodiversity and Ecosystem Functioning: Synthesis and Perspectives*. New York: Oxford University Press.

Rehill, B., A. Class, L. Wieczorek, T. G. Whitham, and R. L. Lindroth. 2005. Foliar phenolic glycosides from *Populus fremontii*, *Populus angustifolia*, and their hybrids. *Biochemical Systematics and Ecology* **33**:125–131.

Schmitz, O. J. 1998. Direct and indirect effects of predation and predation risk in old-field interaction webs. *American Naturalist* **151**:327–342.

Schmitz, O. J., A. P. Beckerman, and K. M. O'Brien. 1997. Behaviorally mediated trophic cascades: effects of predation risk on food web interactions. *Ecology* **78**:1388–1399.

Schmitz, O. J., V. Krivan, and O. Ovadia. 2004. Trophic cascades: the primacy of trait-mediated indirect interactions. *Ecology Letters* **7**:153–163.

Schoenly, K., R. A. Beaver, and T. A. Heumier. 1991. On the trophic relations of insects: a food-web approach. *American Naturalist* **137**:597–638.

Schweitzer, J. A., J. K. Bailey, B. J. Rehill, et al. 2004. Genetically based trait in a dominant tree affects ecosystem processes. *Ecology Letters* **7**:127–134.

Schweitzer, J. A., J. K. Bailey, S. C. Hart, et al. 2005a. The interaction of plant genotype and herbivory decelerate leaf litter decomposition and alter nutrient dynamics. *Oikos* **110**:133–145.

Schweitzer, J. A., J. K. Bailey, S. C. Hart, and T. G. Whitham. 2005b. Nonadditive effects of mixing cottonwood genotypes on litter decomposition and nutrient dynamics. *Ecology* **86**:2834–2840.

Shurin, J. B., E. T. Borer, E. W. Seabloom, *et al.* 2002. A cross-ecosystem comparison of the strength of trophic cascades. *Ecology Letters* **5**:785–791.

Sih, A., G. Englund, and D. Wooster. 1998. Emergent impacts of multiple predators on prey. *Trends in Ecology and Evolution* **13**:350–355.

Simon, M., and M. Hilker. 2003. Herbivores and pathogens on willow: do they affect each other? *Agricultural and Forest Entomology* **5**:275–284.

Stanton, M. L. 2003. Interacting guilds: moving beyond the pairwise perspective on mutualisms. *American Naturalist* **162**:S10–S23.

Strauss, S. Y. 1991. Indirect effects in community ecology: their definition, study and importance. *Trends in Ecology and Evolution* **6**:206–210.

Strauss, S. Y. 1997. Floral characters link herbivores, pollinators, and plant fitness. *Ecology* **78**:1640–1645.

Strauss, S. Y., and A. A. Agrawal. 1999. The ecology and evolution of plant tolerance to herbivory. *Trends in Ecology and Evolution* **14**:179–185.

Strauss, S. Y., and R. E. Irwin. 2004. Ecological and evolutionary consequences of multispecies plant–animal interactions. *Annual Review of Ecology, Evolution, and Systematics* **35**:435–466.

Strong, D. R. 1992. Are trophic cascades all wet? Differentiation and donor-control in speciose ecosystems. *Ecology* **73**:747–754.

Strong, D. R., J. H. Lawton, and R. Southwood. 1984. *Insects on Plants: Community Patterns and Mechanisms.* Oxford, UK: Blackwell Scientific Publications.

Swaty, R. L., R. J. Deckert, T. G. Whitham, and C. A. Gehring. 2004. Ectomycorrhizal abundance and community composition shifts with drought: predictions from tree rings. *Ecology* **85**:1072–1084.

Teder, T., and T. Tammaru. 2002. Cascading effects of variation in plant vigour on the relative performance of insect herbivores and their parasitoids. *Ecological Entomology* **27**:94–104.

Thaler, J. S., M. J. Stout, R. Karban, and S. S. Duffey. 2001. Jasmonate-mediated induced plant resistance affects a community of herbivores. *Ecological Entomology* **26**:312–324.

Thompson, J. N. 1994. *The Coevolutionary Process.* Chicago, IL: University of Chicago Press.

Tilman, D. 1996. Biodiversity: population versus ecosystem stability. *Ecology* **77**:350–363.

Tilman, D. 1999. The ecological consequences of changes in biodiversity: a search for general principles. *Ecology* **80**:1455–1474.

Tilman, D. 2000. Causes, consequences and ethics of biodiversity. *Nature* **405**:208–211.

Tscharntke, T., and B. A. Hawkins (eds.) 2002. *Multitrophic Level Interactions.* Cambridge, UK: Cambridge University Press.

van der Putten, W. H., L. E. M. Vet, J. A. Harvey, and F. L. Wäckers. 2001. Linking above- and belowground multitrophic interactions of plants, herbivores, pathogens, and their antagonists. *Trends in Ecology and Evolution* **16**:547–554.

Van Donk, E., M. Lürling, and W. Lampert. 1999. Consumer-induced changes in phytoplankton: inducibility, costs, benefits, and the impact on grazers, pp. 89–103 in R. Tollrian and C. D. Harvell (eds.) *The Ecology and Evolution of Inducible Defenses*. Princeton, NJ: Princeton University Press.

Wardle, D. A. 2002. *Communities and Ecosystems: Linking the Aboveground and Belowground Components*. Princeton, NJ: Princeton University Press.

Wardle, D. A., G. W. Yeates, W. M. Williamson, K. I. Bonner, and G. M. Barker. 2004. Linking aboveground and belowground communities: the indirect influence of aphid species identity and diversity on a three trophic level soil food web. *Oikos* **107**:283–294.

Werner, E. E., and S. D. Peacor. 2003. A review of trait-mediated indirect interactions in ecological communities. *Ecology* **84**:1083–1100.

White, T. C. R. 1993. *The Inadequate Environment: Nitrogen and the Abundance of Animals*. Berlin, Germany: Springer-Verlag.

Wootton, J. T. 1994. The nature and consequences of indirect effects in ecological communities. *Annual Review of Ecology and Systematics* **25**:443–466.

Wootton, J. T. 2002. Indirect effects in complex ecosystems: recent progress and future challenges. *Journal of Sea Research* **48**:157–172.

Taxonomic index

Acacia spp.
  arthropod leaf constructs 249
  induced extrafloral nectar 195
Aceria tulipae (rust mite) 199
Aculus schlechtendahli (apple rust mite) 199–200
Acyrthosiphum pisum (pea aphid) 200
Agelastica alni (beetle)
  competition with ungulates 110–11
  induced changes in plant morphology 24–5
Agriotes spp. (wireworms), multitrophic interactions case study 154–8, 159
Agriotes lineatus (wireworm)
  competitive interactions 36
  induction of plant allelochemicals 24, 151
Agropyron trachycaulum (grass), effects of earthworms 152–3

Agrostis capillaris (grass), insect–mycorrhiza interactions 126–7
Agrotis ipsilon (black cutworm), endophyte-mediated interactions 174–5
Alces alces (moose), effects of browsing 111
Alnus glutinosa (alder), herbivore interactions 110–11
Alnus incana (grey alder), insect-induced changes in morphology 24–5
Alstroemeria ulula
  effects of herbivory on nectar production 80
  effects of herbivory on pollen quality 80–1
Amblyseius andersoni (predatory mite), plant volatile preferences 199–200
Anomala cupripes, effects of *Eucalyptus* mycorrhizae 134–5

Anthonomus signatus (strawberry bud weevil) 81–2
Aphidius ervi (parasitoid)
  plant volatile preferences 200
  prey identification using plant volatiles 358–9
Aphis fabae (black bean aphid) 200
  effects of root herbivores 150
Aphis farinosa (aphid), on *Salix eriocarpa* 229–30
Aphis nerii (aphid), effects of plant allelochemicals 24
Aphrophora pectoralis (spittlebug), on *Salix miyabeana* 227–8, 229
Arabidopsis thaliana (wall cress)
  herbivore-induced defenses 191
  herbivore-induced plant volatiles 192
  HIPV attraction of predators 197, 360
  susceptibility to herbivores 193

Taxonomic index

Asclepias syriaca (milkweed), insect-induced allelochemicals 24
aspen *see Populus tremuloides*
Asphondylia atriplicis (gall midge), indirect interaction web 342–3
Asphondylia borrichiae (gall midge), indirect interaction web 342
Asphondylia rudbeckiaeconspicua (gall midge), gall size and construction 343
Astermoyia carbonifera (gall midge), indirect interaction web 343
Atractomorpha lata (grasshopper), on *Solidago altissima* 230–2

Battus philenor (swallowtail), larval tolerance of induced allelochemicals 34
Bemisia argentifolii (whitefly) induction of plant allelochemicals 23–4
tolerance of induced defensive proteins 34–5
Betula pendula (birch) herbivore interactions 111
interactions between aphids and leaf miners 55–9
Betula pubescens (mountain birch), carryover effects of induced phenolics 33

Brassica nigra (black mustard), insect-induced changes in morphology 25
Brassica oleracea (cabbage), insect-induced allelochemicals 23–4

Campoletis sonorensis (parasitic wasp) 197
Camponotus spp. (ants), endophyte-mediated effects 180, 181
Capra pyrenaica (Spanish ibex) incidental ingestion of insect herbivores 110, 111–12
interactions with *E. mediohispanicum* and phytophagous insects 113–16
Capsella bursa pastoris, mediator for herbivore interactions 150
Cardamine hirsuta, effects of earthworms 153–4
Cardiochiles nigriceps (parasitoid) attraction to HIPV 365
plant volatile preferences 200
prey identification using plant volatiles 358–9
Carduus nutans (musk thistle), folivore-induced changes in floral traits 25
Castilleja indivisa (hemi-parasite) effects of bud herbivory 78
herbivory impacts on pollination 85–6

Castor canadensis (beaver) ecosystem engineer 316–19
effects of damage to trees 111
effects on arthropod communities 316–19
keystone species 316–19
mosaics of modified and unmodified habitat 297
species of large effect 316–19
Ceratomia catalpae (catalpa sphinx moth), larval feeding-induced attraction of predator 28
Cervus canadensis (elk) aspen–fire–elk interactions 311–14
effects of herbivory on sawfly abundance 314–16
indirect effects on avian community 314–16
species of large effect 311–16
Chaitophorus populicola (aphid) effects of mycorrhizae 134–5
mutualism with *Formica propinqua* (ant) 281–2, 287, 288–95
Chaitophorus salinger (aphid), on *Salix miyabeana* 227–8, 229
Chenopodium album 151
Choristoneura pinus (jack pine budworm), feeding-induced plant susceptibility 29–30

Chromatamyia syngenesiae
(leaf miner)
competitive
interactions 36
effects of mycorrhizae
131, 133–4
effects of root
herbivory 151
parasitism rates and
mycorrhizae 138
Chrysomela confluens
(leaf beetle),
preference for
plant re-growth 111
Cirsium arvense (thistle),
effects of
mycorrhizae on
insect herbivores
130–1, 132
Cirsium palustre (thistle),
effects of root
herbivory 150
Cotesia spp. (parasitoids)
differential responses
to plant volatiles
365–6
phylogenetic survey
of use of HIPV
367–8, 369
Cotesia glomerata (parasitoid)
203–4
olfactory response
selection
experiments 363
Cotesia marginiventris
(parasitoid) 197
Cotesia melanoscelus (galler),
alteration of
phenology by
natural enemies 344
Cotesia plutellae (parasitoid)
203–4
Cotesia rubecula
(endoparasitoid) 197

Cucurbita texana
effects of herbivory
on pollen quantity
and quality 80–1
responses to
herbivory 108
Cynips divisa, nutrient
diversion by
galls 58

Danaus plexippus (monarch
butterfly), larval
feeding-induced
allelochemicals 24
Datura wrightii, interactions
with *Manduca
sexta* 87–9, 90,
91, 92, 93, 94
Delphacodes penedetecta
(planthopper),
competitive
ability 37
Dendrobaena octaedra
(earthworm),
effects on plant
growth 152–3
Depressaria pastinacella
benefits of web
construction 251–2
herbivory and
altered sexual
expression 81
Dichomeris larvae, benefits
and costs of
leaf-tie construction
251–2
Diciphus minimus (mirid bug),
HIPV attraction of
natural enemies 198

Empoasca fabae (potato
leafhopper),
feeding-induced
resistance 28

Encarsia formosa (parasitic
wasp), plant
volatile
preferences 200
Epirrita autumnata (moth
larva), carryover
effects of induced
phenolics 33
Epitrix hirtipennis (flea beetle),
HIPV attraction
of natural
enemies 198
Eriocrania spp. (leaf miners)
impact on the plant
vascular system 65
interactions with
aphids 55–9
Erysimum mediohispanicum,
interactions with
Spanish ibex and
phytophagous
insects 113–16
Eucalyptus, dual mycorrhizal
associations 134–5
Eucalyptus urophylla, effects
of mycorrhizae
on insect
herbivores 134–5
Euceraphis betulae (aphid),
interactions
with leaf miners
55–9, 68–9
Euplectrus (ectoparasites),
endophyte-mediated
interactions 174–5
Eurosta solidaginis
(galling fly)
natural enemies 333
selection by natural
enemies for gall
size 333–41
selection pressures in
different biomes
333–41

414   Taxonomic index

Eurosta solidaginis interaction
　　web 332, 333–4
　　geographical variations
　　　334–41
Eurytoma gigantea (parasitoid)
　　bottom-up cascades
　　　and ovipositor
　　　lengths 336–9
　　geographical variation
　　　in ovipositor length
　　　334–41
　　indirect selection by
　　　bird predation
　　　336–9
　　indirect selection
　　　pressures on 333–41
　　indirect selection
　　　pressures on
　　　ovipositor length
　　　334–41
　　natural enemy of *Eurosta
　　　solidaginis* 333–4
　　selection pressure on
　　　E. solidaginis gall
　　　size 333–41
Eurytoma obtusiventris
　　(parasitoid), natural
　　enemy of *Eurosta
　　solidaginis* 333–4
Euura lasiolepis (galler),
　　vulnerability
　　to parasitoid
　　attack 343–4

Festuca arizonica (Arizona
　　fescue), endophyte
　　associations 168–9
Festuca arundinacea
　　(tall fescue)
　　endophyte associations
　　　168–9
　　endophyte effects
　　　on insect
　　　herbivory 173

Fiorinia externa (scale)
　　competitive benefits of
　　　early colonization 33
　　competitive benefits of
　　　induced nutrient
　　　reduction 33
Forda (gall aphid), negative
　　effects of
　　competition 33
Forda formicaria (aphid)
　　competition with other
　　　phloem feeders 26
　　negative effects of *Geoica*
　　　gall formation 52
　　nutrient diversion by
　　　galls 58
Forelius pruinosus (ant),
　　feeding-induced
　　attraction to prey 28
Formica propinqua (ant),
　　mutualism with
　　*Chaitophorus
　　populicola* (aphid)
　　281–2, 287, 288–95
Frankliniella occidentalis
　　(western flower
　　thrips) 203
　　benefits from plant-
　　　provided foods 205–6
　　benefits of association
　　　with other
　　　herbivores 205–6
　　vector for plant
　　　pathogens 193–4

Galerucella lineola (leaf
　　beetle) 32
　　benefits of occupying leaf
　　　rolls 251–2
Geocoris pallens (insect
　　predator) 198
　　feeding-induced
　　attraction to
　　prey 27–8

　　prey vulnerability
　　　increased by induced
　　　resistance 28–9
Geoica spp. (gall aphids)
　　competitive benefits of
　　　early colonization 33
　　competitive effects of gall
　　　formation 52
　　manipulation of phloem
　　　transport system 26–7
　　nutrient diversion by
　　　other galls 58
Glomus caledonium, arbuscular
　　mycorrhiza 134–5
Gossypium herbaceum (cotton)
　　insect-induced
　　　allelochemicals 24
　　root herbivory-induced
　　　allelochemicals 151

Hayhurstia atriplicis (leaf-
　　galling aphid) 151
Helicoverpa zea (corn earworm,
　　fruitworm) 200
　　effects on plant defenses
　　　193, 206
　　induction of plant
　　　allelochemicals 23–4
Heliothis subflexa (moth),
　　avoidance of
　　plant volatile
　　induction 365
Heliothis virescens (tobacco
　　budworm) 200
　　effects on plant
　　defenses 206
Hepialus californicus (ghost
　　moth), root
　　herbivory by
　　larvae 149
Heteromurus nitidus
　　(collembolan),
　　effects on plant
　　growth 153

Heterorhabditis marelatus (nematode), predator of root herbivore 149
Heterorhabditis megidis (nematode), attraction to root HIPV 204–5
Hordeum, impacts of *Rhinanthus minor* parasitism 69
Hormathophylla spinosa, herbivore interactions 110, 111–12
Hylobius transvittatus (weevil), effects of root herbivory 149, 150
Hyphantria cunea (fall webworm), communal webs 249

Idiocerus spp. (leafhoppers), interspecific feeding associations 31
Iphiseius degenerans (predatory mite)
   benefits from plant-provided foods 205–6
   innate and acquired HIPV responses 203
Ipomopsis aggregata
   herbivory and shifts in flowering phenology 79–80
   overcompensation for herbivory damage 82–3
Ips grandicollis (bark beetle), benefit from deactivated plant defenses 29–30
Isomeris arborea (shrub) effects of herbivory on nectar production 80

   herbivory and altered sexual expression 81
   pollinator avoidance of beetle-damaged flowers 77

Labidomera clivicollis (beetle), effects of plant allelochemicals 24
Laccaria laccata, ectomycorrhiza 134–5
Lactuca, feeding-induced susceptibility 29–30
Lathyrus vernus, pollination and herbivory 84
Leptinotarsa decemlineata (Colorado potato beetle), developmental delay due to induced resistance 28–9
Leucanthemum vulgare (ox-eye daisy)
   effects of mycorrhizae on insect herbivores 131
   rates of parasitism on insect herbivores 138
Liriomyza trifolii (leaf miner)
   negative effects of allelochemicals 23–4
   susceptibility to induced defensive proteins 34
Listronotus bonariensis (Argentine stem weevil), endophyte-mediated interactions 173–5
Lobelia siphilitica, effects of herbivory on pollen quality 80–1

Lolium multiflorum (Italian ryegrass), endophyte-mediated interaction webs 175–81
Lolium perenne (perennial ryegrass)
   endophyte associations 168–9
   endophyte effects on insect natural enemies 170, 173–5
   impact of *Rhinanthus minor* parasitism 60
L. rugulipennis (bug), effects of mycorrhizae 131
Lycopersicon esculentum (tomato)
   attraction of herbivore natural enemies 196–7
   bacterial resistance 193
   herbivore-induced defenses 191
   insect-induced allelochemicals 23–4
Lymantria dispar (gypsy moth)
   feeding-induced attraction of natural enemies 27–8
   feeding-induced changes in plant nutrition 27
Lytrum salicaria (purple loosestrife), effects of root herbivory 149, 150

Macaranga tanarius (macaranga), induced extrafloral nectar production 195

Taxonomic index

Magicicada spp. (periodical cicadas), egg nest inhabitants 250
Malacosoma spp., tent caterpillars 249
Mamestra brassicae (leaf chewer)
  effects of earthworms on development 154
  effects of root herbivory 151
Manduca spp. (tobacco hornworm), plant defensive responses 198–9
Manduca quinquemaculata (tobacco hornworm)
  effects of induced attraction of predator 27–8
  HIPV attraction of natural enemies 198
  induced resistance causes developmental delay 28–9
  vulnerability to predators increased by induced resistance 28–9
Manduca sexta (hawkmoth; tobacco hornworm)
  as weapon in plant competition 194
  herbivorous pollinator 83–4
  HIPV attraction of natural enemies 198
  interactions with *Datura wrightii* 87–9, 90, 91, 92, 93, 94
  plant susceptibility 193
Meligethes rufimanus (beetle)
  damaged flowers avoided by pollinators 77
  effects on plant nectar production 80

Metaseiulus occidentalis (predatory mite), plant volatile preferences 199–200
Metopolophium festucae (aphid), aphid–parasitoid interaction webs 176–80
Microctomus hyperodae (parasitic wasp), endophyte-mediated interactions 173–5
Monochamus carolinensis (pine sawyer), benefit from deactivated plant defenses 29–30
Mononychellus tanajoa (green mite) 204–5
Mordellistena convicta (beetle)
  natural enemy of *Eurosta solidaginis* 333–4
  selection pressure on *E. gigantea* ovipositor length 337–9
  selection pressure on *E. solidaginis* gall size 337–9
Myzus persica (aphid)
  effects of earthworms on reproduction 153–4
  effects of root herbivory 151
  interactions with mycorrhiza and Collembola 137
  interactions with worms and mycorrhiza 137

Neoseiulus cucumeris (predatory mite) 199
  innate and acquired HIPV responses 203

Neoseiulus finlandicus (predatory mite), plant volatile preferences 199–200
Neotyphodium spp. (fungal endophytes) 168–9
  mediation in multitrophic interaction webs 175–81
Neotyphodium lolii (endophyte), effects on insect natural enemies 174–5
Neozygites tanajoae (mite-pathogenic fungus), response to HIPV 204–5
Nephotettix cincticeps (leafhopper), on *Solidago altissima* 230–2
Nicotiana attenuata (wild/native tobacco plant)
  coordination of defensive responses 198–9
  coping with a diversity of attackers 193
  defenses used against neighbouring plants 194
  defensive use of HIPV 198–9
  fitness effects of induced defenses 360
  JA biosynthesis knock-out gene studies 191–2
  nicotine production 191
Nuculaspis tsugae (scale), effects of induced nutrient reduction 33

Octolasion tyrtaeum
    (earthworm),
    multitrophic
    interactions case
    study 154–8, 159
Oecophylla smaragdina
    (weaver ant),
    leaf nest building
    255, 257–8
Oenothera macrocarpa, reduced
    flower size due to
    herbivory 78–9
Onychiurus scotarius
    (collembolan),
    effects on plant
    growth 153
Operophtera brumata,
    susceptibility to
    induced
    allelochemicals 33
Otiorhynchus sulcatus
    (vine weevil)
    effects of mycorrhizae
    133–4
    root herbivory by
    larvae 204–5

Paeonia broteroi, pollination
    and herbivory 84
Panonychus ulmi (fruit tree
    red spider
    mite) 199–200
Papilio canadensis (tiger
    swallowtail)
    effects of induced
    attraction of natural
    enemies 27–8
    effects of induced
    changes in plant
    nutrition 27
Papilio polyxenes
    (swallowtail), larval
    detoxification of
    furanocoumarins 34

Parasaissetia nigra (soft
    scale), on *Solidago*
    altissima 230–2
Pemphigus spp. (gall aphids),
    interspecific
    feeding
    associations 31
Pemphigus batae (root-feeding
    aphid) 151
Phaseolus lunatus (Lima bean),
    herbivore-induced
    plant volatiles 192
Phlogophora meticulosa (moth),
    endophyte
    interactions
    135–6, 137
Phratora vitellinae
    (leaf beetle) 32
    tolerance of induced
    allelochemicals
    33, 34–5
Phyllocolpa spp. (galling sawfly)
    effects on arthropod
    communities
    316–19
    species of large effect
    316–19
Phyllocolpa bozemanii (galling
    sawfly)
    food item for
    insectivorous
    birds 314–16
    impacts on arthropod
    communities 314–16
    species of large effect
    314–16
Phyllonorycter pastorella (leaf
    miner), secondary
    colonization of
    mines 250–1
Phyllopertha horticola
    (chafer larva)
    competitive
    interactions 36

effects on aphid
    performance 150
effects on foliar feeders
    150, 151
Phyllotreta spp. (flea beetles),
    benefits from
    induced plant
    susceptibility 29–30
Phytoseiulus persimilis
    (predatory mite)
    innate and acquired
    HIPV responses
    203
    olfactory response
    selection
    experiments 363–4
    plant volatile
    preferences
    199–200, 202–3
Picoides pubescens (downy
    woodpecker)
    indirect selection
    pressure on
    E. gigantea 336–9
    natural enemy of
    Eurosta
    solidaginis 333–4
Pieris rapae (cabbage
    worm, small
    white cabbage
    butterfly)
    association with other
    herbivore species
    203–4
    caterpillar parasitism 197
    host response to
    herbivory 360
    induced changes in plant
    morphology 25
    induced plant
    allelochemicals 24,
    29–30
    stimulation of plant
    defenses 191

Pinus resinosa (red pine), insect herbivore interactions 52
Pinus sylvestris, effects of mycorrhizae 131
Pistacia palaestina (wild pistachio), effects of phloem-feeding herbivores 26
Pisum sativum (annual pea), insect–mycorrhiza interactions 126
Plagiodera versicolora (leaf beetle)
  on *Salix eriocarpa* 229–30
  on *Salix miyabeana* 227–8, 229
Plantago lanceolata
  effects of earthworms 154
  effects of herbivory on mycorrhizae 126
  effects of mycorrhizae on insect herbivores 130–1
Plutella xylostella (diamondback moth), association with other herbivore species 203–4
Poa annua (grass)
  effects of aphid and hemi-parasite interactions 59–64
  effects of earthworms 153–4
  effects of soil collembolans 153
Podisus maculiventris (stinkbug), predatory benefits of induced resistance 28–9

Poecile atricapillus (chickadee) indirect selection pressure on *E. gigantea* 336–9
natural enemy of *Eurosta solidaginis* 333–4
Popilia japonica (Japanese beetle, root scarab), endophyte-mediated interactions 174, 175
Populus angustifolia (cottonwood)
  effects of beaver damage 111
  species of large effect 316–19
Populus angustifolia (cottonwood) forests
  cottonwood–beaver–sawfly interactions 316–19
  impacts on arthropod communities 316–19
Populus fremontii, effects of beaver damage 111
Populus tremuloides (aspen)
  feeding-induced attraction of insect enemies 27–8
  herbivore-induced changes in nutrition 27
  species of large effect 311–14
Populus tremuloides (aspen) forests
  aspen–fire–elk interactions 311–14
  impacts of galling sawfly 314–16
  impacts on avian community 314–16

influences on biodiversity 311–14
  interactions affecting arthropod communities 311–14
Prokelisia dolus (planthopper)
  competition via induced changes in plant nutrition 26
  competitive superiority over *P. marginata* 34–5
  induced reduction in plant nutrition 53
  interspecific triggering of emigrants 26
  tolerance of induced nitrogen reduction 34
  trade-off between cibarial and flight musculature 37–8
  trade-off between competitive ability and dispersal 37–8
Prokelisia marginata (planthopper)
  dispersal ability 37–8
  effects of induced changes in plant nutrition 26
  interspecific triggering of emigrants 26
  negative effects of induced reduction in nutrition 53
  susceptibility to induced nitrogen reduction 34
Pseudomonas bacterial pathogens, effects on plant defenses 192–4
Psilocorsis quercicella, costs of leaf-tying 252

Taxonomic index   419

Quercus spp.
  interactions within arthropod constructs 258–60
  leaf-tying caterpillars 249
  leafing phenology and insect attack 255
  plant architecture and caterpillar attack 255
  plant traits and gall species richness 254–5
Quercus alba
  gall distribution within the canopy 254
  secondary occupation of leaf mines 250–1
  secondary occupation of spider egg nests 251
Quercus macrocarpa, gall distribution within the canopy 254
Quercus palustris, gall inhabitants 249–50
Quercus robur
  feeding-induced changes in plant nutrition 27
  gall distribution within the canopy 254

Raphanus raphanistrum (wild radish)
  effects of herbivory damage 81–2
  effects of herbivory on nectar production 80
  herbivore damage and flower size 79
  insect-induced allelochemicals 24
  responses to herbivory 108

Raphanus sativa (wild radish), effects of herbivore-induced changes 360
Rhabdophaga rigidae (gall midge), on *Salix eriocarpa* 229–30
Rhabdophaga strobiloides (gall midge), indirect interaction web 342
Rhinanthus minor (hemi-parasitic plant)
  exploitation of host vascular system 59
  impact on aphids 61–2, 63, 68–9
  impact on host plant 59–61, 62–4
  interactions with aphids 59–64
Rhinanthus serotinus (hemi-parasitic plant) 54
Rhinocyllus conicus (weevil), negative effects of changes in floral traits 25
Rhizophora mangle (red mangrove), impacts of stem-borers 250
Rhopalosiphum maidis (aphid), response to induced plant volatiles 190–1
Rhopalosiphum padi (aphid), aphid–parasitoid interaction webs 176–80
Rhyssomatus lineaticollis (stem-boring weevil), competitive interactions 36

Salix (willow) genotypes, conditional susceptibility to insects 308
Salix eriocarpa (willow), plant–insect interactions linkage 229–30, 235, 236
Salix miyabeana (willow)
  competitive effects of shelter-making larvae 52–3
  plant–insect interactions linkage 227–8, 229, 234–5, 236
Salix viminalis, salicylic acid levels and gall midge attack 254–5
Sanicula arctopodes, sexual expression unaffected by herbivory 81
Sinapis arvensis
  effects of root herbivory 150
  multitrophic interactions case study 154–8, 159
Sipha maydis (aphid), endophyte-mediated effects 180, 181
Sitobion avenae (aphid)
  exploitation of plant vascular system 59
  impact on host plant 59–61, 62–4
  interactions with hemi-parasitic plants 59–64, 68–9
Sitonia lineatus (beetle), interactions with mycorrhizae 126
Smaragdina semiaurantiaca (leaf beetle), on *Salix eriocarpa* 229–30

Solidago altissima (goldenrod)
  effects of induced changes in morphology 30
  Eurosta solidaginis interaction web 332, 333–4
  herbivore interactions 66
  impact of indirect evolutionary effects 334
  plant–insect interactions linkage 230–2, 235–7
Spartina alterniflora (cordgrass), herbivore-induced changes in nutrition 26
Spodoptera exigua (beet armyworm)
  competitive interactions 36
  effects of root herbivores 151
  HIPV attraction of parasitoids 196–7
  HIPV blend 202–3
  negative effects of allelochemicals 24
Spodoptera frugiperda (fall armyworm), endophyte-mediated interactions 174–5
Spodoptera littoralis (corn leafworm, cotton leafworm) induced plant responses 190–1

suppression by parasitic wasps 197
Spodoptera ornithogalli (armyworm), benefits from induced plant susceptibility 29–30
Steinernema carpocapsae (parasitic nematode), endophyte-mediated interactions 174–5
Striga hermonthica, parasitic angiosperm 54

Tetranychus urticae (spider mite) 205–6
  HIPV blend 199–200, 202–3
  stimulation of plant defenses 191
  susceptibility of tomato plants 193
Tetranychus viennensis (spider mite) 199–200
Thuja occidentalis, root herbivory-induced chemicals 204–5
Tipula paludosa (root herbivore), interactions with mycorrhizae 126–7
Trialeurodes vaporariorum (whitefly) 200
Trichoplusia ni (cabbage looper) feeding-induced plant susceptibility 29–30

negative effects of allelochemicals 23–4
plant susceptibility 193
susceptibility to induced defensive proteins 34
Trichosirocalus horridus (weevil), induced changes in floral traits 25
Trifolium repens
  effects of earthworms 153–4
  mediation of interactions 54
Tsuga canadensis (hemlock), induced reduction of available nitrogen 33
Tupiocoris notatus (myrid bug)
  feeding-induced attraction of predator 27–8
  feeding-induced proteinase inhibitors 28–9

Uroleucon nigrotuberculatum (aphid), on *Solidago altissima* 230–2
Urophora cardui (gall-forming fly), effects of mycorrhizae 130–1, 132
Urtica dioica (stinging nettle), insect-induced changes in morphology 24–5

Veronica persica, effects of earthworms 154

Author index

Abensperg-Traun, M. 109, 110
Abrahamson, W. G. 66, 255,
  333, 334, 337, 340,
  341, 345, 358, 395
Abrams, P. A. 4, 104, 164,
  165, 166, 167, 179,
  225, 226, 380
Ackerman, J. D. 80
Addicott, J. F. 35, 276, 293
Addison, P. J. 174
Adler, L. S. 75, 77, 78,
  82, 83, 84, 85, 87
Agrawal, A. A. 12, 20, 21, 23,
  24, 25, 27, 32, 33,
  34, 36, 39, 40, 83,
  107, 108, 113, 167,
  188, 202, 206, 222,
  224, 227, 354, 359,
  360, 384
Aizen, M. A. 80, 81
Ajlan, A. M. 36
Allen, W. W. 36
Alphei, J. 153
Amano, H. 343
Ament, K. 191, 192, 195
Ananthakrishnan, T. N.,
  249, 252
Andersen, D. 149
Anstett, M.-C. 83

Aratchige, N. S. 199
Araujo, L. M. 260
Armbruster, W. S. 75
Arnold, S. J. 346
Arroyo, M. T. K. 78
Askew, R. R. 342, 343,
  344, 348
Assis Dansa, C. V. de, 298
Atlegrim, O. 251
Awmack, C. S. 59
Ayres, P. G. 398

Backus, E. A. 37
Bailey, J. K. 306, 307, 310,
  311, 312, 314,
  315, 317, 323
Baines, D. 109, 111, 112
Baker, W. L. 311
Bakker, F. M. 200
Baldwin, I. T. 21-4, 27, 28, 29,
  33, 35, 36, 38, 39, 40,
  41, 51, 105, 107, 110,
  167, 169, 188, 189,
  190, 191, 194, 198,
  205, 221, 222, 237,
  256, 306, 354, 356,
  359, 360, 370, 371,
  383
Ball, O. J. P. 172

Bangert, R. K. 298
Barbosa, P. 261, 331
Bardgett, R. D. 153, 158,
  227, 386, 396, 397
Barker, G. M. 174
Bartos, D. L. 311
Basey, J. M. 311
Bass, K. A. 68
Baur, R. S. 24
Bazely, D. R. 222
Bazzaz, F. A. 86, 96, 222
Behmer, S. T. 132
Belesky, D. P. 168, 170, 172
Bell, G. D. 170, 172, 173, 180
Belliure, B. 194, 206, 398
Belshaw, R. 368
Benrey, B. 28
Benkman, C. W. 332
Berenbaum, M. R. 12, 22,
  34, 251, 283
Bergelson, J. 80, 82, 106
Bergström, R. 107
Berlow, E. L. 221, 234,
  381, 393
Bernard, E. C. 171, 175
Bernasconi, M. L. 190
Bernays, E. 12, 87, 92
Berryman, A. 8
Bertness, M. D. 41, 78

Bestelmeyer, R. 109, 110
Bezemer, T. M. 24, 36, 151
Bi, J. L. 108
Bigger, D. S. 67
Biggs, C. J. 343
Birkett, M. A. 197, 200
Bishop, D. B. 41, 298
Björkman, C. 222
Black, C. A. 283
Bloom, A. J. 86
Blossey, B. 52, 66
Bluthgen, N. 258, 263
Bohlen, P. 152
Bolker, B. 105, 226
Bonkowski, M. 147, 152, 153, 154
Bonos, S. A. 165, 172, 175
Bonsall, M. B. 58, 225, 381, 401
Borer, E. T. 221, 380, 401
Borowicz, V. A. 131, 138, 140
Bostock, R. M. 39, 40
Botrell, D. G. 354, 356
Bouwmeester, H. J. 195
Bower, E. 64, 125
Bowling, D. J. F. 60
Boyce, M. S. 311
Bradley, G. A. 276
Branch, G. M. 380
Breen, J. P. 165, 169, 171, 172, 173, 175
Briand, F. 390
Bristow, C. M. 276, 285, 298
Brodie, E. D. 346
Brody, A. K. 84
Bronstein, J. L. 75, 83, 84, 85, 87, 275, 397, 403
Brooks, D. M. 194
Brown, G. G. 153
Brown, J. H. 96, 105, 165, 166, 307, 309
Brown, J. L. 254
Brown, V. K. 8, 20, 31, 32, 36, 51, 52, 55, 66, 67, 96, 105, 125, 127, 133, 139, 140, 147, 149, 150, 151, 154, 156, 157, 164, 224, 225, 226, 233, 237, 386, 401
Brown, W. L. 331
Bruno, J. F. 233
Bryant, J. P. 310
Buchanan, C. K. 20, 35, 259, 266
Bucheli, E. 84
Buckley, R. C. 276
Budenberg, W. 294
Bultman, T. L. 168, 170, 171, 172, 173, 174, 180, 182, 284
Bush, L. P. 168, 169, 170, 171

Callaham, M. A. 153
Callaway, R. M. 54, 60, 64, 67, 68, 167, 226, 357, 381
Cameron, R. 252
Campbell, C. A. M. 190, 311
Cane, J. T. 334, 337
Cantor, L. F. 311
Cappuccino, N. 30, 31, 225, 233, 251, 259, 264, 306, 310, 317
Cardinale, B. J. 239
Cariveau, D. 84, 86
Carpenter, S. R. 8, 380
Carroll, C. R. 53
Carroll, M. R. 254
Carson, W. P. 106
Carver, M. 260
Casey, T. M. 87
Cebrian, J. 310
Cechin, I. 54
Chabot, B. F. 88
Chadde, S. W. 306, 317
Chamberlin, F. S. 87
Chen, J. L. 96
Cheplick, G. P. 165

Chew, F. S. 23
Choudhury, D. 60
Christensen, K. M. 109
Cipollini, D. 107, 358
Cipollini, D, F. 358
Clark, D. B. 252, 265
Clark, T. L. 169, 252, 265
Clausen, T. P. 23, 311
Clay, K. 64, 165, 168, 169, 170, 171, 172, 173, 276
Clement, S. L. 172
Cloutier, C. 294
Cohen, J. E. 390
Cohen, M. B. 34
Coleman, D. C. 153
Coley, P. D. 166, 222
Colfer, R. G. 202, 206
Collins, C. M. 285, 295
Comstock, J. R. 89
Confer, J. L. 334
Conn, E. E. 23
Connell, J. H. 226
Conner, J. K. 79, 81
Connor, E. F. 55
Constabel, C. P. 22, 23, 188, 370, 383
Cook, A. 32, 36, 37
Cook, J. M. 83
Cornell, H. W. 342, 343
Costa, A. A. 251
Craig, T. P. 227, 253, 333, 340, 342, 343, 348, 395, 396, 403
Crawley, M. J. 20, 59, 67, 82, 106, 128
Crespi, B. J. 249, 346
Cresswell, J. E. 79, 80
Csóka, G. 255
Cuartas, P. 108
Cuba, J. 84
Cui, J. 193, 194
Cullings, K. W. 128, 134
Cunningham, S. A. 77, 84
Currie, A. F. 126

Curry, J. P. 153
Cushman, J. H. 233
Cyr, H. 221, 310, 400

Dafni, A. 150
Dahlman, D. L. 169, 172, 178
Dahlsten, D. L. 55
Dalin, P. 222
Damman, H. 19, 20, 21, 30, 35, 39, 40, 226, 233, 238, 256
Danell, K. 107, 109, 111, 224, 254, 283, 311
Davidowitz, G. 92, 96
Davidson, A. W. 171, 175
Davidson, D. W. 105, 109, 307, 309
Davidor, P. 80
Davies, D. M. 54, 59, 65, 67, 253
Dawson, T. E. 24, 25
De Boer, J. G. 201–3, 206
Debyle, N. V. 311
de Ilarduya, O. M. 36
De Jong, M. C. M. 197
de Mazancourt, C. 83
De Moraes, C. M. 27, 200, 359
de Souza, A. L. T. 249
Dean, W. R. J. 109, 110
DeAngelis, D. L. 83
Del Vecchio, T. A. 126
Delph, L. F. 81, 225
Dennis, P. 109, 110
Denno, R. F. 19, 20, 21, 22, 23, 26, 28, 30, 32, 33, 34, 35, 36, 37, 38, 39, 40, 41, 51, 52, 53, 66, 68, 105, 116, 167, 169, 188, 190, 224, 226, 238, 382, 383, 399
Dettner, K. 294
DeWalt, S. J. 108
Dicke, M. 108, 192, 195, 196, 200, 201, 202, 205, 206, 223, 355, 356, 358, 359, 364, 367
Dickson, L. L. 252, 257, 258, 261, 263, 264, 306, 317
Dill, L. M. 105, 226
Dirzo, R. 310
Dixon, A. F. G. 58, 68, 137, 233
Doares, S. H. 193
Dolch, R. 110
Dorn, S. 201, 363
Drukker, B. 196, 201
Du, Y. 355, 359
Du, Y. J. 195
Duarte Rocha, C. F. 298
Dufaÿ, M. 83
Duffey, S. S. 23, 36, 39, 239
Dukas, R. 78
Dungey, H. S. 277
Dunlop, J. 60
Dussourd, D. E. 22, 30, 53
Dyer, L. A. 166, 167

Edson, J. L. 35
Ehleringer, J. R. 89
Ehrlén, J. 84
Ehrlich, P. R. 12
Eichenseer, H. 178
Einhork, M. 152
Eisikowhich, D. 344
Eisner, T. 30, 53, 294
Eliason, E. A. 250, 254, 259, 261
Elligsen, H. 109
Elton, C. S. 391
Endler, J. A. 356
English-Loeb, G. 81
Enquist, B. J. 96
Eom, A.-H. 127
Esch, T. H. 131
Eubanks, M. D. 333, 395
Evans, T. A. 251

Faeth, S. H. 20, 105, 109, 165, 168, 169, 171, 172, 173, 174, 181, 182, 226, 308, 355
Fagan, W. F. 41
Falconer, D. S. 346
Farmer, E. E. 181
Fasham, M. J. R. 296
Feller, I. C. 250, 254, 255, 260, 264
Fellows, R. J. 58
Felt, E. P. 261
Felton, G. 193
Felton, G. W. 193
Ferdy, J.-B. 83
Fernandes, G. W. 260, 261, 266
Ferrenberg, S. M. 37, 105
Fiala, B. 276
Fiedler, K. 258, 263, 298
Field, C. 89
Fischer, M. K. 285
Fisher, A. E. I. 52, 58, 105
Fitzgerald, T. D. 252
Fleming, T. H. 83
Fontaine, J. 132
Fordyce, J. A. 34
Forkner, R. E. 167, 382
Formusoh, E. S. 31
Foss, L. K. 257
Fournier, V. 249, 261, 264
Fowler, S. V. 233, 251
Fraser, L. H. 60, 63, 66, 67
Freedman, H. I. 288
Freeman, R. S. 79
Freese, G. 344
Fritz, R. S. 105, 308, 319, 396
Fritzsche-Hoballah, M. E., 197, 360
Fukui, A. 30, 31, 52, 224, 225, 251
Fukushima, J. 364
Fuzy, E. 175

Galen, C. 79, 80, 84
Galil, J. 344
Gange, A. 276
Gange, A. C. 8, 51, 64, 67, 125, 126, 127, 128, 130, 131, 132, 133, 134, 137, 138, 139, 140, 147, 149, 154, 156, 164, 224, 231, 237, 276, 284, 386, 401
García-González, R. 108
Gardener, M. C. 106, 276
Gaston, K. J. 253
Gastreich, K. R. 399
Geervliet, J. B. F. 201, 366
Gehring, C. A. 125, 126, 128, 129, 131, 132, 134, 276, 284
Geiger, D. R. 58
Goeden, R. D. 342
Gerard, P. 175
Gilbert, J. E. 24, 222
Glawe, G. A. 358
Gómez, J. M. 40, 84, 105, 106, 107, 108, 109, 110, 111, 112
Gömez, J. M. 222
Godfray, H. C. J. 169, 176, 276, 286, 342, 364
Goheen, J. R. 107, 109
Goldson, S. L. 174
Gols, R. 195, 196, 200
González-Megías, A. 40, 105, 109, 110, 111, 112
Gould, F. 361
Goverde, M. 131, 132
Grace, J. 96
Gratton, C. 383
Graves, J. D. 54, 59, 60, 63, 65, 67, 69
Green, E. S. 252
Greiler, H. J. 110
Grewal, S. 174, 175
Grime, J. P. 275

Gu, H. 363
Guerrieri, E. 195, 200

Haimi, J. 153
Hairston, N. G. 20, 237, 382, 400
Hajek, A. E. 55
Halaj, J. 166
Halldorsson, G. 133
Hamback, P. A. 86
Hamerlynck, E. P. 89
Hamilton, J. G. 132
Hare, J. D. 87
Harman, D. M. 250
Harper, J. L. 78
Harrison, S. 51
Hart, M. 139
Hartley, S. E. 55, 58, 59, 60, 64, 66, 68, 382
Hartvigsen, G. 382
Harvell, C. D. 167
Harvey, P. H. 367
Hassell, M. P. 20, 58, 225, 250, 381, 401
Hatcher, P. E. 398
Haukioja, E. 51, 222
Hause, B. 133
Havill, N. P. 199
Hawkins, B. A. 5, 8, 237, 247, 249, 250, 253, 259, 266, 342, 401
Hübner, G. 295
Heads, P. A. 251
Heard, S. B. 20, 35, 259, 266
Heil, M. 195
Heinrich, B. 87
Helgason, T. 129
Hendrix, S. D. 25, 81, 153
Herms, D. A. 107
Herre, E. A. 83
Herrera, C. M. 84, 86, 147
Heske, E. J. 307, 309
Hicks, D. J. 88
Hildebrand, J. G. 87

Hilker, M. 359
Hinks, J. D. 276
Hjältén, J. 108
Hobbs, R. J. 323
Hölldobler, B. 251, 255, 258
Hopke, J. 195, 196
Hochberg, M. E. 105, 253, 385
Hochwender, C. G. 396
Hodge, A. 124
Hodges, S. A. 80
Hodkinson, I. D. 252
Hoffman, C. A. 53
Holah, J. 168, 276
Holland, J. N. 83, 127
Holopainen, J. K. 197
Holt, R. D. 225, 380, 381
Hooper, D. U. 392, 401
Hopkins, G. W. 298
Horvitz, C. C. 86
Hountondji, F. C. C. 204
Hübner, G. 295
Hudson, E. E. 258
Huenneke, L. F. 323
Huhta, V. 153
Hulme, P. E. 107
Hunter, M. D. 5, 8, 27, 30, 33, 40, 41, 105, 106, 109, 112, 164, 166, 167, 179, 221, 233, 251, 255, 258, 306, 307, 308, 380, 382, 393
Hunt-Joshi, T. R. 52, 66
Huntzinger, M. 108
Hus-Danell, K. 109, 111, 224, 283, 311
Huxman, T. E. 93

Inbar, M. 20, 24, 26, 28, 31, 33, 34, 35, 36, 39, 52, 58, 191, 386
Inouye, B. 113, 114, 396
Irwin, R. E. 82, 357, 397, 401

Ishihara, M. 227
Itami, J. K. 340
Itô, F. 26, 35

Jackson, J. A. 337
Jaeger, N. 83
Jansen, V. A. A. 207
Janssen, A. 196, 197
Janssen, A. R. M. 362
Janzen, D. H. 345
Jaremo, J. 83
Jermy, T. 25
Jeschke, W. D. 54
Jia, F. 201
Jiang, F. 54, 59, 69
Johnson, N. C. 308, 323, 398
Johnson, S. N. 55, 56, 58, 64,
    65, 68, 257, 284, 295
Johnston, C. A. 306, 310, 317
Jones, C. G. 51, 55, 59, 60,
    66, 68, 140, 152,
    154, 225, 227,
    234, 246, 251, 306,
    311, 317, 382
Jordan, G. J. 309
Jousselin, E. 83
Juenger, T. 80, 82, 106

Kagata, H. 225, 251, 284, 383
Kahl, J. 198
Kaitaniemi, P. 33, 41, 254
Kaplan, I. 23, 28–9, 188, 190
Kampichler, C. 254
Kant, M. R. 195
Kapelinksi, A. D. 334
Karban, R. 21–4, 27, 33,
    34, 35, 36, 38, 39,
    40, 41, 51, 52, 77,
    78, 107, 110, 167,
    188, 189, 190, 221,
    222, 237, 256, 306,
    354, 356, 383
Kareiva, P. 20
Katayama, N. 285

Kato, M. 83
Kay, C. E. 306, 311, 317
Kearby, W. H. 254
Kearsley, M. J. C. 309
Kenkel, N. C. 296
Kennedy, J. S. 60
Kerdelhué, C. 344
Kessler, A. 28, 29, 169, 188,
    192, 198, 359
Kidd, N. A. C. 31
Kiss, A. 285
Kitchell, J. F. 8, 380
Kleijn, D. 108
Klinkhamer, P. G. L. 197,
    362, 371
Klironomos, J. N. 139
Kloek, A. P. 194
Kolb, T. E. 128
Knight, F. B. 276
Kondoh, M. 392
Kopelke, J. P. 249
Koppenhöfer, A. 175
Koricheva, J. 20, 21, 40, 222
Korth, K. 193
Krips, O. E. 201
Krishna, K. R. 131
Kruess, A. 109, 110
Krupnick, G. A. 25, 77, 78, 80,
    81, 85
Kruskal, J. B. 296
Kuć, J. 39, 40
Kuikka, K. 128
Kumar, H. 277, 288
Kunkel, B. A. 175
Kunkel, B. N. 195
Kurczewski, F. E. 334, 337

Lacroix, C. R. 343
Lamb, R. J. 26
Lande, R. 346
Langellotto, G. A. 383
Larson, K. C. 26, 27, 31, 52,
    250, 255, 386
Larsson, S. 30, 225, 251, 255

Last, F. T. 140
Latto, J. 343
Lavelle, P. 148
Lawler, S. P. 389, 391, 392
Lawrence, R. 309
Lawton, H. L. 380
Lawton, J. L. 4, 5, 19, 20, 51, 55,
    58, 66, 105, 225, 226,
    237, 251, 307, 381,
    383, 385, 390, 391,
    401
Layne, J. R. Jr. 340
Le Corff, J. 253
Leather, S. R. 59, 283, 285, 295
Leblanc, D. A. 343
Leege, L. M. 81
Lehtilä, K. 25, 78, 79, 150
Leibold, M. A. 164, 166,
    167, 221
Lennartson, T. 82
Lesna, I. 199
Letourneau, D. K. 167, 256,
    261, 331
Lewinsohn, E. 23
Lewis, A. C. 201, 251, 358, 359
Li, C. C. 85
Li, C. Y. 191, 193
Li, L. 191, 194
Liepert, C. 294
Lill, J. T. 10, 19, 30, 31, 225,
    233, 249, 258, 259,
    260, 264
Lindeman, R. L. 382
Lindroth, R. L. 277
Lo Pinto, M. 369
Lodge, D. J. 135
Loeffler, C. C. 251, 252
Lohman, D. J. 78
Loik, M. E. 93
Lombardero, M. J. 250
Loreau, M. 83
Losey, J. E. 399
Louda, S. M. 59, 79,
    106, 108, 251

Louw, S. 53, 252
Lowenberg, G. J. 81
Lürling, M. 400
Lucas, E. 109, 111
Luttbeg, B. 105, 111

Mäntylä, E. 359
Mac Garvin, M. 233, 251
MacArthur, J. W. 297
MacArthur, R. 392
MacArthur, R. W. 297
MacKay, P. A. 26
Mackauer, M. 294
Madden, A. H. 87
Maeda, T. 201
Mahoney, J. H. 316
Mailleux, A.-C. 285
Majerus, M. E. N. 294
Malinowski, D. P. 168, 170, 172
Manninen, A.-M. 131
Mantyla, E. 204
Mapes, C. C. 253
Marchosky, R. J. 342, 343
Margolies, D. C. 201, 363
Marks, S. 276
Maron, J. L. 106, 149
Marquis, R. J. 10, 19, 20, 24, 30, 31, 77, 78, 81, 85, 86, 106, 107, 112, 150, 225, 233, 255, 256, 258, 259, 260, 264
Martin, M. A. 31, 68, 259, 264, 345
Martinez, N. D. 390
Martinsen, G. D. 10, 30, 31, 109, 111, 225, 233, 259, 261, 264, 306, 310, 317
Marvier, M. A. 67
Maschinski, J. 224
Master, G. J. 105, 150

Masters, G. J. 20, 31, 32, 36, 51, 52, 55, 66, 147, 150, 151, 154, 156, 157, 224, 225, 226, 383, 386
Mathis, W. W. 250, 254, 255, 260, 264
Matsumura, M. 26
Matter, S. F. 84
Mattson, W. J. 36, 52, 107, 382
Mauricio, R. 224
May, R. M. 392
Mayer, R. T. 24, 34, 36
McArthur, R. 307
McCann, K. 239
McCann, K. A. 392, 393
McCann, K. S. 392, 393
McClure, M. S. 19, 20, 33, 35, 51, 52
McConn, M. 191
McCrea, K. D. 66
McDade, L. A. 80
McFadden, M. W. 87
McNaughton, S. J. 392
McNeil, S. 59
Mechaber, W. L. 87
Meekijjaroenroj, A. 83
Meiners, T. 359
Menge, B. A. 225, 380
Mercke, P. 195
Merritt, R. W. 109
Mescher, M. C. 365
Meserve, P. L. 164
Messina, F. 169
Meyer, G. A. 66
Michel-Salzat, A. 368
Mikola, J. 158
Milbrath, L. R. 25
Milewski, A. W. 222
Miller, W. E. 249
Minchin, P. R. 296, 312
Mira, A. 87, 92
Miranda, E. 310
Mitchell, R. J. 80, 84, 86

Müller, C. B. 169, 176, 179, 276, 286
Moegenburg, S. M. 25
Moeller, R. 334
Montandon, R. 31
Moon, D. C. 167, 168, 176
Mooney, H. A. 88, 89
Moora, M. 276
Moore, J. C. 153
Mopper, S. 224, 255, 384
Moran, N. A. 20, 151, 154
Mori, N. 206
Morin, P. J. 256, 389, 391, 392
Morris, D. C. 249
Morris, R. F. 249
Morris, F. J. 225, 381
Morse, D. H. 78, 251
Mortimer, S. R. 149
Mothershead, K. 77, 78, 85, 86, 150
Mound, L. A. 249
Mulder, C. P. H. 401
Muller, J. 135
Muñoz, A. A. 78
Murakami, M. 252
Murawski, D. A. 77
Musser, R. O. 193
Mutikainen, P. 54, 81, 225
Myers, J. H. 107, 222, 223, 224, 233, 283, 383

Naiman, R. J. 306, 310, 317
Nakamura, M. 19, 30, 28, 31, 53, 225, 227, 228, 229, 233, 257, 260, 263, 264, 266, 284, 285, 306, 310, 317, 383, 384
Navas, M. L. 96
Nechols, J. R. 25
Nes, W. D. 132
Ness, J. H. 27, 28
Neutel, A. M. 239
Neuvonen, S. 222, 311, 383

Newington, J. 153, 154
Nice, H. E. 131, 132
Nijhout, H. F. 92, 96
Nötzold, R. 149, 150, 154
Noss, R. F. 307
Nozawa, A. 224, 227
Nykänen, H. 20, 21, 40, 222
Nyman, T. 348

O'Dowd, D. I. 195
Odum, E. P. 392
Offenberg, J. 264, 285
Ohgushi, T. 4, 6, 25, 28, 30, 31, 39, 40, 41, 53, 167, 224, 225, 226, 227, 228, 233, 234, 238, 251, 260, 263, 264, 266, 284, 285, 306, 310, 317, 379, 380, 381, 382, 384
Okello, B. D. 108
Oki, Y. 251
Oksanen, L. 164, 166, 167, 179, 221, 391
Olff, L. 237
Oliveira, P. S. 108
Oliver, K. M. 164
Ollerstam, O. 255
Olmstead, K. L. 26, 34
Omacini, M. 26, 34, 164, 171, 172, 173, 175, 176, 179, 180, 181, 182, 284
Omer, A. D. 283
Oppenheim, S. A. 361
Orians, C. M. 277, 308, 319
Orloci, L. 296
Osier, T. L. 277
Ott, J. R. 251
Ouellet, H. R. 337
Ozawa, R. 192, 195, 196

Pace, M. L. 4, 5, 221, 225, 310, 380, 400, 401

Pacovsky, R. S. 131, 132
Pagel, M. D. 367
Paicos, P. 334
Paige, K. N. 82
Paine, E. O. 344
Paine, R. T. 221, 380
Paine, T. D. 344
Pallini, A. 205
Palmisano, S. 107
Papaj, D. R. 201
Parkinson, D. 152, 153, 154
Paré, P. W. 27, 195
Passos, L. 108
Pattrasudhi, P. 20
Paul, N. D. 133, 205
Peacor, S. D. 4, 104, 105, 111, 164, 166, 169, 174, 188, 226, 227, 234, 381, 400
Pellmyr, O. 83, 150
Pennings, S. C. 54, 60, 64, 67, 68
Petersen, M. K. 52
Pieterse, C. M. J. 192, 193
Pilson, D. 12, 30, 79, 80, 233, 257, 283, 295
Pimm, S. L. 104, 234, 382, 389, 390, 391, 392
Ping, C. 333
Plakidas, J. D. 343
Plantard, O. 253
Polis, G. A. 164, 166, 221, 225, 234, 238, 310, 380, 381, 392, 399, 400, 401
Poorter, H. 96
Popay, A. J. 165, 171, 172, 175
Post, B. J. 224
Potter, D. A. 36, 171, 175, 250, 254, 259, 261
Potvin, M. A. 79, 106
Porazinska, D. L. 147
Poveda, K. 150, 155, 156
Powell, W. 364, 368

Power, M. E. 234, 310, 317
Preisser, E. L. 149
Press, M. C. 53, 54, 59, 65, 67
Prestidge, R. A. 59
Prestige, R. A. 172
Preston, C. A. 193
Preszler, R. W. 223
Price, P. W. 5, 8, 19, 41, 53, 80, 82, 86, 108, 130, 164, 166, 167, 169, 171, 179, 189, 221, 223, 238, 252, 254, 255, 261, 264, 276, 307, 331, 342, 347, 348, 380, 381, 382, 383, 393
Primack, R. B. 78
Prittinen, K. 107
Pullin, A. S. 24, 222
Puustinen, S. 54, 59
Pywell, R. F. 59, 60, 64, 67

Quesada, M. 78, 80, 108, 224, 225
Quested, H. M. 54
Quicke, D. L. J. 368

Rabin, L. B. 131, 132, 276
Raffa, K. F. 22, 29, 199
Raffaele, E. 80, 81
Raffaelli, D. 401
Raguso, R. A. 87
Raimondi, P. T. 105
Rambo, J. L. 109
Ramlan, M. F. 60
Rasplus, J. Y. 83, 344
Rathcke, B. J. 19, 78
Rathcke, R. J. 250
Raupp, M. J. 21, 33
Rausher, M. D. 20, 295
Raven, J. A. 34, 36, 55, 69
Raven, P. H. 12
Read, D. J. 124, 125, 127, 130
Recher, H. F. 297

Redfern, M. 252
Redman, A. M. 20, 27, 28, 29
Rehill, B. J. 254, 397
Retamosa, E. C. 111, 112
Reymond, P. 191, 195
Reynolds, J. F. 96
Reznik, S. Y. 110
Rhoades, D. F. 132
Richards, A. J. 85
Richmond, D. S. 175
Rieske, L. K. 257
Riihimaki, J. 254, 258
Ringel, M. S. 288
Roche, B. M. 105
Roda, A. L. 370
Rodman, J. E. 108
Roderick, G. K. 20, 26, 35, 52
Rohlf, F. J. 85, 319
Roininen, H. 254, 259
Roitberg, B. 201
Rojo, E. 194
Roland, J. 252, 342
Romero, G. Q. 251, 258, 263, 264
Romme, W. H. 311
Rood, S. B. 316
Root, R. B. 307
Rossi, A. M. 40, 342, 343
Rovira, A. D. 127
Rudgers, J. A. 276
Ruess, L. 152
Russell, L. M. 250
Rush, S. 79, 81
Rush, S. L. 79, 81
Rutter, M. T. 108
Rutledge, C. E. 84, 366, 368
Ruuhola, T. 32, 33
Ryan, C. A. 191

Sabelis, M. W. 189, 195, 196, 197, 200, 201, 205, 207
Sagers, C. L. 251, 283

Saikkonen, K. 128, 129, 135, 168, 169, 172, 173
Salonen, V. 59
Salyk, R. P. 31, 35
Sandberg, S. L. 251, 283
Sanders, C. J. 276
Sandstrom, J. P. 52
Salt, D. T. 20
Sanver, D. 247, 249, 250, 253, 259, 266
Schardl, C. 165, 168, 169, 171
Scheu, S. 137, 147, 148, 152, 153, 154
Schlichter, L. 334
Schluter, D. 331, 346
Schlutz, J. C. 254
Schmidt, O. 153
Schmitz, O. 104
Schmitz, O. J. 8, 41, 164, 166, 188, 226, 381, 399, 400
Schneider, M. 192
Schoener, T. W. 19, 226
Schoenly, T. 390, 391
Schonrogge, K. 259, 261
Schoonhoven, L. M. 105, 116, 125, 130, 131, 138, 139
Schupp, E. W. 276
Schweitzer, J. A. 397
Scriber, J. M. 20, 27, 28, 29, 224
Scutareanu, P. 195, 196
Sears, M. K. 32, 36, 39
Seel, W. E. 54, 59, 63, 65, 67
Seibert, T. F. 253
Sessions, L. A. 107
Setälä, H. 147, 148, 153
Seyffarth, J. A. S. 255
Seymour, C. L. 109, 110
Sharaf, K. E. 80, 82, 86
Shaw, M. R. 344
Shearer, J. W. 31, 35
Shimoda, T. 202, 206

Shiojiri, K. 204
Shorthouse, J. D. 55, 253
Shurin, J. B. 380, 399
Shykoff, J. A. 84
Siegel, M. R. 180
Sih, A. 381, 399, 401
Simberloff, D. 84, 238, 255
Simms, E. L. 106
Slansky, F. J. 224
Sloggett, J. J. 294
Smith, F. A. 86
Smith, S. E. 124, 125, 127, 130
Smith, S. M. 337
Sokal, R. R. 85, 319
Sopow, S. L. 253
Spain, A. V. 148, 153
Speight, M. R. 105, 111, 116
Spoel, S. H. 192
Staddon, P. L. 128
Stadler, B. 295
Stanton, M. L. 5, 12, 227, 401
Stanton, N. L. 149
Steidle, J. L. M. 367
Stein, J. S. 311
Steinger, T. 108
Stewart, A. J. A. 105, 116
Stiling, P. 20, 167, 258, 342, 343
Stinchcombe, J. R. 113, 114, 396
Stoetzel, M. B. 250
Stone, G. N. 342
Storeck, A. 201
Stout, M. J. 23, 36, 39, 40
Stowe, K. A. 81
Strack, D. 133
Strain, B. R. 86
Strand, M. 109
Strauss, S. 150
Strauss, S. Y. 24, 25, 67, 75, 77, 78, 79, 80, 81, 82, 85, 108, 224, 225, 227, 233, 308, 311, 357, 380, 384, 397, 401

Strong, D. R. 19, 20, 52, 116, 149, 164, 166, 380, 382, 383, 392, 401
Strong, D. R. J. 226, 237
Strong, D. R. Jr. 226, 237
Sugiura, S. 253
Sullivan, D. J. 31, 35
Sullivan, T. J. 169
Suominen, O. 109
Suzuki, N. 285
Suzuki, Y. 26
Swaty, R. L. 398
Szentesi, A. 25

Tabuchi, K. 343
Takabayashi, J. 108, 195, 200, 201, 365
Tallamy, D. W. 21, 33
Tamaki, G. 35
Tammaru, T. 383
Taverner, M. P. 55
Teder, T. 383
Teschner, M. 254
Tester, J. R. 335, 337
Thacker, J. I. 298
Thaler, J. S. 27, 29, 133, 191, 193, 195, 196, 199, 200, 202, 206, 283, 361, 396
Thalmann, C. 25
Thogerson, M. T. 334
Thompson, D. B. 307
Thompson, J. N. 10, 12, 238, 335, 383, 396, 401
Tilles, D. A. 276
Tilman, D. 239, 310, 392
Tjallingii, W. F. 131
Tollrian, R. 167
Tomlin, E. S. 32, 36, 39
Trapp, E. J. 25
Traw, M. V. 24, 25
Trussell, G. C. 105
Tsarouhas, V. 277

Tscharntke, T. 5, 8, 105, 107, 109, 110, 111, 112, 237, 401
Tumlinson, J. H. 195
Tuomi, J. 224
Turlings, T. C. J. 27, 190, 195, 196, 197, 201, 223, 354, 358, 360, 364, 367

Uhler, L. D. 333, 334
Underwood, N. 41
Uriarte. M. 41

Vázquez, D. P. 84, 238
Vail, S. G. 83
Van Baalen, M. 207
Van Dam, N. M. 194, 199, 207
Van der Baan, H. E. 199, 200
van der Heijden, M. G. A., 140, 197, 275
van der Meijden, E. 197, 362, 371
van der Putten, W. 147
van der Putten, W. H. 133, 138, 140, 227, 386, 396
Van Donk, E. 400
Van Hezewijk, B. H. 342
Van Loon, J. J. A. 192, 193, 360, 364, 367
Van Ommeren, R. J. 308
Van Tol, R. W. H. M. 204, 359
Van Wijk, B. H. 202
Van Zandt, P. A. 12, 20, 24, 36, 39, 40, 113, 190, 227, 266, 359
Varanda, E. M. 251
Vasconcellos-Neto, J. 251, 258, 263, 264
Vaughn, T. T. 369
Vet, L. E. M. 201, 223, 354, 355, 356, 358, 359, 364, 367

via, S. 358
Vicari, M. 132, 135
Völkl, W. 276, 285, 294, 295
Vos, M. 60, 204

Wallin, K. F. 29
WallisDeVries, M. F. 108
Waloff, N. 19
Waltz, A. M. 10, 31
Wamberg, C. 126, 127, 128, 130, 132
Wang, Q. 201, 363
Wangai, A. W. 60
Wardle, D. 147, 149, 152
Wardle, D. A. 148, 152, 158, 227, 386, 396
Warren, P. H. 390
Warwick, R. M. 312
Waser, N. M. 80
Washburn, J. O. 261, 343
Way, M. J. 26, 35, 276
Weiblen, G. D. 344, 348
Wiedenmann, R. N. 366, 368
Wiens, J. A. 109, 110
Weinburg, A. M. 28, 359
Weis, A. E. 77, 81, 85, 252, 261, 333, 334, 335, 337, 339, 340, 341, 343, 345, 346, 358
Weiss, M. R. 252
Weisser, W. W. 286
Werner, E. E. 4, 104, 105, 111, 164, 165, 166, 169, 174, 188, 226, 227, 234, 381, 386, 400
West, C. 130, 131, 132, 133, 276
Westercamp, C. 83, 85
Whelan, C. J. 256
White, T. C. R. 53, 130, 153, 155, 382

Whitfield, J. B. 368
Whitham, T. G. 5, 10, 20,
    27, 31, 52, 82, 109,
    125, 126, 127, 128,
    129, 131, 132, 134,
    151, 154, 223, 224,
    233, 252, 254, 255,
    257, 258, 261, 263,
    264, 276, 284, 286,
    288, 289, 296, 298,
    306, 307, 308, 309,
    310, 311, 312, 314,
    315, 317, 323, 345,
    362, 386
Williams, A. G. 223
Williams, K. S. 224, 233
Williamson, G. B. 288
Willmer, P. G. 251
Willott, S. J. 25

Wilson, D. 105, 253, 255,
    307, 331
Wimp, G. M. 31, 233, 276, 286,
    288, 289, 296, 298,
    306, 307, 312, 315
Winemiller, K. O. 221,
    234, 381
Wise, M. J. 28, 166, 359
Wold, E. N. 20, 24
Wolf, J. B. 311
Wolfe, D. W. 86
Wolfe, L. M. 79, 81
Wood, D. L. 276
Woodman, R. L. 261, 264
Wool, D. 53, 55
Wootton, J. T. 51, 104,
    164, 165, 166, 174,
    180, 225, 308, 311,
    380, 403

Worm, B. 239
Wright, J. T. 297, 317, 368
Wurst, S. 137, 153, 154

Yamamura, N. 83, 86
Yamauchi, A. 83, 86
Young, G. R. 249
Young, T. P. 107, 108, 261
Yu, D. W. 83
Yue, Q. 171

Zamora, R. 106, 107,
    108, 222
Zangerl, A. R. 12, 22, 23, 34,
    67, 84, 370
Zera, A. J. 37
Zhao, Y. F. 193, 194
Zimmerman, M. 80
Zobel, M. 276

Subject index

aboveground–belowground
   interactions
      147–8, 158–9
   interface in indirect
      interaction
      webs 396–7
   interface in plant-
      mediated indirect
      effects 396–7
   multitrophic interactions
      case study
      154–8, 159
allogenic engineering
   see ecosystem
      engineers
ant–aphid mutualism 275–7
   effects of ecosystem
      engineers 283–4
   effects of host plant
      traits 277–86, 287
   effects of mycorrhizae
      284–5
   effects of spatial
      variation on
      biodiversity 295–7
   effects on aphid
      predators and
      parasites 281–2,
      289–95
   effects on community
      stability 286–9, 290
   effects on community
      structure 232–3,
      281–2, 289–95
   endophyte-mediated
      effects 180,
      181, 284
   influence on other
      herbivores 260
   mosaic of occupied
      and unoccupied
      habitat 296–7
   on *Salix eriocarpa* 229–30
   on *Salix miyabeana*
      227–8, 229
   on *Solidago altissima*
      230–2
   spatial variations in
      host plant quality
      295
ant-mediated indirect
   effects 232–3
ant–rodent interactions,
   timescale-related
   effects 309
ant–scale insect mutualism,
   on *Solidago*
   altissima 231, 232
ants
   as keystone species 232–3
   influence on ecosystem
      engineers 260–1
   see also ant–aphid
      mutualism *and*
      specific ant species
Aphidiinae, phylogenetic
   patterns of HIPV
   use 368–9, 370
aphids
   aggregation and
      nutrient sink
      creation 35
   aphid–parasitoid
      interaction webs
      175–81
   asymmetry of indirect
      interactions 66–7
   effects of earthworms
      153–4
   effects of ecosystem
      engineers 283–4
   effects of endophytic
      fungi 284
   effects of host plant
      traits 277–86, 287
   effects of mycorrhizae
      284–5

aphids (cont.)
    effects of root
        herbivores 150–2
    interactions based on
        the plant vascular
        system 65–6
    interspecific association
        to induce nutrient
        sinks 31–2
    interspecific triggering
        of emigrants 26
    mechanical vs. chemical
        effects of
        interactions 68–9
    on *Salix eriocarpa* 229–30
    on *Salix miyabeana*
        227–8, 229
    on *Solidago altissima*
        230–2
    response to induced
        plant volatiles 190–1
    see also ant–aphid
        mutualism *and*
        specific aphid species
aphids and hemi-parasitic
    plants
    comparison of impacts
        on host plant 67–8
    exploitation of host
        vascular system 59
    impact of hemi-parasites
        on aphid
        performance
        61–2, 68–9
    impacts on host vascular
        system 62–4
    impacts on the shared
        host 59–61, 62–4
    interactions based on
        vascular system
        resources 59–64
    interactions between
        different
        kingdoms 67–8

interactions on
    Poa annua 59
aphids and leaf miners
    impact of leaf miners on
        aphid performance
        56–8, 68–9
    impact of leaf miners
        on the vascular
        system 58–9
    interactions based on
        vascular system
        changes 55–9
    interactions on *Betula*
        pendula 55
aquatic systems
    plant-mediated indirect
        effects 400
    trait-mediated indirect
        interactions 238
arbuscular mycorrhizae
    124–5
    effects of insect
        herbivory 125–30
    effects on insect
        herbivores 130–5
    Glomus caledonium 134–5
arthropod communities
    effects of aspen–fire–elk
        interactions 311–14
    effects of beaver
        316–19
    effects of cottonwood–
        beaver–sawfly
        interactions 316–19
    effects of induced
        changes in plants
        188–9
    effects of galling sawfly
        (*Phyllocolpa* spp.)
        314–19
    in aspen forests
        311–14
    in cottonwood forests
        316–19

arthropod construct builders
    see ecosystem
        engineers
avian community
    effects of elk herbivory
        314–16
    effects of forest fire
        314–16
    predation on galling
        sawfly 314–16
    responses to HIPV 204–5
bat pollinators, avoidance
    of katydid
    damage 77
beaver see *Castor canadensis*
bee pollinators
    avoidance of herbivore
        damage 77
    herbivorous
        pollinators 83
beetles, as herbivorous
    pollinators 83
belowground–aboveground
    interactions
    147–8, 158–9
    interface in indirect
        interaction
        webs 396–7
    interface in plant-
        mediated indirect
        effects 396–7
    multitrophic
        interactions
        case study
        154–8, 159
belowground biota 148
belowground processes
    see also decomposers;
        root herbivores
biocontrol of insect
    herbivores,
    effects
    of HIPV 196–8

Subject index

biodiversity
    and indirect interaction webs 389–94
    effects of conditional responses 309–10
    effects of numbers of interactions (review) 319–23
    effects of spatial variation in ant–aphid mutualisms 295–7
    importance of plant-mediated indirect effects 10–12, 389–94
    interaction effects and pattern reversals (review) 319–23
    of species and interactions 238–9
birds see avian community
bottom-up cascades 166–8
    endophyte-induced 169–72
    plant-mediated 169–72
bottom-up indirect effects, endophytic fungi 164–5
bottom-up influences
    on community organization 382–4
    on community structure 4–5, 164–5
butterfly pollinators, avoidance of tephritid fly damage 77

collembolans
    effects on plant growth 153
    interactions with mycorrhizae 137

community composition
    plant-mediated effects of herbivores 306–7
    plant trait-mediated effects 310–11
community dynamics
    holistic approach 40–1
    need for multiple species interaction studies 40–1
    spatial and temporal scaling 41
community interactions
    conditional responses 307–10
    spatial and temporal changes in responses 307–10
community species richness, effects of ecosystem engineers 261–3
community stability
    and indirect interaction webs 389–94
    and plant-mediated indirect effects 389–94
    effects of ant–aphid mutualisms 286–9, 290
community structure
    and indirect interaction webs 389–94
    and plant mediated indirect effects 389–94
    bottom-up influences 4–5
    effects of induced changes in plants 188–9
    effects of insect–mycorrhiza interactions 139–40

impacts of ant–aphid mutualism 232–3
impacts of ecosystem engineers 256–63
important roles of plant-mediated indirect effects 399–400
influence of plant-mediated indirect effects 232–3
plant-mediated indirect effects 10–12
role of competition 19–21
top-down influences 8–10
see also arthropod communities
competition
    between organisms from different kingdoms 67–8
    conditions for 68
competition theory
    asymmetric plant-mediated interactions 39
    challenges from plant-mediated interactions 38–40
    herbivore densities and plant-mediated competition 38–9
    niche divergence and resource partitioning 39
    phylogenetic relatedness and competition 39–40
conditional responses
    effects and pattern reversals (review) 319–23
    effects on biodiversity 309–10

conditional responses (*cont.*)
  in community
    interactions 307–10
  mistletoe–juniper
    relationship 308
  mycorrhizae response
    to fertilizer 308–9
  negative effects of
    fertilizers 308–9
  ontological changes
    in plant resistance
    and susceptibility
    309
  timescale-related
    rodent–ant
    interactions 309
  willow genotypes'
    susceptibility to
    insects 308
conifers, co-evolution with
  seed predators 332
cottonwood *see Populus
  angustifolia*

decomposers 148, 152–4
  effects on
    herbivores 153–4
  effects on plant
    growth 152–3
  multitrophic
    interactions case
    study 154–8, 159
  nutrient mineralization
    process 152
density-mediated indirect
  interactions (DMIIs)
  3–4, 5, 104–5,
  165–6, 380–1
  mammalian and insect
    herbivores 109–10
direct interactions 104–5
  mammalian and insect
    herbivores 110,
    111–12

DMIIs *see* density-mediated
  indirect interactions
domatia 195

earthworms
  effects on aboveground
    herbivores 153–4
  effects on plant
    growth 152–3
  in indirect interaction
    webs 137
  multitrophic interactions
    case study 154–8, 159
ecosystem engineers
  beaver (*Castor canadensis*)
    316–19
  benefits and costs of
    secondary
    inhabitation 253–4
  benefits of construct
    formation 251–2
  consequences for
    plant fitness 257,
    262, 263–5
  costs of construct
    formation 253
  definition 246
  distribution of constructs
    within plants 254
  effects of leafing
    phenology 255
  effects on aphids 283–4
  effects on community
    species richness
    261–3
  galls and gall-makers
    249–50
  habitat construction
    on plants 225
  impacts on arthropod
    communities
    246–7, 256–63
  influence of abiotic
    factors 255

influence of natural
  enemies 256,
  257, 260–1
influence of plant
  traits 254–6
influence on value
  of nonengineered
  habitat 257, 260
inquilines 247,
  249–50, 253,
interactions between
  secondary
  inhabitants and
  other herbivores
  257, 260
interactions with other
  herbivores 256–8
interactions within
  constructs 258–60
leaf constructs and
  webs 249
leaf mines 250–1
plant architecture
  and attack
  levels 255
plant vigor and gall
  attack 254–5
potential keystone
  species 234
predator constructs 251
secondary inhabitants
  of constructs
  247–51, 257, 260
species of large
  effect 306–7
stem-bored cavities 250
types of constructs and
  their builders 247–51
ectomycorrhizae 124–5
  effects of insect
    herbivory 125–30
  effects on insect
    herbivores 130–5
  Laccaria laccata 134–5

Subject index 435

elk *see Cervus canadensis*
emigrants (macropterous forms, alates), triggered by induced changes in plant nutrition 26
endophyte–grass symbiosis 168–9
endophyte-induced bottom-up cascades 169–72
endophytes (endophytic fungi) 168–9
  bottom-up indirect effects 164–5
  effects on aphids 284
  effects on grass–insect interactions 172–3
  effects on higher trophic levels 170, 173–5
  effects on insect natural enemies 170, 173–5
  in indirect interaction webs 135–6, 137
  in multitrophic interaction webs 175–81
endophytic grasses, allelochemical modifications 170–1
entomopathogens, use of HIPV as a signal 204–5
evolutionary changes
  consequences of plant-mediated indirect effects 12–13
  evidence for herbivore-induced origins 361, 368, 369, 370
  from ecological effects 354–6
  from herbivore-induced changes 354–6

future research on
  herbivore-induced impacts 371–2
  implications of indirect interaction webs 394–6
  implications of plant-mediated indirect effects 394–6
  induced changes linked with presence of herbivory 358–9
  investigation by comparisons of related species 365–7
  investigation by selection experiments 362
  multitrophic fitness effects from induced changes 360–1
  multitrophic level detection of induced changes 359
  phylogenetic distribution of defensive plant genes 370–1
  phylogenetic surveys across taxa 367–71
  prerequisites for 356–61
  TMIIs between mammalian and insect herbivores 112–13
  see also indirect evolutionary effects
evolutionary interaction web 331–3
extrafloral nectar (EFN) production, herbivore-induced 195

fallow deer, incidental ingestion of insect herbivores 110, 111–12
fertilizers, conditional negative responses to 308–9
figs and fig wasps 83
  indirect evolutionary interactions 344
fire (forest fire)
  impacts on arthropod communities 311–14
  impacts on avian community 314–16
floral traits, effects of root herbivory 149–50
flower numbers, reduced due to herbivory 78
flower size, reduced due to herbivory 78–9
food webs 4–5
  comparison with indirect interaction webs 229, 230, 231, 234–7
  incorporation of nontrophic indirect interactions 234
  interactions absent from 229, 230, 231, 234–7
  theory of 381–2
fungal endophytes *see* endophytic fungi

gall-centered interaction webs 342–4 *see also Eurosta solidaginis*
gall formation
  effects on plant-mediated interactions 52–3
  effects on plant vasculature 53
galls and gall-makers 249–50
genes involved in plant defenses 191–2

grass–endophyte symbiosis
168–9
endophyte-produced
allelochemicals
170–1
grass–insect interactions,
effects of fungal
endophytes 172–3
grasslands, role of parasitic
plants in community
dynamics 64

hemi-parasitic plants
effects on host
biomass 54
exploitation of
host vascular
system 53–4
host-mediated
interactions with
other organisms 54
interactions based on
the host vascular
system 65–6
role in community
dynamics 64
see also aphids and
hemi-parasitic
plants
herbivore-induced changes
in plants 10–12
changes to plant
distribution
pattern 108–9
chemical changes 188–9
detection at other
trophic levels 359
distribution of defensive
plant genes 370–1
ecological effects and
evolutionary
change 354–6
effects on plant
phenotype 107–9

evidence for evolutionary
change as a result
361, 368, 369, 370
extrafloral nectar
production 195
fitness effects at other
trophic levels 360–1
future research on
evolutionary
impacts 371–2
impacts on plant
fitness 360
indirect evolutionary
connections 354–6
induced changes linked
with presence of
herbivory 358–9
induced resistance 107–8
mediation of interaction
linkages 232–3
negative fitness effects
for herbivores 361
nonadaptive phenotypic
plasticity 108
nonlethal effects 382–4
phenotypic plasticity
of plants 382–4
predator associative
learning abilities 364
predator olfactory
response
experiments 362
predator responses to
plant volatiles 365–7
prerequisites for
evolutionary
effects 356–61
tolerance to damage 108
herbivore-induced
indirect effects
benefits of sharing the
same host 226
community structural
organization 227

frequent temporal
and spatial
separation 226
multiple interaction
linkage 227
widespread and common
occurrence 226
herbivore-induced plant
responses 222–5
effects of leaf herbivory
on floral traits 224–5
effects on arthropod
communities 221–2
endophyte-induced
resistance 170–1
enhanced growth
following
herbivory 224
habitat construction
by ecosystem
engineers 225
induced defense 222–3
mediation of multiple
arthropod
interactions 221–2
nutrients in damaged
plants 224
release of plant
volatiles 222–3
via nontrophic
linkages 221–2
herbivore-induced plant
volatiles see HIPV
herbivore-induced shelter
provisioning 195
herbivore natural enemies,
attraction to
HIPV 196–8
herbivore vectors for
plant pathogens
193–4
herbivores
benefits from HIPV for
other species 205–6

Subject index

effects on plant
community
diversity 106–7
effects on plant
performance 106
effects on plant
population
densities 106
impact of mammalian
versus insect
herbivory 107
plant-mediated effects
on community
composition 306–7,
310–11
species of large effect
306–7
see also insect and
mammalian
herbivores
herbivorous insects
alteration of phloem
source–sink
dynamics 26–7
as plant weapon
against
competitors 206–7
association with other
herbivore species
203–4
asymmetry in plant-
mediated
interactions 52
benefits from induced
changes in plant
morphology 30–1
changes in mycorrhizal
species composition
128–30
competition between
phloem feeders 26–7
competition through
effects on plant
nutrition 26–7

competitive benefits of
aggregation 35
competitive benefits of
early breaking of
diapause 33
competitive benefits of
early colonization 33
competitive trade-offs
and constraints 37–8
cooperation with plant
pathogens 206–7
creators of interactions
linkages 225–7
developmental delay
caused by induced
resistance 28–9
differential tolerance of
allelochemicals 33–5
dispersal capability
and competitive
ability 35–6
effects of altered flower
size and number 25
effects of changes in
leaf-trichome
density 24–5
effects of mycorrhizae
130–5
effects of root
herbivory 150–2
effects on
mycorrhizae 125–30
feeding guild and
competitive
ability 36
feeding-induced
attraction of natural
enemies 27–8
frequency of plant-
mediated
interactions 52
importance of plant-
mediated
interactions 52

incidental predation
by mammals
110, 111–12
indirect interaction web
with mycorrhizae
and plants 135–8
induced allelochemistry
and plant
susceptibility 29–30
induced changes in plant
morphology 24–5
induced changes to plant
morphology benefit
predators 28
induced defenses
and reduced
enemy attack 32
induced plant
allelochemistry 22–4
induced resistance and
plant-mediated
competition 22–9
induced susceptibility
and plant mediated
facilitation 29–32
inter-annual carryover
of induced effects 33
interactions based
on the plant vascular
system 65–6
interactions with
E. mediohispanicum
and Spanish
ibex 113–16
interactions with
parasitic
angiosperms 54
interspecific
interactions 19–21
manipulation of plant
nutrients 52
manipulation of the
plant vascular
system 53–4

herbivorous insects (cont.)
  mechanisms of
    plant-mediated
    interactions 21–32
  on endophytic
    grasses 172–3
  phloem feeders change
    plant source-sink
    dynamics 31–2
  plant-mediated
    competitive effects
    20–1, 51–3
  plant-mediated indirect
    interactions 225–6
  plant vein cutting and
    trenching behaviors
    29–30
  positive interspecific
    association to
    induce nutrient
    sinks 31–2
  preference for
    nitrogen-rich
    plants 111
  preference for plant
    re-growth 111
  role of competition in
    structuring
    communities 19–21
  shelter construction by
    "ecosystem
    engineers" 30–1
  structural modification
    of plant tissues 52–3
  top-down influences
    on performance 138
  trade-off between
    competitive ability
    and dispersal 37–8
  trait-mediated indirect
    interactions 225–6
  traits for plant-mediated
    competitive
    superiority 32–6

underestimation of
  interspecific
  interactions 237–8
vulnerability to
  predators increased
  by induced
  resistance 28–9
herbivory and pollination
  alteration of flowering
    phenology 79–80
  alteration of plant
    sexual expression 81
  altered plant investment
    in reproduction
    78–81
  buffers against herbivory
    effects 81–2
  consequences of anti-
    herbivore traits 82
  effects of pollination
    on herbivory 84
  effects of root herbivory
    149–50
  effects on nectar quantity
    and quality 80
  effects on pollen quantity
    and quality 80–1
  herbivorous pollinator
    interactions 83–4
  increase in pollination
    success 82–4
  linkages and interactions
    75–6
  Manduca/Datura
    interactions 87–8
  mechanisms for reduced
    plant reproductive
    success 76–7
  pollinator avoidance
    mechanisms 77–8
  plant compensation
    for damage 81–2
  plant-mediated effects on
    reproduction 78–81

plant overcompensation
  for damage 82–3
pollinator avoidance of
  herbivore damage 77
pollinator avoidance of
  herbivores 77–8
pollinator avoidance
  of reduced size
  flowers 78–9
pollinators which are
  not influenced 81–2
reduced flower
  numbers 78
reduced flower
  size 78–9
resource allocation
  model (Manduca/
  Datura interactions)
  89, 90, 91, 92,
  93, 94, 88
review of interactions
  76–84
towards a predictive
  framework 84–6
herbivory and population
  ecology, interface
  in plant-mediated
  indirect interactions
  397–8
HIPV (herbivore-induced
  plant volatiles)
  192–4, 195–205
  attraction of herbivore
    natural enemies
    196–8
  benefits to other
    herbivores 205–6
  coordination of plant
    responses 198–9
  evidence for active
    role of plants 195
  evolution of plant
    responses 197
  herbivore avoidance 198

Subject index 439

herbivore specificity 199–200
herbivore suppression by parasitic wasps 196–7
innate and acquired predator responses 203
learned predator responses 200–4
phylogenetic patterns of use in Aphidiinae 368–9, 370
phylogenetic survey of use by *Cotesia* spp. 367–8, 369
predator associative learning abilities 364
predator olfactory response experiments 362
predator preferences 199–200
reduced induction by herbivores 206
responses of birds and mammals 204–5
specificity of mixtures of volatiles 199–200
use by entomopathogens 204–5
use by nonarthropod insectivores 204–5
variability of blends 200–4

indirect evolutionary interactions 331–3
conditions for development 341–2
E. solidaginis interaction web 332, 333–4
evaluation tools 346–7
evolution of natural enemies 341–4

gall-centered interaction webs 342–4
mediation by plant traits 347–9
problems with detection and evaluation 344–6
spatial and temporal separation 344–6
widespread potential for development 341–4
indirect interaction webs among spatially isolated species 385–7
among taxonomically distant species 385
among temporally separated species 387–9
and community stability 389–94
and community structure and biodiversity 389–94
collembolan interactions 137
complexity of 389–94
concept 4–5
earthworm interactions 137
endophyte interactions 135–6, 137
evolutionary implications 394–6
ideas on future studies 400–3
in aquatic systems 400
insect herbivores–mycorrhizae–plants 135–8
interactions absent from food webs 229, 230, 231, 234–7

interface of below- and aboveground ecology 396–7
interface of herbivory and population ecology 397–8
interface of insect and microbial ecology 398–9
interface of separated study disciplines 396–9
nontrophic (within-trophic-level) links 234–9
plant-feeding nematodes 137–8
richness of interactions 384–9
spatial mosaics 385–7
within-trophic-level (nontrophic) links 234–9
indirect interactions and community composition 306–7
and food web regulation 165–8
between highly dissimilar herbivores 105–6
bottom–up cascades 166–8
evolution of 331–3
top–down forces 8–10, 166–8
trophic cascades 8–10, 166–8, 380–1
see also density-mediated indirect interactions; trait-mediated indirect interactions

induced resistance in
plants 107–8
insect and mammalian
herbivores 107
density-mediated
indirect interactions
109–10, 110
direct interactions 111–12
incidental predation
by mammals 111–12
interactions between
109–12
prospects for future
research 116–17
trait-mediated indirect
interactions 110–11
insect and microbial ecology,
interface in plant-
mediated indirect
effects 398–9
insect herbivores *see*
herbivorous insects
insect–mycorrhiza
interactions 125
effects on community
structure 136,
139–40
insectivores (nonarthropod),
use of HIPV as a
signal 204–5
interaction biodiversity
10–12, 238–9
interaction webs, gall-
centered 342–4
see also indirect
interaction webs

JA (jasmonic acid), 191–4

katydid damage, avoidance
by bat pollinators 77
keystone species 306–7
beaver (*Castor canadensis*)
316–19

galling sawfly
(*Phyllocolpa*) 314–16

leaf miners, impact on
the plant vascular
system 65
lodgepole pine, co-evolution
with seed
predators 332

mammalian and insect
herbivores 107
density-mediated
indirect
interactions 109–10
direct interactions
111–12
evolutionary
consequences of
TMIIs 112–13
incidental predation by
mammals 111–12
interactions between
109–12
interactions case study
113–16
prospects for future
research 116–17
trait-mediated indirect
interactions 110–11
mammals, responses to
HIPV 204–5
microbial and insect
ecology, interface
in plant-mediated
indirect effects
398–9
microbial symbionts 167–8
see also endophytic
fungi
mistletoe-juniper
relationship,
conditional
effects 308

multitrophic belowground–
aboveground
interactions (case
study) 154–8, 159
multitrophic interaction
webs, endophyte-
driven 175–81
mutualisms, mediation by
host plants 275–7
see also ant–aphid
mutualism
mycorrhiza-forming
fungi 124–5
mycorrhiza-insect
interactions, effects
on community
structure 139–40
mycorrhizae
conditional response to
fertilizer 308–9
dual mycorrhizal
associations 134–5
effects of insect
herbivory 125–30
effects of insect herbivory
on species
composition 128–30
effects on aphids 284–5
effects on insect
herbivores 130–5
indirect interaction
web with insect
herbivores and
plants 135–8
interactions with
Collembola 137
interactions with
earthworms 137
interactions with
pathogenic
fungi 137–8
interactions with
plant-feeding
nematodes 137–8

types and functions
124–5
see also arbuscular
mycorrhizae;
ectomycorrhizae

natural enemies
alteration of host
phenology 344
indirect evolutionary
interactions 341–4
of *Eurosta solidaginis* 333
selection for gall size
in *E. solidaginis*
333–41
nectar quantity and
quality, effects of
herbivory 80
nematodes, in indirect
interaction
webs 137–8
nontrophic indirect
interaction
webs 234–9
incorporation into
food web
structure 234
nontrophic linkages in
herbivore-induced
plant responses
221–2

parasitic angiosperms
exploitation of host
vascular system
53–4
role in community
dynamics 64
see also hemi-parasitic
plants
parasitic wasps, herbivore
suppression by
HIPV attraction
196–7

parasitoids
feeding-induced
attraction to
prey 27–8
indirect selection
pressures among 344
see also specific species
phenotypic plasticity of
plants 382–4
physical ecosystem engineers
see ecosystem
engineers
phytophagous insects see
herbivorous insects
plant allelochemicals
"activated synthesis"
of chemicals 23
alkaloids 22
and induced
susceptibility 29–30
defensive proteins
22, 23–4
herbivore-induced
chemical
changes 188–9
in endophytic
grasses 170–1
indole glucosinoloates
22–3, 24
induction by root
herbivory 151
insect feeding-induced
22–4
phenolics 22–3
"preformed" chemical
release 22–3
terpenoids 22–3, 24
tolerance by herbivorous
insects 33–5
plant-based interaction
webs, in terrestrial
systems 237–9
plant constructors see
ecosystem engineers

plant defense elicitors
herbivore-derived 190–1
jasmonic acid 191–2
phytohormones 191–2
plant-specific elicitors
191–2
plant defenses
benefits to other
herbivores 205–6
changes in composition
of signal traits 207
co-regulation of
signaling
pathways 192–4
coordination of defense
responses 198–9
coordination of HIPV
response 198–9
coping with a diversity
of attackers 192–4
effects of plant
pathogens 192–4
genes involved 191–2
herbivore-induced
plant volatiles
(HIPV) 195–205
herbivore suppression
by parasitic wasps
196–7
herbivore vectors for
plant pathogens
193–4
herbivores as weapon
against plant
competitors 206–7
octadecanoid pathway
191–2
prioritizing defenses
192–4
protection from
herbivore natural
enemies 207
reduced induction by
herbivores 206

442 Subject index

plant defenses (*cont.*)
  salicylic acid
    phytohormone
    192–4
  shikimate pathway 192–4
  used against neighboring
    plants 194
plant fitness, consequences
  of ecosystem
  engineering 257,
  262, 263–5
plant growth, effects of
  root herbivory 149
plant–herbivore interactions
  10–12
  conditional interactions
    (review) 319–23
  effects and pattern
    reversals (review)
    319–23
plant–insect interactions
  case studies 227–32,
    234–7
  indirect interaction
    web with insect
    herbivores and
    mycorrhizae 135–8
  interaction linkage on
    Salix eriocarpa
    (willow) 229–30,
    235, 236
  interaction linkage on
    Salix miyabeana
    (willow) 227–8, 229,
    234–5, 236
  interaction linkage on
    Solidago altissima
    (goldenrod) 230–2,
    235–7
plant-mediated effects
  at the second trophic
    level 190–4
  at the third trophic
    level 195–205

bottom–up cascades
  169–72
on ant–aphid mutualisms
  277–86, 287
on community
  composition 306–7
responses to induced
  plant volatiles 190–1
see also plant trait-
  mediated effects
plant-mediated indirect
  effects
among spatially isolated
  species 385–7
among taxonomically
  distant species 385
among temporally
  separated
  species 387–9
complexity of
  interactions 8–10,
  389–94
creation of spatial
  mosaics 385–7
evolutionary
  consequences
  12–13, 394–6
ideas on future
  studies 400–3
in aquatic systems 400
in multitrophic
  systems 8–10
influence on
  biodiversity 10–12
influence on community
  stability 389–94
influence on community
  structure 10–12,
  188–9, 232–3, 389–94
interaction diversity
  10–12
interface of below- and
  aboveground
  ecology 396–7

interface of herbivory
  and population
  ecology 397–8
interface of insect and
  microbial ecology
  398–9
interface of separated
  study disciplines
  396–9
novel interaction
  linkages 5–8
richness of interaction
  webs 384–9
when and where they are
  important 399–400
see also plant trait-
  mediated effects
plant-mediated interactions
  asymmetry of
    interactions 67,
    66–7
  based on changes to
    the vascular system
    53–4, 55–9
  based on resources in
    the vascular
    system 59–64
  effects of gall formation
    52–3
  frequency among insect
    herbivores 52
  importance for insect
    herbivores 52
  structural modification
    of plant tissues 52–3
plant morphology
  changes induced
    by herbivorous
    insects 24–5
  folivory-induced changes
    in floral traits 25
  folivory-induced changes
    in leaf-trichome
    density 24–5

Subject index   443

induced changes
  benefit herbivorous
    insects 30–1
  induced changes benefit
    predators 28
plant nutrition
  changes induced
    by herbivore
    feeding 26–7
  induced changes trigger
    dispersal 26
plant pathogens
  cooperation with insect
    herbivores 206–7
  effects on plant
    defenses 192–4
  herbivore vectors 193–4
plant–predator
  communication,
    temporal and
    spatial signal
    changes 207
plant responses to herbivory
  changes to distribution
    pattern 108–9
  induced resistance 107–8
  non-adaptive phenotypic
    plasticity 108
  re-allocation of
    resources 108
  tolerance to damage 108
plant susceptibility
  feeding-induced
    deactivation of
    defenses 29–30
  induced changes to
    source–sink
    dynamics 31–2
plant-trait-mediated effects
  3–4, 5–6
  indirect evolutionary
    interactions 347–9
  on community
    composition 310–11

see also plant-mediated
  effects; plant-
  mediated indirect
  effects
plant traits
  phenotypic plasticity
    382–4
  ontological changes in
    resistance and
    susceptibility 309
plant vascular system
  effects of gall
    formation 53
  exploitation by parasitic
    angiosperms 53–4
  range of interactions
    associated with 65–6
pollen quantity and
  quality, effects of
  herbivory 80–1
pollination and herbivory
  see herbivory and
  pollination
population ecology and
  herbivory, interface
  in plant-mediated
  indirect
  interactions 397–8
predator arthropods,
  learned responses
  to HIPV 200–4
predators, feeding-induced
  attraction to
  insect prey 27–8
psyllids
  control using natural
    predators 196
  HIPV attraction of natural
    predators 196

red deer, incidental
  ingestion of insect
  herbivores 110,
  111–12

resource allocation model
  (*Manduca/Datura*
  interactions) 88, 89,
  90, 91, 92, 93, 94
rodent–ant interactions,
  timescale-related
  effects 309
root herbivores 148, 149–52
  effects on aboveground
    herbivores and their
    parasitism 150–2
  effects on floral traits
    and pollination
    149–50
  effects on foliar
    feeders 150–2
  effects on plant
    growth 149
  endophyte-induced
    resistance 170–1
  endophyte-mediated
    interactions 175
  induction of
    secondary plant
    compounds 151
  multitrophic
    interactions case
    study 154–8, 159

SA (salicylic acid)
sexual expression of plants,
  alteration by
  herbivory 81
shelter provisioning by
  plants 195
Spanish ibex see *Capra
  pyrenaica*
species of large effect 306–7
  Castor canadensis (beaver)
    316–19
  Cervus canadensis (elk)
    311–16
  Phyllocolpa spp. (galling
    sawfly), 316–19

Subject index

species of large effect (*cont.*)
   Phyllocolpa bozemanii
      (galling sawfly)
      314–16
   Populus angustifolia
      (cottonwood) 316–19
   Populus tremuloides (aspen)
      311–14
strawberries, effects of
   strawberry bud
      weevil 81–2
symbiotic microbes 167–8 *see also* endophytic fungi

tephritid flies, effects of root herbivores 150
tephritid fly damage, avoidance by butterfly pollinators 77
TMIIs *see* trait-mediated indirect interactions
tobacco mosaic virus 193
tobacco plants *see Nicotiana attenuata*
tomato-spotted wilt virus 193–4
top-down forces 166–8
   influences on community structure 8–10
trait-mediated indirect interactions (TMIIs) 3–4, 5–6, 104–5, 165–6, 381
   evolutionary consequences 112–13

   in aquatic systems 238
   in terrestrial systems 238
   mammalian and insect herbivores 110–11
trophic cascades 8–10, 166–8, 380–1
tulip bulbs, coordination of defensive responses 199

ungulates, interactions with *E. mediohispanicum* and phytophagous insects 113–16

willow *see Salix*

yuccas and yucca moths 83